PROJECTIVE GEOMETRY AND ITS APPLICATIONS TO COMPUTER GRAPHICS

Michael A. Penna
and
Richard R. Patterson
Indiana University—Purdue University at Indianapolis

Prentice-Hall, Englewood Cliffs, New Jersey 07632

Library of Congress Cataloging-in-Publication Data

PENNA, MICHAEL A. (date)
  Projective geometry and its applications to computer graphics.

  Bibliography: p.
  Includes index.
  1. Geometry, Projective. 2. Computer graphics.
  I. Patterson, Richard R. (date). II. Title.
  QA471.P395 1986   516′.5   85-16690
  ISBN 0-13-730649-0

Editorial/production supervision
  and interior design: **Diana Drew**
Cover design: **Photo Plus Art**
Manufacturing buyer: **Gordon Osbourne**

The author and publisher of this book have used their best efforts in preparing this book. These efforts include the development, research, and testing of the theories and programs to determine their effectiveness. The author and publisher make no warranty of any kind, expressed or implied, with regard to these programs or the documentation contained in this book. The author and publisher shall not be liable in any event for incidental or consequential damages in connection with, or arising out of, the furnishing, performance, or use of these programs.

© 1986 by Prentice-Hall
A Division of Simon & Schuster, Inc.
Englewood Cliffs, New Jersey 07632

All rights reserved. No part of this book
may be reproduced, in any form or by any means,
without permission in writing from the publisher.

Printed in the United States of America

10 9 8 7 6 5 4 3 2 1

0-13-730649-0 025

Prentice-Hall International, Inc., *London*
Prentice-Hall of Australia Pty. Limited, *Sydney*
Editora Prentice-Hall do Brasil, Ltda., *Rio de Janeiro*
Prentice-Hall Canada Inc., *Toronto*
Prentice-Hall Hispanoamericana, S.A., *Mexico*
Prentice-Hall of India Private Limited, *New Delhi*
Prentice-Hall of Japan, Inc., *Tokyo*
Prentice-Hall of Southeast Asia Pte. Ltd., *Singapore*
Whitehall Books Limited, *Wellington, New Zealand*

To our wives,
without whose support
this would never have been written

Contents

Preface — ix

1
Perspective and Projective Geometry — 1

Introduction — 1
Section 1: The Study of Geometry — 2
Section 2: A Model for Perspective — 7
Section 3: The Projective Plane — 15
Section 4: A Topological Model of the Projective Plane — 26
Section 5: The Standard Embedded Projective Plane — 32
Section 6: Some Synthetic Projective Geometry — 39
Section 7: Remarks — 49

2
Analytic Projective Geometry — 54

Introduction — 54
Section 1: Homogeneous Coordinates — 55
Section 2: Projective Lines (Nonparametric Description) — 63
Section 3: Projective Planes (Nonparametric Description) — 77
Section 4: Projective Lines (Parametric Description) — 86

3
Projective Transformations — 97

Introduction — 97
Section 1: Transformations of Projective Lines — 101
Section 2: Transformations of Projective Planes — 116
Section 3: The Geometry of Projective Transformations — 126
Section 4: General Homogeneous Coordinates — 149
Section 5: Coordinate Transformations — 159
Section 6: The Cross Ratio — 162
Section 7: An Important Result — 171

4
Two-Dimensional Graphics — 176

Introduction — 176
Section 1: The Complete Viewing Operation in Two Dimensions — 177
Section 2: A Simple Two-Dimensional Graphics Library — 189
Section 3: A Two-Dimensional Graphics Display Program — 201

5
Three-Dimensional Graphics — 220

Introduction — 220
Section 1: Some Concepts and Terminology — 221
Section 2: The Complete Viewing Operation in Three Dimensions — 226
Section 3: A Simple Three-Dimensional Graphics Library — 240
Section 4: A Three-Dimensional Graphics Display Program — 244
Section 5: Graphing Functions of Two Variables — 255
Section 6: The Aesthetics of Perspective and Parallel Imaging — 270
Section 7: Depth Clipping — 296
Section 8: An Attitude Indicator Simulation — 303

6
Reconstruction — 322

Introduction — 322
Section 1: Least Squares and the Right Inverse — 323
Section 2: Object Reconstruction — 329
Section 3: Camera Calibration — 336
Section 4: Photogrammetric Reconstruction — 341

7
Conics in Design — 357

Introduction — 357
Section 1: The Projective Theory of Conics — 357
Section 2: The Conic through Five Points — 369
Section 3: Additional Conic-Fitting Problems — 377

Bibliography — 394

Index — 396

Preface

This book is about projective geometry and its applications to computer graphics. It is intended to help people who design computer graphics software learn, understand, and use projective geometry.

In this book we interpret the term *computer graphics* in a liberal sense to include not only two- and three-dimensional graphics (in which the goal is to obtain an image of a real or imaginary object) but also vision systems (in which the goal is to reconstruct an object from a given image or images of the object) and design.

We are, in fact, concerned only with computer graphics systems in which both the objects and their images exist as mathematical models or numerical abstractions. In such systems there are three major elements. The first element is the mathematical model, or numerical abstraction, of the object. The second element is the image of the object. The third element is the transformation process. The transformation process can be viewed in two ways: First, given an object, that object can be used to construct the image. Second, given the image, that image can be used to reconstruct the original object.

The goal of this book is to support the design of solutions to computer graphics problems through an understanding of the underlying elements of projective geometry. Some of the topics discussed in this book include

- How to view the structure of two- and three-dimensional graphics libraries
- How to create two-dimensional perspective and parallel images of a three-dimensional object
- How to draw the graph of a function of two real variables

- How to perform depth clipping
- How to draw shadows and reflections
- How to construct stereoscopic images
- How to compute the coordinates of vanishing points and the equations of vanishing lines
- How to reconstruct a three-dimensional object and how to reconstruct a three-dimensional viewing mechanism from images of the object
- How to determine the equation of a conic meeting various incidence requirements

Some of the specific issues that can be elucidated by an understanding of the underlying elements of projective geometry include

- How perspective transformations can be represented by 3×3, 4×4, 3×4, and 4×3 matrix multiplications, and the relations between these representations
- How and why points in a plane (that is two-dimensional) are represented by homogeneous coordinates (that have three components)
- How projective geometry eliminates having to worry about special cases and ad hoc techniques (such as those that arise in producing perspective and parallel images of three-dimensional objects)
- How homogeneous coordinates are limited in relation to certain graphics constructs (such as drawing line segments)
- How formulas that are not "canned" can be obtained efficiently
- How to interpret the feedback provided by an improperly written program so that bugs can be isolated.

This book is not intended to be hardware- or software-specific. (Programs in this book are written in pseudocode, although all have been implemented in Pascal, FORTRAN, and BASIC as well.) The primary reason for this is that a wide variety of software in which programs can be written is presently available for many types of hardware. Addressing specific implementations would take us too far from our intended goal.

The computer prerequisite for this book is a first course in computer programming (minimally, knowing something about pseudocode). The mathematical prerequisites for this book are analytic geometry in the Euclidean plane and in Euclidean space (the geometry normally covered in three semesters of calculus with analytic geometry) and matrix theory. Thus this book is suitable for use at the upper undergraduate level and beyond.

This book has three major parts: The first part (Chapter 1) is a brief introduction to perspective and projective geometry. The second part (Chapters 2 and 3) is devoted to analytic projective geometry. The third part (Chapters 4, 5, 6, and 7) focuses on application of the preceding material to solutions of problems in computer graphics. Chapter 4 spotlights two-dimensional graphics, Chapter 5 presents three-dimensional graphics, Chapter 6 deals with reconstruction (vision systems), and Chapter 7 addresses design.

The first three chapters, as well as parts of Chapters 4 and 5, were developed by the first author as an upper-level undergraduate course in projective geometry and its applications. The remaining material was later added by both authors in a number of other courses taught from material in this book.

Results and examples are numbered consecutively within each section, by section and item number. The chapter number is included only when reference is made outside the current chapter. For example, Proposition 1.2.3 refers to the third proposition of Section 2 in Chapter 1, and this reference is being made in a chapter other than Chapter 1. The section number of an exercise is included only when reference is made outside the current section; the chapter number is included only when reference is made outside the current chapter. For example, Exercise 1.3 refers to the third exercise in Section 1 of the chapter you are reading, and Exercise 3 refers to the third exercise in the section you are reading. Exercises marked with a dagger (†) are of particular importance because they are used later in the book. Figures are numbered consecutively throughout each chapter.

We would like to thank Owen Burkinshaw and Jim Harvey for some critical help with programming, and Ingrid Carlbom for her assistance in helping us obtain an illustration we otherwise could not have located.

1

Perspective and Projective Geometry

INTRODUCTION

This chapter provides a brief introduction to perspective and projective geometry. You are assumed to have had little prior experience with or exposure to either. Since this chapter contains important content (particularly in statements of fundamental definitions and results), you are advised to read it even if you have had some experience.

Section 1 is an informal orientation to the study of Euclidean and projective geometries. Section 2 is an informal introduction to perspective using Euclidean geometry, while Section 3 is a more formal introduction to perspective using projective geometry. Section 4 is a discussion of some of the topological aspects of projective geometry. The standard embedded projective plane is defined in Section 5. Section 6 is a description of some of the main results of synthetic projective geometry. Finally, Section 7 is a brief discussion of the history of perspective and projective geometry, and the axiomatic foundations of projective geometry.

The primary purpose of this chapter is to introduce projective geometry and to discuss it in relation to Euclidean geometry. The reasons for doing this are twofold. First, Euclidean geometry is well known and hence is a good foundation for the discussion of a "new" geometry. Second, the geometry of real objects is Euclidean, while the geometry of imaging an object is projective; hence the study of computer graphics naturally involves both geometries.

Projective geometry is, more accurately, an extension of Euclidean geometry. As a consequence, the techniques of projective geometry are not only more powerful than those of Euclidean geometry, but they are just as efficient and easy to use. In fact, the reason projective geometry is worth learning is that it is often more efficient and easier to use than Euclidean geometry.

SECTION 1: THE STUDY OF GEOMETRY

If the subject of geometry is mentioned, you might automatically think of Euclidean geometry. Most people are surprised to learn that there are actually many geometries. The primary topic of this book is a geometry other than Euclidean geometry; the topic is projective geometry.

In Euclidean geometry one studies the actual shapes of objects. More accurately, one studies those properties of (planar and solid) objects that are unchanged by rotations, translations, and reflections of the objects. Such properties include congruence of segments (being of equal length), congruence of angles (being of equal angle measure), and parallelism. In projective geometry one studies the way objects are seen.

The following examples are intended to clarify the distinction between the actual shape of an object and the way the object is seen.

Consider first a set of railroad tracks in the desert that disappear off into the horizon. If you were standing in the middle of the tracks and looking down the tracks, you would see what is illustrated in Figure 1.1. Observe that although the tracks may actually be parallel, they seem to meet on the horizon. Further, although the lengths of all the railroad ties may be equal, their lengths appear to diminish as you look down the tracks. And although the railroad ties may be spaced a uniform distance apart, the distance from one tie to the next appears to diminish as you look down the tracks. Finally, although four (congruent) right angles are determined by any given railroad tie and the tracks (two interior right angles at each end of the tie), the angles you see do not appear to be equal.

The rim of a coffee cup may be perfectly circular, and it may, in fact, look perfectly circular when viewed from above. But from the side the rim will appear to be elliptical (see Figure 1.2).

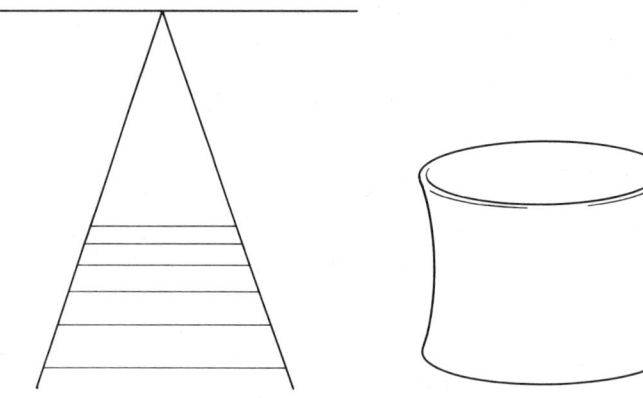

Figure 1.1 Railroad tracks.   **Figure 1.2** A coffee cup.

Thus lengths, angles, parallelism, and shapes can become distorted when we view objects. Here are two more examples that reinforce this observation.

Look at the corner of a room (see Figure 1.3). Although we know that three right angles meet at the corner forming a (three-dimensional) geometric figure consisting of $3 \times 90° = 270°$, what we see (or what we see in a picture of the corner) is a planar point about which we know there are 360°. Thus in what we see, at least one of the three right angles must be distorted.

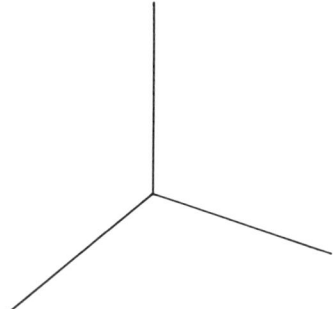

Figure 1.3 The corner of a room.

Consider the checkerboard pattern illustrated in Figure 1.4. Observe that—as with the railroad tracks—even though this figure communicates the idea of a checkerboard, the lengths, angles, and parallelism of a true checkerboard have all been distorted. (The checkerboard pattern is important in art since it is a framework that allows an artist to copy a picture identically, to a different scale, or in perspective: By superimposing checkerboards on both the original picture and on the canvas of the new picture, an artist can easily reproduce the picture square by square using proportionality. See Exercise 3.15.)

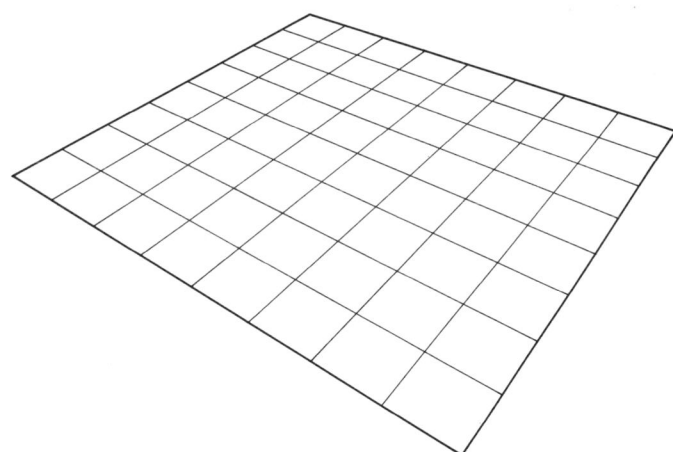

Figure 1.4 A checkerboard.

The preceding examples have illustrated that Euclidean properties of an object need not be present in pictures of the object. The next example illustrates, conversely, that a property present in a picture need not reflect a Euclidean property.

On the horizon you see mountains A, B, and C (see Figure 1.5). Although B may appear to be between A and C, a topographical map might reveal that B is not really between A and C at all (see Figure 1.6). It may just be that you are looking at things from point P; if you were to look at things from point Q instead, then C would appear to be between A and B!

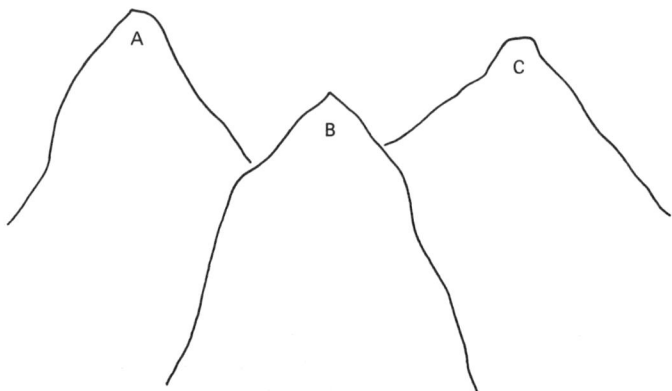

Figure 1.5 Three mountains.

These examples all illustrate two different aspects of visual communication—the first aspect is *what* a picture (for example) is trying to communicate (this is the realm of Euclidean geometry), and the second aspect is *how* the picture communicates (this is the realm of projective geometry).

The primary topics of this book are Euclidean geometry, projective geometry, and the relationships between the two. These subjects are the mathematical foundation for the geometry of computer graphics. Knowing the geometric theory behind computer graphics is essential to the production of quality graphics products.

There were other reasons that originally led to this study. In particular, understanding the universe in which we live is, to a large extent, based on observation; when the interpretation of observations is ambiguous, clear statements about how objects are seen are essential to understanding. For example, the concept introduced in the example of the three mountains—namely, the apparent displacement of objects due to a change in position of the observer—is called *parallax shift* by astronomers. Astronomers use parallax shift to measure distances between heavenly bodies. In fact, the Greeks knew that if the earth revolved around the sun, then there should be a parallax shift in the stars; this was the principal phenomenon for which the Danish astronomer and nobleman Tycho Brahe was looking (one of Brahe's goals was verification of the heliocentric theory). Unfortunately, his instruments and those of the Greeks were not accurate

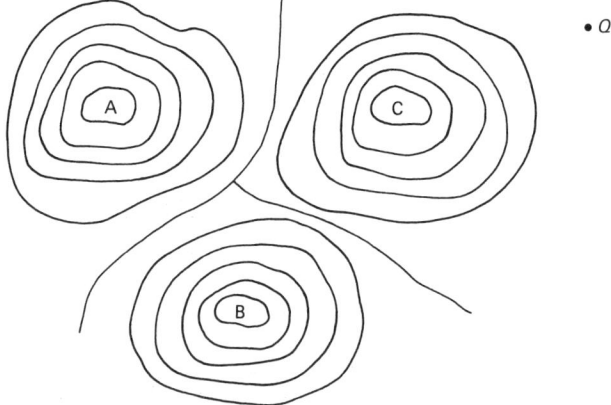

Figure 1.6 A topographical map of the three mountains.

enough to detect the small parallax shifts that even the closest stars display. Brahe's observations were not in vain, however; his young German assistant Johann Kepler used Brahe's observations to deduce three famous laws of planetary motion.

Since input/output devices for computer graphics systems are oriented toward numbers (coordinates of points), our approach will be primarily analytic. There are two basic approaches to geometry: synthetic (the qualitative approach) and analytic (the quantitative approach). Synthetic geometry concentrates on proving qualitative results via axioms. Analytic geometry is based on the association of numbers to points (in the form of coordinates) and ultimately the use of algebraic techniques to solve geometric problems. Historically, the use of algebraic techniques to solve geometric problems meant proving results. Today it means establishing communication with computer graphics devices as well.

One final point: An interesting aspect of our study will be self-referencing—in the rest of this book we use pictures to illustrate how we illustrate in pictures. In other words, pictures play two closely related, yet different, roles for us: first as objects of our study, and second as a means for communicating how we see what we see.

SECTION 1 EXERCISES

1. Describe scenes (other than those given in the text) illustrating that perspective distorts lengths, angles, parallelism, betweenness, and area. Are there any other such Euclidean properties distorted by perspective? (*Hint:* Consider photography, for example.)
2. Are there any Euclidean properties that are not distorted by perspective?
3. Cut out six pictures from newspapers or magazines that illustrate the distortions described in this section. In particular, find some pictures that illustrate the distortion of parallelism. (Save these pictures; you will need them in the next section.)

4. What properties contribute to a well-drawn picture?
5. Carefully use the checkerboard-pattern algorithm described in this section to reproduce a picture in each of the following ways.
   (a) Identically
   (b) Magnified by a factor of 2
   (c) Contracted by a factor of 2
   (After seeing how to draw an accurate picture of a checkerboard in one- and two-point perspective in the exercises of Section 3, you will also be able to reproduce your picture in one- and two-point perspective.)
6. What is wrong in Figure 1.7?

Figure 1.7 William Hogarth, *False Perspective* (reproduced by courtesy of the Trustees of the British Museum).

SECTION 2: A MODEL FOR PERSPECTIVE

In order to motivate the theory of (analytic) projective geometry further, let us take a brief look at the theory of perspective. (We initially concentrate our attention on perspective. Parallel projection—for those who are already familiar with it—will be discussed later as an extended form of perspective.)

In perspective (see Figure 1.8) the artist's canvas is thought of as a transparent screen through which the artist views a scene. Light rays from each point in the scene pass through the screen and enter the eye. The set of points in which these light rays (or *projectors*) meet the screen is what the artist draws. As illustrated in the previous section, perspective distorts lengths, angles, parallelism, and shapes. Yet the eye and mind automatically accept and synthesize this perspective into an impression of reality. (An artist drawing in perspective uses a variety of skills—some inherent and some learned, some quantitative and some qualitative. All of this must be made routine in computer graphics if the same results are to be achieved.)

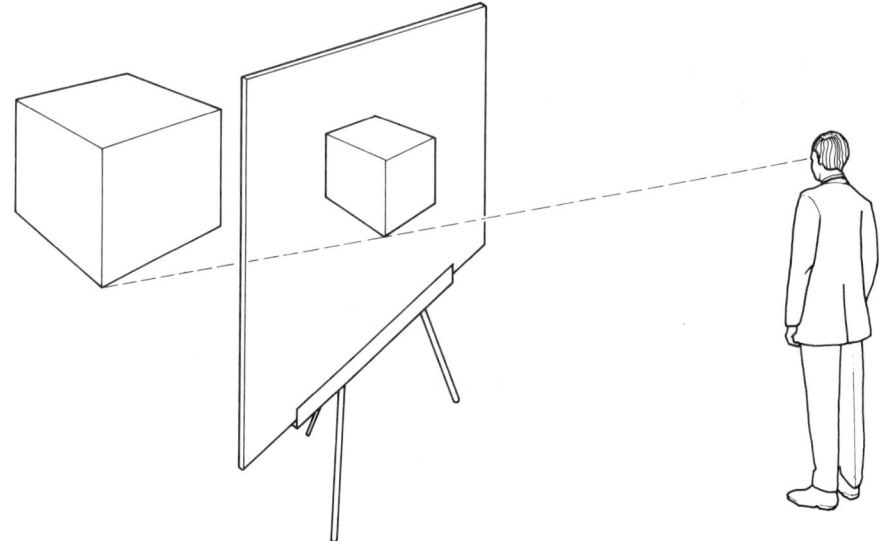

Figure 1.8 Viewing in perspective.

A physical model for the study of perspective (see Figure 1.9) might thus consist of an *object* in space (normally this object would be two- or three-dimensional), an *image plane,* and an *eye* on the opposite side of the image plane from the object. Requiring the eye to lie on the opposite side of the image plane from the object avoids image reversal, as occurs in photography (see Figure 1.10); this is not a necessary assumption (in fact reversal may be a desired effect)—it is just being made temporarily for simplicity.

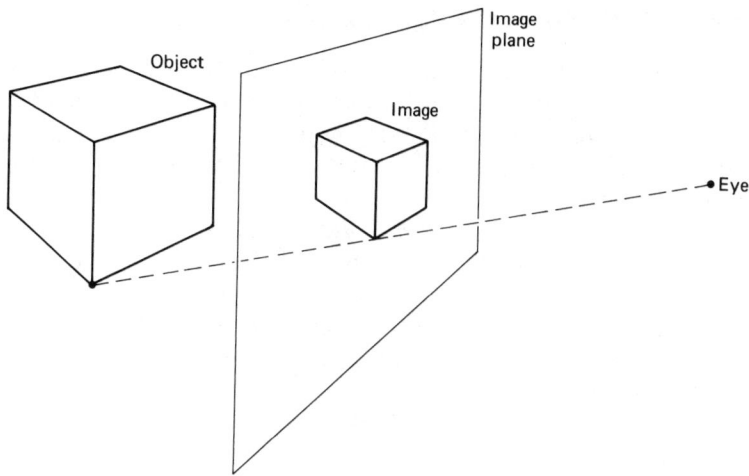

Figure 1.9 A physical model for viewing in perspective.

Actually, what we refer to as perspective is more accurately referred to as *linear perspective*. Theories of perspective other than linear perspective—as discussed in this book—can be obtained by replacing the viewplane by another surface (such as part of a sphere, for example) or by replacing linear projectors by curved projectors. We will pursue these alternate theories no further than stating their existence.

There is a natural correspondence between the points of the object and points of the image, which is established by associating to each point of the object the point of intersection of the image plane with the line containing the object point and the eye.

Although the object of interest may be three-dimensional, it is planar, or two-dimensional, perspective that will be of fundamental interest to us. The reason for this is that solid, or three-dimensional, objects are bounded by two-dimensional surfaces. A perspective drawing of a solid can thus be considered

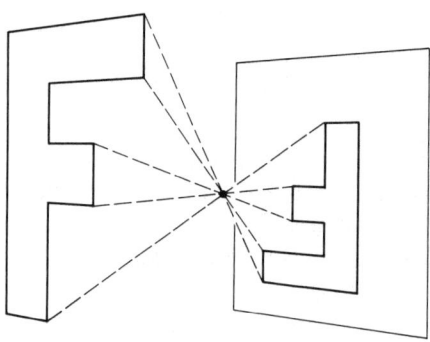

Figure 1.10 Image reversal.

as a composite of perspective drawings of the surfaces bounding the object. The simplest possible such drawing is a planar perspective drawing.

Consequently, a better physical model for the study of perspective (see Figure 1.11) might consist of a two-dimensional *object* in an *object plane,* an *image plane*, and an *eye* on the opposite side of the image plane from the object, which lies neither in the object plane nor in the image plane. Specifying that the eye lies neither in the object plane nor in the image plane eliminates the singular situations of viewing a planar figure on its edge, and of obtaining a point image of a two-dimensional figure. Again, requiring the eye to lie on the opposite side of the image plane from the object avoids image reversal. Also note that the object and image planes need not be perpendicular—there is no reason to expect that the plane containing a face of a three-dimensional object in which we are interested, be perpendicular to the image plane; in fact it will not be in general.

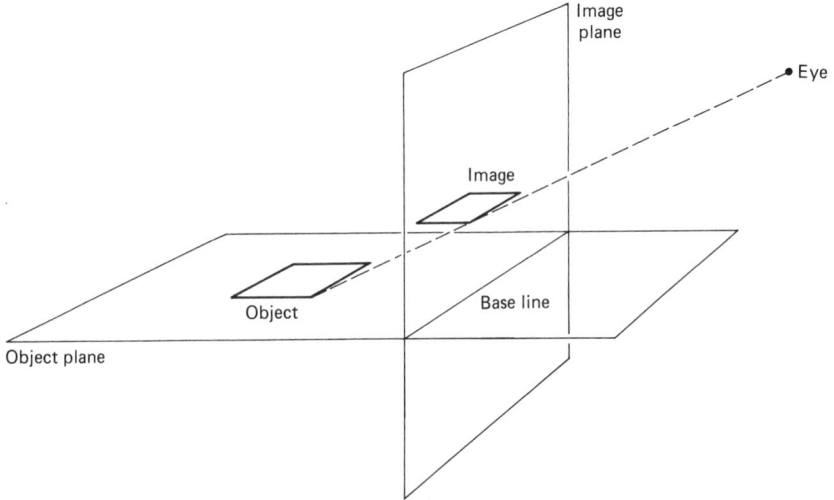

Figure 1.11 A planar model for perspective.

There is a one-to-one correspondence between points of the object and points of the image, which is established by associating to each point of the object the point of intersection of the image plane with the line containing the object point and the eye.

To add realism to a picture of a scene containing, for example, a pair of railroad tracks that disappear off into the distance (see Figure 1.12), an artist adds a vanishing point to the picture. A *vanishing point,* in general, is that point in a picture at which two parallel lines (or line segments; see Figure 1.25) in the scene appear to meet.

In our model of perspective the vanishing point is the limit of the images of the object points on both parallel lines as the object points move further and further away from the eye. This vanishing point can be located geometrically

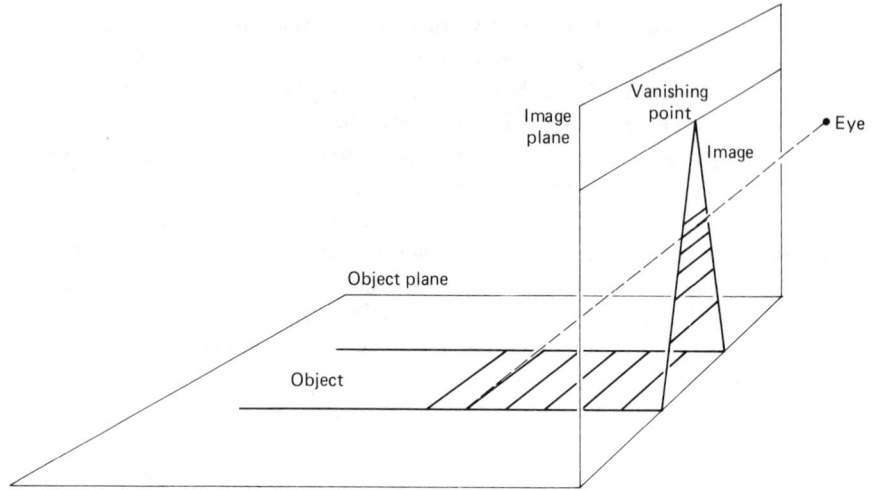

Figure 1.12 The vanishing point for a picture of railroad tracks.

as follows (see Figure 1.13): A line and a point not on the line determine a unique plane in Euclidean three-space. Thus for each of the parallel lines l_1 and l_2 in the object plane, there is a unique plane—π_1 and π_2, respectively—containing it and the eye. Since π_1 and π_2 contain a common point (namely, the eye), their intersection is a line l; l is parallel to l_1, l_2, and the object plane. The point at which l meets the image plane is the vanishing point associated to l_1 and l_2. The

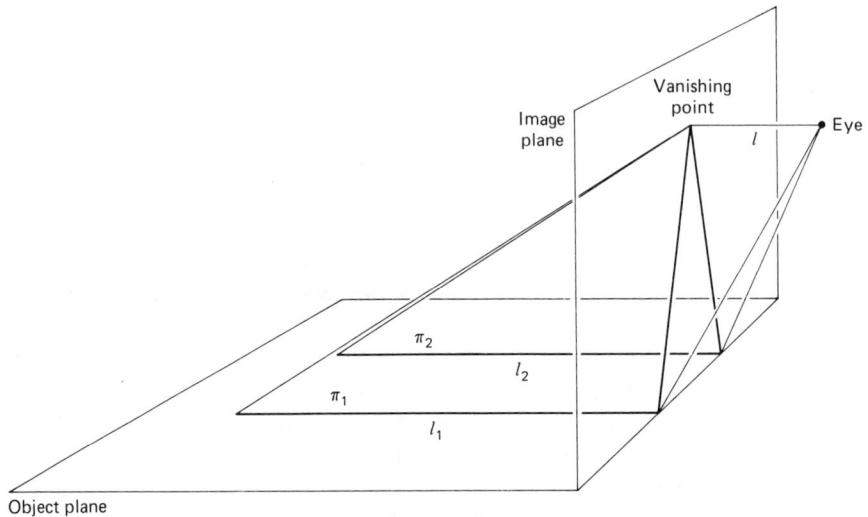

Figure 1.13 Locating the vanishing point.

reason for this is that all the lines of sight between object points P along l_1 and the eye lie in π_1, and all the lines of sight between object points P along l_2 and the eye lie in π_2. Thus the limiting position of the lines of sight to object points along both lines (as the object points move further and further away from the eye) must be the intersection l of the planes π_1 and π_2 containing the object lines and the eye.

Actually it is not necessary to have two parallel lines to determine a vanishing point; only one line is necessary (see Figure 1.14). Given any object line l_0,

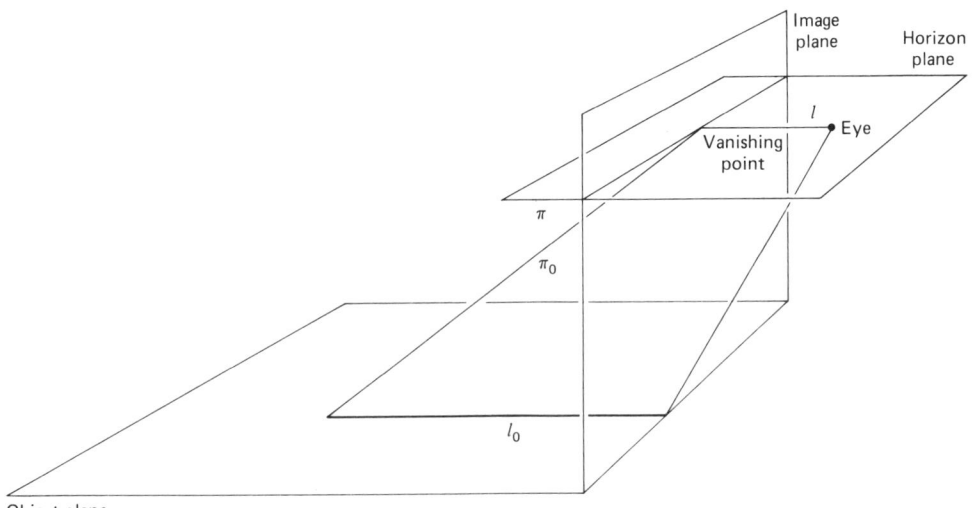

Figure 1.14 Locating the vanishing point of a line.

consider the plane π_0 containing it and the eye. This plane meets the plane π, which contains the eye and is parallel to the object plane—the *horizon plane*—in a line l, and l meets the image plane in a point. This point is the vanishing point for the original object line l_0 and also for any other line in the object plane parallel to l_0. Indeed, it is exactly the vanishing point for any pair of lines in the object plane parallel to l_0 as constructed above.

We should note that the object lines in Figures 1.13 and 1.14 are drawn perpendicular to the image plane. This perpendicularity has not been used nor is it necessary (see Figure 1.15). In fact as the lines in the object plane take on all possible inclinations (relative to the image plane), a line of vanishing points in the image plane is described; this is the *vanishing line,* or *horizon.* It is the intersection of the image plane with the horizon plane.

Even though the object and image planes of Figure 1.15 are perpendicular, it is not necessary for the object and image planes to be perpendicular to locate

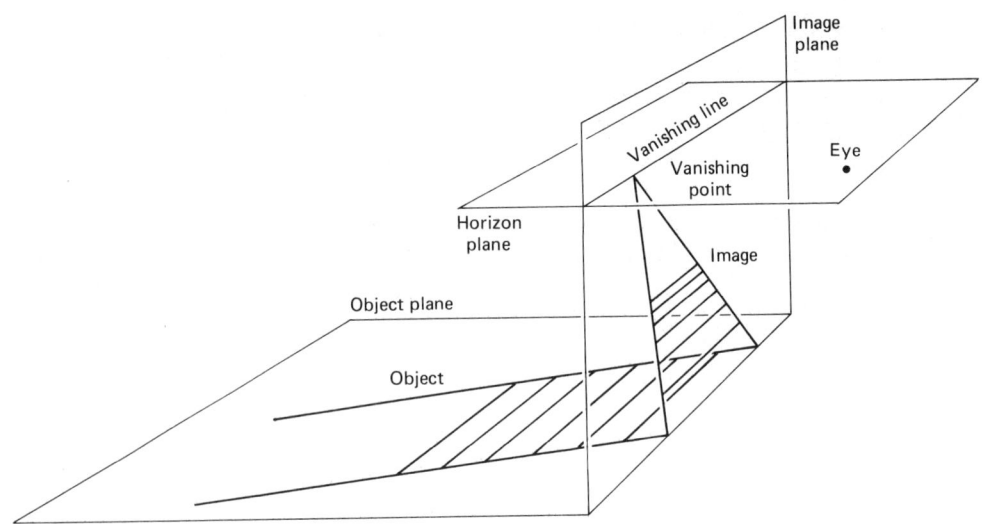

Figure 1.15 The vanishing line.

vanishing points and the vanishing line (see Figure 1.16). In particular, even if the object and image planes are not perpendicular, the vanishing line is still the line of intersection of the image plane with the plane that contains the eye and is parallel to the object plane, and the vanishing point for a line in the object plane is the intersection of the vanishing line and the plane containing both the original line and the eye.

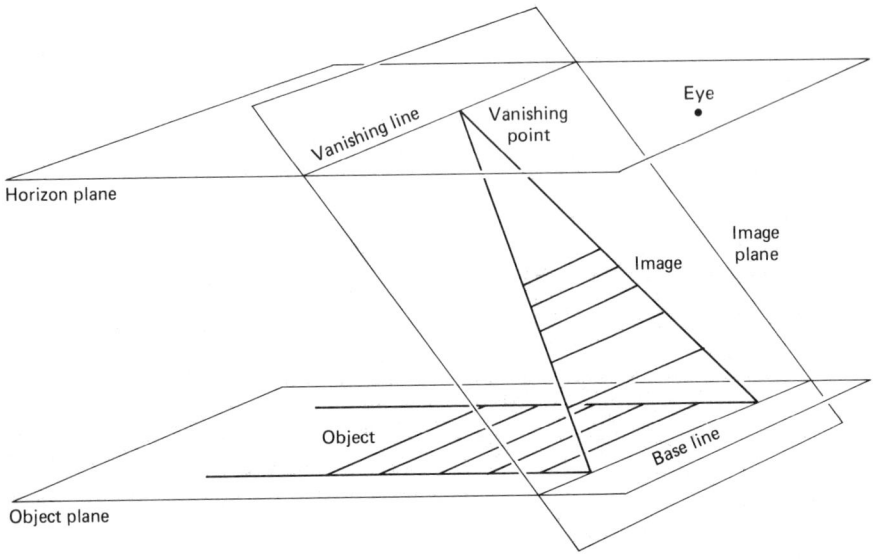

Figure 1.16 A general configuration for vanishing points and vanishing lines.

(If you ever thought about why you see a horizon when you look out across the Earth, you might have thought that it was because of the spherical shape of the Earth (see Figure 1.17(a)). However that is not the case (see Figure 1.17(b))— you would see a horizon even if the Earth were flat! The spherical shape of the Earth *can* be detected by observing that at sea, for example, the mast of a distant ship is visible before the body of the ship is visible; *this* would not be so on a flat Earth.)

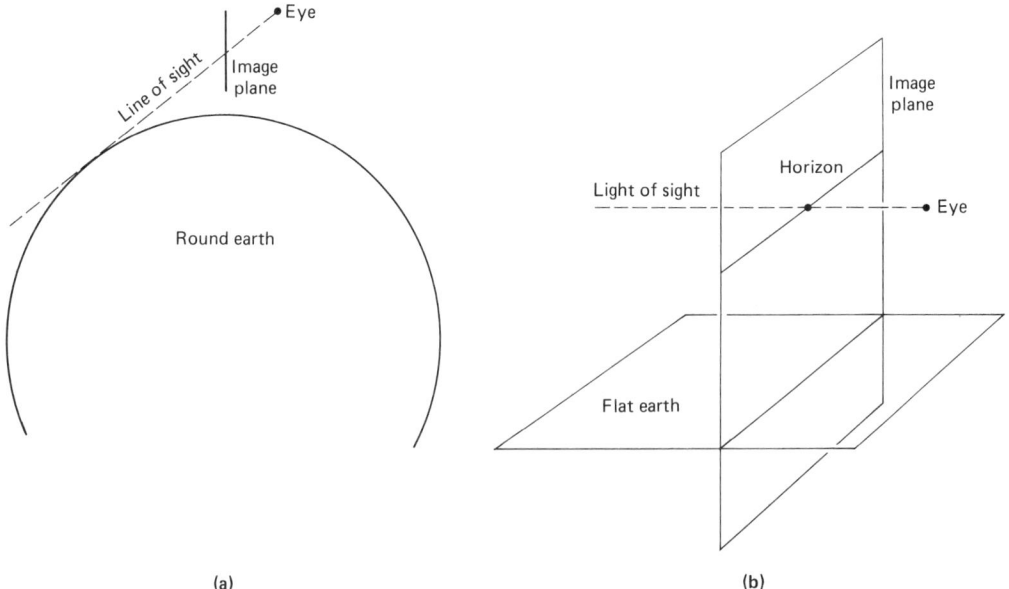

Figure 1.17 Why you see a horizon.

One final (and *very* important) point: There are two philosophically different ways to look at geometry—extrinsically and intrinsically. Extrinsic geometry relies on the existence of a larger ambient space; intrinsic geometry does not. For example, in this section we looked at models for perspective (see Figure 1.8, for example) as external observers; in these models we see the image plane as *a* Euclidean plane embedded in Euclidean three-space. The observer in the model, however, sees the image plane as *the* Euclidean plane. The geometry of *a* Euclidean plane embedded in Euclidean three-space (see Figure 1.18(a)) is *extrinsic* Euclidean geometry; the geometry of *the* Euclidean plane (see Figure 1.18(b)) is *intrinsic* Euclidean geometry. The difference between intrinsic and extrinsic geometry is precisely the difference between pictures as an object of our study and pictures as a means for communicating how we see what we see, as mentioned at the end of Section 1. Throughout this book extrinsic and intrinsic Euclidean planes are distinguished from each other by the use of the indefinite articles *a* and *an*, and the definite article *the*, respectively.

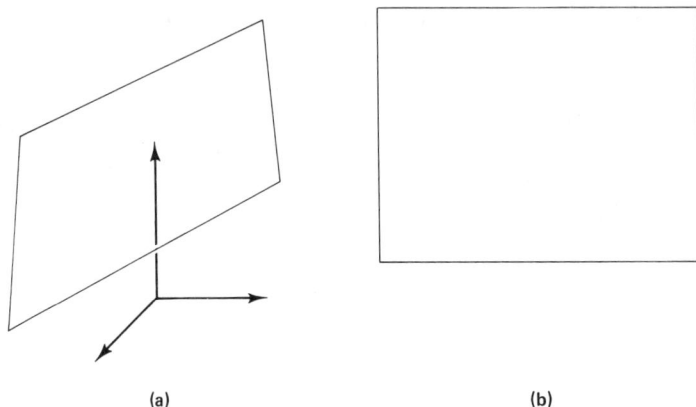

Figure 1.18 (a) *A* Euclidean plane (extrinsic) and (b) *the* Euclidean plane (intrinsic).

SECTION 2 EXERCISES

1. Using a straightedge, locate the vanishing points for the pictures you cut out in Exercise 1.3, which illustrate the distortion of parallelism in perspective. (The vanishing point associated with a pair of parallel line segments is, by definition, the vanishing point associated with the parallel lines containing the segments. Thus you may need to attach the pictures to larger auxiliary sheets of paper to do this.) Observe that there can be more than one vanishing point per picture; why is this so? Do any relations exist between the various vanishing points?
2. (This exercise requires an acetate sheet and a grease pen or other suitable marker.) Tape an acetate sheet to a window, locate an object on the other side of the window that has parallel lines (or line segments) in it that meet in a reasonably close vanishing point, and sketch the object.
3. Draw freehand pictures of the following objects (using at least one vanishing point in parts (a), (b), and (d)).
   (a) A square in perspective
   (b) A set of railroad tracks that disappear off into the horizon
   (c) A coffee cup in perspective
   (d) A checkerboard in perspective
   (e) A cube in perspective
4. In a perspective picture of a checkerboard, what can be said about the vanishing points of the diagonals?
5. Why is it possible to have several different vanishing points and vanishing lines in a single picture?
6. Draw a freehand picture of your home in perspective.
7. Where might image reversal be a desired effect?
†8. Explain why specifying that the eye lies neither in the object plane nor the image plane eliminates the singular situations of viewing a planar figure on its edge and of obtaining a point image of a two-dimensional figure.
9. Is the study of the Euclidean line extrinsic or intrinsic Euclidean geometry? Is the study of a Euclidean line embedded in the Euclidean plane extrinsic or intrinsic?

10. Discuss some possible types and applications of nonlinear perspective. (*Hint:* See Figure 1.19)

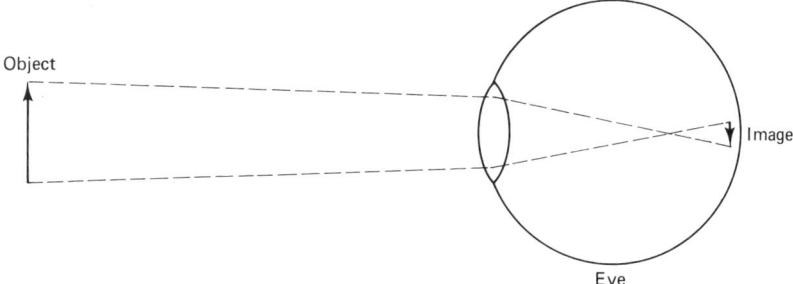

Figure 1.19 Nonlinear perspective.

11. What specific artistic skills must be made routine in computer graphics if the results of an artist are to be achieved?

SECTION 3: THE PROJECTIVE PLANE

As indicated in the previous section, planar, or two-dimensional, perspective is the fundamental form of perspective drawing. The reason for this is that solid, or three-dimensional, objects are bounded by two-dimensional surfaces; thus a perspective drawing of a solid can be considered as a composite of perspective drawings of surfaces bounding the object, and the simplest possible such drawing is a planar perspective drawing.

In planar perspective drawing, the scene of interest normally lies in a half-plane on the opposite side of the image plane from the eye, the image lies in a band between the *base line* (that is, the intersection of the object plane and the image plane) and the horizon, and there is a one-to-one correspondence between the set of object points and the set of image points. This one-to-one correspondence can be regarded as a one-to-one transformation (or function) from the set of object points to the set of image points.

A geometric theory is the study of a set of objects and one-to-one transformations of those objects. (For example, planar Euclidean geometry is the study of the Euclidean plane and Euclidean transformations—rotations, translations, and reflections—of the Euclidean plane.) It is unnatural to build such a theory with objects as different as a half-plane (such as the half of the object plane in which the scene of interest lies) and an infinite band (such as the band in the image plane between the base line and horizon).

It would be natural to try to drop the restrictions that the scene of interest lies in a half-plane and that its image lies in an infinite band and instead make Euclidean planes the objects of our study. However, this creates problems since there is no natural way to extend the one-to-one correspondence between the set of object points and the set of image points to a one-to-one correspondence between the set of *all* points in the object plane and the set of *all* points in the image plane. In particular, suppose we were to try to define such a one-to-one correspondence by associating to each point of the object plane the point of intersection of the image plane and the line containing the object point and the eye (see Figure 1.20). Unfortunately, this correspondence is not defined for all points in the object plane—if m denotes the line of intersection of the object plane with the plane that contains the eye and is parallel to the image plane, then for any point P on m, the line containing P and the eye does not meet the image plane. So the correspondence is not defined for points on m. Similarly, no point in the object plane corresponds to a point on the vanishing line.

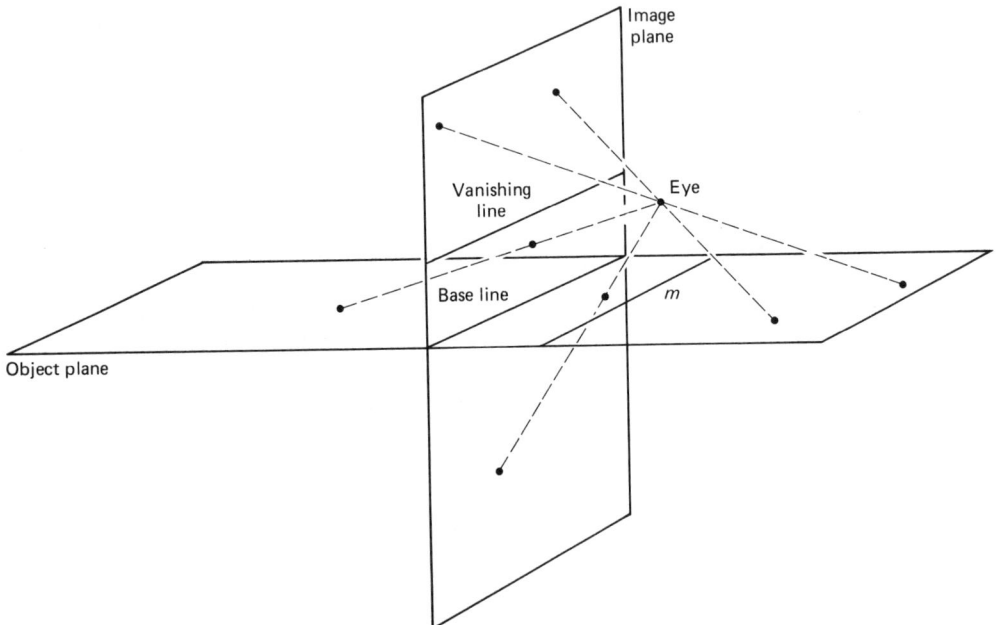

Figure 1.20 The extended correspondence.

This dilemma is resolved by introducing ideal points, or points at infinity, in the object plane, and ideal points, or points at infinity, in the image plane. These points are introduced in the same way in both planes: We group the lines in either plane into families, all the lines in each family being parallel to one another. For each family of lines, a single *ideal point* is attached to the plane. This ideal point is attached to each parallel line in the family at infinity; in other words, it is attached in such a manner that all parallel lines in the family intersect at this ideal point. More precisely—see Section 4 for more details—if the Euclidean plane is thought of as corresponding to the *interior* of a circular disk (see Figure 1.21), then attaching the ideal points to the Euclidean plane corresponds to attaching the boundary points to the disk in such a manner that antipodal (or

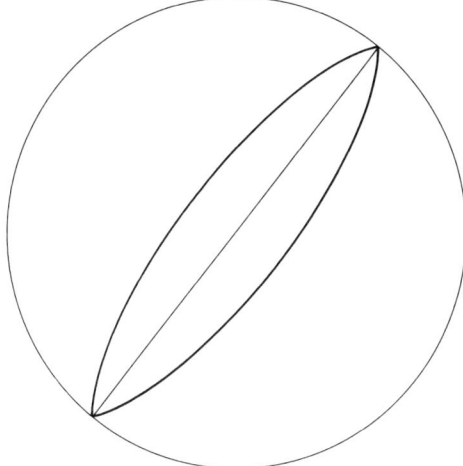

Figure 1.21 A model of the Euclidean plane.

diametrically opposite) points are identified (or considered the same). See Figure 1.22.

Having added ideal points to both the object and image planes, the partial correspondence between the object and image planes constructed earlier becomes a one-to-one correspondence between the object plane completed with its ideal points and the image plane completed with its ideal points. This is because parallel lines in the object plane now meet in the object plane (in an ideal point), while their images (as before) meet in the image plane (at a vanishing point); these points correspond to one another by the very definition of ideal points in

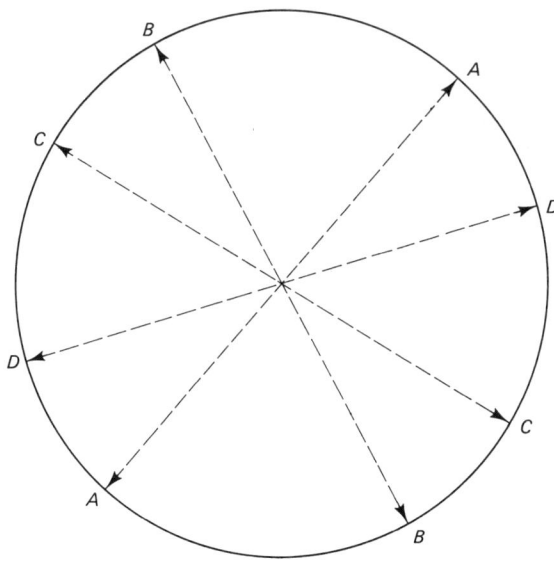

Figure 1.22 Ideal points being attached to the model of the Euclidean plane.

Sec. 3 The Projective Plane

the object plane and the way in which ideal points were attached to the object plane. Similar remarks pertain to the ideal points introduced to the image plane.

Ideal points are a mathematical invention; they are invented to achieve a one-to-one correspondence as indicated above. The ideal points attached to a Euclidean plane should be viewed as abstract points, each representing a direction, which are added to the plane, as opposed to Euclidean points that have somehow been moved or become attached to the plane. (Again, see Section 4 for more details.)

The object plane completed with its ideal points, the image plane completed with its ideal points, or any other Euclidean plane completed with its ideal points is called a *projective plane*. Although we have thus far been concerned primarily with Euclidean planes embedded in Euclidean three-space, "any other" also includes *the* (intrinsic) Euclidean plane: the projective plane obtained by completing *the* Euclidean plane with its ideal points is *the* projective plane. (In our model of perspective, the observer sees the image plane as *the* projective plane.) A projective plane obtained by completing an embedded Euclidean plane with its ideal points is an embedded projective plane. (The space—projective three-space—in which this projective plane is embedded is discussed in Section 5.) As mentioned previously, throughout this book extrinsic and intrinsic projective planes are distinguished from each other by the use of the indefinite articles *a* and *an* and the definite article *the*, respectively.

A Euclidean line completed with its ideal point is called a *projective line*. If a Euclidean line is embedded in the Euclidean plane, the resulting projective line is embedded in the projective plane. The projective line obtained by completing a Euclidean line embedded in Euclidean space is an embedded projective line, too. (The space—projective three-space—in which this projective line is embedded is discussed in Section 5.) The projective line obtained by completing *the* Euclidean line is *the* projective line. Later we will see that the set of all ideal points in a projective plane looks like a projective line. Hence the set of all ideal points in a projective plane is said to form a line, the *ideal line* of the projective plane.

Projective lines are of central importance in projective geometry. One reason for this is that we can use them to reason by analogy when dimensions get too large. In Section 6, for example, we introduce the concept of duality in projective geometry. Although duality for planes may ultimately be more important to us than duality for lines, it is difficult to address duality for planes immediately, because doing so requires that we visualize four-dimensional (!) projective space. Under these circumstances it is easier to address duality for lines (which requires only that we visualize projective three-space; see Section 5) and then reason one dimension higher by analogy. In general, the principle is: If you have trouble visualizing the geometry of a problem, drop things down one dimension, if possible, solve this reduced problem, and then reason up one dimension by analogy.

A one-to-one correspondence (or transformation) between projective planes as constructed above is called a perspective transformation. That is, a *perspective transformation* is a transformation that associates to each point in one plane the point of intersection of the other plane and the projective line containing the given point and a fixed *eye*; the eye is also referred to as the *center of perspectivity*. More generally, a *projective transformation* is either a perspective transformation or a composite of perspective transformations. We will speak of perspectivities and projectivities, respectively, for short.

Composites of perspective transformations arise in the consideration of shadows and reflections. For example, in the case of shadows (see Figure 1.23), a light source perspectively projects an object to its shadow, which lies in a background plane; thus in drawing a picture of a shadow, we are drawing a twofold composite of perspectivities of the original object—the first perspectivity, which has the light source as its center of perspectivity, takes the object to its shadow in the background plane, and the second perspectivity, which has the viewing point as its center of perspectivity, takes the shadow in the background plane to its image in the image plane.

Note that Figure 1.23(b) is drawn in the projective plane; but since ideal points are attached to the Euclidean plane at infinity, it is impossible to tell just from the picture whether the picture is drawn in the projective plane or the Euclidean plane. In general, it is only in context that a picture (such as Figure 1.23(b)) is identifiable as a picture in the projective plane or the Euclidean plane.

Earlier, projective geometry was described as the geometry of objects the way they are seen as opposed to the geometry of objects the way they actually are. We can now refine this description: planar Euclidean geometry is the study of those properties of Euclidean planes that are unchanged under Euclidean transformations (rotations, translations, and reflections). Such properties are *Euclidean properties* and consist of incidence (or intersection) of lines, congruence of segments, congruence of angles, and parallelism. Projective geometry is the

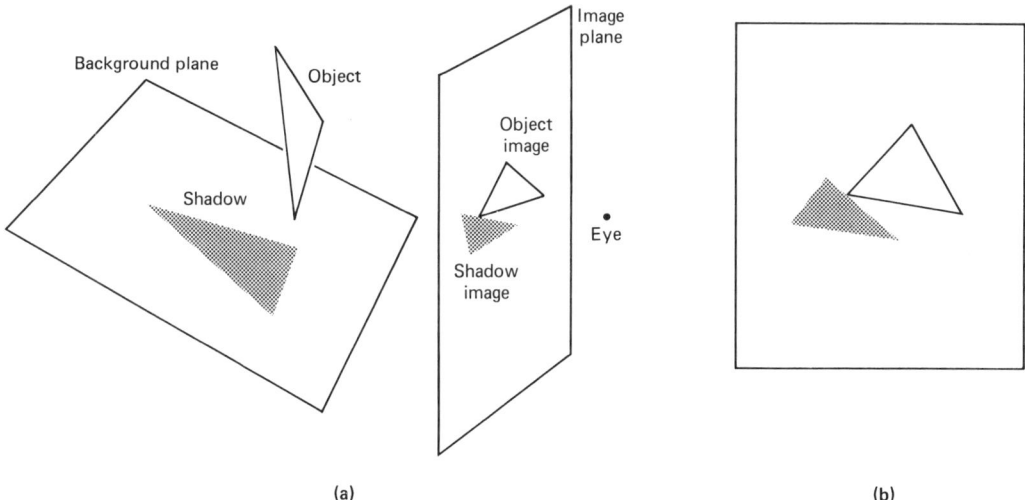

Figure 1.23 Drawing shadows.

study of those properties of projective planes that are unchanged under projective transformations. Such properties are *projective properties*. Incidence of lines is a projective property; congruence of segments, congruence of angles, and parallelism are not projective properties. The discussion of projective properties is pursued in later sections.

SECTION 3 EXERCISES

1. What is the difference between an ideal point and a vanishing point?

In Exercises 2–6, you are asked to construct perspectivities. A *construction* consists of positioning the object, image, and eye in space so that the desired condition is met.

2. Suppose two lengths are arbitrarily given. Construct a perspectivity that takes a segment (any segment) of one of these lengths to a segment (any segment) of the other length.
3. Suppose two angle measures (other than 0° or 180°) are arbitrarily given. Construct a perspectivity that takes an angle (any angle) of one of these angle measures to an angle (any angle) of the other angle measure.
4. Suppose a pair of intersecting lines is given. Construct a perspectivity that takes this pair of lines to a pair of parallel lines. (*Hint:* How could you reverse the process of drawing a picture of railroad tracks in perspective?)
5. Suppose a pair of intersecting lines is given. Construct a perspectivity that takes this pair of lines to a pair of perpendicular lines.
†6. Suppose any two triangles are given. Show that (after possibly a rotation and translation in Euclidean space) there is a perspectivity taking one triangle to the other. (*Hint:* Position the triangles in such a manner that both have a vertex in common and the sides opposite these vertices are parallel. Parallel lines are coplanar.)
†7. Justify each step in the following proof that if an arbitrary quadrilateral *ABCD* is given, then there is a perspectivity that takes this quadrilateral to a square.

Proof. (See Figure 1.24(a).) Let P and Q be the points of intersection of the lines determined by the opposite sides of the quadrilateral. Let l denote the line determined by P and Q, and let R and S denote the points of intersection of the diagonals of the quadrilateral with l. Draw two circles in the plane π_{object} of the quadrilateral—one with P and Q as endpoints of a diameter, and one with R and S as endpoints of a diameter; let O denote one of the two points of intersection of these circles. Let $\pi_{horizon}$ denote any other plane that passes through l, and E denote the point on $\pi_{horizon}$ that is the image of O under a rotation of π_{object} to $\pi_{horizon}$ through l. Let π_{image} denote any plane other than $\pi_{horizon}$ that is parallel to $\pi_{horizon}$. The image in π_{image} of the original quadrilateral under the perspectivity whose center is E is a square. (*Hint:* It suffices to show that the image of the quadrilateral is a rectangle whose diagonals are perpendicular. To do this, first verify that the image is a parallelogram. To show that this parallelogram is a rectangle (see Figure 1.24(b)), use the fact that angle $\angle PEQ$ is a right angle to show that the parallelogram has an interior right angle. To show that the diagonals are perpendicular, use the fact that angle $\angle RES$ is a right angle.)

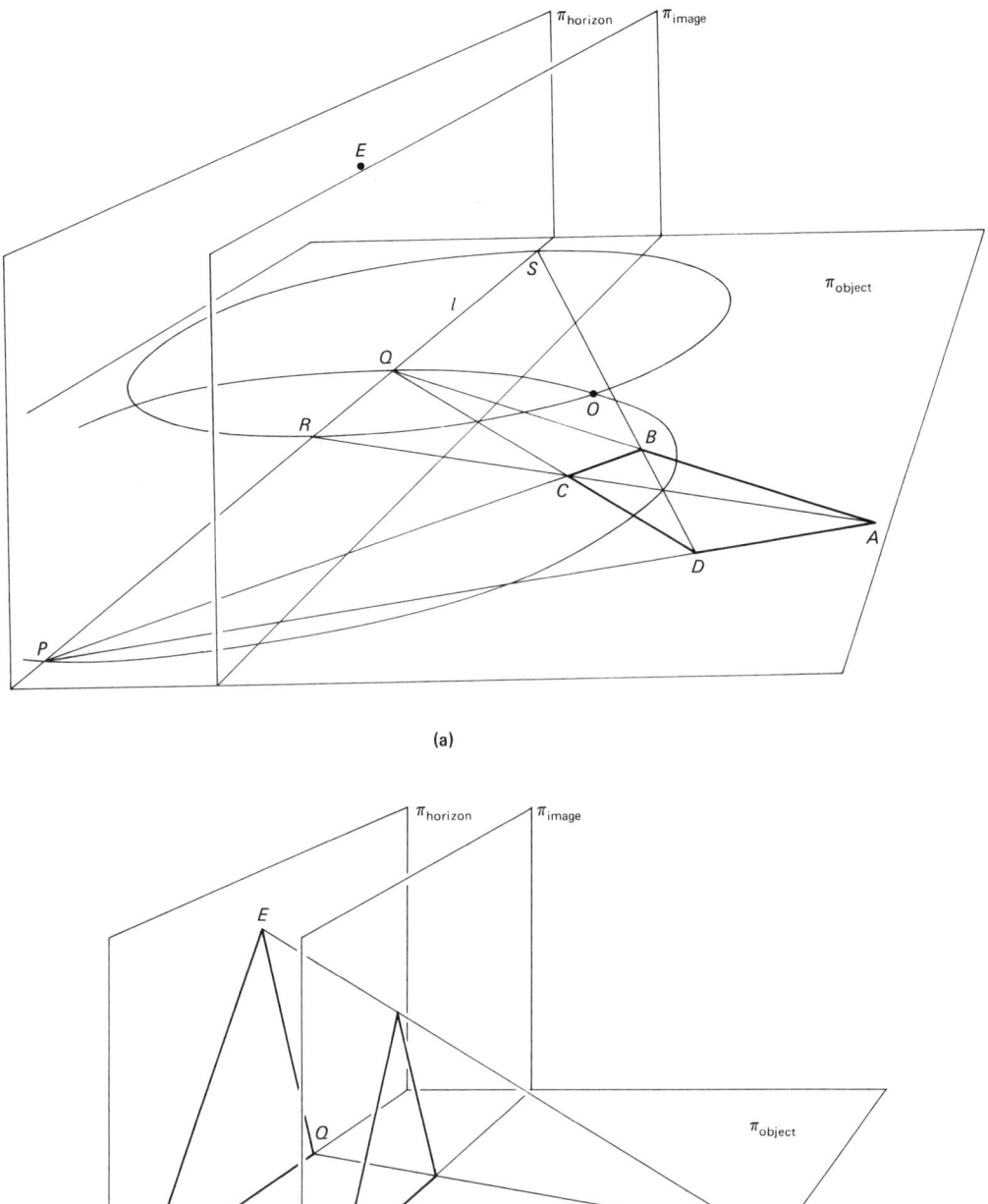

Figure 1.24 Any quadrilateral is a square in perspective.

Sec. 3 The Projective Plane

A square is in *one-* or *two-point perspective* if it has one or two Euclidean vanishing points, respectively.

In Exercises 8–15, use a straightedge to obtain accurate pictures.

8. Use the following algorithm to draw a picture of a square in two-point perspective (see Figure 1.25).
   (a) Draw a horizon l.
   (b) Choose two vanishing points VP_1 and VP_2 on l and the near vertex A of the square so that it does not lie on the horizon.
   (c) Draw line m through VP_1 and A, and line n through VP_2 and A.
   (d) Draw a line p through VP_1, which intersects n between A and VP_2 at B, and a line q through VP_2, which intersects m between A and VP_1 at C. Let D be the intersection of p and q.

   The points A, B, C, and D are the vertices of the desired figure.

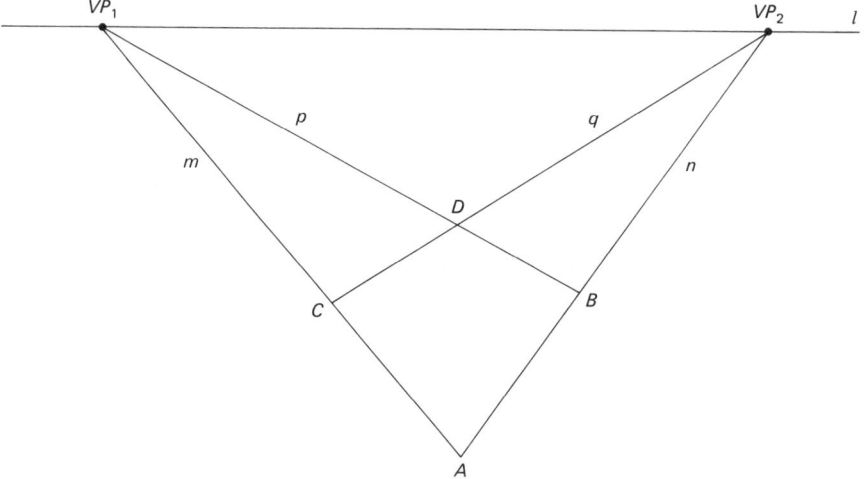

Figure 1.25 A square in two-point perspective.

9. Draw a picture of a square in one-point perspective.
10. Use the following algorithm to duplicate in perspective the picture of the square drawn in Exercise 8 (see Figure 1.26).
    (a) Locate the intersection E of the diagonals of quadrilateral $ABCD$.
    (b) Locate the intersection F of line q and the line through VP_1 and E.
    (c) Locate the intersection G of line m and the line through B and F, and the intersection H of line p and the line through A and F.

    The points C, D, H, and G are the vertices of the desired perspective duplication.
11. Perspectively subdivide the picture of the square drawn in Exercise 8 into four equal parts.
12. Draw an accurate picture of a set of railroad tracks that disappear off into the horizon.
13. Draw an accurate picture of a checkerboard in two-point perspective.

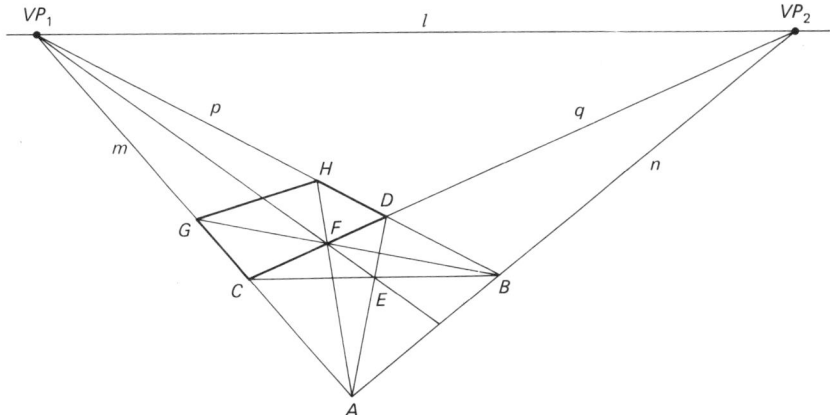

Figure 1.26 Duplicating the square.

14. Draw an accurate picture of a checkerboard in one-point perspective.
15. If a square is subdivided as illustrated in Figure 1.27(a), the 12 labeled points all come close to lying on an inscribed circle (see Figure 1.27(b)). Use this fact to produce a picture of a circle in perspective (see Figure 1.28) by locating the appropriate 12 points in a perspective square and drawing a smooth curve through them.

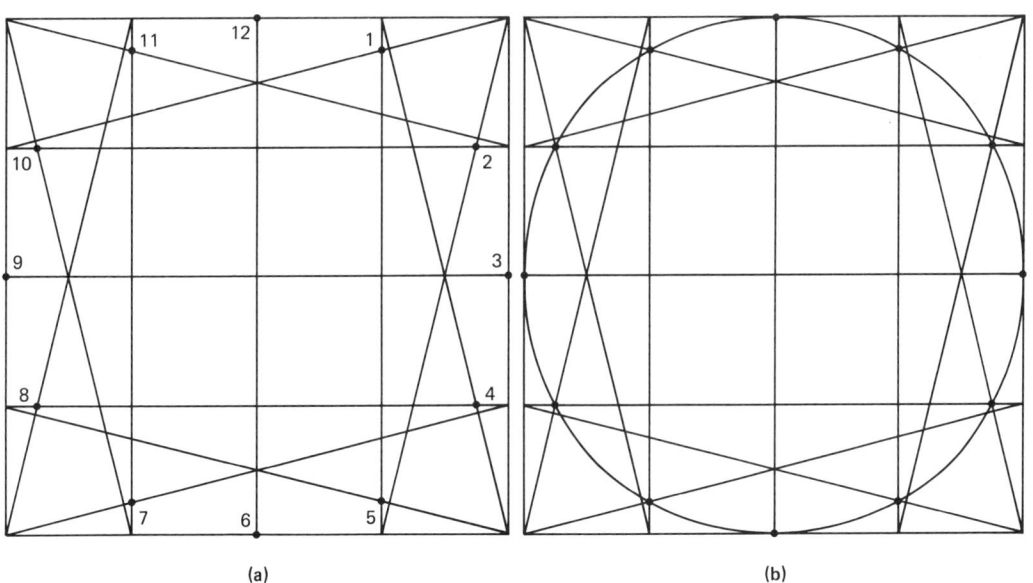

Figure 1.27 The Twelve-Point Circle.

Sec. 3 The Projective Plane

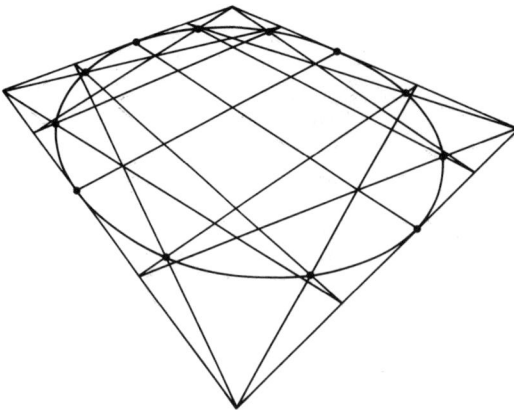

Figure 1.28 Drawing a circle in perspective.

16. Explain why there is only one ideal point associated to each pair of parallel lines in a plane (instead of two, one at each end of the lines). (*Hint:* See Figure 1.29.)

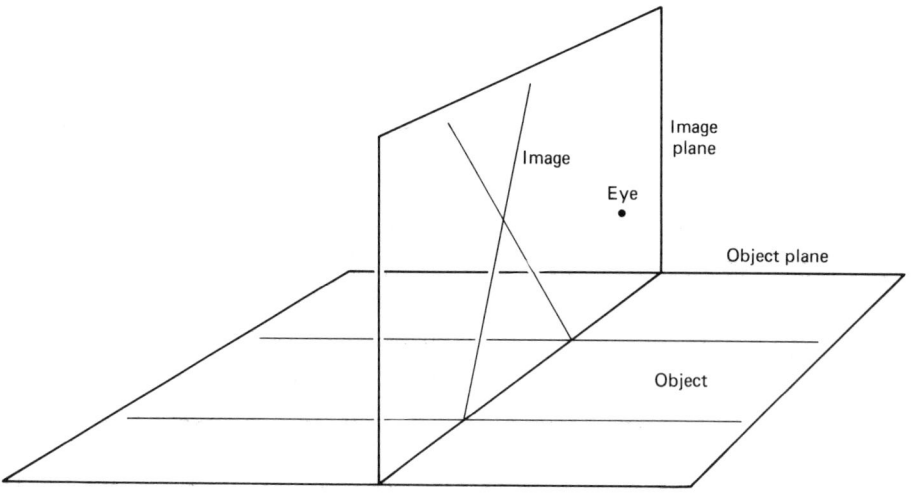

Figure 1.29 A pair of parallel lines determines a single vanishing point.

17. Given a quadrilateral (or, equivalently, a square in perspective), let A and B be the points of intersection of the opposite sides (see Figure 1.30). The diagonals of the quadrilateral meet the (horizon) line l determined by A and B in two points, C and D. The points A, B, C, and D are said to form a *harmonic quadrad*.
    (a) Find the harmonic quadrads associated to the quadrilaterals illustrated in Figure 1.31.
    (b) For each set of three points A, B, and C in Figure 1.32, find the fourth harmonic point D. (You will have to make some choices to do this. The point D is independent of these choices. Why?)

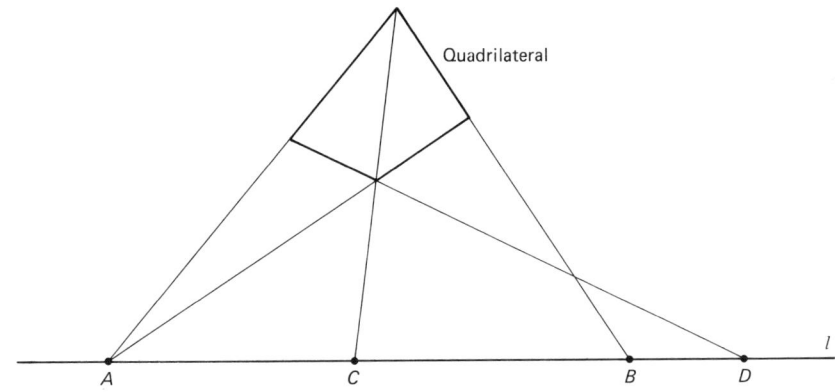

Figure 1.30 A harmonic quadrad.

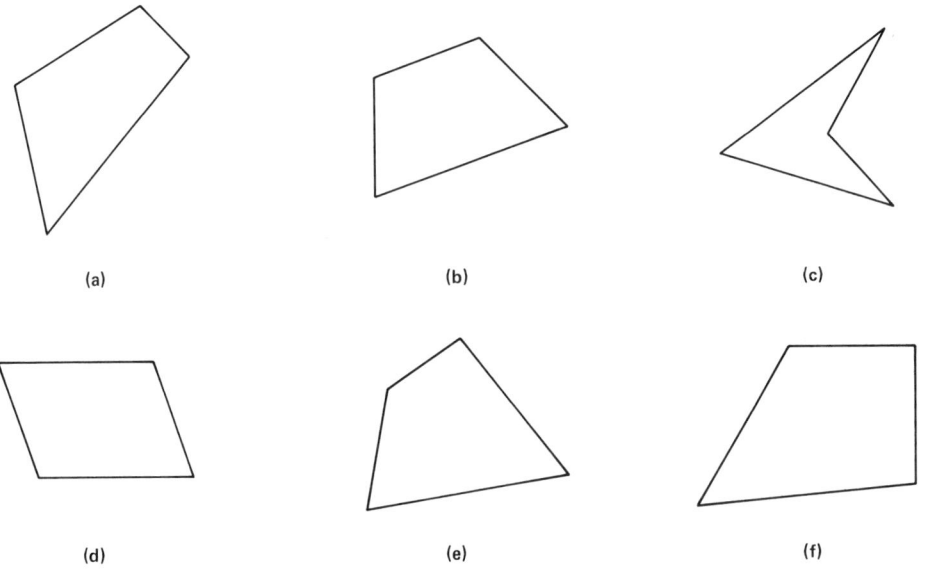

Figure 1.31 Arbitrary quadrilaterals.

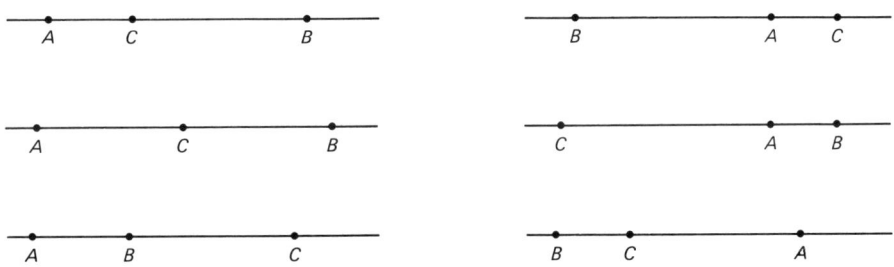

Figure 1.32 Sets of three given points.

Sec. 3   The Projective Plane   25

SECTION 4: A TOPOLOGICAL MODEL OF THE PROJECTIVE PLANE

With the introduction of ideal points, the following question arises: Since a Euclidean plane is of infinite extent, where and how are its ideal points to be attached? One answer to this question is supplied by the subject of topology. Unfortunately, this answer is not useful in studying projective geometry, and in the end we will see that it is perhaps best to think of an ideal point in the projective plane as a direction (the direction of the set of parallel lines to which the ideal point corresponds).

Topology, like geometry, is a study of spaces and shapes. Two objects are topologically equivalent to each other if one can be obtained from the other by a continuous transformation. For example, if we were to take a rubber sheet and distort it (by stretching, bending, and poking it), then as long as we did not tear it, it would be topologically equivalent to itself in its original undistorted state (see Figure 1.33).

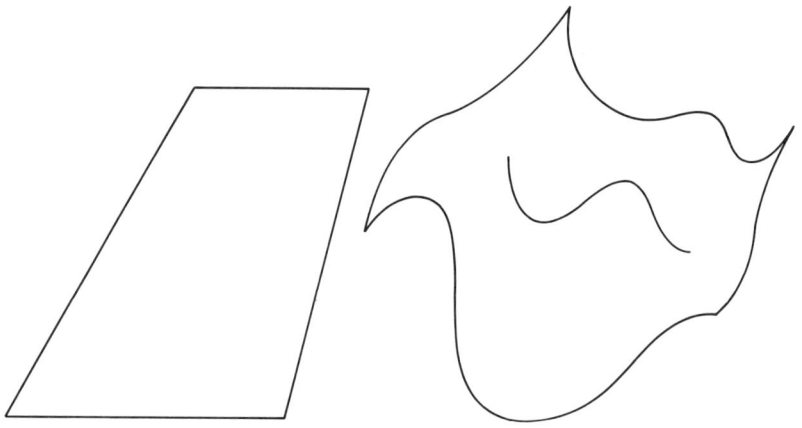

Figure 1.33 Topologically equivalent surfaces.

The properties studied in topology are generally much weaker than the properties studied even in projective geometry. If, for example, a line were drawn on the rubber sheet discussed above, then it and its image in the distorted state would always be topologically equivalent. However if the line after distortion were no longer a straight line, the line and its image would not be projectively equivalent (the property of being a line is a projective property).

Topologically, a Euclidean plane is equivalent to the interior of a circular disk of radius 1. Such an equivalence is described graphically in Figure 1.34: Two families of curves in the Euclidean plane—namely, a family of concentric circles and a family of radial lines through the common center of these circles—are drawn, and the corresponding images of these curves in the disk are also

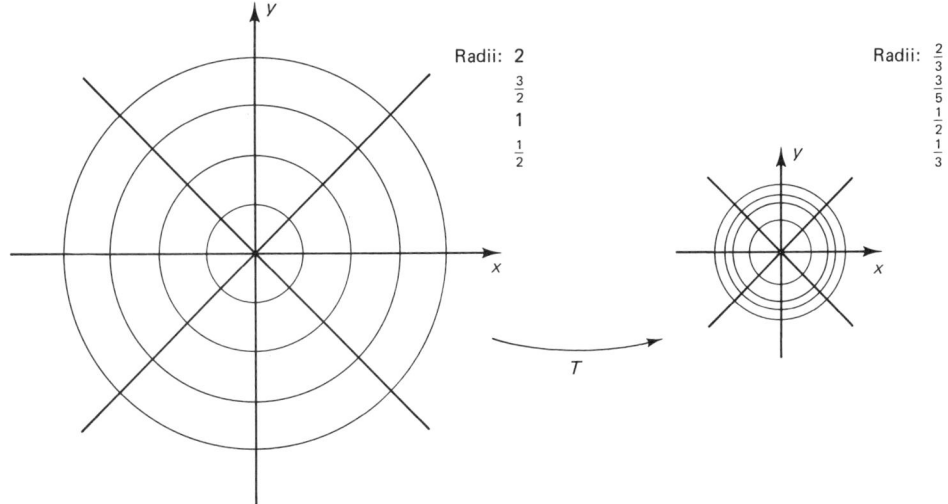

Figure 1.34 A model of the Euclidean plane.

drawn. The transformation (from the Euclidean plane to the disk) describing this equivalence is radial contraction (given in polar coordinates by $T(r, \theta) = (r/(r + 1), \theta)$). This equivalence could alternately be described geometrically by drawing the horizontal and vertical grid lines in the Euclidean plane and their images in the disk (see Figure 1.35). Note, however, that now the only lines in the plane whose images are still true lines in the disk are those lines that pass through the common center; all other lines in the plane correspond to curved lines in the disk.

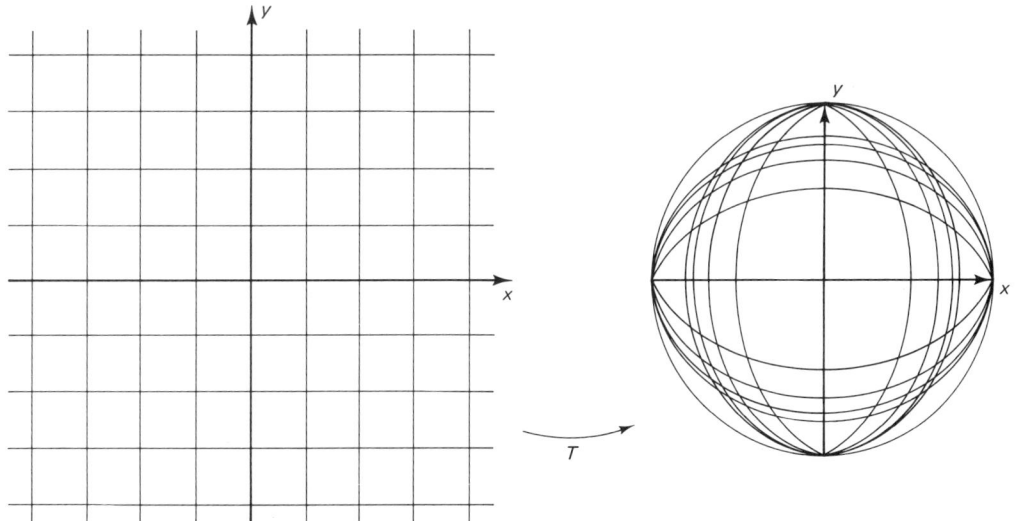

Figure 1.35 The model of the Euclidean plane again.

Sec. 4   A Topological Model of the Projective Plane

When we attach ideal points to a Euclidean plane, we are attaching the boundary points to the circular disk topologically in a very special way: Each family of parallel lines in the Euclidean plane has a unique representative that passes through the origin; the image of this representative is a diameter of the disk. Since we attach only one ideal point to the Euclidean plane for every set of parallel lines (see Exercise 3.16), we must identify as one point the antipodal (or diametrically opposite) endpoints of each and every such diameter (see Figure 1.36).

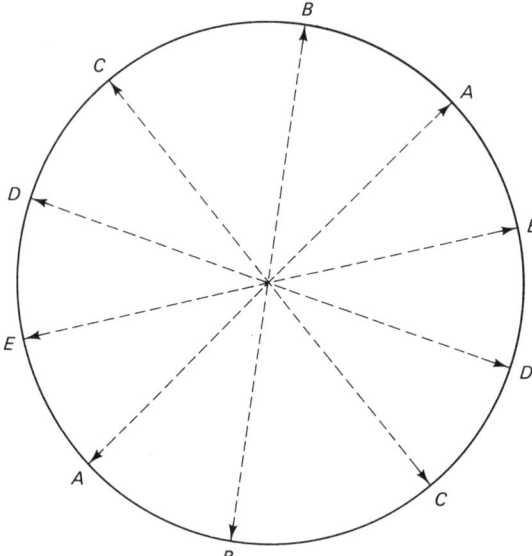

Figure 1.36 An identification model of the projective plane.

It can be shown that it is impossible to sew the boundary circle of the disk up in this manner without having the resulting surface intersect itself. If we are willing to allow such self-intersections, though, then this sewing up process can be performed as in Figure 1.37. Note the curve of self-intersection points in the final closed surface.

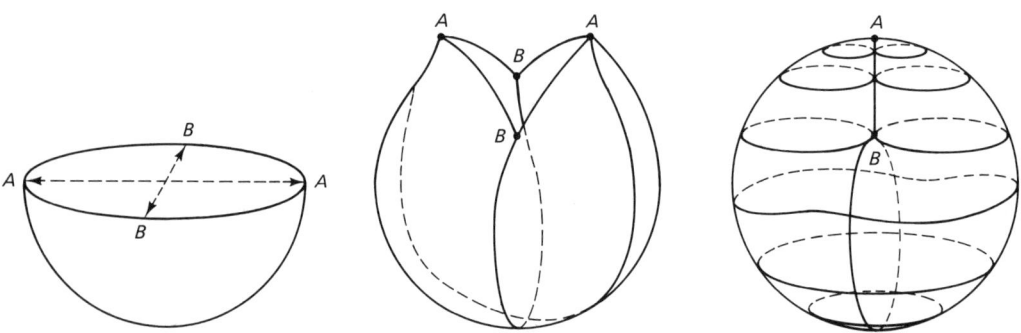

Figure 1.37 Sewing up the boundary of the disk.

Again, this is only a topological model for the projective plane and not a model convenient for doing projective geometry: Although lines in this model can be viewed as some sort of special curve, it is not easy to identify these curves and further pursue geometry on this surface. This surface was introduced only to describe how ideal points can be attached to a Euclidean plane.

The analogous model for the projective line is obtained by first transforming the Euclidean line to a bounded open (line) segment (see Figure 1.38) and then joining both ends of this segment with a single point, the ideal point of the line. This model of a projective line is topologically equivalent to a circle. Note that this is consistent with our model for the projective plane in that if we identify

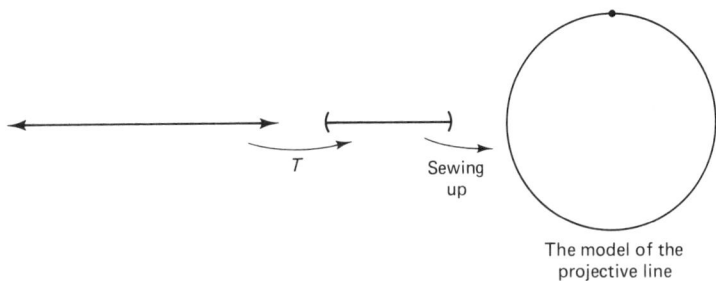

Figure 1.38 A model of the projective line.

the antipodal points of any diameter of the disk (as illustrated in Figure 1.36), we obtain the same model for the projective line as a subset of the projective plane.

Using the topological model of the projective plane, it is also easy to see that the projective plane is a nonorientable surface (see Figure 1.39): If we place

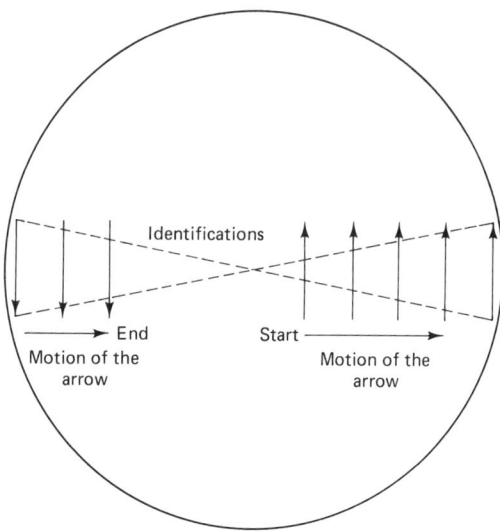

Figure 1.39 The projective plane is nonorientable.

Sec. 4   A Topological Model of the Projective Plane

a small arrow at the center of the disk and move the arrow out beyond the boundary of the disk (the ideal line), the arrow will eventually return to the center of the disk, but its orientation will be reversed.

SECTION 4 EXERCISES

1. By considering the topological model of the projective plane presented above, show that the ideal line topologically looks like a Euclidean line completed with its ideal point.
2. A Moebius band is a single-sided surface obtained from a rectangular strip by giving the strip a half twist and then pasting the two opposite edges of the strip together (see Figure 1.40). Show that the topological model of the projective plane constructed above can be obtained by sewing a closed disk onto a Moebius band, the boundary of the disk being sewn to the boundary of the Moebius band. (*Hint:* Cut up the disk illustrated in Figure 1.37 as illustrated in Figure 1.41 before forming the boundary identification.)

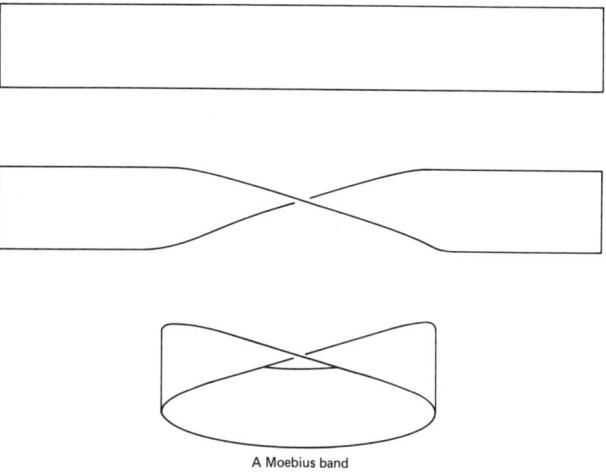

Figure 1.40 Making a Moebius band.

3. In computer graphics, *clipping* is the process of truncating those parts of a picture that lie outside the window of a graphics display device (see Figure 1.42). In Chapter 2, coordinates will be introduced on the projective line and on the projective plane. Without knowing anything specific about these coordinates at this time, what difficulty might one have in using them to clip line segments in the projective plane? (*Hint:* A common algorithm for this type of clipping involves a binary search. Consider the topological nature of a line in the projective plane and the concept of betweenness.)
4. Explain why the concept of a segment is *not* a projective concept. In other words (see Figure 1.43), explain why it does not make sense to talk about a segment AB in the projective plane.

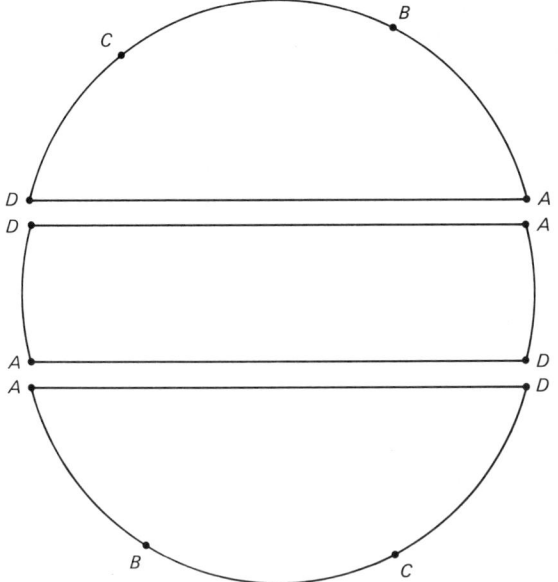

Figure 1.41 The projective plane is a Moebius band plus a closed disk.

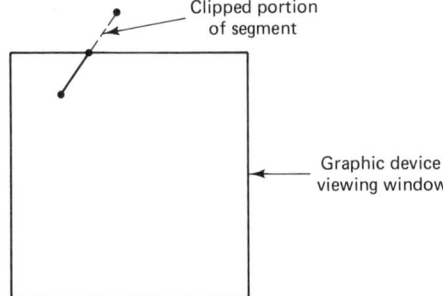

Figure 1.42 Clipping.

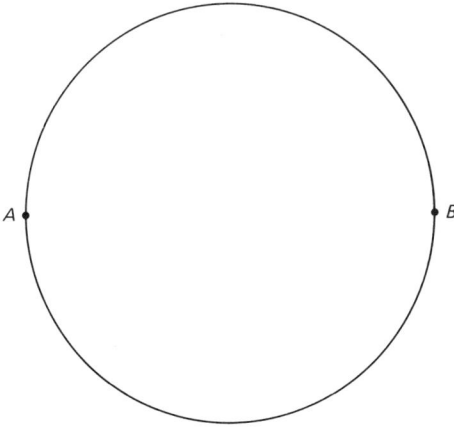

Figure 1.43 Segment is *not* a projective concept.

Sec. 4   A Topological Model of the Projective Plane

SECTION 5: THE STANDARD EMBEDDED PROJECTIVE PLANE

Until now we have concentrated on planar, or two-dimensional, perspective. We have indicated that in two-dimensional perspective, the object and image planes need not be perpendicular; indeed the angle between these planes can be arbitrary. This generality was in anticipation of solid, or three-dimensional, perspective.

We henceforth take as *the standard embedded projective plane* the Euclidean plane $z = 1$ in Euclidean xyz-coordinate space, completed with its ideal line (the space in which this projective plane is embedded will soon be discussed). We take as *the standard model for perspective* the standard embedded projective plane as the image plane with the origin $O(0, 0, 0)$ as the center of perspectivity (see Figure 1.44).

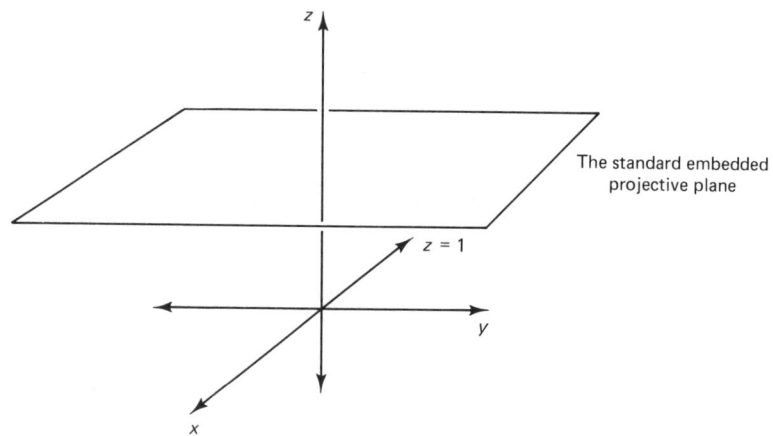

Figure 1.44 The standard model for perspective.

Note that we have not specified the object plane in this model. This is because we will eventually apply the theory of two-dimensional perspective to drawing three-dimensional objects in perspective. To do this we treat three-dimensional objects as being bounded by two-dimensional faces. Each of these two-dimensional faces is treated by two-dimensional perspective; what ties everything together is that the edges (that is, the boundaries between the two-dimensional faces) are treated consistently.

For example, in a typical view of a cube we see three faces (see Figure 1.45). A perspective view of a cube is accomplished by combining three perspective transformations: each transformation has the same image plane and center of perspectivity, and each has as an object plane one of the three planes determined by the three visible faces of the cube. Each object plane has a corresponding vanishing line in the image plane, and the three transformations are consistent in the sense that if an edge is common to two visible faces then the corresponding vanishing points of the two respective transformations are the same (namely,

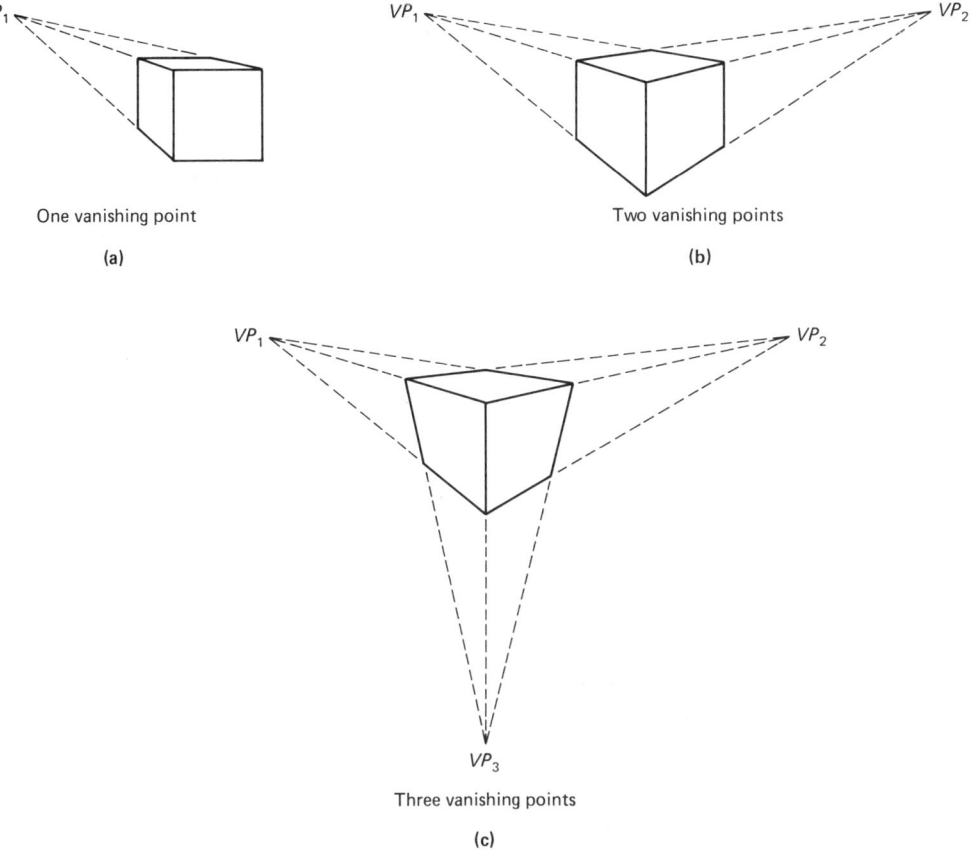

Figure 1.45 The three possible perspective views of a cube.

the point of intersection of the respective vanishing lines). It follows that a perspective view of a cube must have either one, two, or three vanishing points, depending on where the vanishing lines intersect.

The combination of an image plane and a center of perspectivity is a *viewing mechanism*. We will later see that the standard model for perspective is a natural (or standard) viewing mechanism.

As indicated earlier, any Euclidean plane can be completed with its set of ideal points to obtain a projective plane. *Projective three-space*—or, simply, *projective space*—can be defined in a similar manner: We group the lines in Euclidean three-space into families, all the lines in each family being parallel to one another. For each family of lines, a single *ideal point* is attached to Euclidean three-space. This ideal point is attached to each parallel line in the family at infinity; in other words, it is attached in such a manner that all parallel lines in the family intersect at this ideal point. More precisely, if Euclidean three-space is thought of as corresponding to the interior of a three-dimensional spherical Euclidean ball (see Figure 1.46), then attaching the ideal points to Euclidean three-space corresponds to attaching the boundary points to the ball in such a

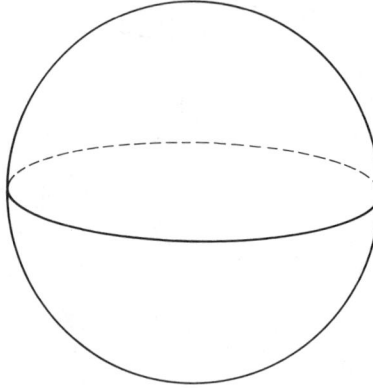

Figure 1.46 A three-dimensional spherical Euclidean ball.

manner that antipodal (or diametrically opposite) points are identified (or considered the same). (See Figure 1.47.) Although we can represent the projective plane as a surface in Euclidean three-space (see Section 4), it is impossible to do anything similar for projective three-space. The best way to visualize projective three-space is as constructed above.

From another point of view, when we construct a projective plane, we add a set of ideal points to a Euclidean plane. When we do this for each Euclidean plane in Euclidean three-space, we are actually adding a plane of ideal points, an *ideal plane,* to Euclidean three-space to obtain projective three-space. (Later we will see that the set of all ideal points added to Euclidean three-space looks like a projective plane.) Although this may seem like an odd way to view things, it should be pointed out that the introduction of the concept of ideal points in projective geometry is generally attributed to Johann Kepler. Kepler thought of the stars as points at infinity in the celestial sphere: The stars are so far away that light rays from them are virtually parallel at any two points on Earth.

Unfortunately, the correspondence between Kepler's notion of ideal point and our notion of ideal point is not perfect: We would want to identify antipodal points on the celestial sphere. Kepler would say that we can only see half of the celestial sphere at any given time; he might be inclined to attach *two* ideal

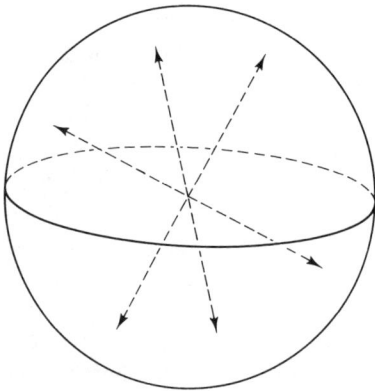

Figure 1.47 A model of projective three-space.

points to every line, one at each end of the line. In other words, if we think about ideal points as directions, there are two different senses in which the word *direction* can be used: in the sense of the direction of a vector (Kepler's concept of ideal point), and in the sense of the direction of a line (our concept of ideal point).

For us, projective three-space will provide a convenient context in which to study planar projective geometry. In particular, if a Euclidean plane is embedded in Euclidean three-space, the projective plane obtained by completing this Euclidean plane with its line of ideal points is embedded in projective three-space. We, as external observers, view this projective plane as a projective plane embedded in projective three-space; we, as observers in projective three-space, view the same projective plane as *the* projective plane. (In other words, the window seen by graphics devices is part of *the* projective plane being viewed intrinsically.) The geometry of an embedded projective plane is extrinsic projective geometry; the geometry of *the* projective plane is intrinsic projective geometry.

There are two types of embedded projective planes: Euclidean planes completed with their lines of ideal points, and the ideal plane. The lines in projective three-space and in projective planes are embedded projective lines. There are two different types of embedded projective lines: Euclidean lines each completed with its ideal point, and ideal lines (one for completing each Euclidean plane). *The standard embedded projective line* is the line $y = 1$ in the (Euclidean) xy-coordinate plane, completed with its ideal point. Projective lines, projective planes, and projective space are the central objects of study in projective geometry.

SECTION 5 EXERCISES

1. What conditions on the vanishing lines for a picture of a cube determine whether the picture has one, two, or three vanishing points?
2. Use the following algorithm to draw a picture of a cube in one-point perspective (see Figure 1.48).

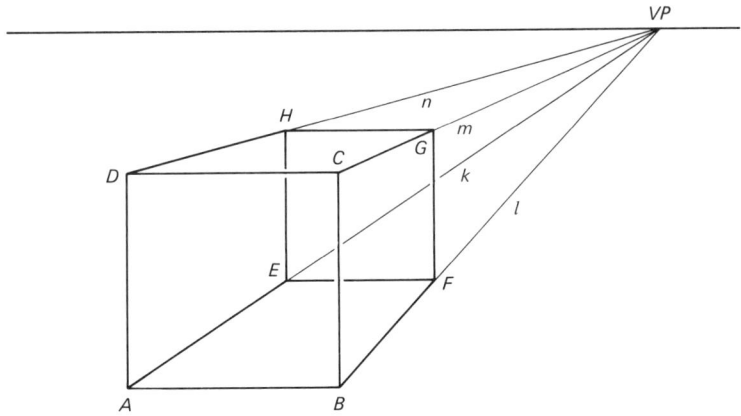

Figure 1.48 A cube in one-point perspective.

(a) Draw a horizon and choose a vanishing point, *VP*, on it.
(b) Draw a square *ABCD*, none of whose vertices lie on the horizon, two of whose sides are parallel to the horizon, and whose other sides are perpendicular to the horizon.
(c) Draw lines *k*, *l*, *m*, and *n* from *VP* to *A*, *B*, *C*, and *D*, respectively.
(d) Choose any point *E* on *k*, and draw another square *EFGH*, in which two sides are parallel to the horizon, in which the other sides are perpendicular to the horizon, and for which *F*, *G*, and *H* lie on lines *l*, *m*, and *n*, respectively.

The points *A*, *B*, *C*, *D*, *E*, *F*, *G*, and *H* are the vertices of the desired figure.

3. Use the following algorithm to draw a picture of a cube in two-point perspective (see Figure 1.49); P-Q denotes the line through points *P* and *Q*.
   (a) Draw a horizon and choose vanishing points VP_1 and VP_2 on it.
   (b) Locate the midpoint *M* of the line segment determined by VP_1 and VP_2, and draw a semicircle below the horizon with center *M* and that passes through VP_1 and VP_2.
   (c) Choose a point *A* between VP_1 and VP_2, and draw a line *l* through *A* perpendicular to the horizon. Let *B* denote the point of intersection of *l* with the semicircle.
   (d) Choose points *C* and *D* between *A* and *B*, *C* below *D*. Through *C*, draw a line parallel to the horizon, and locate the points *E* and *F* on this line for which the distances from *C* to *E* and from *C* to *F* are both the same as the distance from *C* to *D*.
   (e) Locate the points *G* and *H* on the horizon for which the distance from VP_1 to *G* is the same as the distance from VP_1 to *B* and for which the distance from VP_2 to *H* is the same as the distance from VP_2 to *B*.
   (f) Locate *I*, the point of intersection of C-VP_2 and *F*-*H*, and *J*, the point of intersection of C-VP_1 and *E*-*G*.

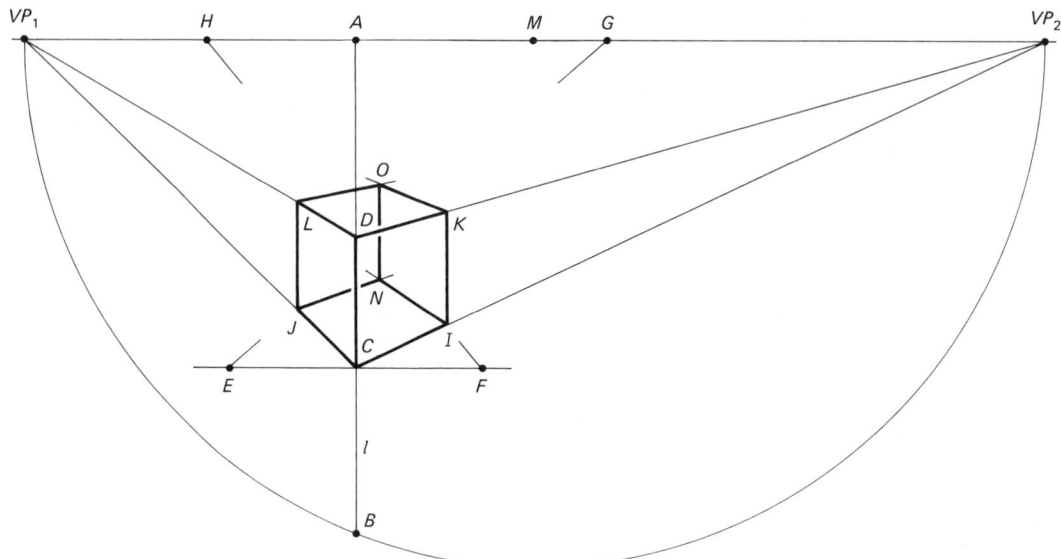

Figure 1.49 A cube in two-point perspective.

(g) Draw lines through I and J parallel to l locating K and L, the points of intersection of these lines with D-VP_2 and D-VP_1, respectively.

(h) Locate N, the point of intersection of I-VP_1 and J-VP_2, and O, the point of intersection of K-VP_1 and L-VP_2.

The points C, D, I, K, N, O, J, and L are the vertices of the desired figure.

4. Use the following algorithm to draw a picture of a cube in three-point perspective (see Figure 1.50); P-Q denotes the line through points P and Q:

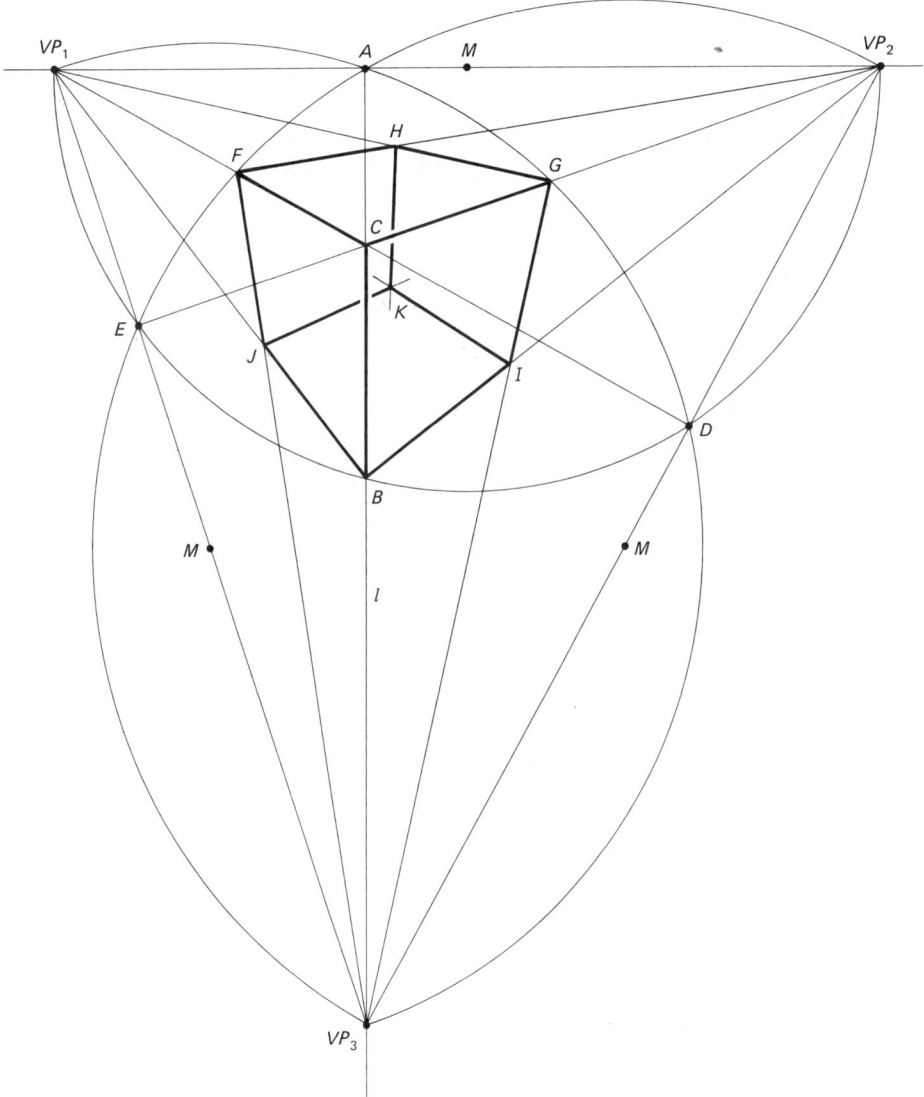

Figure 1.50 A cube in three-point perspective.

Sec. 5 The Standard Embedded Projective Plane

(a) Draw a horizon and choose vanishing points VP_1 and VP_2 on it.

(b) Locate the midpoint M of the line segment determined by VP_1 and VP_2, and draw a semicircle below the horizon with center M and that passes through VP_1 and VP_2.

(c) Choose a point A between VP_1 and VP_2, and draw a line l through A perpendicular to the horizon. Let B denote the intersection of l with the semicircle.

(d) Choose a point C between A and B, and locate D and E, the points of intersection of C-VP_1 and C-VP_2, respectively, with the semicircle.

(e) Locate the third vanishing point VP_3 on l as the point of intersection of E-VP_1 and D-VP_2.

(f) Locate point F by finding the midpoint M of the line segment determined by VP_2 and VP_3, drawing a semicircle with center M that passes through VP_2 and VP_3, and finding the intersection of this semicircle with D-VP_1. Locate the point G similarly.

(g) Locate H, the point of intersection of G-VP_1 and F-VP_2, I, the point of intersection of B-VP_2 and G-VP_3, and J, the point of intersection of F-VP_3 and B-VP_1.

(h) Locate K, the point of intersection of I-VP_1, J-VP_2, and H-VP_3.

The points B, C, I, G, K, H, J, and F are the vertices of the desired figure.

5. Draw a picture of a cube in two-point perspective, starting with the base of the cube (a square in two-point perspective).

6. Draw a picture of a cube in three-point perspective, starting with the three vanishing points.

7. Perspectively duplicate a picture of a cube drawn in
   (a) One-point perspective
   (b) Two-point perspective
   (c) Three-point perspective
   (*Hint:* See Exercise 3.10.)

8. Perspectively subdivide into eight equal parts a cube drawn in
   (a) One-point perspective
   (b) Two-point perspective
   (c) Three-point perspective
   (*Hint:* See Exercise 3.11.)

9. Using a straightedge, draw an accurate picture of your home in perspective. (Use vanishing points properly.)

10. Explain why the vanishing point of the image of a line l under a perspectivity is the point of intersection of the image plane and the line determined by the vanishing point of l and the center of perspectivity.

11. Draw Figure 1.9: a picture of a perspectivity of a cube. (*Hint:* First construct a cube in three-point perspective, making one vanishing line horizontal. This will be your image of the object. Next choose the image E' in your viewplane of the center E of the perspectivity that you are viewing. Now use an extrinsic argument to show that for each pair of parallel edges of the object, the image in your viewplane of the perspectivity's vanishing point is collinear with E' and your vanishing point. This may help you locate the images in your viewplane of the vanishing points of the perspectivity's image. The frame of the perspectivity's image can be chosen arbitrarily, but for your picture to be realistic it is advisable for the vanishing point of the top and bottom of the image of the perspectivity's frame to be on the image in your viewplane of the vanishing line in the perspectivity's viewplane corresponding to your horizontal vanishing line.)

SECTION 6: SOME SYNTHETIC PROJECTIVE GEOMETRY

We now briefly discuss some of the fundamental results of synthetic projective geometry. In the statement of these results

1. *Points* refer to the points in projective three-space—either Euclidean points or ideal points.
2. *Lines* refer to the lines in projective three-space—either Euclidean lines, each completed with its ideal point, or ideal lines.
3. *Planes* refer to planes in projective three-space—either Euclidean planes, each completed with its ideal line, or the ideal plane.

In Euclidean geometry we have one set of points, lines, and planes, and in projective geometry we have a different set. Although some of our first results are stated in terms of points, lines, and planes, so they are *statements* in both Euclidean and projective geometry, they are all *true statements* only in projective geometry.

In the first two results, we restrict our attention to a fixed projective plane, *any* projective plane. (Without loss of generality, we might as well assume it is *the* projective plane.)

6.1 Proposition.   Two distinct points lie on one and only one line.

This result is valid in both Euclidean and projective geometry. The difference is that as a result in projective geometry, either one or both points can be ideal: If one point is Euclidean and the other ideal, we can find one and only one Euclidean line passing through the given Euclidean point whose direction is that associated to the given ideal point. By completing this line with its ideal point—the given ideal point—we obtain the desired line. If both points are ideal, then both lie on the ideal line; the result follows since there is only one line—namely, the ideal line—which contains more than one ideal point.

6.2 Proposition.   Two distinct lines meet in one and only one point.

This result is *not* valid in Euclidean geometry: Two parallel Euclidean lines do not meet in a point in the Euclidean plane. It is only after ideal points have been added to the Euclidean plane—that is, that we move into projective geometry—that this result is valid: Two parallel Euclidean lines meet at their common ideal point. Two lines that intersect in Euclidean geometry intersect—or, more accurately, their extensions in the ideal plane intersect—in projective geometry in a unique Euclidean point. And a Euclidean line completed with its ideal point intersects the ideal line at the ideal point of the Euclidean line and only at that point.

Propositions 6.1 and 6.2 are the fundamental incidence results of planar projective geometry. Observe that these two results are closely related, at least formally, in the sense that if we interchange the words *point* and *line* and the phrases *lies on* and *meets in* in either result, we obtain the other result. This phenomenon is known as the *duality* of planar projective geometry (see Figure 1.51).

More generally, we have the following result. (We will not prove this result, since doing so would be beyond the scope of this book. The statement of this result, however, will be used in the sequel.)

Concept	Dual concept
Point	Line
Line	Point
Lies on	Contains
Contains	Lies On

A point *lies on* a line.
A line *contains* a point.

Figure 1.51 The duality of planar projective geometry.

6.3 Theorem. (Duality of Planar Projective Geometry) In projective geometry the dual of every valid result is a valid result.

We have only introduced duality formally here. Later we will describe duality both analytically and geometrically as well.

The next several results are stated for projective three-space. Consideration of why these results are true and whether or not they are valid in Euclidean geometry is left to the exercises.

6.4 Proposition. Two distinct planes meet in one and only one line.

6.5 Proposition. Two distinct points lie on one and only one line.

6.6 Proposition. Three distinct noncollinear points lie on one and only one plane.

6.7 Proposition. Three distinct planes not containing a common line meet in one and only one point.

6.8 Proposition. A line and a plane not containing the line meet in one and only one point.

6.9 Proposition. A line and a point not on the line are contained in one and only one plane.

Again, there is a duality in spatial projective geometry that is slightly different than duality in planar projective geometry (see Figure 1.52). Observe that results 6.4 and 6.5, 6.6 and 6.7, and 6.8 and 6.9 are duals of each other.

The following definition extends our earlier definitions of perspectivity and projectivity. In this definition, *lines* are lines either both in the projective plane or both in projective three-space (and *planes* are planes in projective three-space).

Concept	Dual concept
Point	Plane
Line	Line
Plane	Point
Lies on: point on line	Contains: plane contains line
point on plane	plane contains point
line on plane	line contains point
Contains: line contains point	Lies on: line on plane
plane contains point	point on plane
plane contains line	point on line

A point *lies on* a line.
A point *lies on* a plane.
A line *lies on* a plane.
A plane *contains* a point.
A plane *contains* a line.
A line *contains* a point.

Figure 1.52 The duality of spatial projective geometry.

6.10 Definition. A *perspective transformation*, or a *perspectivity*, *of lines* (see Figure 1.53(a)) is a transformation taking the points of a fixed line, the *object line*, onto the points of another fixed line, the *image line*, in such a way that each pair of corresponding points is collinear with a fixed point, the *center of perspectivity*, or *eye*, which lies neither on the object nor the image line. A *perspective transformation*, or a *perspectivity*, *of planes* (see Figure 1.53(b)) is a transformation taking the points of a fixed plane, the *object plane*, onto the points of another fixed plane, the *image plane*, in such a manner that each pair of corresponding points is collinear with a fixed point, the *center of perspectivity*, or *eye*, which lies neither on the object nor the image plane. A *projective*

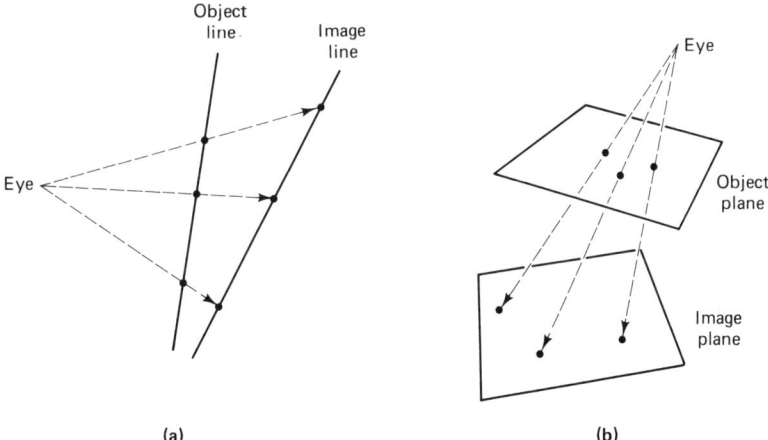

Figure 1.53 Perspectivities (a) of lines and (b) of planes.

transformation, or a *projectivity*, *of lines* is a perspectivity of lines or a composite of perspectivities of lines (see Figure 1.54(a)), and a *projective transformation*, or a *projectivity*, *of planes* is a perspectivity of planes or a composite of perspectivities of planes (see Figure 1.54(b)).

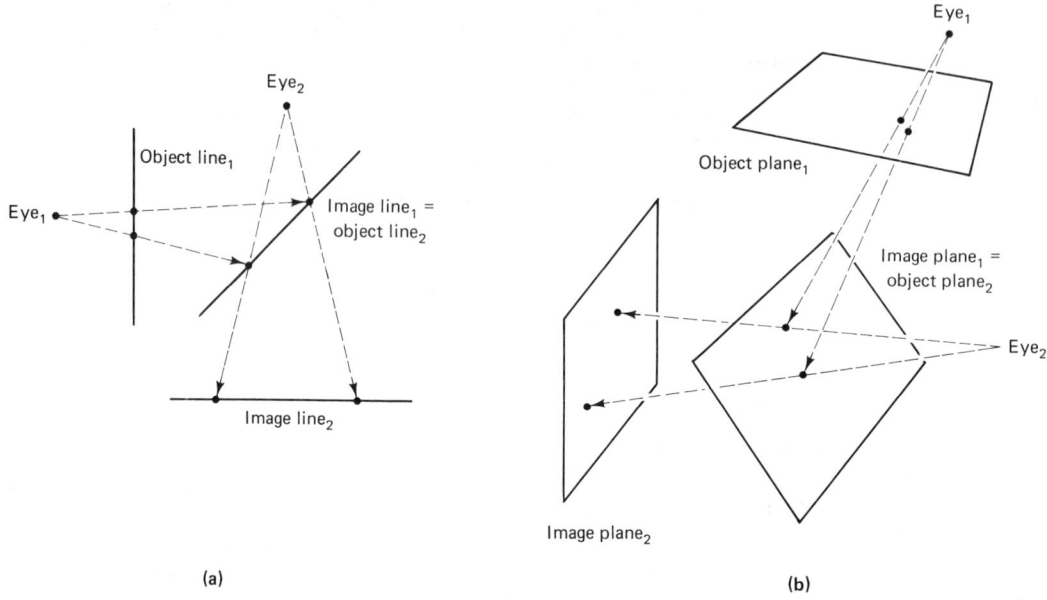

Figure 1.54 Projectivities (examples of twofold perspectivities) (a) of lines and (b) of planes.

This definition extends our earlier definition(s) in two respects. The first is that it extends our discussion to perspectivities and projectivities of lines (before we only considered planes). The second is that it permits the use of ideal points as centers of perspectivity; a perspectivity whose center is an ideal point is a *parallel projection* (see Figure 1.55) because lines along which points are projected are all parallel and all have the same ideal point—namely, the center of perspectivity. Pictures obtained through parallel projection are *plan views*; this type of picture includes floor plans and other layout views used by designers, architects, builders, and engineers.

(*Note:* The term *perspective* is typically used in two slightly different ways: One way is as introduced here, in which case a parallel projection is a special form of perspectivity. The other demands that the center of perspectivity be a Euclidean point, so that perspectivities and parallel projections are mutually exclusive. We will use the term perspective as given in Definition 6.10, since the formulas of analytic projective geometry to be developed in this book are all valid whether the center of perspectivity is Euclidean or ideal; this convention thus avoids an unnecessary and awkward wording of results. When clarification

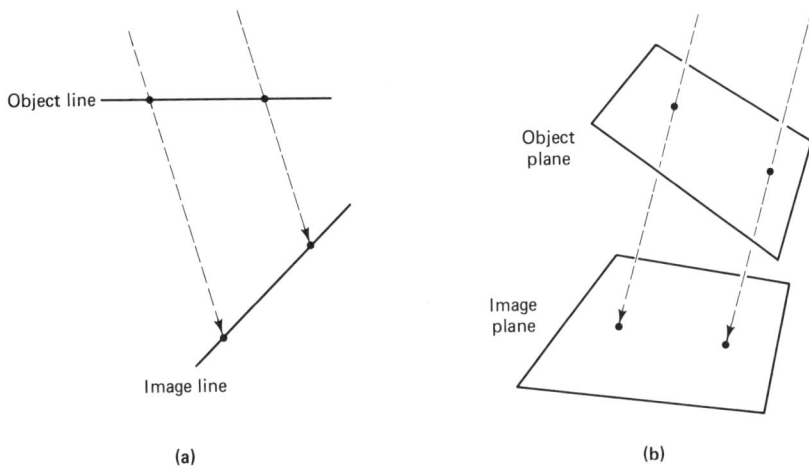

Figure 1.55 Parallel projection (a) of lines and (b) of planes.

is necessary, one refers to perspectivities for which the center of perspectivity is Euclidean as *Euclidean perspectivities* and perspectivities for which the center of perspectivity is ideal as *parallel projections*.)

As a consequence of Propositions 6.1–6.9, perspectivities and projectivities of lines (and planes) are one-to-one transformations and hence have inverses. We will exploit this fact later.

Even simple examples of projectivities of lines reveal some surprising geometry (see Figure 1.56; observe what happens to both order and scale in these examples). Such examples lead us to ask what flexibility, or latitude, there is to projectivities in general. The following result answers this question.

6.11 Theorem. (The Fundamental Theorem of Projective Geometry for Lines) Let l and m be lines either both in the projective plane or both in projective three-space. If A, B, and C are three distinct points on l, and P, Q, and R are three distinct points on m, then there is one and only one projective transformation taking l to m and A to P, B to Q, and C to R.

We do not attempt a thorough proof of this result here; a relatively simple proof will be given later as an application of analytic projective geometry. We *do* now briefly describe a proof that introduces an important concept in projective geometry—the *cross ratio*.

Until now, our study of perspective and projective geometry has been highly synthetic (or qualitative). The importance of the cross ratio is that it provides a perspective scaling in projective geometry. In other words, the cross ratio is an analytic (a metric or quantitative) concept in projective geometry.

The cross ratio is a Euclidean concept (it is defined in terms of distances) which extends to projective geometry: If A, B, C, and D are any four distinct

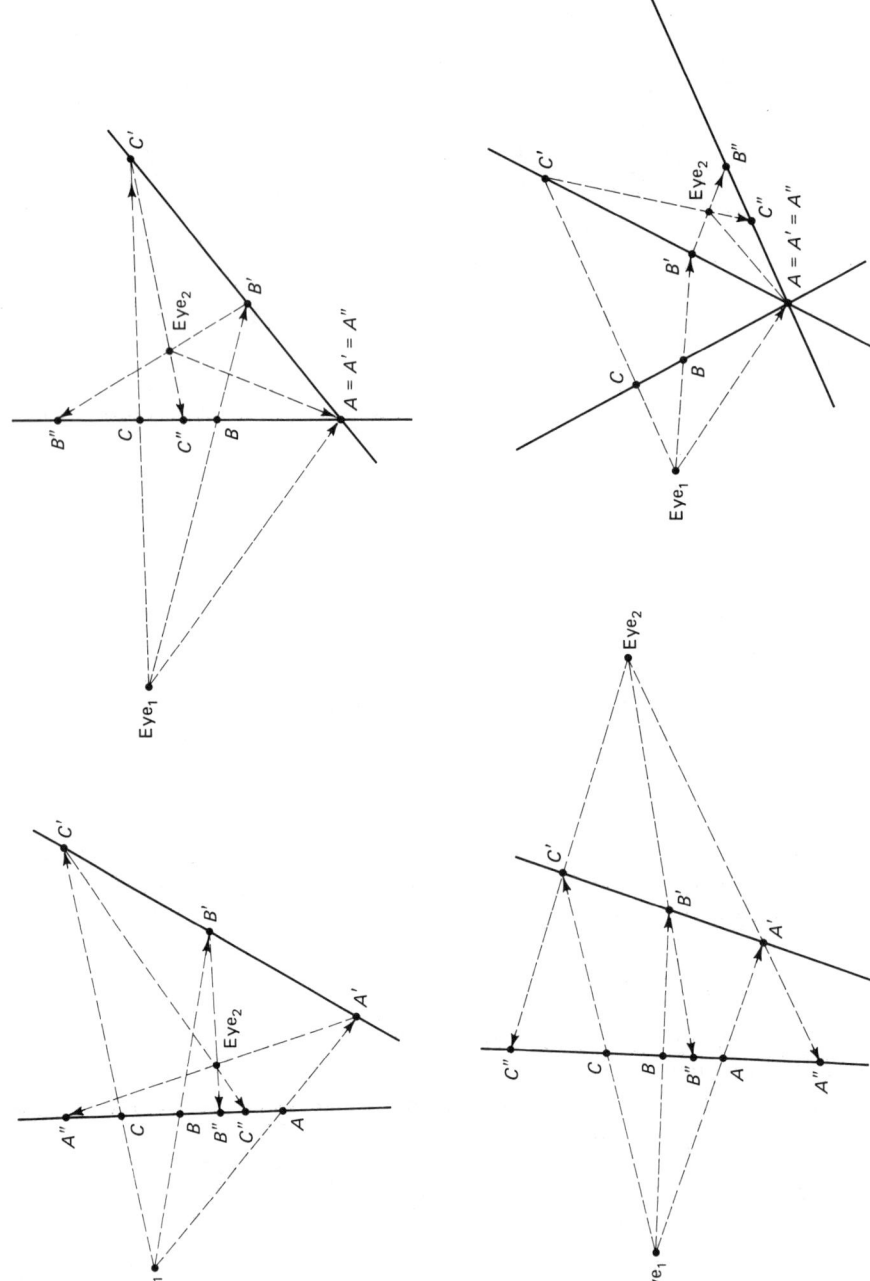

Figure 1.56 Some projectivities of lines.

points that lie on the Euclidean line (or on a line in the Euclidean plane or Euclidean three-space), we define the real number $R(A, B, C, D)$ by

$$R(A, B, C, D) = \frac{\text{dist}(A, C) \times \text{dist}(B, D)}{\text{dist}(B, C) \times \text{dist}(A, D)}$$

where dist(P, Q) denotes the directed Euclidean distance from point P to point Q. This expression is a well-defined extended real number even if only three of the points A, B, C, and D are distinct. (An extended real number is a finite real number or $\pm\infty$. The inclusion of $\pm\infty$ is necessitated by the event that either $A = D$ or $B = C$.) Indeed, this expression is also a naturally well-defined extended real number if any of A, B, C, and D are allowed to be ideal points. In other words, the extended real number $R(A, B, C, D)$ can be defined for any four points A, B, C, and D, at least three of which are distinct, on the (or a) projective line.

6.12 Definition. For any four points A, B, C, and D, at least three of which are distinct, on the projective line (or four collinear points A, B, C, and D, at least three of which are distinct, in the projective plane or in projective three-space) the extended real number $R(A, B, C, D)$ is the *cross ratio* associated to A, B, C, and D.

The cross ratio satisfies the following two properties: First, it is invariant under projective transformations in the sense that if T is any projective transformation from a projective line l to a projective line m, then

$$R(A, B, C, D) = R(T(A), T(B), T(C), T(D))$$

Furthermore, if T is a transformation from l to m for which

$$R(A, B, C, D) = R(T(A), T(B), T(C), T(D))$$

for all sets of four points A, B, C, and D on l, then T is a projective transformation. Second, given distinct collinear points A, B, and C, $R(A, B, C, -)$ as a function of one variable from the set of all points D collinear with A, B, and C to the set of extended real numbers (with $+\infty$ identified, or set equal to, $-\infty$) is one-to-one, onto, and continuous.

Using the cross ratio, the Fundamental Theorem of Projective Geometry can be proved as follows: Given points A, B, and C on l, and points P, Q, and R on m, we define T for A, B, and C by

$$T(A) = P, \qquad T(B) = Q, \qquad T(C) = R$$

For any other point D on l we compute $R(A, B, C, D)$ and then find the unique point S for which

$$R(A, B, C, D) = R(P, Q, R, S)$$

We then let $T(D) = S$. It follows that T is a projectivity, and it is unique.

In other words, the cross ratio can be thought of as a perspective scaling. To clarify this, let us consider the problem of drawing a perspective view of railroad tracks (see Figure 1.1). The Fundamental Theorem of Projective Geometry states that after drawing the two tracks, only *three* railroad ties can be placed arbitrarily; the relative positions of all the other ties are well determined by proper perspective or, more accurately, the cross ratio. (The positioning of the first three ties is not totally arbitrary—the distance between the ties must decrease in the direction of the vanishing point.)

To be more specific (see Figures 1.57 and 1.58), if the first tie is actually located at A and its image is located at P, the second tie is actually located at B and its image is located at Q, and the third tie is located at C and its image is located at R, if D_n denotes the actual position of the nth tie and S_n the position of its image in our picture, and if distances are as labeled in Figures 1.57 and 1.58 (the scales in both parts of Figure 1.57 are Euclidean), then

$$R(A, B, C, D_n) = \frac{\text{dist}(A, C) \times \text{dist}(B, D_n)}{\text{dist}(B, C) \times \text{dist}(A, D_n)} = \frac{2(n-1)}{1 \times n}$$

$$R(P, Q, R, S_n) = \frac{\text{dist}(P, R) \times \text{dist}(Q, S_n)}{\text{dist}(Q, R) \times \text{dist}(P, S_n)} = \frac{(\frac{20}{11})s_{n-1}}{(\frac{9}{11})s_n}$$

and, since the picture of the railroad tracks is to be in perspective,

$$R(A, B, C, D_n) = R(P, Q, R, S_n)$$

It follows that

$$s_n = \frac{10n}{n+9}$$

Thus, for example, the $n = 3$ tie should be drawn $s_3 = 10 \times 3/(3+9) = 2.5$

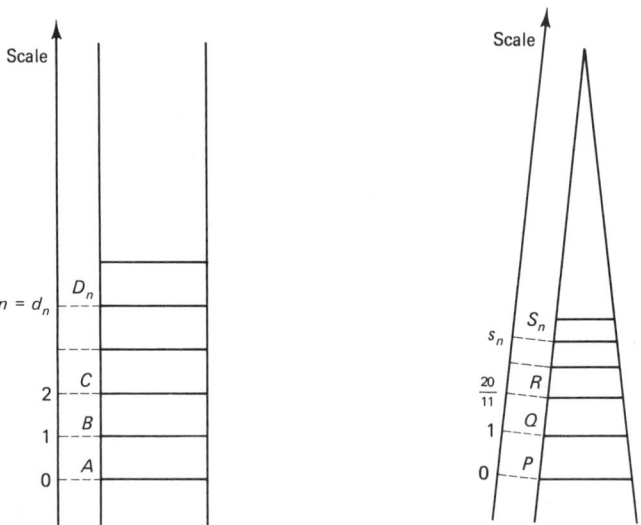

Figure 1.57 Drawing railroad tracks.

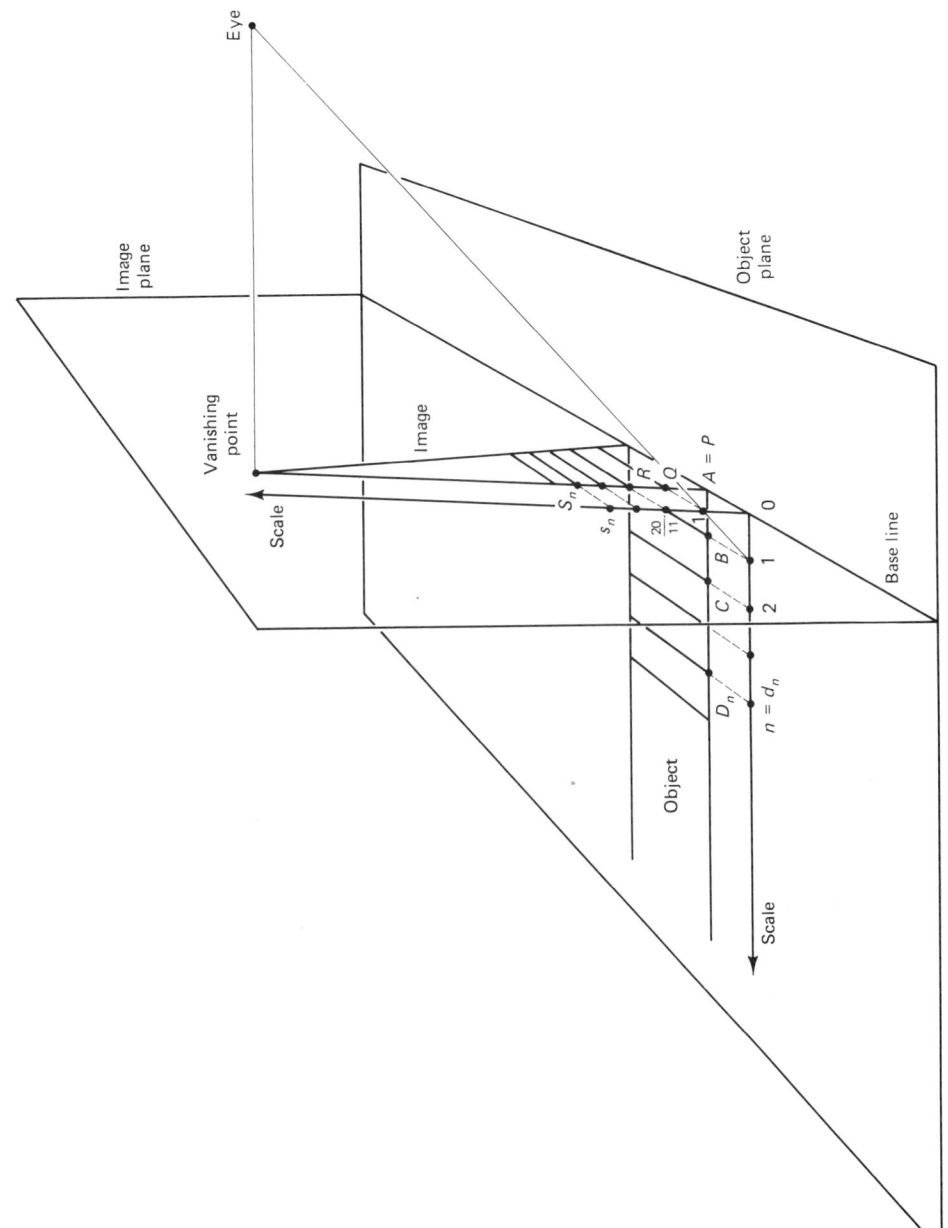

Figure 1.58 Drawing railroad tracks (in three dimensions).

units down the axis from P, the $n = 4$ tie should be drawn $s_4 = 10 \times 4/(4 + 9) = \frac{40}{13}$ units down the axis from P, and so on. In particular, the vanishing point—which corresponds to $n = \infty$—should be drawn

$$\lim_{n \to \infty} s_n = \lim_{n \to \infty} \frac{10n}{n + 9} = \lim_{n \to \infty} \frac{10}{1 + 9/n} = 10$$

units down the axis from P. Note that there is nothing special about the placement of the first three ties. In particular, we could have positioned the first two ties arbitrarily and the third tie at ∞ (so its image would correspond to the vanishing point) and then used the cross ratio to position the remaining ties.

We return to the cross ratio in Section 3.6.

There is an analog of the Fundamental Theorem of Projective Geometry for lines which applies to planes.

6.13 Theorem. (The Fundamental Theorem of Projective Geometry for Planes) Let π and μ be planes in projective three-space. If A, B, C, and D are four points on π, no three of which are collinear, and P, Q, R, and S are four points on μ, no three of which are collinear, then there is one and only one projectivity taking π to μ and A to P, B to Q, C to R, and D to S.

SECTION 6 EXERCISES

1. (a) Explain why Propositions 6.4–6.9 are valid in projective space.
   (b) Which of Propositions 6.4–6.9 is valid in Euclidean space as well as projective space? For those that are valid, explain why. For those that are not, provide a counterexample.
2. Does every pair of lines in the projective plane intersect? How about in projective three-space?
3. Determine the placement of the railroad ties in a perspective picture of a set of railroad tracks (as was done in the text) if the first three ties are located at positions corresponding to distances of 0, 1, and 1.5 units along a Euclidean scale in the image plane. Where is the vanishing point?
4. Determine the placement of the railroad ties in a perspective picture of a set of railroad tracks if the first two ties are located at positions corresponding to distances of 0 and 1 units along a Euclidean scale in the image plane and the vanishing point is located at a position corresponding to a distance of 12 units along the same scale.
5. Show that if railroad ties are located at positions A, B, C, and D in a perspective picture of railroad tracks and if the lengths of these ties are a, b, c, and d, respectively, then

$$R(A, B, C, D) = \frac{(c - a)(d - b)}{(d - a)(c - b)}$$

   (*Hint:* Use similar triangles.)
6. Consider the statement: Given a line and a point not on the line, distinct lines through the given point meet the given line in distinct points.

(a) Give the planar dual of this statement.

(b) Give the spatial dual of this statement.

7. What can be said about a perspectivity from a plane to itself?

†8. Outline how the result proved in Exercise 3.7 could be used to prove the existence of the projectivity described in the Fundamental Theorem of Projective Geometry for Planes.

9. (a) Show that the identity transformation on any line (or plane) can be written as a composite of perspectivities in infinitely many ways.

(b) Use part (a) to show that any projective transformation of lines (or planes) can be written as a composite of perspectivities in infinitely many ways.

10. What differences are there between two- and three-dimensional computer graphics? In particular, what would you interpret one-dimensional graphics to be?

SECTION 7: REMARKS

In this section we briefly discuss the history of projective geometry (for more details, see [Carlbom and Paciorek] and [Wylie]), and the axiomatic foundations of projective geometry (for more details see [Fishback], [Wylie], or [Adler]).

Projective geometry originated with the study of Euclidean perspective and plan views. The history of projective geometry as a practical science can be traced back as far as 2150 B.C. to a floor plan view of a building carved on a stone tablet, which is part of a statue of King Gudea of the city of Lagash in Mesopotamia (see Figure 1.59). The Egyptians, although displaying no sense of perspective in their art, must have also used plan views in their engineering and architecture. Later the painters of Greek antiquity (roughly in the fifth century B.C.) displayed some knowledge of perspective. But perspective is an applied science, and Greek mathematicians took a dim view of applied science; thus while the Greeks devoted much effort to an axiomatic development of Euclidean geometry, projective geometry remained a collection of rules of thumb used by artists, architects, and builders. The major contribution of the Romans came in

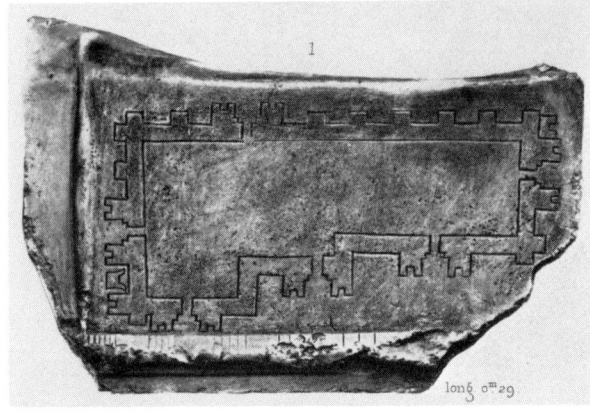

Figure 1.59 Plan view of a building, part of statue of Gudea, from Lagash, Mesopotamia, c. 2150 B.C. (Ernest de Sarzec, *Découvertes en Chaldée*, 1891, Sterling Memorial Library, Yale University).

approximately 14 B.C. with the publication of *De Architectura* by the architect and engineer Vitruvius; however it would eventually take 1200 years until the beginning of the Italian Renaissance for projective geometry, in the form of perspective, to move forward appreciably.

Before the Renaissance, art was thematically religious and symbolic; pictures often looked flat, or two-dimensional (see Figure 1.60). Artists of the Renaissance

Figure 1.60 *Madonna and Child on a Curved Throne*, Byzantine School, XIII Century (the National Gallery of Art, Washington, D.C., Andrew W. Mellon Collection, 1937).

became concerned with making their art more realistic. Early attempts made in the thirteenth century by Duccio di Buoninsegna, for example, were not wholly successful (see Figure 1.61). Later attempts in the fifteenth century by Filippo Brunelleschi, Leone Battista Alberti, and Leonardo Da Vinci (see Figure 1.62) were much more successful; these were true Renaissance artists—they were amateur mathematicians as well as artists and architects. (Da Vinci also wrote *A Treatise on Painting*, in which he stressed the importance of the scientific method and mathematical principles.) In the early sixteenth century, the German artist and mathematician Albrecht Dürer traveled to Italy to learn perspective and succeeded in improving on what he learned (see Figure 1.63).

Figure 1.61 Duccio di Buoninsegna, *The Last Supper*, Opera del Duomo, Siena, fresco, c. 1308 (Alinari/Art Resource).

In the seventeenth century, while artists continued developing their use of perspective, Gerard Desargues (a French architect and engineer) and his student Blaise Pascal (a French mathematician, physicist, and philosopher) initiated a systematic study of perspective. This study contributed largely to the development of synthetic projective geometry but was overshadowed at the time by the invention of Cartesian coordinates by René Descartes and the related study of Euclidean geometry. This was also the period during which the astronomer Johann Kepler made various contributions to projective geometry (such as the concept of ideal

Figure 1.62 Engraving of Leonardo da Vinci's fresco *The Last Supper*, Santa Maria delle Grazie, Milan, c. 1495–1498 (The Bettmann Archive).

point). In the late eighteenth century, Gaspard Monge, motivated by applications to engineering, developed mechanical drawing as it is known today (his major work on the subject was *An Introduction to Perspective and Projective Geometry*). And in the early nineteenth century, Jean Victor Poncelet started the systematic axiomatic study of projective geometry.

Euclid's great contribution to science was the axiomatic method, not (perhaps surprisingly) Euclidean geometry. The axiomatic method is the development of a subject that involves starting with certain rules of logic (or reasoning), certain undefined terms (such as *point*, *line*, *plane*, and *lies on*—as in "a point lies on a line") which are terms that we accept without question as to their meaning, and certain postulates or axioms, which are basic results that we accept without question, from which further results of the subject are deduced. Although Euclid introduced Euclidean geometry using this method, his execution of it is, by modern standards, somewhat less than perfect. (Today the axiomatic treatment used in most courses in synthetic Euclidean geometry taught in high schools and colleges is due to David Hilbert. For further details see [Greenberg].)

Just as synthetic Euclidean geometry can be developed axiomatically, so can synthetic projective geometry. We take *point*, *line*, *plane*, and *lies on* (with meanings as indicated in Section 6 in mind) as undefined concepts and appropriately worded versions of the results stated in Section 6 (together with certain existence postulates) as postulates. All results from synthetic projective geometry can be obtained in this framework. This is the development started by Poncelet.

Figure 1.63 Albrecht Dürer, *Artist Drawing a Lute*, woodcut from *Unterweysung der Messung mit dem Zyrkel und Rychtscheyd*, 1525 (The Metropolitan Museum of Art, Harris Brisbane Dick Fund, 1941).

Our goal is analytic projective geometry, though, since it is necessary to use numbers in working with computer graphics systems. Hence we now abandon synthetic projective geometry and move on to analytic projective geometry.

2

Analytic Projective Geometry

INTRODUCTION

In this chapter we begin the coordinate, or analytic, study of projective geometry. In Section 1 we associate coordinates—homogeneous coordinates—to points in the projective plane. Just as the introduction of Cartesian coordinates in Euclidean geometry allows us to relate algebra to synthetic Euclidean geometry, the introduction of homogeneous coordinates in projective geometry allows us to relate algebra to synthetic projective geometry. Homogeneous coordinates are thus the key to the quantitative study of projective geometry; and this quantitative approach is precisely what is necessary in the study of computer graphics.

Homogeneous coordinates, moreover, have special attributes. In Chapter 3, for example, we see that homogeneous coordinates facilitate the development of transformation formulas (the transformation formulas of central importance to computer graphics). Furthermore, points and directions can be treated equally if homogeneous coordinates are used; consequently, anomalies that arise with the use of Cartesian coordinates can be handled more readily with homogeneous coordinates.

The nonparametric equation of a line in the projective plane is described in Section 2, and the corresponding nonparametric equation of a plane in projective three-space is described in Section 3. The parametric equations of lines in the projective plane and in projective three-space are described in Section 4. In Chapter 3 we return to the parametric representation of lines and also discuss the parametric representation of planes.

SECTION 1: HOMOGENEOUS COORDINATES

The goal of this section is to define coordinates—homogeneous coordinates—on the projective plane.

Recall that the projective plane is the Euclidean plane completed with its ideal line. Although the Euclidean points of this plane can be described by Cartesian coordinates ($P = P(x, y)$), the ideal points cannot. Hence new coordinates are necessary to describe *all* points in the projective plane. Homogeneous coordinates do this while accomplishing two further ends at the same time. First, all points in the projective plane (both Euclidean and ideal) are algebraically treated equally—this is the homogeneity of homogeneous coordinates. Second, it is immediately and easily determined from the homogeneous coordinates of a point whether the point is Euclidean or ideal.

We begin by defining the important analytic set RP^2. As far as homogeneous coordinates are concerned, RP^2 plays a role similar to the role R^2 plays in coordinatizing the Euclidean plane: A set of homogeneous coordinates on the projective plane is a one-to-one correspondence between this plane and RP^2.

The set $R^3 - \{\langle 0, 0, 0\rangle\}$ consists of all nonzero vectors $\mathbf{p} = \langle p_1, p_2, p_3\rangle$ in R^3.

1.1 Definition. The set RP^2 is the set of all triples $[\mathbf{p}] = [p_1, p_2, p_3]$, where $\mathbf{p} = \langle p_1, p_2, p_3\rangle$ is in $R^3 - \{\langle 0, 0, 0\rangle\}$. Two such triples, $[p_1, p_2, p_3]$ and $[q_1, q_2, q_3]$, are considered equal if and only if there is a nonzero real constant k for which

$$p_1 = kq_1, \quad p_2 = kq_2, \quad \text{and} \quad p_3 = kq_3$$

For example, $[4, -8, 12] = [1, -2, 3]$, since (letting $k = 4$)

$$4 = 4 \times 1, \quad -8 = 4 \times (-2), \quad \text{and} \quad 12 = 4 \times 3$$

However, $[4, -8, 12] \neq [1, 5, 3]$, since there is no real number k for which

$$4 = k \times 1, \quad -8 = k \times 5, \quad \text{and} \quad 12 = k \times 3$$

(The first equality implies that $k = 4$, but $k = 4$ does not satisfy the second equality.)

The point of these examples is to emphasize, first, that just as fractions have multiple representations—such as

$$\tfrac{1}{2} = \tfrac{2}{4} = \tfrac{3}{6}$$

—elements of RP^2 also have multiple representations—for instance,

$$[4, -8, 12] = [1, -2, 3] = [5, -10, 15]$$

Second, *all three* of the equalities

$$p_1 = kq_1, \quad p_2 = kq_2, \quad \text{and} \quad p_3 = kq_3$$

must hold for some (single) k in order that $[p_1, p_2, p_3] = [q_1, q_2, q_3]$. Finally, $[0, 0, 0]$ is meaningless since $\mathbf{0} = \langle 0, 0, 0\rangle$ is not in $R^3 - \{\langle 0, 0, 0\rangle\}$.

Before defining homogeneous coordinates, we make four remarks.

First, the term representative will be useful in the rest of this book.

1.2 Definition. A vector $\langle q_1, q_2, q_3 \rangle$ is a *representative* of $[p_1, p_2, p_3]$ if and only if $[p_1, p_2, p_3] = [q_1, q_2, q_3]$.

For example, $\langle 4, -8, 12 \rangle$ is a representative of $[1, -2, 3]$ since $[4, -8, 12] = [1, -2, 3]$.

The following result follows immediately from Definition 1.1.

1.3 Proposition The vector $\langle q_1, q_2, q_3 \rangle$ is a representative of $[p_1, p_2, p_3]$ if and only if there is a nonzero real constant k for which $\langle p_1, p_2, p_3 \rangle = k \langle q_1, q_2, q_3 \rangle$.

Second, for those who know about equivalence relations, another way to look at RP^2 is as follows: If $\equiv$ denotes the relation on $R^3 - \{\langle 0, 0, 0\rangle\}$ given by

$$\langle p_1, p_2, p_3 \rangle \equiv \langle q_1, q_2, q_3 \rangle$$

if and only if there is a nonzero real constant k for which

$$p_1 = kq_1, p_2 = kq_2, \text{ and } p_3 = kq_3,$$

then $\equiv$ is an equivalence relation on $R^3 - \{\langle 0, 0, 0\rangle\}$, and RP^2 is the set of equivalence classes of $R^3 - \{\langle 0, 0, 0\rangle\}$ with respect to $\equiv$.

Third, in the rest of this book homogeneous coordinates will be denoted by square brackets, $[-,-,-]$, and Cartesian coordinates will be denoted by curved brackets, $(-,-,-)$. Although these notations may be similar in appearance, analytically they are really very different. One major difference between RP^2 and R^2 is that elements of RP^2 do not have unique representatives, and consequently RP^2 does not possess a vector space structure similar to that of R^2. If, for example, we were to try to define addition in RP^2 by

$$[p_1, p_2, p_3] + [q_1, q_2, q_3] = [p_1 + q_1, p_2 + q_2, p_3 + q_3]$$

we would be faced with the following type of inconsistency: $[1, -2, 3] = [4, -8, 12]$, so that $[1, 1, 1] + [1, -2, 3] = [2, -1, 4]$ should be the same as $[1, 1, 1] + [4, -8, 12] = [5, -7, 13]$; but $[2, -1, 4] \neq [5, -7, 13]$. Any other type of addition in RP^2 would lead to similar problems: There is no consistent form of addition in RP^2.

Fourth, and finally, in order to define homogeneous coordinates we need the following result.

1.4 Proposition. There is a natural decomposition of RP^2 into two disjoint subsets

$$RP^2 = \{[p_1, p_2, 1] \in RP^2\} \cup \{[p_1, p_2, 0] \in RP^2\}$$

Proof. For any $[q_1, q_2, q_3]$ in RP^2, either $q_3 = 0$ or $q_3 \neq 0$. If $q_3 = 0$, then

$$[q_1, q_2, q_3] = [q_1, q_2, 0] \in \{[p_1, p_2, 0] \in RP^2\}$$

If $q_3 \neq 0$, then

$$[q_1, q_2, q_3] = \left[\frac{q_1}{q_3}, \frac{q_2}{q_3}, 1\right] \in \{[p_1, p_2, 1] \in RP^2\} \qquad \blacksquare$$

This is not the only natural decomposition of RP^2. There are others, and we address these in Section 3.4 when we consider general homogeneous coordinates.

We now define standard homogeneous coordinates on the projective plane: Again, these coordinates can be thought of as a one-to-one correspondence H between the projective plane and RP^2.

Think of the projective plane as the standard embedded projective plane (recall—see Section 1.5—that we, as external observers, view the standard embedded projective plane as a projective plane embedded in projective three-space, while we, as observers in projective three-space, view the same plane as *the* projective plane).

There is a one-to-one correspondence H between the Euclidean plane $z = 1$ and $\{[p_1, p_2, 1] \in RP^2\}$ given as follows (see Figure 2.1): Each point P in this plane has Cartesian coordinates $(x, y, 1)$. We define the correspondence H by associating to P the element $H(P) = [x, y, 1]$ of RP^2. We denote by $P[x, y, 1]$ the Euclidean point in the standard embedded projective plane whose Cartesian coordinates are $(x, y, 1)$.

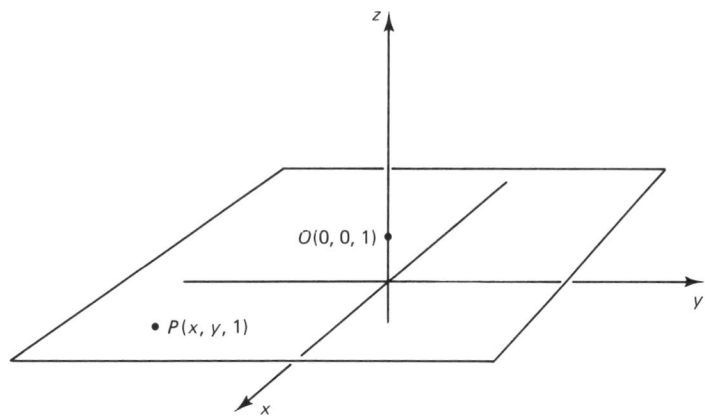

Figure 2.1 Coordinates on the Euclidean plane $z = 1$.

This correspondence H naturally extends to a one-to-one correspondence H between the set of ideal points added to the Euclidean plane $z = 1$ to obtain the standard embedded projective plane, and $\{[p_1, p_2, 0] \in RP^2\}$: For a specific ideal point P, let l denote the line in the Euclidean plane $z = 1$, which passes through the "origin" $O(0, 0, 1)$ and whose direction is P (see Figure 2.2). Then l can be described parametrically as

$$\{P_t = P_t(at, bt, 1) \in R^3 \mid t \in R\}$$

where $a, b, 0$ are direction numbers for l. For each t,

$$H(P_t) = [at, bt, 1] = \left[a, b, \frac{1}{t}\right]$$

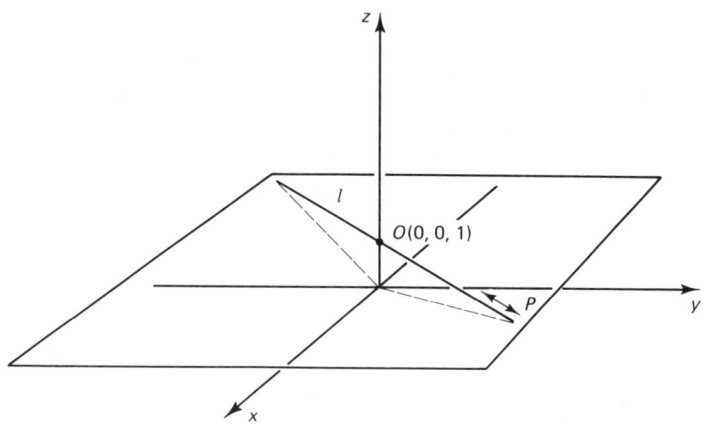

Figure 2.2 A line in the Euclidean plane $z = 1$ passing through the "origin."

and P is the limit of the Euclidean points P_t on l: $P = \lim_{t \to \infty} P_t$. If H (that is, the association of standard homogeneous coordinates) is to respect the manner in which ideal points are attached to the Euclidean plane $z = 1$ (respect means in the sense that $H(P) = H(\lim_{t \to \infty} P_t) = \lim_{t \to \infty} H(P_t)$), it should follow that

$$H(P) = H(\lim_{t \to \infty} P_t) = \lim_{t \to \infty} H(P_t) = \lim_{t \to \infty} \left[a, b, \frac{1}{t} \right] = [a, b, 0]$$

In other words, the standard homogeneous coordinates of P should be $[a, b, 0]$.

This is supported by the fact that the assignment of the standard homogeneous coordinates $[a, b, 0]$ to P does not depend on the choice of the line in the Euclidean plane $z = 1$ whose direction is P (see Figure 2.3). Let m be any other line in this plane that is parallel to l. Then m can be described parametrically as

$$\{P_t = P_t(at + x_0, bt + y_0, 1) \in R^3 \mid t \in R\}$$

where $a, b, 0$ are the direction numbers of both l and m and $P_0(x_0, y_0, 1)$ is a point that lies on m. Again, since for each t

$$H(P_t) = [at + x_0, bt + y_0, 1] = \left[a + \frac{x_0}{t}, b + \frac{y_0}{t}, \frac{1}{t} \right]$$

and P is the limit of the Euclidean points P_t on m, $P = \lim_{t \to \infty} P_t$, it should follow that

$$H(P) = H(\lim_{t \to \infty} P_t) = \lim_{t \to \infty} H(P_t) = \lim_{t \to \infty} \left[a + \frac{x_0}{t}, b + \frac{y_0}{t}, \frac{1}{t} \right] = [a, b, 0]$$

We denote by $P[a, b, 0]$ the ideal point in the standard embedded projective plane that is attached to all lines in the Euclidean plane $z = 1$ whose direction numbers are $a, b, 0$.

In summary we have the following definition.

1.5 Definition. The *standard homogeneous coordinates* $[p_1, p_2, p_3]$ of a point

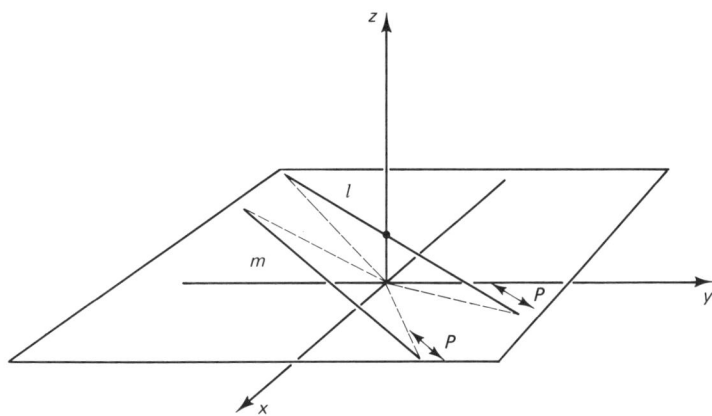

Figure 2.3 An arbitrary line in the Euclidean plane $z = 1$.

P in the projective plane (thought of as the standard embedded projective plane), written $P[p_1, p_2, p_3]$, are of the form

(a) $[x, y, 1]$ if P is a point in the Euclidean plane $z = 1$ whose Cartesian coordinates are $(x, y, 1)$

(b) $[a, b, 0]$ if P is the ideal point associated to all lines in the Euclidean plane $z = 1$ with direction numbers $a, b, 0$

Thus the decomposition of RP^2 described in Proposition 1.4 corresponds to

projective plane = {Euclidean points} ∪ {ideal points}

Note that it is easy to determine whether a point $P[p_1, p_2, p_3]$ is Euclidean ($p_3 \neq 0$) or ideal ($p_3 = 0$), but the notation $[p_1, p_2, p_3]$ for the standard homogeneous coordinates of a point downplays the distinction between Euclidean and ideal points; this is the homogeneity of homogeneous coordinates.

Some points in the projective plane together with their standard homogeneous coordinates are illustrated in Figure 2.4. (Figure 2.4 is a picture of *the* projective plane. Recall that we, as external observers, view the standard embedded projective plane as a projective plane embedded in projective three-space, while we, as observers in projective three-space, view the same projective plane as *the* projective plane.) Note that the ideal points in Figure 2.4 are represented by directions, and since we mean direction in the sense of the direction of a line (as opposed to the direction of a vector), directions are described by double-headed arrows. (In particular, *both* ends of the coordinate axes are labeled by arrows.) Also observe that since homogeneous coordinates are not unique, rational homogeneous coordinates can always be replaced by integral homogeneous coordinates. (In Chapter 3 we see that when considering projective transformations this feature lets us replace rational arithmetic with integral arithmetic.)

One last point: Just as standard homogeneous coordinates have been attached to the projective plane, standard homogeneous coordinates can be attached to the projective line and projective three-space.

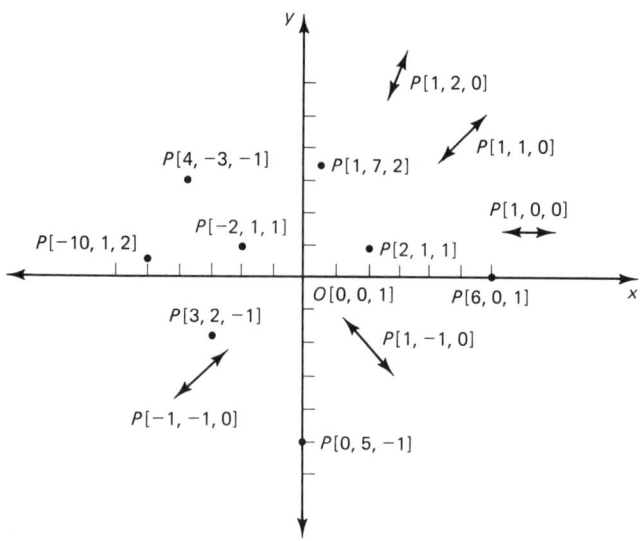

Figure 2.4 Standard homogeneous coordinates of points.

For the projective line (see Figure 2.5), standard homogeneous coordinates are of the form $[p_1, p_2]$, where $\langle p_1, p_2 \rangle$ is an element of $R^2 - \{\langle 0, 0 \rangle\}$, with $[p_1, p_2]$ equal to $[q_1, q_2]$ if and only if there is a nonzero real constant k for which $p_1 = kq_1$ and $p_2 = kq_2$. The standard homogeneous coordinates of a Euclidean point $P(x)$ are of the form $[x, 1]$, and the standard homogeneous coordinates of the ideal point are $[1, 0]$. The symbol $[0, 0]$ is meaningless.

For projective three-space (see Figure 2.6), standard homogeneous coordinates are of the form $[p_1, p_2, p_3, p_4]$, where $\langle p_1, p_2, p_3, p_4 \rangle$ is an element of $R^4 - \{\langle 0, 0, 0, 0 \rangle\}$, with $[p_1, p_2, p_3, p_4]$ equal to $[q_1, q_2, q_3, q_4]$ if and only if there is a nonzero real constant k for which $p_1 = kq_1$, $p_2 = kq_2$, $p_3 = kq_3$, and

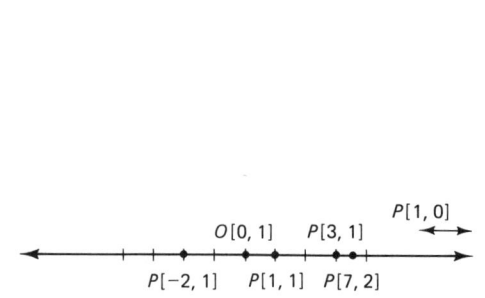

Figure 2.5 Standard homogeneous coordinates on the projective line.

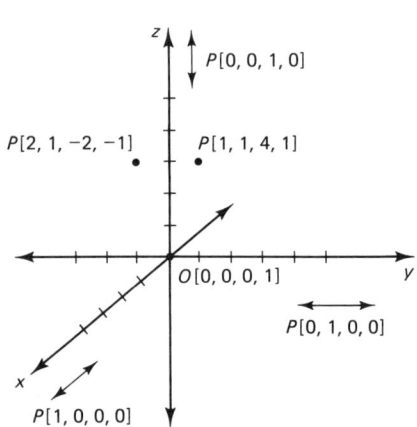

Figure 2.6 Standard homogeneous coordinates on projective three-space.

$p_4 = kq_4$. The standard homogeneous coordinates of a Euclidean point $P(x, y, z)$ are of the form $[x, y, z, 1]$, and the standard homogeneous coordinates of the ideal point associated to the direction numbers a, b, c are $[a, b, c, 0]$. Again, the symbol $[0, 0, 0, 0]$ is meaningless.

SECTION 1 EXERCISES

1. List three different representatives for each of the following.
   (a) $[1, -2, 3]$ (b) $[3, 0, -2]$
   (c) $[3, 5]$ (d) $[-2, 3]$
   (e) $[1, 0, -6, 3]$ (f) $[5, 1, 7, -2]$
2. State whether or not the following points are the same and explain why.
   (a) $A[2, -1, 3], B[4, -2, 6]$ (b) $A(2, 1, 3), B[2, 1, 3]$
   (c) $A[\sqrt{2}/2, -1, 0], B[1, -\sqrt{2}, 0]$ (d) $A[2, 7], B[4, 5]$
   (e) $A[3, 1, 7, 6], B[-6, -2, -14, 10]$ (f) $A[2, -1, 3], B[10, -5, 0]$
3. Find standard homogeneous coordinates for the points illustrated in Figure 2.7.

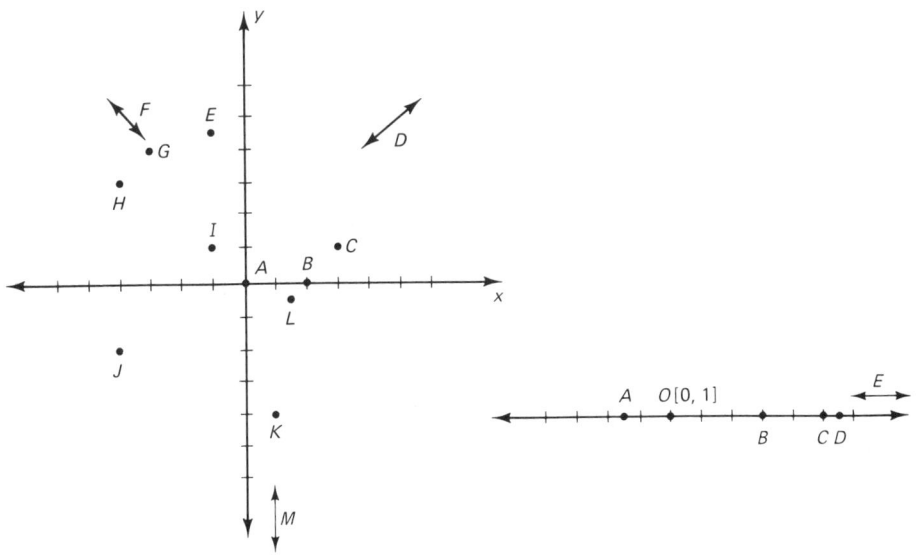

Figure 2.7 Points on the projective plane and the projective line.

4. Plot the points with the following standard homogeneous coordinates in the projective plane.
   (a) $A[0, 0, 1]$ (b) $B[0, 1, 1]$
   (c) $C[1, 0, 1]$ (d) $D[1, 1, 1]$
   (e) $E[1, 2, 1]$ (f) $F[1, 2, 2]$
   (g) $G[1, 2, 4]$ (h) $H[1, 2, \tfrac{1}{2}]$
   (i) $I[1, 0, 0]$ (j) $J[0, 1, 0]$

5. Plot the points with the following standard homogeneous coordinates on the projective line.
   (a) $A[0, 1]$
   (b) $B[1, 1]$
   (c) $C[1, 0]$
   (d) $D[\frac{1}{2}, 1]$
   (e) $E[-\frac{1}{2}, 1]$
   (f) $F[\frac{1}{2}, -1]$
6. In projective three-space, what are the standard homogeneous coordinates of the origin and the ideal points determined by the intersections of the extensions of the coordinate axes and the ideal plane?
7. Give two reasons—one analytic and one geometric—why the points $P[p_1, p_2, p_3]$, $p_3 \neq 0$, and $Q[q_1, q_2, 0]$ can never be the same.
8. For what values of t are the points $P[2t, t + 1, 3t - 4]$ and $Q[t, -t + 2, 3t + 2]$ ideal?
9. (a) For what value of t does $[2t, t + 1, 3t - 4] = [2, 3, -5]$?
   (b) For what value of t does $[t, -t + 2, 3t + 2] = [2, -1, 7]$?
10. (a) The hyperbola $x^2/a^2 - y^2/b^2 = 1$ can be parametrized by $x = a \sec \theta$, $y = b \tan \theta$. Find the standard homogeneous coordinates of the ideal points on the curve in the projective plane determined by this hyperbola. Sketch this hyperbola in the identification model of the projective plane illustrated in Figure 1.36. (*Hint:* Use the asymptotes of the hyperbola.)
    (b) Find the ideal point(s) on the curve in the projective plane determined by the graph of $y = \tan x$.
†11. Explain why the following statements are true, or provide counterexamples to show why they are false.
    (a) If $\mathbf{u} = \mathbf{v}$, then $[\mathbf{u}] = [\mathbf{v}]$.
    (b) If $[\mathbf{u}] = [\mathbf{v}]$, then $\mathbf{u} = \mathbf{v}$.
    (c) If $\mathbf{u} = \mathbf{v} + \mathbf{w}$, then $[\mathbf{u}] = [\mathbf{v}] + [\mathbf{w}]$.
    (d) If $\mathbf{u} + \mathbf{v} + \mathbf{w} = \mathbf{0}$, then $[\mathbf{u}] = [\mathbf{v} + \mathbf{w}]$.
12. Verify that the relation $\equiv$ defined in this section is an equivalence relation.
13. Describe two natural decompositions of RP^2 other than the one given in Proposition 1.4. To what viewing mechanisms do these decompositions correspond?

The *rank* of a matrix M is the size of the largest invertible submatrix of M.

†14. (a) Verify that two points $P[p_1, p_2, p_3]$ and $Q[q_1, q_2, q_3]$ in the projective plane are distinct if and only if the matrix
$$\begin{pmatrix} p_1 & p_2 & p_3 \\ q_1 & q_2 & q_3 \end{pmatrix}$$
has rank 2.
   (b) State and verify comparable results for points on the projective line and points in projective three-space.
   (c) How are these three results related geometrically?
†15. Consider the perspectivity in *Euclidean* three-space whose center is the origin and whose image plane is the Euclidean plane $z = 1$. Show that the image of the point $P(x, y, z)$, with $z \neq 0$, has Cartesian coordinates $(x/z, y/z, 1)$ under this perspectivity. Thus show that two distinct Euclidean points $P'(x', y', z')$ and $P''(x'', y'', z'')$ have the same image under this perspectivity if and only if $[x', y', z'] = [x'', y'', z'']$. (This is motivation for Definition 1.1 and for the terminology "the standard embedded projective plane.")

16. Assuming that it takes two bytes to store a real number and 1 byte to store an integer, what percentage savings in space can be made by storing the coordinates of a point in the plane as three integers (homogeneous coordinates) as opposed to two real numbers?
17. Describe how homogeneous coordinates might be used on a computer to store the coordinates of points that have at least one coordinate larger than the largest number that the computer can handle or at least one coordinate smaller than the smallest number that the computer can handle. Explain in what sense resolution becomes an issue.
18. In Section 1.6 we discussed extended real numbers informally. In what sense is RP^1 a concrete model of the extended real numbers?
19. (a) Using homogeneous coordinates in the projective plane, show that the set of ideal points in the projective plane looks like a projective line.
    (b) Using homogeneous coordinates in projective three-space, show that the set of ideal points in projective three-space looks like a projective plane.

SECTION 2: PROJECTIVE LINES (NONPARAMETRIC DESCRIPTION)

In this section we discuss lines in the projective plane analytically. We begin with a preliminary result.

2.1 Lemma. If the equation
$$l_1 p_1 + l_2 p_2 + l_3 p_3 = 0$$
is satisfied by the coordinates of one representative $\langle p_1, p_2, p_3 \rangle$ of an element $[p_1, p_2, p_3]$ of RP^2, then it is satisfied by the coordinates of all representatives of $[p_1, p_2, p_3]$.

Proof. Any other representative of $[p_1, p_2, p_3]$ is of the form $k\langle p_1, p_2, p_3 \rangle$, and
$$l_1(kp_1) + l_2(kp_2) + l_3(kp_3) = k(l_1 p_1 + l_2 p_2 + l_3 p_3) = 0 \qquad \blacksquare$$

This result motivates the following terminology.

2.2 Definition. The equation of Lemma 2.1 is *satisfied by* $[p_1, p_2, p_3]$ if and only if it is satisfied by the coordinates of any representative (hence all representatives) $\langle p_1, p_2, p_3 \rangle$ of $[p_1, p_2, p_3]$.

Recall that in the projective plane there are two different types of lines (see Figure 2.8): Euclidean lines completed with their ideal points and the ideal line. (The "ideal line" of Figure 2.8 is, at best, an artistic suggestion of the ideal line. To draw the actual ideal line, we would need to appeal to a model of the projective plane, as described in Section 1.4.)

2.3 Proposition. For every line l in the projective plane, there are real constants l_1, l_2, and l_3, not all of which are zero, such that $P[p_1, p_2, p_3]$ lies on l if and only if the equation
$$l_1 p_1 + l_2 p_2 + l_3 p_3 = 0$$

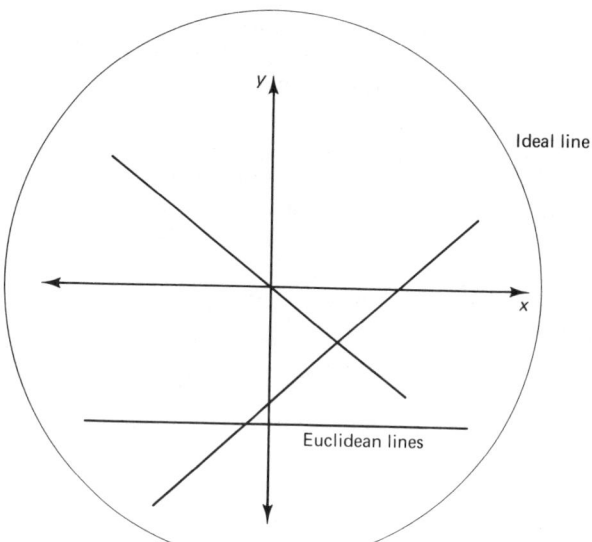

Figure 2.8 Lines in the projective plane.

is satisfied by the standard homogeneous coordinates $[p_1, p_2, p_3]$ of P. Conversely, if l_1, l_2, and l_3 are real constants, not all of which are zero, then the set of points $P[p_1, p_2, p_3]$ whose standard homogeneous coordinates satisfy this equation is a line in the projective plane.

Proof. Think of the projective plane as the standard embedded projective plane.

To prove the first part of this result, first consider a Euclidean line completed with its ideal point. The set of Euclidean points on this line is the intersection of the Euclidean plane $z = 1$ and a Euclidean plane in Euclidean three-space passing through the origin, whose equation is of the form $l_1 x + l_2 y + l_3 z = 0$ for some real constants l_1, l_2, and l_3, where l_1 and l_2 are not both zero (see Figure 2.9). The Euclidean point Q lies on l if and only if its Cartesian coordinates $(q_1, q_2, 1)$ satisfy the equation $l_1 x + l_2 y + l_3 z = 0$; thus the Euclidean point $Q[q_1, q_2, 1]$ lies on l if and only if its standard homogeneous coordinates $[q_1, q_2, 1]$ satisfy the corresponding equation $l_1 p_1 + l_2 p_2 + l_3 p_3 = 0$. Furthermore, the standard homogeneous coordinates of the ideal point $Q[-l_2, l_1, 0]$ of l satisfy the equation $l_1 p_1 + l_2 p_2 + l_3 p_3 = 0$:

$$l_1(-l_2) + l_2(l_1) + l_3(0) = 0$$

and this is the only ideal point whose standard homogeneous coordinates do so.

If l is the ideal line, then $Q[q_1, q_2, q_3]$ lies on l if and only if $q_3 = 0$ for every representative of $[q_1, q_2, q_3]$. In this case $l_1 = 0$, $l_2 = 0$, and $l_3 = 1$.

To prove the second part of the result, observe that l_1, l_2, and l_3 determine the equation $l_1 x + l_2 y + l_3 z = 0$ of a Euclidean plane that passes through the origin in Euclidean three-space (again see Figure 2.9). The intersection (or nonintersection) of this plane and the Euclidean plane $z = 1$ determines a line

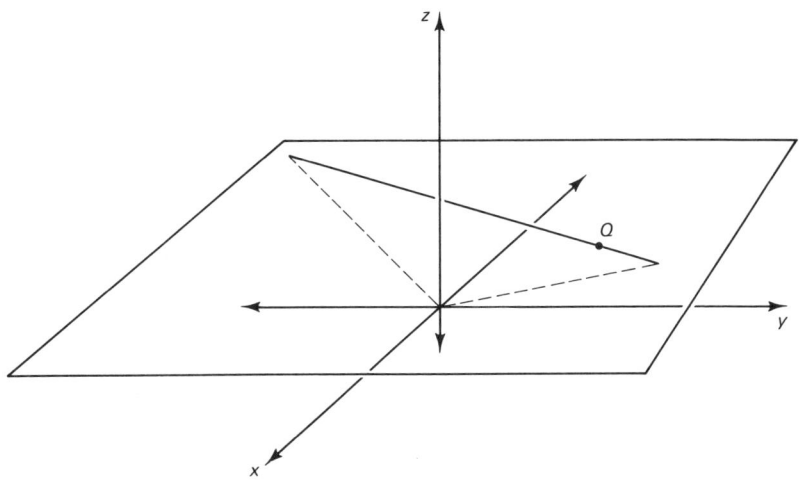

Figure 2.9 A line in the standard embedded projective plane.

in the standard embedded projective plane: This is a Euclidean line completed with its ideal point if l_1 or l_2 is nonzero and the ideal line if l_1 and l_2 are both zero (and l_3 is nonzero). ■

2.4 Definition. The equation associated to the line l in Proposition 2.3 is the *equation of l*.

As an immediate consequence of Proposition 2.3, it follows that if l_1 or l_2 is nonzero, then l is a Euclidean line completed with its ideal point; and if l_1 and l_2 are both zero (and l_3 is nonzero), then l is the ideal line. Furthermore, a point $P[p_1, p_2, p_3]$ lies on a line l if and only if the standard homogeneous coordinates of P satisfy the equation of l (l_1, l_2, and l_3 being treated as known and p_1, p_2, and p_3 being treated as unknown).

2.5 Example. The set of points $P[p_1, p_2, p_3]$ for which $3p_1 - p_2 + 2p_3 = 0$ is a line. The point $A[1, 3, 0]$ lies on this line, since $3(1) - 1(3) + 2(0) = 0$. However, the point $B[2, 3, 6]$ does not lie on this line, since $3(2) - 1(3) + 2(6) = 15 \neq 0$.

To graph the line l in the projective plane whose equation is $l_1 p_1 + l_2 p_2 + l_3 p_3 = 0$, first determine whether l is a Euclidean line completed with its ideal point or the ideal line. If l is the ideal line, its graph is the set of ideal points. If l is a Euclidean line completed with its ideal point, it suffices to find two points that lie on l to draw the graph of l.

2.6 Example. The points $A[-2, 0, 1]$ and $B[0, \frac{2}{3}, 1]$ lie on the line whose equation is $p_1 - 3p_2 + 2p_3 = 0$. To graph l, plot A and B and draw the line that they determine (see Figure 2.10).

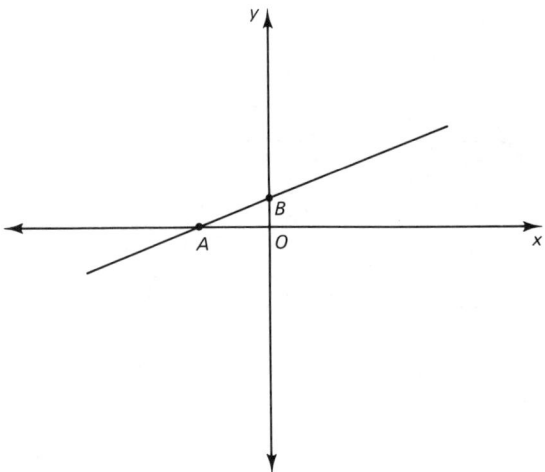

Figure 2.10 Graphing a line in the projective plane.

The next result is an analytic formulation of two dual axioms of planar projective geometry. To illustrate the theory, we both verify and derive this result.

2.7 Proposition

(a) The equation of the line l that passes through the (distinct) points $Q[q_1, q_2, q_3]$ and $R[r_1, r_2, r_3]$ (see Figure 2.11) is $l_1 p_1 + l_2 p_2 + l_3 p_3 = 0$ where

$$l_1 = \begin{vmatrix} q_2 & q_3 \\ r_2 & r_3 \end{vmatrix}, \quad l_2 = -\begin{vmatrix} q_1 & q_3 \\ r_1 & r_3 \end{vmatrix}, \quad l_3 = \begin{vmatrix} q_1 & q_2 \\ r_1 & r_2 \end{vmatrix}$$

(b) The point P of intersection of the distinct lines whose equations are

$$l_1 p_1 + l_2 p_2 + l_3 p_3 = 0$$
$$m_1 p_1 + m_2 p_2 + m_3 p_3 = 0$$

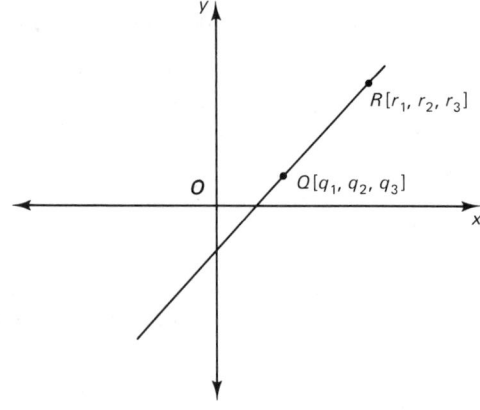

Figure 2.11 The line through two distinct points.

(see Figure 2.12) is $P[p_1, p_2, p_3]$, where

$$p_1 = \begin{vmatrix} l_2 & l_3 \\ m_2 & m_3 \end{vmatrix}, \quad p_2 = -\begin{vmatrix} l_1 & l_3 \\ m_1 & m_3 \end{vmatrix}, \quad p_3 = \begin{vmatrix} l_1 & l_2 \\ m_1 & m_2 \end{vmatrix}$$

Proof. To verify part (a) (part (b) is similar), first observe that l_1, l_2, and l_3 are not all zero by Exercise 1.14(a). Since the graph of the equation of part (a) is thus a line, it suffices to show that the standard homogeneous coordinates of Q and R satisfy this equation. For example,

$$l_1 q_1 + l_2 q_2 + l_3 q_3 = \begin{vmatrix} q_2 & q_3 \\ r_2 & r_3 \end{vmatrix} q_1 - \begin{vmatrix} q_1 & q_3 \\ r_1 & r_3 \end{vmatrix} q_2 + \begin{vmatrix} q_1 & q_2 \\ r_1 & r_2 \end{vmatrix} q_3$$

$$= (q_2 r_3 - r_2 q_3) q_1 - (q_1 r_3 - r_1 q_3) q_2 + (q_1 r_2 - r_1 q_2) q_3 = 0$$

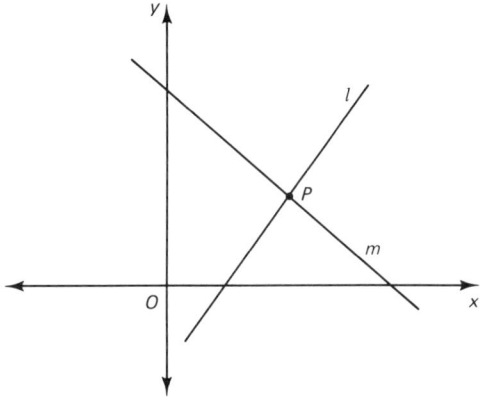

Figure 2.12 The point of intersection of two distinct lines.

Both parts of this result can be derived as follows: For part (b) for example (part (a) is similar), first observe that since P lies on l and m,

$$p_1 l_1 + p_2 l_2 + p_3 l_3 = 0$$
$$p_1 m_1 + p_2 m_2 + p_3 m_3 = 0$$

Since l and m are distinct, the matrix

$$\begin{pmatrix} l_1 & l_2 & l_3 \\ m_1 & m_2 & m_3 \end{pmatrix}$$

has rank 2 (see Exercise 1.14) and hence has an invertible 2×2 submatrix. If, for example, this submatrix is

$$\begin{pmatrix} l_1 & l_2 \\ m_1 & m_2 \end{pmatrix}$$

(the other two possible cases are treated similarly), then applying Cramer's Rule to the system

$$p_1 l_1 + p_2 l_2 = -p_3 l_3$$
$$p_1 m_1 + p_2 m_2 = -p_3 m_3$$

(treating p_1 and p_2 as unknowns), we obtain

$$p_1 = p_3 \frac{\begin{vmatrix} -l_3 & l_2 \\ -m_3 & m_2 \end{vmatrix}}{\begin{vmatrix} l_1 & l_2 \\ m_1 & m_2 \end{vmatrix}} \qquad p_2 = p_3 \frac{\begin{vmatrix} l_1 & -l_3 \\ m_1 & -m_3 \end{vmatrix}}{\begin{vmatrix} l_1 & l_2 \\ m_1 & m_2 \end{vmatrix}}$$

Now p_3 cannot be zero (if it were then p_1 and p_2 would also be zero, so $[p_1, p_2, p_3]$ would be $[0, 0, 0]$, and $[0, 0, 0]$ is meaningless). Thus

$$[p_1, p_2, p_3] = \left[p_3 \frac{\begin{vmatrix} l_2 & l_3 \\ m_2 & m_3 \end{vmatrix}}{\begin{vmatrix} l_1 & l_2 \\ m_1 & m_2 \end{vmatrix}}, -p_3 \frac{\begin{vmatrix} l_1 & l_3 \\ m_1 & m_3 \end{vmatrix}}{\begin{vmatrix} l_1 & l_2 \\ m_1 & m_2 \end{vmatrix}}, p_3 \right]$$

Simplifying this expression yields the desired result. ∎

2.8 Example. To find the equation of the line passing through the points $A[1, 0, 2]$ and $B[3, 1, 2]$, compute

$$l_1 = \begin{vmatrix} 0 & 2 \\ 1 & 2 \end{vmatrix} = -2, \qquad l_2 = -\begin{vmatrix} 1 & 2 \\ 3 & 2 \end{vmatrix} = 4, \qquad l_3 = \begin{vmatrix} 1 & 0 \\ 3 & 1 \end{vmatrix} = 1$$

Thus the equation of l is $-2p_1 + 4p_2 + p_3 = 0$.

2.9 Example. The standard homogeneous coordinates of the point P of intersection of the lines $2p_1 - p_2 + 3p_3 = 0$ and $p_1 + 3p_2 + p_3 = 0$ are given by

$$p_1 = \begin{vmatrix} -1 & 3 \\ 3 & 1 \end{vmatrix} = -10, \qquad p_2 = -\begin{vmatrix} 2 & 3 \\ 1 & 1 \end{vmatrix} = 1, \qquad p_3 = \begin{vmatrix} 2 & -1 \\ 1 & 3 \end{vmatrix} = 7$$

Thus the point of intersection is $P[-10, 1, 7]$.

The constants l_1, l_2, and l_3 that determine the equation of a line are unique only up to nonzero scalar multiples: If k is any nonzero real constant, the equation determined by l_1, l_2, l_3,

$$l_1 p_1 + l_2 p_2 + l_3 p_3 = 0$$

is the same as the equation determined by kl_1, kl_2, kl_3,

$$(kl_1)p_1 + (kl_2)p_2 + (kl_3)p_3 = k(l_1 p_1 + l_2 p_2 + l_3 p_3) = 0$$

Thus associated to each line l, there is a vector $\langle l_1, l_2, l_3 \rangle$ uniquely defined up to scalar multiples—something that looks very much like an element of RP^2. This leads to a concrete analytic description of planar duality (see Section 1.6).

To begin with, there is a one-to-one correspondence H between the set of all lines in the projective plane and RP^2, which associates to each line l whose equation is $l_1 p_1 + l_2 p_2 + l_3 p_3 = 0$ the element $H(l) = [l_1, l_2, l_3]$ of RP^2.

2.10 Definition. The *standard homogeneous coordinates* of a line l in the projective plane are given by $H(l) = [l_1, l_2, l_3]$, where H is the one-to-one correspondence described above. The line l whose standard homogeneous coordinates are $[l_1, l_2, l_3]$ is denoted by $l[l_1, l_2, l_3]$.

Combining the one-to-one correspondence between the points in the projective plane and RP^2 described in Section 1 and the one-to-one correspondence between the lines in the projective plane and RP^2 described above, we obtain the following result.

2.11 Theorem. (Analytic Duality) There is a one-to-one correspondence between the points in the projective plane and the lines in the projective plane that associates to the point P with standard homogeneous coordinates $[x, y, z]$ the line l with standard homogeneous coordinates $[x, y, z]$.

For example, the point $P[1, -2, 3]$ and the line $l[1, -2, 3]$ are dual to each other.

Planar duality means that points and lines in the projective plane can be treated symmetrically. Analytically, for example, this means that points have equations: The equation of a point $P[p_1, p_2, p_3]$ is $l_1p_1 + l_2p_2 + l_3p_3 = 0$ (p_1, p_2, and p_3 being treated as known, and l_1, l_2, and l_3 being treated as unknown); that is, the point P is (the intersection of) the set of all lines $l[l_1, l_2, l_3]$ that pass through P. In fact, our notation has been chosen so that the equation $l_1p_1 + l_2p_2 + l_3p_3 = 0$ can be treated symmetrically in $[l_1, l_2, l_3]$ and $[p_1, p_2, p_3]$. (Notice, in particular, the duality between parts (a) and (b) of Proposition 2.7.)

Geometrically, duality can be viewed as follows (see Figure 2.13): Think of the projective plane as the standard embedded projective plane. Every line $l[l_1, l_2, l_3]$ in the standard embedded projective plane is determined by the intersection (or nonintersection) of the Euclidean plane $z = 1$ with a Euclidean plane in Euclidean three-space that passes through the origin and whose equation is of the form $l_1x + l_2y + l_3z = 0$. Since the position vector $\langle l_1, l_2, l_3 \rangle$ in R^3 is

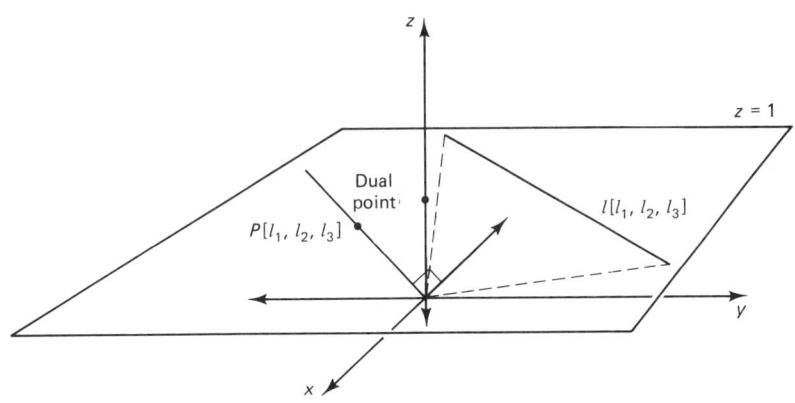

Figure 2.13 A geometric interpretation of planar duality.

perpendicular to the Euclidean plane $l_1 x + l_2 y + l_3 z = 0$, the point with standard homogeneous coordinates $[l_1, l_2, l_3]$—the point dual to l—is the point of intersection of the standard embedded projective plane with the line in projective three-space determined by the Euclidean line in Euclidean three-space that passes through the origin and is perpendicular to the Euclidean plane $l_1 x + l_2 y + l_3 z = 0$. (Notice, in particular, the similarity between the formulas for l_1, l_2, and l_3 given in Proposition 2.7(a) and the formulas for the components of the cross product **OQ** × **OR** of the position vectors **OQ** and **OR**.)

This geometric description of Theorem 2.11 reveals an important principle: There are two different ways to think about duality. One way (see Theorem 1.6.3) is as a concept related to the formal statement of results. This is an abstract, or formal, duality. The other way (Theorem 2.11) is as a specific one-to-one correspondence between objects (such as points and lines) and incidence relations. This is concrete duality. Establishing concrete duality requires that choices be made. In Theorem 2.11 the choices were those that determine standard homogeneous coordinates. In Section 3.4 we discuss general homogeneous coordinates, and at that time we observe that there are other forms of concrete duality. This is not the case for formal duality.

We now consider conditions for collinearity.

2.12 Proposition. The following are equivalent.

(a) The points $P[p_1, p_2, p_3]$, $Q[q_1, q_2, q_3]$, and $R[r_1, r_2, r_3]$ are collinear.
(b) The determinant

$$\begin{vmatrix} p_1 & p_2 & p_3 \\ q_1 & q_2 & q_3 \\ r_1 & r_2 & r_3 \end{vmatrix}$$

is zero.

(c) Given representatives $\langle p_1, p_2, p_3 \rangle$ of $[p_1, p_2, p_3]$, $\langle q_1, q_2, q_3 \rangle$ of $[q_1, q_2, q_3]$, and $\langle r_1, r_2, r_3 \rangle$ of $[r_1, r_2, r_3]$, there are constants k_1, k_2, and k_3, not all of which are zero, for which

$$k_1 \langle p_1, p_2, p_3 \rangle + k_2 \langle q_1, q_2, q_3 \rangle + k_3 \langle r_1, r_2, r_3 \rangle = \langle 0, 0, 0 \rangle$$

Proof. It is not difficult to verify this result if P, Q, and R are not distinct. So suppose P, Q, and R are distinct.

If P, Q, and R are collinear, then they lie on a line $l = l[l_1, l_2, l_3]$; that is, there are real constants l_1, l_2, and l_3, not all of which are zero, for which

$$l_1 p_1 + l_2 p_2 + l_3 p_3 = 0$$
$$l_1 q_1 + l_2 q_2 + l_3 q_3 = 0$$
$$l_1 r_1 + l_2 r_2 + l_3 r_3 = 0$$

or, in matrix form,

$$\begin{pmatrix} p_1 & p_2 & p_3 \\ q_1 & q_2 & q_3 \\ r_1 & r_2 & r_3 \end{pmatrix} \begin{pmatrix} l_1 \\ l_2 \\ l_3 \end{pmatrix} = \begin{pmatrix} 0 \\ 0 \\ 0 \end{pmatrix}$$

Since this system has a nontrivial solution, the determinant of the coefficient matrix must be zero. Thus (a) implies (b).

The fact that (b) implies (c) is a standard result in linear algebra.

To see that (c) implies (a), observe that none of the constants k_1, k_2, or k_3 can be zero (if two constants were zero, then the point corresponding to the third constant would have standard homogeneous coordinates [0, 0, 0], which is impossible; if one constant were zero, then the points corresponding to the other two constants would be the same, that is, not distinct). In particular, k_1 is not zero, and

$$\langle p_1, p_2, p_3 \rangle = -\frac{k_2}{k_1}\langle q_1, q_2, q_3 \rangle - \frac{k_3}{k_1}\langle r_1, r_2, r_3 \rangle$$

or

$$p_1 = \frac{k_2}{k_1}q_1 - \frac{k_3}{k_1}r_1$$

$$p_2 = \frac{k_2}{k_1}q_2 - \frac{k_3}{k_1}r_2$$

$$p_3 = \frac{k_2}{k_1}q_3 - \frac{k_3}{k_1}r_3$$

If $l[l_1, l_2, l_3]$ is the line determined by Q and R (see Proposition 2.7), then by directly substituting these values of p_1, p_2, and p_3 into the equation $l_1 p_1 + l_2 p_2 + l_3 p_3 = 0$ of l, it follows that P lies on l. ∎

2.13 Example. The points $A[2, 1, 3]$, $B[1, 2, -1]$, and $C[1, -4, 9]$ are collinear since

$$\begin{vmatrix} 2 & 1 & 3 \\ 1 & 2 & -1 \\ 1 & -4 & 9 \end{vmatrix} = 0$$

However the points $A[1, 2, 1]$, $B[2, 0, 1]$, and $C[1, 1, 3]$ are not collinear since

$$\begin{vmatrix} 1 & 2 & 1 \\ 2 & 0 & 1 \\ 1 & 1 & 3 \end{vmatrix} = -9 \neq 0$$

From the proof of Proposition 2.12 it follows that if P, Q, and R are distinct then the constants k_1, k_2, and k_3 are all nonzero. In this case these constants can be found either by using Gauss–Jordan Elimination or by using right inverses: If M is an $m \times n$ matrix, $m \leq n$, of rank m then the matrix MM^T is invertible and the matrix M^+ is defined by $M^+ = M^T(MM^T)^{-1}$. The matrix M^+ is a right inverse of M in the sense that $MM^+ = MM^T(MM^T)^{-1} = I$.

2.14 Example. Since $A[2, 1, 3]$, $B[1, 2, -1]$, and $C[1, -4, 9]$ are collinear, given representatives $\langle 2, 1, 3 \rangle$ of $[2, 1, 3]$, $\langle 1, 2, -1 \rangle$ of $[1, 2, -1]$, and $\langle 1, -4, 9 \rangle$ of $[1, -4, 9]$, there must be constants k_1, k_2, and k_3 for which

$$k_1\langle 2, 1, 3 \rangle + k_2\langle 1, 2, -1 \rangle + k_3\langle 1, -4, 9 \rangle = \langle 0, 0, 0 \rangle$$

To find these constants, observe that solving
$$\langle 2k_1 + k_2 + k_3, k_1 + 2k_2 - 4k_3, 3k_1 - k_2 + 9k_3 \rangle = \langle 0, 0, 0 \rangle$$
for k_1, k_2, and k_3 is equivalent to solving
$$2k_1 + k_2 + k_3 = 0$$
$$k_1 + 2k_2 - 4k_3 = 0$$
$$3k_1 - k_2 + 9k_3 = 0$$
We can use Gauss–Jordan Elimination to solve this system. We find $k_1 = -2$, $k_2 = 3$, and $k_3 = 1$:
$$(-2)\langle 2, 1, 3 \rangle + (3)\langle 1, 2, -1 \rangle + (1)\langle 1, -4, 9 \rangle = \langle 0, 0, 0 \rangle$$

Alternately, we can rewrite the original system as a system in k_1/k_3 and k_2/k_3:
$$\langle k_1/k_3, k_2/k_3 \rangle \begin{pmatrix} 2 & 1 & 3 \\ 1 & 2 & -1 \end{pmatrix} = \langle 1, -4, 9 \rangle$$

If
$$M = \begin{pmatrix} 2 & 1 & 3 \\ 1 & 2 & -1 \end{pmatrix}$$
then
$$M^+ = \begin{pmatrix} 11/83 & 12/83 \\ 4/83 & 27/83 \\ 19/83 & -17/83 \end{pmatrix}$$
so
$$\langle k_1/k_3, k_2/k_3 \rangle = \langle k_1/k_3, k_2/k_3 \rangle MM^+ = \langle 1, -4, 9 \rangle M^+ = \langle -2, 3 \rangle$$
Thus if $k_3 = 1$, then $k_1 = -2$ and $k_2 = 3$. (Depending on how the original system of equations is rewritten, there are three different ways to apply the right inverse technique. We obtain equivalent solutions to the original system using any of these three possible alternatives.)

If P, Q, and R are distinct, so that the constants k_1, k_2, and k_3 are all nonzero, we can absorb these constants into the representatives of $[p_1, p_2, p_3]$, $[q_1, q_2, q_3]$, and $[r_1, r_2, r_3]$. This gives us the following result.

2.15 Corollary. The distinct points P, Q, and R are collinear if and only if there are representatives $\langle p_1, p_2, p_3 \rangle$ of $[p_1, p_2, p_3]$, $\langle q_1, q_2, q_3 \rangle$ of $[q_1, q_2, q_3]$, and $\langle r_1, r_2, r_3 \rangle$ of $[r_1, r_2, r_3]$, for which
$$\langle p_1, p_2, p_3 \rangle + \langle q_1, q_2, q_3 \rangle + \langle r_1, r_2, r_3 \rangle = \langle 0, 0, 0 \rangle$$

2.16 Example. The points $A[2, 1, 3]$, $B[1, 2, -1]$, and $C[1, -4, 9]$ are collinear (see Example 2.14) and
$$\langle -4, -2, -6 \rangle + \langle 3, 6, -3 \rangle + \langle 1, -4, 9 \rangle = \langle 0, 0, 0 \rangle$$

As an application of Theorem 2.11 and Proposition 2.12, we now prove a fundamental result of projective geometry known as Desargues' Theorem.

In order to state Desargues' Theorem, recall (see Section 1.6) that the following pairs of terms are planar duals of each other: *point* and *line*, and *lie(s) on* and *meet(s) in*. We let *P-Q* (or *PQ* for short) denote the line determined by the points *P* and *Q*, and *l-m* (or *lm* for short) denote the point of intersection of the lines *l* and *m*.

2.17 Theorem. (Desargues' Theorem) Consider triangles △*ABC* and △*A'B'C'* (see Figure 2.14). If the lines *A-A'*, *B-B'*, and *C-C'* meet in a common point *P*, then the points *AB-A'B'*, *AC-A'C'*, and *BC-B'C'* lie on a common line *l*. And conversely, if the points *AB-A'B'*, *AC-A'C'*, and *BC-B'C'* lie on a common line *l*, then the lines *A-A'*, *B-B'*, and *C-C'* meet in a common point *P*. (*Note:* Our notation here has been chosen to emphasize the planar duality of the two parts of Desargues' Theorem.)

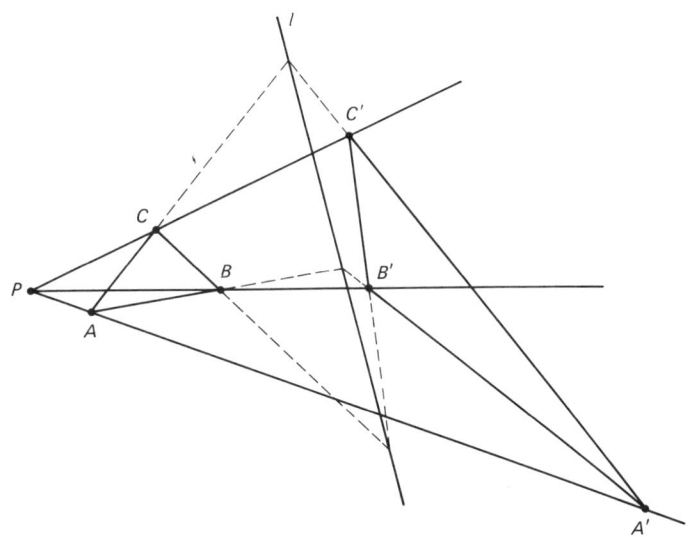

Figure 2.14 Desargues' Theorem.

Proof. Assume that *A*, *B*, *C*, *A'*, *B'*, and *C'* are all distinct (the result follows in the event *A*, *B*, *C*, *A'*, *B'*, and *C'* are not all distinct as a special case of what we will do below).

If *A-A'*, *B-B'*, and *C-C'* meet in a common point *P*, then *P*, *A*, and *A'* are collinear, *P*, *B*, and *B'* are collinear, and *P*, *C*, and *C'* are collinear. Having chosen a representative $\mathbf{p} = \langle p_1, p_2, p_3 \rangle$ of the standard homogeneous coordinates of $P[\mathbf{p}] = P[p_1, p_2, p_3]$, we can find representatives **a** of the standard homogeneous coordinates of *A*[**a**], **b** of *B*[**b**], **c** of *C*[**c**], **a'** of *A'*[**a'**], **b'** of *B'*[**b'**], and **c'** of *C'*[**c'**] for which

$$\mathbf{p} = \mathbf{a} + \mathbf{a'}, \mathbf{p} = \mathbf{b} + \mathbf{b'}, \text{ and } \mathbf{p} = \mathbf{c} + \mathbf{c'}$$

But then

$$\mathbf{a} + \mathbf{a'} = \mathbf{b} + \mathbf{b'}, \mathbf{b} + \mathbf{b'} = \mathbf{c} + \mathbf{c'}, \text{ and } \mathbf{a} + \mathbf{a'} = \mathbf{c} + \mathbf{c'}$$

so that

$$\mathbf{a} - \mathbf{b} = \mathbf{b}' - \mathbf{a}', \mathbf{b} - \mathbf{c} = \mathbf{c}' - \mathbf{b}', \text{ and } \mathbf{c} - \mathbf{a} = \mathbf{a}' - \mathbf{c}'$$

Thus $[\mathbf{a} - \mathbf{b}] = [\mathbf{b}' - \mathbf{a}']$. Since $A[\mathbf{a}]$, $B[\mathbf{b}]$, and $C''[\mathbf{a} - \mathbf{b}]$ are collinear, C'' lies on AB; and since $A'[\mathbf{a}']$ $B'[\mathbf{b}']$, and $C''[\mathbf{b}' - \mathbf{a}']$ are collinear, C'' also lies on $A'B'$. Thus C'' must be AB-$A'B'$. Similarly $B''[\mathbf{c} - \mathbf{a}] = B''[\mathbf{a}' - \mathbf{c}']$ must be AC-$A'C'$, and $A''[\mathbf{b} - \mathbf{c}] = A''[\mathbf{c}' - \mathbf{b}']$ must be BC-$B'C'$. The conclusion follows, since

$$(\mathbf{a} - \mathbf{b}) + (\mathbf{b} - \mathbf{c}) + (\mathbf{c} - \mathbf{a}) = \langle 0, 0, 0 \rangle$$

implies that $A''[\mathbf{b} - \mathbf{c}]$, $B''[\mathbf{c} - \mathbf{a}]$, and $C''[\mathbf{a} - \mathbf{b}]$ are collinear.

Since, in projective geometry, the dual of every valid result is a valid result (see Theorem 1.6.3), since the first part of Theorem 2.17 is valid and since the second part of Theorem 2.17 is the dual of the first part of Theorem 2.17, the second part of Theorem 2.17 is valid. ■

The proof of Desargues' Theorem illustrates a standard technique for proving results in (both planar and spatial) projective geometry: The technique is to prove the dual of the desired result and then to appeal to duality (Theorem 1.6.3).

SECTION 2 EXERCISES

1. Which of the following points lie on the line whose equation is $3p_1 - 2p_2 + 5p_3 = 0$?
    (a) $A[1, 1, 2]$ (b) $B[4, 1, -2]$
    (c) $C[2, 3, 0]$ (d) $D[-5, 0, 3]$

2. Find the equations of the lines that are the extensions to the projective plane of the following Euclidean lines.
    (a) $3x + 2y = 6$ (b) $4x + 5y + 7 = 0$
    (c) $x = 0$ (d) $y = 0$
    (e) $y = x$ (f) $y = 3x + 2$

3. Sketch each line in the projective plane whose equation is given.
    (a) $2p_1 + 3p_2 + 5p_3 = 0$ (b) $3p_1 - 2p_2 - p_3 = 0$
    (c) $p_1 = 0$ (d) $p_1 + 3p_2 = 0$
    (e) $p_1 - 2p_3 = 0$ (f) $p_2 = 0$

4. Find the equations of the lines illustrated in Figure 2.15.

5. In each of the following cases, sketch the line determined by the two given points; then find the equation of the line.
    (a) $A[3, 1, 2], B[1, 2, -1]$ (b) $A[2, 1, 3], B[1, 2, 0]$
    (c) $A[1, 1, 0], B[3, 2, 0]$ (d) $A[5, 2, 1], B[-1, 1, 3]$
    (e) $A[2, 1, 2], B[1, 3, 5]$ (f) $A[1, 2, 1], B[-1, 3, 0]$

6. Find the equation of the line determined by the points $A[1, -2, 3]$ and $B[-4, 5, 6]$ twice, each time using two different representatives for A and B. The equations you obtain should be the same. Why should this be true in general?

7. Find the standard homogeneous coordinates of the point of intersection for each pair of lines.

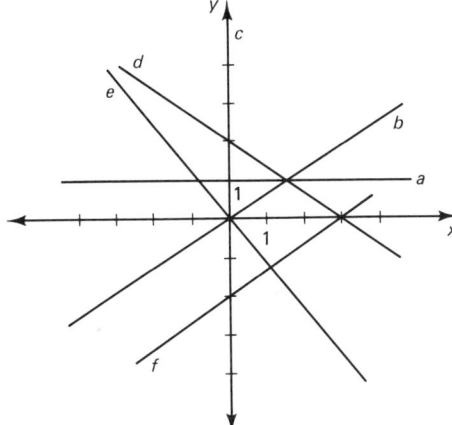

Figure 2.15 Lines in the projective plane.

(a) $p_1 + p_2 - 2p_3 = 0$, $3p_1 + p_2 + 4p_3 = 0$
(b) $p_1 + p_2 = 0$, $4p_1 - 2p_2 + p_3 = 0$
(c) $3p_1 + 2p_2 + p_3 = 0$, $p_3 = 0$
(d) $l[1, 2, 1]$, $m[-1, 1, -1]$
(e) $l[2, -3, 1]$, $m[3, 1, 5]$
(f) $l[0, 0, 1]$, $m[3, -2, 4]$

8. Find the standard homogeneous coordinates of the lines in Exercises 2 and 3.
9. What are the equations of the lines whose standard homogeneous coordinates are given?
   (a) $a[1, 3, -4]$ (b) $b[0, 0, 1]$
   (c) $c[1, 0, 0]$ (d) $d[0, 2, -1]$
   (e) $e[1, 0, -2]$ (f) $f[-2, 3, 1]$
10. Show axiomatically that if $A[a_1, a_2, a_3]$ and $B[b_1, b_2, b_3]$ are two distinct points in the projective plane, then

$$\begin{vmatrix} p_1 & p_2 & p_3 \\ a_1 & a_2 & a_3 \\ b_1 & b_2 & b_3 \end{vmatrix} = 0$$

is the equation of the line through A and B. (*Hint:* Since this is a linear equation in p_1, p_2, and p_3, since the graph of a linear equation is a line, and since a line is uniquely determined by two points that lie on it, it suffices to show that the coordinates of A and B satisfy this equation.) Use this result to verify Proposition 2.7.

11. Verify that if PQ denotes the line determined by $P = P[\mathbf{p}]$ and $Q = Q[\mathbf{q}]$, then $PQ = PQ[\mathbf{p} \times \mathbf{q}]$; also verify that if lm denotes the point of intersection of the lines $l = l[\mathbf{l}]$ and $m = m[\mathbf{m}]$, then $lm = lm[\mathbf{l} \times \mathbf{m}]$.

12. Find M^+ for each of the following matrices M.
   (a) $\begin{pmatrix} 1 & 0 & 1 \\ -1 & -1 & 1 \end{pmatrix}$ (b) $\begin{pmatrix} -2 & 1 & 1 \\ 3 & -1 & 2 \end{pmatrix}$
   (c) $\begin{pmatrix} 2/\sqrt{14} & 1/\sqrt{14} & 3/\sqrt{14} \\ -1/\sqrt{3} & -1/\sqrt{3} & 1/\sqrt{3} \end{pmatrix}$ (d) $(4\ 3\ 7)$

(e) $\begin{pmatrix} 2 & 1 & 3 & -4 \\ 3 & 2 & 3 & 5 \\ -1 & -8 & 3 & 3 \end{pmatrix}$ (f) $\begin{pmatrix} 3 & 0 & -1 & 2 \\ 1 & 1 & 3 & 0 \end{pmatrix}$

†13. (a) Show that if M is a square matrix whose determinant is nonzero then M^+ exists and $M^+ = M^{-1}$.
  (b) Show that if M is an $m \times n$ matrix, $m \le n$, whose rows are orthonormal (that is, whose rows are of length 1 and perpendicular to one another) then M^+ exists and $M^+ = M^T$.

14. Determine which of the following sets of three points are collinear. For those that are, obtain equations of the form described in Proposition 2.12(c) and Corollary 2.15.
  (a) $A[1, 2, 1]$, $B[0, 1, 3]$, $C[2, 1, 1]$
  (b) $A[1, 2, 3]$, $B[2, 4, 3]$, $C[1, 2, -2]$
  (c) $A[2, 1, -3]$, $B[4, -2, 4]$, $C[10, -1, 0]$
  (d) $A[1, 1, 0]$, $B[1, 1, 1]$, $C[3, 3, 1]$

15. State the dual of Proposition 2.12. (By Theorem 1.6.3, the dual of Proposition 2.12 is automatically true.)

16. Using the result stated in Exercise 15, determine which of the following sets of three lines meet in a point. For those that do, obtain equations of the form described in part (c) of this same result.
  (a) $l[1, 0, 1]$, $m[1, 1, 0]$, $n[0, 1, -1]$ (b) $l[1, 0, -1]$, $m[1, -2, 1]$, $n[3, -2, -1]$
  (c) $l[1, 1, 1]$, $m[1, 2, 0]$, $n[1, -1, 3]$ (d) $l[1, 2, 1]$, $m[3, 1, 0]$, $n[1, 7, 4]$

17. Find the equations of the following points.
  (a) $A[1, -2, 3]$ (b) $B[1, 0, 3]$

18. Show that the point $P[a, b, c]$ can never lie on the line $l[a, b, c]$.

19. Explain why the following statement is true or provide a counterexample to show why it is false: If $\mathbf{u} = k_1\mathbf{v} + k_2\mathbf{w}$, then $[\mathbf{u}]$, $[\mathbf{v}]$, and $[\mathbf{w}]$ are collinear.

†20. Verify that the equation of any line can be written in the matrix form

$$(p_1 \ p_2 \ p_3) \begin{pmatrix} l_1 \\ l_2 \\ l_3 \end{pmatrix} = 0$$

Write the equation of the line $3p_1 - 2p_2 + 5p_3 = 0$ in this form.

†21. Equations of the form

$$3p_1 - p_2 + 5p_3 = 0$$
$$p_1 p_2 - 2p_2^2 - p_2 p_3 = 0$$
$$3p_1^2 p_2 - 4p_1 p_2^2 - p_3^3 = 0$$

in which all terms have the same degree, are called *homogeneous*. This is the fundamental type of equation that appears in analytic projective geometry. Why? (*Hint:* See Lemma 2.1.)

22. In this section, planar duality was described extrinsically—it used the fact that the standard embedded projective plane is a subspace of a larger space (namely, projective three-space) in which we think of ourselves existing (see Figure 2.13). Starting from this extrinsic description, verify—both synthetically and analytically—the following intrinsic description of planar duality. Again, intrinsic means that it does not rely on the existence of an ambient space.

(a) The line dual to the origin O of the Euclidean plane is the ideal line.

(b) The line l dual to a Euclidean point P other than the origin is determined as follows (see Figure 2.16): draw a Euclidean line m through P and O, and determine the point Q on m on the opposite side of O from P for which dist$(O, Q) = 1/$dist(O, P); l is the line determined by the Euclidean line that passes through Q and is perpendicular to m.

(c) The dual line l of an ideal point P (thought of as a direction) is the line determined by the Euclidean line passing through the origin that is perpendicular to the direction of P (see Figure 2.17).

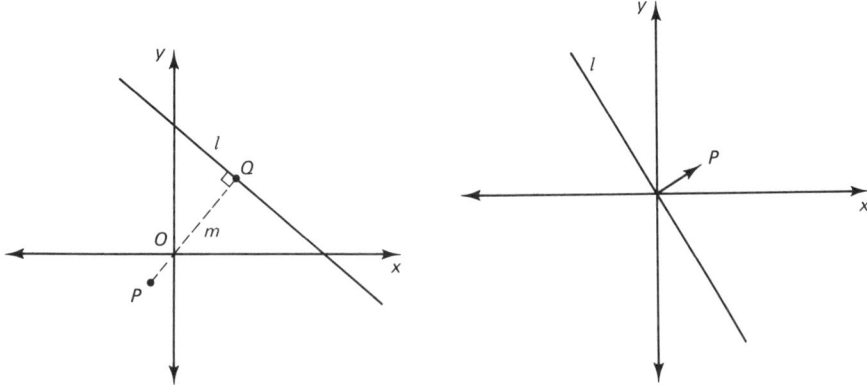

Figure 2.16 Intrinsic planar duality for Euclidean points.

Figure 2.17 Intrinsic planar duality for ideal points.

23. (Continuation of Exercise 22) Intrinsically describe the dual points of the following lines.
    (a) A line determined by a Euclidean line that passes through the origin
    (b) A line determined by a Euclidean line that does not pass through the origin
    (c) The ideal line

24. Verify Proposition 2.12 in the event that P, Q, and R are not distinct.

25. How is Desargues' Theorem related to Exercise 1.3.6?

26. If $A[\mathbf{a}]$ and $B[\mathbf{b}]$ are distinct, explain why each of the following is true.
    (a) A, B, and $C'''[\mathbf{a} - \mathbf{b}]$ are collinear.
    (b) A, B, and $C'''[\mathbf{b} - \mathbf{a}]$ are collinear.

27. The sides of the line in the Euclidean plane whose equation is $ax + by + c = 0$ are distinguished by the sign of the quantity $ax + by + c$ for points $P(x, y)$ not on the line. Explain why this characterization breaks down analytically in the projective plane. (The projective plane is nonorientable: There are no sides of a line in the projective plane; see Section 1.4.)

SECTION 3: PROJECTIVE PLANES (NONPARAMETRIC DESCRIPTION)

In this section planes in projective three-space are discussed in a manner parallel to the manner in which lines in the projective plane are discussed in Section 2.

Since this parallelism is analytically so straightforward, proofs in this section are minimal.

In this section we also begin to use the notation [**p**] instead of component notation $[p_1, p_2, p_3, p_4]$ (or $[p_1, p_2, p_3]$) for elements of RP^3 (or RP^2) more frequently. Both notations are useful in different ways at different times: The notation [**p**] is more compact, and component notation $[p_1, p_2, p_3, p_4]$ (or $[p_1, p_2, p_3]$) is more explicit.

If the equation

$$\pi_1 p_1 + \pi_2 p_2 + \pi_3 p_3 + \pi_4 p_4 = 0$$

is satisfied by the coordinates of one representative $\mathbf{p} = \langle p_1, p_2, p_3, p_4 \rangle$ of an element $[\mathbf{p}] = [p_1, p_2, p_3, p_4]$ in RP^3, then it is satisfied by the coordinates of all representatives of [**p**]. This equation is *satisfied by* [**p**] if and only if it is satisfied by the coordinates of any representative (hence all representatives) **p** of [**p**].

Recall that there are two different types of planes in projective three-space (see Figure 2.18): Euclidean planes completed with their ideal lines and the ideal plane.

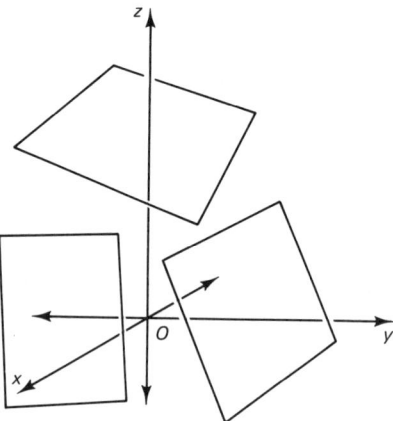

Figure 2.18 Planes in projective three-space.

3.1 Proposition. For every plane π in projective three-space, there are real constants π_1, π_2, π_3, and π_4, not all of which are zero, such that $P[\mathbf{p}]$ lies on π if and only if the equation

$$\pi_1 p_1 + \pi_2 p_2 + \pi_3 p_3 + \pi_4 p_4 = 0$$

is satisfied by the standard homogeneous coordinates $[\mathbf{p}] = [p_1, p_2, p_3, p_4]$ of P. Conversely, if π_1, π_2, π_3, and π_4 are real constants, not all of which are zero, then the set of points $P[p_1, p_2, p_3, p_4]$, whose standard homogeneous coordinates satisfy this equation, is a plane in projective three-space.

3.2 Definition. The equation associated to the plane π in Proposition 3.1 is the *equation of* π.

As an immediate consequence of Proposition 3.1, it follows that if π_1, π_2, or π_3 is nonzero, then π is a Euclidean plane completed with its ideal line; if π_1, π_2, and π_3 are all zero (and π_4 is nonzero), then π is the ideal plane. Furthermore, a point $P[\mathbf{p}] = P[p_1, p_2, p_3, p_4]$ lies on a plane π if and only if the standard homogeneous coordinates of P satisfy the equation of π (π_1, π_2, π_3, and π_4 being treated as known, and p_1, p_2, p_3, and p_4 being treated as unknown).

3.3 Example. The set of all points $P[\mathbf{p}]$ for which $3p_1 - p_2 + 2p_3 + p_4 = 0$ is a plane. The point $A[2, -1, 0, -7]$ lies on this plane since $3(2) - 1(-1) + 2(0) + 1(-7) = 0$. However, the point $B[2, 3, 4, 1]$ does not lie on this plane, since $3(2) - 1(3) + 2(4) + 1(1) = 12 \neq 0$.

The next result is an analytic formulation of two dual axioms of spatial projective geometry.

3.4 Proposition

(a) If the (distinct) points $Q[\mathbf{q}]$, $R[\mathbf{r}]$, and $S[\mathbf{s}]$ are noncollinear, then the equation of the plane they determine (see Figure 2.19) is $\pi_1 p_1 + \pi_2 p_2 + \pi_3 p_3 + \pi_4 p_4 = 0$ where

$$\pi_1 = \begin{vmatrix} q_2 & q_3 & q_4 \\ r_2 & r_3 & r_4 \\ s_2 & s_3 & s_4 \end{vmatrix} \qquad \pi_2 = -\begin{vmatrix} q_1 & q_3 & q_4 \\ r_1 & r_3 & r_4 \\ s_1 & s_3 & s_4 \end{vmatrix}$$

$$\pi_3 = \begin{vmatrix} q_1 & q_2 & q_4 \\ r_1 & r_2 & r_4 \\ s_1 & s_2 & s_4 \end{vmatrix} \qquad \pi_4 = -\begin{vmatrix} q_1 & q_2 & q_3 \\ r_1 & r_2 & r_3 \\ s_1 & s_2 & s_3 \end{vmatrix}$$

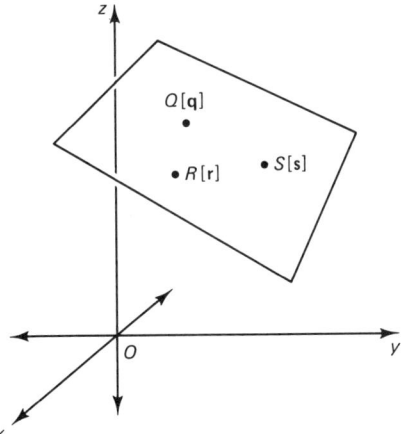

Figure 2.19 The plane determined by three distinct points.

Sec. 3   Projective Planes (Nonparametric Description)

(b) If the planes whose equations are
$$\pi_1 p_1 + \pi_2 p_2 + \pi_3 p_3 + \pi_4 p_4 = 0$$
$$\mu_1 p_1 + \mu_2 p_2 + \mu_3 p_3 + \mu_4 p_4 = 0$$
$$\nu_1 p_1 + \nu_2 p_2 + \nu_3 p_3 + \nu_4 p_4 = 0$$
do not meet in a common line, they meet in the point $P[p_1, p_2, p_3, p_4]$ (see Figure 2.20), where

$$p_1 = \begin{vmatrix} \pi_2 & \pi_3 & \pi_4 \\ \mu_2 & \mu_3 & \mu_4 \\ \nu_2 & \nu_3 & \nu_4 \end{vmatrix} \quad p_2 = -\begin{vmatrix} \pi_1 & \pi_3 & \pi_4 \\ \mu_1 & \mu_3 & \mu_4 \\ \nu_1 & \nu_3 & \nu_4 \end{vmatrix}$$

$$p_3 = \begin{vmatrix} \pi_1 & \pi_2 & \pi_4 \\ \mu_1 & \mu_2 & \mu_4 \\ \nu_1 & \nu_2 & \nu_4 \end{vmatrix} \quad p_4 = -\begin{vmatrix} \pi_1 & \pi_2 & \pi_3 \\ \mu_1 & \mu_2 & \mu_3 \\ \nu_1 & \nu_2 & \nu_3 \end{vmatrix}$$

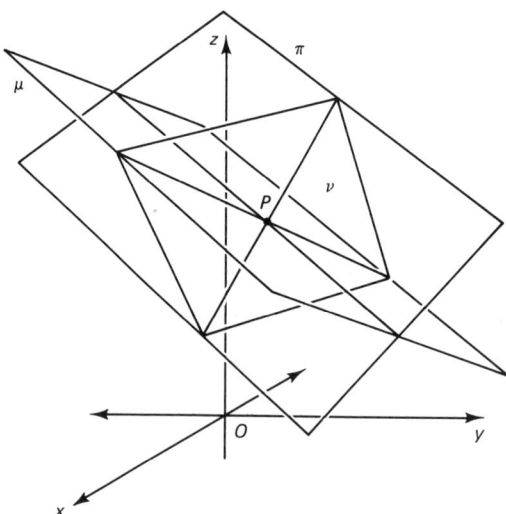

Figure 2.20 The point determined by three planes.

The statement of Proposition 3.4 reflects special conditions that are geometrically singular: three (distinct) points determine a unique plane if and only if they are noncollinear, and three (distinct) planes determine a unique point if and only if they do not meet in a common line. In Section 4 we see that the points $Q[\mathbf{q}]$, $R[\mathbf{r}]$, and $S[\mathbf{s}]$ are collinear if and only if the four determinants of Proposition 3.4(a) are all zero. Thus we have a test for collinearity (Are the four determinants of Proposition 3.4(a) all zero?) which, if it fails, automatically gives the equation of the plane determined by Q, R, and S. Similarly, the planes whose equations are

$$\pi_1 p_1 + \pi_2 p_2 + \pi_3 p_3 + \pi_4 p_4 = 0$$
$$\mu_1 p_1 + \mu_2 p_2 + \mu_3 p_3 + \mu_4 p_4 = 0$$
$$\nu_1 p_1 + \nu_2 p_2 + \nu_3 p_3 + \nu_4 p_4 = 0$$

meet in a common line if and only if the four determinants of Proposition 3.4(b) are all zero; so again we have a test for collinearity which, if it fails, automatically gives a set of standard homogeneous coordinates for the point of intersection of the three planes.

3.5 Example. To find the equation of the plane passing through the points $A[1, 0, 2, -1]$, $B[3, 1, 2, 0]$, and $C[-1, 2, 2, 1]$, compute

$$\pi_1 = \begin{vmatrix} 0 & 2 & -1 \\ 1 & 2 & 0 \\ 2 & 2 & 1 \end{vmatrix} = 0, \qquad \pi_2 = -\begin{vmatrix} 1 & 2 & -1 \\ 3 & 2 & 0 \\ -1 & 2 & 1 \end{vmatrix} = 12$$

$$\pi_3 = \begin{vmatrix} 1 & 0 & -1 \\ 3 & 1 & 0 \\ -1 & 2 & 1 \end{vmatrix} = -6, \qquad \pi_4 = -\begin{vmatrix} 1 & 0 & 2 \\ 3 & 1 & 2 \\ -1 & 2 & 2 \end{vmatrix} = -12$$

Thus A, B, and C determine a plane π, and the equation of π is $12p_2 - 6p_3 - 12p_4 = 0$.

3.6 Example. Given the plane π whose equation is $2p_1 - p_2 + 3p_3 + p_4 = 0$, μ whose equation is $p_1 + 3p_2 + p_3 = 0$, and ν whose equation is $3p_1 - p_2 + 2p_4 = 0$,

$$p_1 = \begin{vmatrix} -1 & 3 & 1 \\ 3 & 1 & 0 \\ -1 & 0 & 2 \end{vmatrix} = -19, \qquad p_2 = -\begin{vmatrix} 2 & 3 & 1 \\ 1 & 1 & 0 \\ 3 & 0 & 2 \end{vmatrix} = 5$$

$$p_3 = \begin{vmatrix} 2 & -1 & 1 \\ 1 & 3 & 0 \\ 3 & -1 & 2 \end{vmatrix} = 4, \qquad p_4 = -\begin{vmatrix} 2 & -1 & 3 \\ 1 & 3 & 1 \\ 3 & -1 & 0 \end{vmatrix} = 31$$

Thus π, μ, and ν meet in a single point P, and $P = P[-19, 5, 4, 31]$.

The constants π_1, π_2, π_3, and π_4 that determine the equation of a plane are unique only up to nonzero scalar multiples: If k is any nonzero real constant, the equation determined by $\pi_1, \pi_2, \pi_3, \pi_4$,

$$\pi_1 p_1 + \pi_2 p_2 + \pi_3 p_3 + \pi_4 p_4 = 0$$

is the same as the equation determined by $k\pi_1, k\pi_2, k\pi_3, k\pi_4$,

$$(k\pi_1)p_1 + (k\pi_2)p_2 + (k\pi_3)p_3 + (k\pi_4)p_4 = k(\pi_1 p_1 + \pi_2 p_2 + \pi_3 p_3 + \pi_4 p_4) = 0$$

Thus associated to each plane π, there is a vector $\langle \pi_1, \pi_2, \pi_3, \pi_4 \rangle$ uniquely defined up to scalar multiples—something that looks very much like an element of RP^3. This leads to a concrete analytic description of spatial duality (see Section 1.6).

There is a one-to-one correspondence H between the set of all planes in projective three-space and RP^3 that associates to each plane π whose equation is $\pi_1 p_1 + \pi_2 p_2 + \pi_3 p_3 + \pi_4 p_4 = 0$ the element $H(\pi) = [\boldsymbol{\pi}] = [\pi_1, \pi_2, \pi_3, \pi_4]$ of RP^3.

3.7 Definition. The *standard homogeneous coordinates* of a plane π in projective three-space are given by $H(\pi) = [\boldsymbol{\pi}] = [\pi_1, \pi_2, \pi_3, \pi_4]$, where H is the one-to-one correspondence described above. The plane π whose standard homogeneous coordinates are $[\boldsymbol{\pi}]$ is denoted by $\pi[\boldsymbol{\pi}]$.

3.8 Theorem. (Analytic Duality) There is a one-to-one correspondence between the points in projective three-space and the planes in projective three-space that associates to the point P with standard homogeneous coordinates $[w, x, y, z]$ the plane π with standard homogeneous coordinates $[w, x, y, z]$. There is a one-to-one self-correspondence defined on the lines in projective three-space that associates the line of intersection of the (distinct) dual planes π and μ to the line passing through the distinct points P and Q.

For example, the point $P[4, -2, 3, 1]$ and the plane $\pi[4, -2, 3, 1]$ are dual to each other. And the dual of the line passing through the points $P[1, 4, -3, 2]$ and $Q[2, 3, 1, 0]$ is the line of intersection of the planes $\pi[1, 4, -3, 2]$ and $\mu[2, 3, 1, 0]$.

Spatial duality means that points and planes in projective three-space can be treated symmetrically. Analytically, for example, this means that points have equations: The equation of a point $P[\mathbf{p}] = P[p_1, p_2, p_3, p_4]$ is $\pi_1 p_1 + \pi_2 p_2 + \pi_3 p_3 + \pi_4 p_4 = 0$ (p_1, p_2, p_3, and p_4 being treated as known and π_1, π_2, π_3, and π_4 being treated as unknown); that is, the point P is the (intersection of the) set of planes $\pi[\pi_1, \pi_2, \pi_3, \pi_4]$ that pass through P. Again, our notation has been chosen so that this last equation can be treated symmetrically in $[\pi_1, \pi_2, \pi_3, \pi_4]$ and $[p_1, p_2, p_3, p_4]$.

As with planar duality, there are two different ways to think about spatial duality—formally and concretely. Concrete duality is well defined subject to certain choices being made. Formal duality is independent of such choices. The geometry of spatial duality is pursued in Exercises 8, 9, and 10.

We now consider conditions for coplanarity.

3.9 Proposition. The following are equivalent.

(a) The points $P[\mathbf{p}], Q[\mathbf{q}], R[\mathbf{r}]$, and $S[\mathbf{s}]$ are coplanar.

(b) The determinant

$$\begin{vmatrix} p_1 & p_2 & p_3 & p_4 \\ q_1 & q_2 & q_3 & q_4 \\ r_1 & r_2 & r_3 & r_4 \\ s_1 & s_2 & s_3 & s_4 \end{vmatrix}$$

is zero.

(c) Given representatives **p** of [p], **q** of [q], **r** of [r], and **s** of [s], there are constants k_1, k_2, k_3, and k_4 for which
$$k_1\mathbf{p} + k_2\mathbf{q} + k_3\mathbf{r} + k_4\mathbf{s} = \langle 0, 0, 0, 0\rangle$$

3.10 Example. The points $A[2, 1, 3, 0]$, $B[1, 2, -1, 2]$, $C[1, -4, 9, 3]$, and $D[-5, -7, 0, 3]$ are coplanar, since

$$\begin{vmatrix} 2 & 1 & 3 & 0 \\ 1 & 2 & -1 & 2 \\ 1 & -4 & 9 & 3 \\ -5 & -7 & 0 & 3 \end{vmatrix} = 0$$

However, the points $A[1, 2, 1, 0]$, $B[2, 0, 1, 1]$, $C[1, 1, 3, 0]$, and $D[0, 1, 3, 0]$ are not coplanar, since

$$\begin{vmatrix} 1 & 2 & 1 & 0 \\ 2 & 0 & 1 & 1 \\ 1 & 1 & 3 & 0 \\ 0 & 1 & 3 & 0 \end{vmatrix} = -5 \neq 0$$

3.11 Example. Since $A[2, 1, 3, 0]$, $B[1, 2, -1, 2]$, $C[1, -4, 9, 3]$, and $D[-5, -7, 0, 3]$ are coplanar, given representatives $\langle 2, 1, 3, 0\rangle$ of $[2, 1, 3, 0]$, $\langle 1, 2, -1, 2\rangle$ of $[1, 2, -1, 2]$, $\langle 1, -4, 9, 3\rangle$ of $[1, -4, 9, 3]$, and $\langle -5, -7, 0, 3\rangle$ of $[-5, -7, 0, 3]$, there must be constants k_1, k_2, k_3, and k_4 for which
$$k_1\langle 2, 1, 3, 0\rangle + k_2\langle 1, 2, -1, 2\rangle + k_3\langle 1, -4, 9, 3\rangle + k_4\langle -5, -7, 0, 3\rangle = \langle 0, 0, 0, 0\rangle$$
We find (see Example 2.14) that $k_1 = -3$, $k_2 = 0$, $k_3 = 1$, and $k_4 = -1$:
$$(-3)\langle 2,1,3,0\rangle + (0)\langle 1,2,-1,2\rangle + (1)\langle 1,-4,9,3\rangle + (-1)\langle -5,-7,0,3\rangle = \langle 0,0,0,0\rangle$$

This last example illustrates that distinct points P, Q, R, and S are coplanar if there are representatives **p** of [p], **q** of [q], **r** of [r], and **s** of [s] for which
$$\mathbf{p} + \mathbf{q} + \mathbf{r} + \mathbf{s} = \langle 0, 0, 0, 0\rangle$$
but even though the distinct points P, Q, R, and S are coplanar, there need not be representatives **p** of [p], **q** of [q], **r** of [r], and **s** of [s] for which
$$\mathbf{p} + \mathbf{q} + \mathbf{r} + \mathbf{s} = \langle 0, 0, 0, 0\rangle$$
Thus even though there are potentially four ways to apply right inverses to find k_1, k_2, k_3, and k_4, not all of them always work. In this example, if we rewrite the original system as a system in k_1/k_2, k_3/k_2, and k_4/k_2, the matrix M has rank 2 (not rank 3) so M^+ does not exist.

SECTION 3 EXERCISES

1. Which of the following points lie on the plane whose equation is $p_1 - 3p_2 + 2p_3 - p_4 = 0$?
   (a) $A[1, 1, 1, 0]$ (b) $B[1, 0, 1, 1]$
   (c) $C[-2, 1, 1, 3]$ (d) $D[2, 1, 1, 1]$

2. Find the equation and standard homogeneous coordinates for the planes determined by the extension to projective three-space of the following Euclidean planes.
   (a) $3x - 2y + z = 4$ (b) $5x + 3y - 2z = 0$
   (c) $x = 1$ (d) $y = z$
   (e) $x + 2z = 3$ (f) $z = 1$

3. In each case, state whether the set of three points determines a plane, and find the equation of the plane if it exists.
   (a) $A[1, 2, 1, 1]$, $B[0, 3, -2, 1]$, $C[1, 2, 0, 3]$
   (b) $A[-2, 1, 3, 1]$, $B[3, 1, -1, 0]$, $C[-1, 1, 0, 1]$
   (c) $A[0, 1, 1, 0]$, $B[2, 1, 3, 1]$, $C[-4, 0, 1, 6]$
   (d) $A[0, 0, 0, 1]$, $B[-1, 2, 3, 1]$, $C[3, 2, 1, 1]$

4. In each of the following cases, state whether the set of three planes intersect in a single point, and find the standard homogeneous coordinates of that point if it exists.
   (a) $\pi[1, 2, 2, 1]$, $\mu[2, -1, 1, 3]$, $\nu[4, 1, 2, 0]$
   (b) $\pi[2, 3, 1, 4]$, $\mu[3, 0, -1, -3]$, $\nu[1, -2, 2, -5]$
   (c) $\pi[1, 1, 1, 2]$, $\mu[1, 0, -1, 1]$, $\nu[1, 1, 0, 1]$
   (d) $\pi[2, 3, 1, 2]$, $\mu[-1, 2, 3, -1]$, $\nu[-3, -3, 1, 0]$

5. In each of the following cases, describe the plane with the given standard homogeneous coordinates analytically (give its equation) and geometrically.
   (a) $a[1, 0, 0, 1]$ (b) $b[2, 3, 0, 1]$
   (c) $c[0, 0, 1, -1]$ (d) $d[1, 1, 1, -1]$
   (e) $e[2, -3, -4, -1]$ (f) $f[1, 2, 3, 0]$

6. Determine which of the following sets of four points is coplanar. For those that are, obtain an equation of the form described in Proposition 3.9(c) using Gauss–Jordan Elimination and the right inverse technique.
   (a) $A[-2, 1, -7, 0]$, $B[1, 3, 0, 0]$, $C[0, 2, -3, -1]$, $D[1, -1, -1, -5]$
   (b) $A[3, 1, 2, -1]$, $B[2, -1, 3, -1]$, $C[2, 1, 1, -1]$, $D[1, -3, 4, 0]$
   (c) $A[1, 2, 3, 4]$, $B[2, 3, 4, 5]$, $C[3, 4, 5, 6]$, $D[4, 5, 6, 7]$
   (d) $A[0, 2, 1, 1]$, $B[1, 2, 1, 0]$, $C[1, -1, 0, 1]$, $D[1, 4, 1, -4]$

†7. Verify that the equation of a plane can be written in the matrix form

$$(p_1 \ p_2 \ p_3 \ p_4) \begin{pmatrix} \pi_1 \\ \pi_2 \\ \pi_3 \\ \pi_4 \end{pmatrix} = 0$$

Write the equation of the plane $2p_1 - 3p_2 + 4p_3 + p_4 = 0$ in this form.

8. Give an analytic verification of the following intrinsic description of spatial duality (see Exercises 2.22 and 2.23):
   (a) The plane dual to the origin O of Euclidean three-space is the ideal plane.
   (b) The plane π dual to a Euclidean point P other than the origin is determined as follows (see Figure 2.21): Draw a Euclidean line m through P and O and determine the point Q on m on the opposite side of O from P for which $\text{dist}(O, Q) = 1/\text{dist}(O, P)$; π is the plane determined by the Euclidean plane that passes through Q and is perpendicualr to m.
   (c) The dual plane π of an ideal point P (thought of as a direction) is the plane determined by the Euclidean plane passing through the origin that is perpendicular to the direction of P (see Figure 2.22).

9. (Continuation of Exercise 8) Give an intrinsic description of the dual points of the following planes.

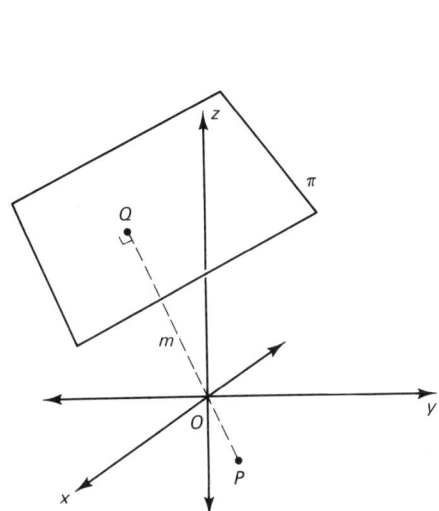

Figure 2.21 Intrinsic spatial duality for Euclidean points.

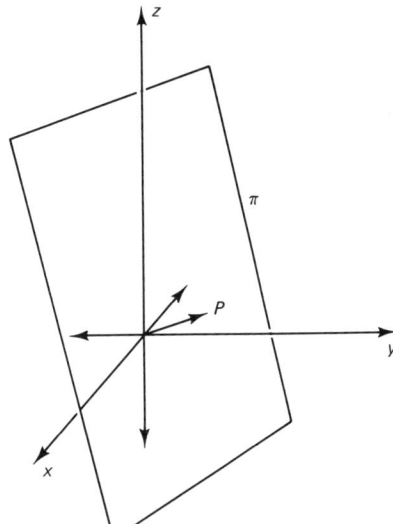

Figure 2.22 Intrinsic spatial duality for ideal points.

   (a) A plane determined by a Euclidean plane that passes through the origin
   (b) A plane determined by a Euclidean plane that does not pass through the origin
   (c) The ideal plane
10. (Continuation of Exercise 9) Lines in projective three-space are dual to lines in projective three-space. Describe the duals of the following lines in projective three-space.
   (a) A line determined by a Euclidean line that passes through the origin
   (b) A line determined by a Euclidean line that does not pass through the origin
   (c) An ideal line
   (*Hint:* Duality of lines is consistent with the duality of points and planes in projective three-space in the sense that if P and Q are points that lie on the line l, then the dual of l is the line of intersection of the planes dual to P and Q.)
†11. Show that the following conditions are equivalent.
   (a) The points $P[p_1, p_2, p_3, p_4]$, $Q[q_1, q_2, q_3, q_4]$, and $R[r_1, r_2, r_3, r_4]$ in projective three-space are collinear.
   (b) The matrix
   $$\begin{pmatrix} p_1 & p_2 & p_3 & p_4 \\ q_1 & q_2 & q_3 & q_4 \\ r_1 & r_2 & r_3 & r_4 \end{pmatrix}$$
   has rank 2.
   (c) Given representatives $\langle p_1, p_2, p_3, p_4 \rangle$ of $[p_1, p_2, p_3, p_4]$, $\langle q_1, q_2, q_3, q_4 \rangle$ of $[q_1, q_2, q_3, q_4]$, and $\langle r_1, r_2, r_3, r_4 \rangle$ of $[r_1, r_2, r_3, r_4]$, there are constants k_1, k_2, and k_3, not all of which are zero, for which
   $$k_1 \langle p_1, p_2, p_3, p_4 \rangle + k_2 \langle q_1, q_2, q_3, q_4 \rangle + k_3 \langle r_1, r_2, r_3, r_4 \rangle = \langle 0, 0, 0, 0 \rangle$$
12. Use Exercise 11 to verify that each of the following sets of three points in projective

three-space is a collinear set. In each case, obtain an equation of the form described in part (c) using both Gauss–Jordan Elimination and the right inverse technique.
(a) $P[1, 0, 0, 0]$, $Q[1, 1, 0, 0]$, $R[1, 2, 0, 0]$
(b) $P[1, 2, 1, 1]$, $Q[1, 2, -1, 1]$, $R[1, 2, 4, 1]$
(c) $P[1, 2, -1, 4]$, $Q[1, 2, -1, 0]$, $R[1, 2, -1, -1]$
(d) $P[0, 1, 2, -1]$, $Q[1, 2, 3, -2]$, $R[2, 1, 0, -1]$

SECTION 4: PROJECTIVE LINES (PARAMETRIC DESCRIPTION)

In Exercises 2.20 and 3.7 we observed that the equation of a line in the projective plane can be written in matrix form

$$(p_1 \;\; p_2 \;\; p_3) \begin{pmatrix} l_1 \\ l_2 \\ l_3 \end{pmatrix} = 0$$

or $\mathbf{pl} = 0$ for short, and that the equation of a plane in projective three-space can be written in matrix form

$$(p_1 \;\; p_2 \;\; p_3 \;\; p_4) \begin{pmatrix} \pi_1 \\ \pi_2 \\ \pi_3 \\ \pi_4 \end{pmatrix} = 0$$

or $\mathbf{p}\pi = 0$ for short. In this section we begin taking advantage of these matrix forms.

We begin by parametrizing lines in the projective plane.

4.1 Proposition. The line l in the projective plane that passes through the (distinct) points $Q[\mathbf{q}]$ and $R[\mathbf{r}]$ consists of all points of the form $P[t\mathbf{q} + (1 - t)\mathbf{r}]$, for $t \in R$, together with the point $P[\mathbf{q} - \mathbf{r}]$:

$$\{P[t\mathbf{q} + (1 - t)\mathbf{r}] \mid t \in R\} \cup \{P[\mathbf{q} - \mathbf{r}]\}$$

Proof. First, each point in this set lies on l: If $\mathbf{pl} = 0$ is the equation of l and $Q[\mathbf{q}]$ and $R[\mathbf{r}]$ lie on l, then $\mathbf{ql} = 0$ and $\mathbf{rl} = 0$. If $P = P[t\mathbf{q} + (1 - t)\mathbf{r}]$, then since

$$(t\mathbf{q} + (1 - t)\mathbf{r})\mathbf{l} = t\mathbf{ql} + (1 - t)\mathbf{rl} = 0$$

P lies on l. And if $P = P[\mathbf{q} - \mathbf{r}]$, then since

$$(\mathbf{q} - \mathbf{r})\mathbf{l} = \mathbf{ql} - \mathbf{rl} = 0$$

P lies on l.

Second, every point that lies on l is in this set: Since $Q[\mathbf{q}]$ and $R[\mathbf{r}]$ determine l, for every point $P[\mathbf{p}]$ on l, there are constants k_1, k_2, and k_3 for which

$$k_1\mathbf{p} + k_2\mathbf{q} + k_3\mathbf{r} = \mathbf{0}$$

Since Q and R are distinct, k_1 must be nonzero and

$$-k_1\mathbf{p} = k_2\mathbf{q} + k_3\mathbf{r}$$

If $k_2 + k_3 = 0$, then $-k_1\mathbf{p} = k_2(\mathbf{q} - \mathbf{r})$, k_2 must be nonzero, and $P = P[\mathbf{p}] = P[\mathbf{q} - \mathbf{r}]$ is in this set. If $k_2 + k_3 \neq 0$, then

$$-\frac{k_1}{k_2 + k_3}\mathbf{p} = \frac{k_2}{k_2 + k_3}\mathbf{q} + \left(1 - \frac{k_2}{k_2 + k_3}\right)\mathbf{r}$$

So letting $t = k_2/(k_2 + k_3)$, $P[\mathbf{p}] = P[t\mathbf{q} + (1 - t)\mathbf{r}]$ is in this set. ∎

This result has two interesting facets.

First, Proposition 4.1 illustrates how a projective line is parametrized. Parametrizations of projective lines are generically different than parametrizations of Euclidean lines: The set R of real numbers can be used to parametrize a Euclidean line, but it cannot be used to parametrize a projective line. Topologically this is because a projective line (which is a circle—see Section 1.4) is different than a Euclidean line. Analytically, let l, for example, be the parametrized projective line given in Proposition 4.1. If $t = 0$, then

$$P[t\mathbf{q} + (1 - t)\mathbf{r}] = P[\mathbf{r}] = R[\mathbf{r}]$$

(and if $t = 1$, then $P[t\mathbf{q} + (1 - t)\mathbf{r}] = P[\mathbf{q}] = Q[\mathbf{q}]$). If $t \neq 0$, then

$$P[t\mathbf{q} + (1 - t)\mathbf{r}] = P[t(\mathbf{q} - \mathbf{r}) + \mathbf{r}] = P\left[\mathbf{q} - \mathbf{r} + \left(\frac{\mathbf{r}}{t}\right)\right]$$

so

$$\lim_{t \to \infty} P[t\mathbf{q} + (1 - t)\mathbf{r}] = P[\mathbf{q} - \mathbf{r}]$$

Thus the point $P[\mathbf{q} - \mathbf{r}]$ cannot be included in this parametrization of l by R. It is (in a sense to be clarified in Chapter 3) the ideal point of l with respect to this parametrization.

Second, the Euclidean line passing through the Euclidean points $Q(q_1, q_2, 1)$ and $R(r_1, r_2, 1)$ can be parametrized in Cartesian coordinates by

$$\{P[tq_1 + (1 - t)r_1, tq_2 + (1 - t)r_2, 1] \mid t \in R\}$$

and the corresponding set of points in the projective plane is

$$\{P[tq_1 + (1 - t)r_1, tq_2 + (1 - t)r_2, 1] \mid t \in R\}$$

Although there is nothing too surprising about this, what is surprising is that according to Proposition 4.1, part of a line (part of a projective line that topologically looks like a Euclidean line) can be parametrized in standard homogeneous coordinates by

$$\{P[tq_1 + (1 - t)r_1, tq_2 + (1 - t)r_2, tq_3 + (1 - t)r_3] \mid t \in R\}$$
$$= \left\{P\left[\frac{tq_1 + (1 - t)r_1}{tq_3 + (1 - t)r_3}, \frac{tq_2 + (1 - t)r_2}{tq_3 + (1 - t)r_3}, 1\right] \mid t \in R\right\}$$

and the first and second coordinate functions in this parametrization are not linear in t! The geometric implications of this are discussed in the next example.

4.2 Example. The line l passing through the points $A[4, 4, 1]$ and $B[-2, 0, 2]$ is

$$\{P[t\langle 4, 4, 1\rangle + (1 - t)\langle -2, 0, 2\rangle] \mid t \in R\} \cup \{P[\langle 4, 4, 1\rangle - \langle -2, 0, 2\rangle]\}$$
$$= \{P[6t - 2, 4t, -t + 2] \mid t \in R\} \cup \{P[6, 4, -1]\}$$

Several points in the projective plane determined by this parametrization are plotted in Figure 2.23. Observe that if $t = 0$, then
$$P[6t - 2, 4t, -t + 2] = P[-2, 0, 2] = R[-2, 0, 2]$$
and that if $t = 1$, then
$$P[6t - 2, 4t, -t + 2] = P[4, 4, 1] = Q[4, 4, 1]$$
Also observe that even though the spacing between parameter values is constant, the geometric spacing between points plotted is not constant; this is because of the nonlinearity of the first two component functions in the normalized form of the parametrizations. (This is the same phenomenon we observe, for example, when looking at a train moving down a set of railroad tracks into the distance: Even though the speed of the train may be constant, it appears to move slower and slower as it moves further and further away.)

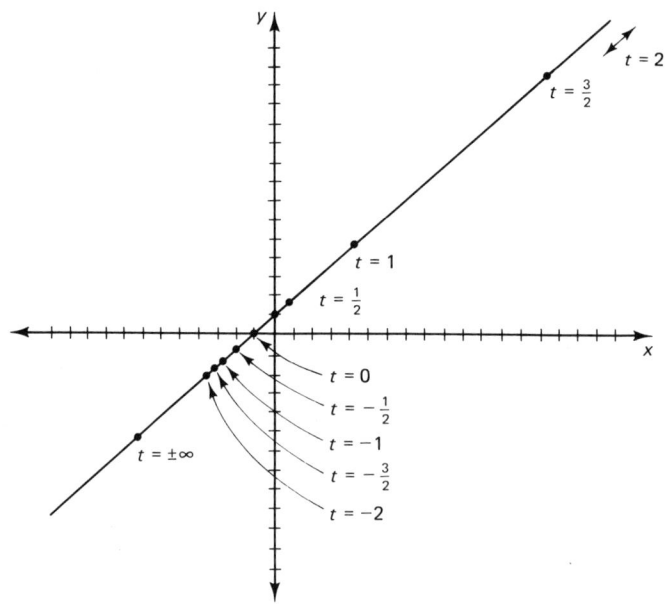

Figure 2.23 Points in the projective plane determined by the parametrization of a line.

Parametrizations of lines are not unique. If, in particular, two different sets of representatives are chosen for $[\mathbf{q}]$ and $[\mathbf{r}]$, two different parametrizations of l might well result.

4.3 Example. The line l passing through the points $A[4, 4, 1]$ and $B[-2, 0, 2] = C[-1, 0, 1]$ (see Example 4.2) may also be written
$$\{P[t\langle 4, 4, 1\rangle + (1 - t)\langle -1, 0, 1\rangle] \mid t \in R\} \cup \{P[\langle 4, 4, 1\rangle - \langle -1, 0, 1\rangle]\}$$
$$= \{P[5t - 1, 4t, 1] \mid t \in R\} \cup \{P[5, 4, 0]\}$$

Observe that if $t = 0$, then
$$P[5t - 1, 4t, 1] = P[-1, 0, 1] = R[-2, 0, 2]$$
and that if $t = 1$, then
$$P[5t - 1, 4t, 1] = P[4, 4, 1] = Q[4, 4, 1]$$
However, the ideal point of this parametrization is $P[5, 4, 0]$, and this is not the same as the ideal point $P[6, 4, -1]$ of the parametrization given in Example 4.2. (See Figure 2.24.)

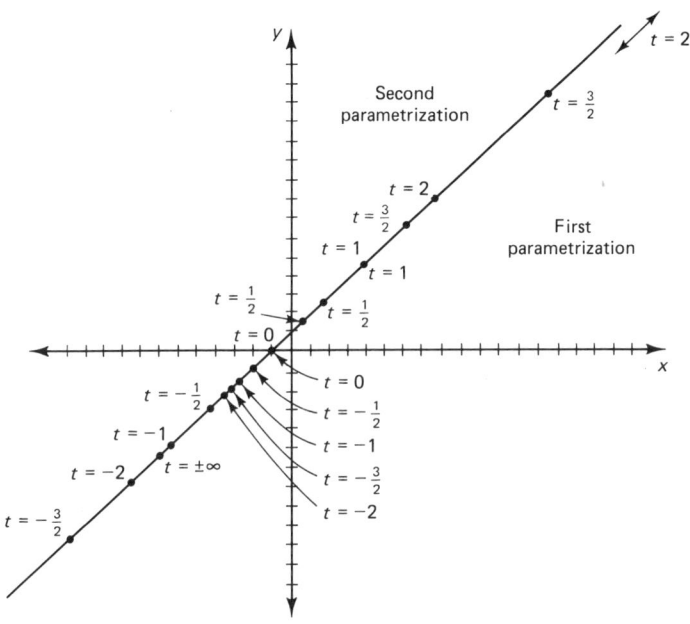

Figure 2.24 A comparison of two parametrizations of a line.

The nonuniqueness of parametrizations is pursued further in Chapter 3.

4.4 Example. To find the intersection of the lines
$$l_1 = \{P[3t - 1, -2t + 1, -4t + 1] \mid t \in R\} \cup \{P[3, -2, -4]\}$$
and
$$l_2 = \{P[-t + 2, t, 2t + 3] \mid t \in R\} \cup \{P[-1, 1, 2]\}$$
first observe that $P[3, -2, -4]$ does not lie on l_2: $[3, -2, -4] \neq [-1, 1, 2]$, and if $[3, -2, -4]$ were of the form $[-t + 2, t, 2t + 3]$ for some $t \in R$, then there would be a nonzero real constant k for which
$$3 = k(-t + 2), \quad -2 = k(t), \quad -4 = k(2t + 3)$$
The first two equations imply $t = \tfrac{5}{2}$ and $k = -\tfrac{4}{5}$, but these values do not satisfy the third equation. Similarly $P[-1, 1, 2]$ does not lie on l_1. Thus there must be parameter values t and s for which

Sec. 4   Projective Lines (Parametric Description)

$$[3t - 1, -2t + 1, -4t + 1] = [-s + 2, s, 2s + 3]$$

(The parameters of l_1 and l_2 are independent of each other.) So for some nonzero real constant k,

$$\langle 3t - 1, -2t + 1, -4t + 1 \rangle = k \langle -s + 2, s, 2s + 3 \rangle$$

or

$$3t - 1 = k(-s + 2), \quad -2t + 1 = k(s), \quad -4t + 1 = k(2s + 3)$$

Solving these equations, we find $t = -\frac{2}{3}$, $s = -7$, and $k = -\frac{1}{3}$. Thus the point of intersection is

$$P[3(-\tfrac{2}{3}) - 1, -2(-\tfrac{2}{3}) + 1, -4(-\tfrac{2}{3}) + 1] = P[-3, \tfrac{7}{3}, \tfrac{11}{3}]$$
$$= P[9, -7, -11] = P[-(-7) + 2, -7, 2(-7) + 3]$$

We now turn our attention to lines in projective three-space. In order to obtain a representation similar to that of Proposition 4.1, we need the following version of Proposition 2.12.

4.5 Proposition. The following are equivalent.

(a) The points $P[p_1, p_2, p_3, p_4]$, $Q[q_1, q_2, q_3, q_4]$, and $R[r_1, r_2, r_3, r_4]$ are collinear.
(b) The matrix

$$\begin{pmatrix} p_1 & p_2 & p_3 & p_4 \\ q_1 & q_2 & q_3 & q_4 \\ r_1 & r_2 & r_3 & r_4 \end{pmatrix}$$

has rank 2.
(c) Given representatives $\mathbf{p} = \langle p_1, p_2, p_3, p_4 \rangle$ of $[p_1, p_2, p_3, p_4]$, $\mathbf{q} = \langle q_1, q_2, q_3, q_4 \rangle$ of $[q_1, q_2, q_3, q_4]$, and $\mathbf{r} = \langle r_1, r_2, r_3, r_4 \rangle$ of $[r_1, r_2, r_3, r_4]$, there are constants k_1, k_2, and k_3 for which

$$k_1 \mathbf{p} + k_2 \mathbf{q} + k_3 \mathbf{r} = \langle 0, 0, 0, 0 \rangle$$

Proof. It is not difficult to verify this result if P, Q, and R are not distinct. So suppose P, Q, and R are distinct.

First, we show that (a) implies (b): If P, Q, and R are collinear, then they lie on a line l; that is, there are distinct planes $\pi[\boldsymbol{\pi}]$ and $\mu[\boldsymbol{\mu}]$ on which P, Q, and R all lie:

$$\mathbf{p}\boldsymbol{\pi} = 0 \qquad \mathbf{p}\boldsymbol{\mu} = 0$$
$$\mathbf{q}\boldsymbol{\pi} = 0 \qquad \mathbf{q}\boldsymbol{\mu} = 0$$
$$\mathbf{r}\boldsymbol{\pi} = 0 \qquad \mathbf{r}\boldsymbol{\mu} = 0$$

or, in matrix form,

$$\begin{pmatrix} p_1 & p_2 & p_3 & p_4 \\ q_1 & q_2 & q_3 & q_4 \\ r_1 & r_2 & r_3 & r_4 \end{pmatrix} \begin{pmatrix} \pi_1 & \mu_1 \\ \pi_2 & \mu_2 \\ \pi_3 & \mu_3 \\ \pi_4 & \mu_4 \end{pmatrix} = \begin{pmatrix} 0 & 0 \\ 0 & 0 \\ 0 & 0 \end{pmatrix}$$

Since P, Q, and R are distinct, the coefficient matrix

$$\begin{pmatrix} p_1 & p_2 & p_3 & p_4 \\ q_1 & q_2 & q_3 & q_4 \\ r_1 & r_2 & r_3 & r_4 \end{pmatrix}$$

must have rank at least 2. (If it were to have rank 1 then $P[\mathbf{p}]$, $Q[\mathbf{q}]$, and $R[\mathbf{r}]$ would all be the same point.) If the rank of the coefficient matrix were 3, the coefficient matrix would contain an invertible 3×3 submatrix. If, for example, this submatrix were

$$\begin{pmatrix} p_1 & p_2 & p_3 \\ q_1 & q_2 & q_3 \\ r_1 & r_2 & r_3 \end{pmatrix}$$

then

$$\begin{pmatrix} p_1 & p_2 & p_3 \\ q_1 & q_2 & q_3 \\ r_1 & r_2 & r_3 \end{pmatrix} \begin{pmatrix} \pi_1 & \mu_1 \\ \pi_2 & \mu_2 \\ \pi_3 & \mu_3 \end{pmatrix} = \begin{pmatrix} 0 & 0 \\ 0 & 0 \\ 0 & 0 \end{pmatrix}$$

But since the 3×3 coefficient matrix in this system is invertible, it must follow that

$$\begin{pmatrix} \pi_1 & \mu_1 \\ \pi_2 & \mu_2 \\ \pi_3 & \mu_3 \end{pmatrix} = \begin{pmatrix} 0 & 0 \\ 0 & 0 \\ 0 & 0 \end{pmatrix}$$

If this were true, then $\pi[0, 0, 0, \pi_4] = \pi[0, 0, 0, 1]$ and $\mu[0, 0, 0, \mu_4] = \mu[0, 0, 0, 1]$ would be the same. This contradicts our choice of π and μ (they were to be distinct).

The fact that (b) implies (c) is a standard result in linear algebra.

To see that (c) implies (a), observe that (just as in Proposition 2.12), none of the constants k_1, k_2, or k_3 can be zero. In particular, k_1 is not zero and

$$\mathbf{p} = -\frac{k_2}{k_1}\mathbf{q} - \frac{k_3}{k_1}\mathbf{r}$$

So if π and μ are distinct planes on which both Q and R lie,

$$\mathbf{q}\pi = 0 \quad \mathbf{q}\mu = 0$$
$$\mathbf{r}\pi = 0 \quad \mathbf{r}\mu = 0$$

then

$$\mathbf{p}\pi = \left(-\frac{k_2}{k_1}\mathbf{q} - \frac{k_3}{k_1}\mathbf{r}\right)\pi = 0$$

and

$$\mathbf{p}\mu = \left(-\frac{k_2}{k_1}\mathbf{q} - \frac{k_3}{k_1}\mathbf{r}\right)\mu = 0$$

so that P lies on π and μ. Since P lies on the line of intersection of π and μ, and since this line is the line determined by Q and R, P is collinear with Q and R. ■

4.6 Example. The points $A[2, -1, 3, 1]$, $B[2, 3, -5, -7]$, and $C[4, 0, 2, -2]$ are collinear, since the matrix

$$\begin{pmatrix} 2 & -1 & 3 & 1 \\ 2 & 3 & -5 & -7 \\ 4 & 0 & 2 & -2 \end{pmatrix}$$

has rank 2. However the points $A[1, 2, 0, -1]$, $B[3, 2, 1, 6]$, and $C[0, 1, 1, 1]$ are not collinear, since the matrix

$$\begin{pmatrix} 1 & 2 & 0 & -1 \\ 3 & 2 & 1 & 6 \\ 0 & 1 & 1 & 1 \end{pmatrix}$$

has rank 3.

4.7 Example. Since the points $A[2, -1, 3, 1]$, $B[2, 3, -5, -7]$, and $C[4, 0, 2, -2]$ are collinear, given representatives $\langle 2, -1, 3, 1 \rangle$ of $[2, -1, 3, 1]$, $\langle 2, 3, -5, -7 \rangle$ of $[2, 3, -5, -7]$, and $\langle 4, 0, 2, -2 \rangle$ of $[4, 0, 2, -2]$, there must be constants k_1, k_2, and k_3 for which

$$k_1 \langle 2, -1, 3, 1 \rangle + k_2 \langle 2, 3, -5, -7 \rangle + k_3 \langle 4, 0, 2, -2 \rangle = \langle 0, 0, 0, 0 \rangle$$

To find these constants we can use Gauss–Jordan Elimination or the right inverse technique to find that $k_1 = -3$, $k_2 = -1$, and $k_3 = 2$:

$$(-3)\langle 2, -1, 3, 1 \rangle + (-1)\langle 2, 3, -5, -7 \rangle + (2)\langle 4, 0, 2, -2 \rangle = \langle 0, 0, 0, 0 \rangle$$

There is an analog to Corollary 2.15.

4.8 Corollary. The distinct points P, Q, and R are collinear if and only if there are representatives $\mathbf{p}$ of $[\mathbf{p}]$, $\mathbf{q}$ of $[\mathbf{q}]$, and $\mathbf{r}$ of $[\mathbf{r}]$ for which

$$\mathbf{p} + \mathbf{q} + \mathbf{r} = \langle 0, 0, 0, 0 \rangle$$

4.9 Example. The points $A[2, -1, 3, 1]$, $B[2, 3, -5, -7]$, and $C[4, 0, 2, -2]$ are collinear (see Example 4.6), and

$$\langle -6, 3, -9, -3 \rangle + \langle -2, -3, 5, 7 \rangle + \langle 8, 0, 4, -4 \rangle = \langle 0, 0, 0, 0 \rangle$$

As a consequence of Proposition 4.5, we have the following parametric representation of lines in projective three-space.

4.10 Proposition. The line l in projective three-space that passes through the (distinct) points $Q[\mathbf{q}]$ and $R[\mathbf{r}]$ is

$$\{P[t\mathbf{q} + (1 - t)\mathbf{r}] \mid t \in R\} \cup \{P[\mathbf{q} - \mathbf{r}]\}$$

The proof of this result is completely analogous to that of Proposition 4.1.

4.11 Example. The line passing through the point $A[3, -1, 2, 1]$ and $B[2, 1, 3, 1]$ is

$$\{P[t\langle 3, -1, 2, 1 \rangle + (1 - t)\langle 2, 1, 3, 1 \rangle] \mid t \in R\} \cup \{P[\langle 3, -1, 2, 1 \rangle - \langle 2, 1, 3, 1 \rangle]\}$$
$$= \{P[t + 2, -2t + 1, -t + 3, 1] \mid t \in R\} \cup \{P[1, -2, -1, 0]\}$$

Various synthetic results concerning lines in projective three-space can now be treated analytically. Here we mention just one result that will be important to us in the rest of the book. To illustrate the theory we both verify and derive this result.

4.12 Proposition. If the (distinct) points $Q[\mathbf{q}]$ and $R[\mathbf{r}]$ do not both lie on the plane $\pi[\boldsymbol{\pi}]$, then the point P of intersection of the line determined by Q and R, and π is $P[(\mathbf{q}\boldsymbol{\pi})\mathbf{r} - (\mathbf{r}\boldsymbol{\pi})\mathbf{q}]$.

Proof. To verify this result simply observe that the line determined by Q and R must intersect π in some point, and that since

$$((\mathbf{q}\boldsymbol{\pi})\mathbf{r} - (\mathbf{r}\boldsymbol{\pi})\mathbf{q})\boldsymbol{\pi} = (\mathbf{q}\boldsymbol{\pi})(\mathbf{r}\boldsymbol{\pi}) - (\mathbf{r}\boldsymbol{\pi})(\mathbf{q}\boldsymbol{\pi}) = 0$$

$P[(\mathbf{q}\boldsymbol{\pi})\mathbf{r} - (\mathbf{r}\boldsymbol{\pi})\mathbf{q}]$ must be that point.

To derive this result, observe that the line determined by Q and R can be written parametrically

$$\{P[t\mathbf{q} + (1 - t)\mathbf{r}] \mid t \in R\} \cup \{P[\mathbf{q} - \mathbf{r}]\}$$

If $(\mathbf{q} - \mathbf{r})\boldsymbol{\pi} = 0$, then $P = P[\mathbf{q} - \mathbf{r}]$; if $(\mathbf{q} - \mathbf{r})\boldsymbol{\pi} \neq 0$, in which case the condition that $P[t\mathbf{q} + (1 - t)\mathbf{r}]$ lies on π,

$$(t\mathbf{q} + (1 - t)\mathbf{r})\boldsymbol{\pi} = 0$$

implies that $t = \mathbf{r}\boldsymbol{\pi} / (\mathbf{r}\boldsymbol{\pi} - \mathbf{q}\boldsymbol{\pi})$, then $P = P[t\mathbf{q} + (1 - t)\mathbf{r}] = P[(\mathbf{q}\boldsymbol{\pi})\mathbf{r} - (\mathbf{r}\boldsymbol{\pi})\mathbf{q}]$. ■

4.13 Example. Neither of the points $A[2, 0, -1, 3]$ nor $B[1, 2, 1, 1]$ lies on the plane $\pi[3, -2, 0, -1]$ since

$$\mathbf{q}\boldsymbol{\pi} = (2 \quad 0 \quad -1 \quad 3) \begin{pmatrix} 3 \\ -2 \\ 0 \\ -1 \end{pmatrix} = 3 \neq 0$$

and

$$\mathbf{r}\boldsymbol{\pi} = (1 \quad 2 \quad 1 \quad 1) \begin{pmatrix} 3 \\ -2 \\ 0 \\ -1 \end{pmatrix} = -2 \neq 0$$

The point of intersection of the line determined by A and B and the plane π is

$$P[(3)\langle 1, 2, 1, 1 \rangle - (-2)\langle 2, 0, -1, 3 \rangle] = P[7, 6, 1, 9]$$

The following result is the dual of Proposition 4.12.

4.14 Proposition. If the point $P[\mathbf{p}]$ does not lie on either of the distinct planes $\mu[\boldsymbol{\mu}]$ or $\nu[\boldsymbol{\nu}]$, then the plane π determined by P and the line of intersection of μ and ν is $\pi[(\mathbf{p}\boldsymbol{\mu})\boldsymbol{\nu} - (\mathbf{p}\boldsymbol{\nu})\boldsymbol{\mu}]$.

4.15 Example. The point $P[0, 1, 2, 1]$ lies neither on the plane $\mu[3, 2, 1, -1]$

nor on the plane $\nu[1, 2, 0, 0]$, since $\mathbf{p}\boldsymbol{\mu} = 3$ and $\mathbf{p}\boldsymbol{\nu} = 2$. The plane determined by P and the line of intersection of μ and ν is

$$\pi[(3)\langle 1, 2, 0, 0\rangle - (2)\langle 3, 2, 1, -1\rangle] = \pi[-3, 2, -2, 2]$$

SECTION 4 EXERCISES

1. Find parametric representations of the lines in the projective plane determined by the following points. Plot the points in the projective plane given by each parametrization for $t = 0, 1, 2, 3, \ldots$.
   (a) $A[2, -1, 3]$, $B[1, 2, 1]$ (b) $A[1, 0, 2]$, $B[-2, 3, 1]$
   (c) $A[2, 3, 1]$, $B[0, 0, 1]$ (d) $A[2, 3, 1]$, $B[4, 5, 0]$

2. The parametrization $\{P[2t + 1, 3t - 2, t + 3] \mid t \in R\}$ describes all the points on a line in the projective plane except for one. What is that point?

3. In Proposition 4.1 (and 4.10) it was tacitly assumed that if $Q[\mathbf{q}]$ and $R[\mathbf{r}]$ are distinct points in the projective plane (or projective three-space), then $[t\mathbf{q} + (1 - t)\mathbf{r}]$, for each t, and $[\mathbf{q} - \mathbf{r}]$ are standard homogeneous coordinates of points (that is, $t\mathbf{q} + (1 - t)\mathbf{r}$, for each t, and $\mathbf{q} - \mathbf{r}$ are nonzero). Explain why this is in fact true.

4. Find the point of intersection of the following lines in the projective plane.
   (a) $\{P[3t + 1, 2t - 2, t + 3] \mid t \in R\} \cup \{P[3, 2, 1]\}$,
      $\{P[2t, t, 2t - 3] \mid t \in R\} \cup \{P[2, 1, 2]\}$
   (b) $\{P[t + 2, 3t, 2t + 1] \mid t \in R\} \cup \{P[1, 3, 2]\}$,
      $\{P[3t + 2, 5t + 6, 5t + 2] \mid t \in R\} \cup \{P[3, 5, 5]\}$
   (c) $\{P[2t, t + 1, 3t - 2] \mid t \in R\} \cup \{P[2, 1, 3]\}$,
      $\{P[3t + 2, 2t, 4t + 4] \mid t \in R\} \cup \{P[3, 2, 4]\}$
   (d) $\{P[2t + 1, 3t, -t - 4] \mid t \in R\} \cup \{P[2, 3, -1]\}$,
      $\{P[t - 5, 4t - 4, -4t + 2] \mid t \in R\} \cup \{P[1, 4, -4]\}$

5. If $m[\mathbf{m}]$ and $n[\mathbf{n}]$ are distinct lines in the projective plane, verify that

$$\{l[t\mathbf{m} + (1 - t)\mathbf{n}] \mid t \in R\} \cup \{l[\mathbf{m} - \mathbf{n}]\}$$

is the set of lines in the projective plane that pass through the point of intersection of m and n.

6. Determine which of the following sets of three points are collinear. For those that are, obtain equations of the form described in Proposition 4.5(c) and Corollary 4.8.
   (a) $A[1, 3, -2, 1]$, $B[2, 1, -3, 1]$, $C[0, 5, -1, 1]$
   (b) $A[4, -1, -2, 3]$, $B[1, 1, -3, -3]$, $C[1, -1, 1, 3]$
   (c) $A[1, 2, 1, -1]$, $B[3, 2, 1, -1]$, $C[4, 6, 3, -3]$
   (d) $A[2, 1, 0, 3]$, $B[1, 2, 0, 1]$, $C[1, -4, 0, 3]$

7. Find parametric representations of the lines in projective three-space determined by the following points.
   (a) $A[2, 1, -2, -1]$, $B[1, 3, 4, 2]$ (b) $A[1, 2, 1, 3]$, $B[3, 1, -1, -2]$
   (c) $A[-1, 0, 1, 1]$, $B[2, 3, 1, -4]$ (d) $A[3, -2, -1, -3]$, $B[0, 0, 0, 1]$

8. Show that in projective three-space, the line PQ determined by the points $P[\mathbf{p}]$ and $Q[\mathbf{q}]$ and the line RS determined by the points $R[\mathbf{r}]$ and $S[\mathbf{s}]$ intersect if and only if the determinant

$$\begin{vmatrix} p_1 & p_2 & p_3 & p_4 \\ q_1 & q_2 & q_3 & q_4 \\ r_1 & r_2 & r_3 & r_4 \\ s_1 & s_2 & s_3 & s_4 \end{vmatrix}$$

is zero. Why is this also the condition for the two lines being coplanar?

9. Use Exercise 8 to determine whether the lines AB and CD determined by the following pairs of points intersect. Find the standard homogeneous coordinates of the point of intersection if it exists.
   (a) $A[1, 3, 2, 0]$, $B[2, 1, 1, 0]$, $C[-1, 2, 0, 1]$, $D[3, 4, 2, 1]$
   (b) $A[1, 2, -4, -3]$, $B[1, -2, 3, 0]$, $C[1, -1, 5, -2]$, $D[1, 0, -2, -1]$
   (c) $A[2, -5, 2, 2]$, $B[-2, 3, 0, -4]$, $C[1, 0, -1, 1]$, $D[-2, 1, 0, 4]$
   (d) $A[3, 2, 0, 1]$, $B[-4, 2, 4, 3]$, $C[0, 1, 1, 1]$, $D[1, 2, 0, 1]$

10. Find the point of intersection of the line AB in projective three-space determined by the points A and B and the plane π.
    (a) $A[1, 2, 0, 1]$, $B[2, -1, 3, 1]$, $\pi[1, 0, -2, 2]$
    (b) $A[0, 1, 3, 2]$, $B[1, -1, 1, 1]$, $\pi[2, 1, -1, 3]$
    (c) $A[-2, 1, 3, 3]$, $B[2, 1, 0, -1]$, $\pi[3, -2, 0, 4]$
    (d) $A[1, 3, 0, 2]$, $B[2, 1, 1, 1]$, $\pi[1, 2, 3, -2]$

11. If $\mu[\mu]$ and $\nu[\nu]$ are distinct planes in projective three-space, verify that
    $$\{\pi[t\mu + (1-t)\nu] \mid t \in R\} \cup \{\pi[\mu - \nu]\}$$
    is the set of planes in projective three-space that pass through the line of intersection of μ and ν.

12. Find the plane determined by the line of intersection of the planes μ and ν and the point P.
    (a) $\mu[1, 2, -1, 3]$, $\nu[2, 0, 0, 1]$, $P[1, -1, 1, 0]$
    (b) $\mu[3, 1, 2, 0]$, $\nu[1, -2, 1, 1]$, $P[1, 0, -2, 3]$
    (c) $\mu[4, 1, 2, 1]$, $\nu[2, 2, 3, 5]$, $P[1, 0, 0, -1]$
    (d) $\mu[2, 3, 1, -1]$, $\nu[1, 5, -1, -2]$, $P[2, 1, -1, 3]$

In the next two problems, $\pi[\pi]$ and $\mu[\mu]$ are planes in projective three-space, and p_{ij}, for $i, j = 1, \ldots, 4$, $i \neq j$, denotes the determinant of the 2×2 submatrix of
$$\begin{pmatrix} \pi_1 & \pi_2 & \pi_3 & \pi_4 \\ \mu_1 & \mu_2 & \mu_3 & \mu_4 \end{pmatrix}$$
given by
$$p_{ij} = \begin{vmatrix} \pi_i & \pi_j \\ \mu_i & \mu_j \end{vmatrix}$$

13. Verify each of the following.
    (a) $p_{ij} = -p_{ji}$
    (b) If π and μ are distinct, then not all the p_{ij}'s are zero.
    (c) $p_{12}p_{34} + p_{14}p_{23} + p_{13}p_{24} = 0$

14. Verify that if π and μ are distinct, then at least two of the four vectors
    $$\langle 0, -p_{34}, p_{24}, -p_{23} \rangle$$
    $$\langle -p_{34}, 0, -p_{14}, p_{13} \rangle$$
    $$\langle p_{24}, -p_{14}, 0, -p_{12} \rangle$$
    $$\langle -p_{23}, p_{13}, -p_{12}, 0 \rangle$$

are nonzero, and the points in projective three-space whose homogeneous coordinates are determined by these (nonzero) vectors lie on the line of intersection of π and μ. Geometrically explain why one or two of these vectors may be zero, but at least two of them must be nonzero. (*Hint:* The points determined in this manner are the points of intersection of l with the planes determined by the three Euclidean coordinate planes and the ideal plane.)

15. Using Exercise 14, find parametric representations of the lines of intersection of the following planes.
    (a) $\pi[2, -1, 3, 1]$, $\mu[0, 1, 2, 2]$ (b) $\pi[3, 1, -1, 2]$, $\mu[-1, 2, 0, 3]$
    (c) $\pi[-1, 3, 1, 2]$, $\mu[2, 1, 2, 0]$ (d) $\pi[1, -1, 0, 3]$, $\mu[2, 0, -2, 1]$

16. If $Q[\mathbf{q}]$ and $R[\mathbf{r}]$ are distinct points in the projective plane, show that the parameter value t corresponding to the point of intersection of the parametrized line
$$\{P[t\mathbf{q} + (1 - t)\mathbf{r}] \mid t \in R\} \cup \{P[\mathbf{q} - \mathbf{r}]\}$$
and the ideal line is $\pm\infty$ if $q_3 = r_3$, and $r_3/(r_3 - q_3)$ if $q_3 \neq r_3$. (Observe that this value of t depends on the specific representatives $\mathbf{q}$ of $[\mathbf{q}]$ and $\mathbf{r}$ of $[\mathbf{r}]$ used in the parametrization.)

17. Find the points of intersection of the following parametrized lines and the ideal line.
    (a) $\{P[t\langle 1, 2, 3\rangle + (1 - t)\langle 2, 1, 1\rangle]\} \cup \{P[1, -1, -2]\}$
    (b) $\{P[t\langle 0, 1, 2\rangle + (1 - t)\langle 1, 2, 1\rangle]\} \cup \{P[1, 1, -1]\}$
    (c) $\{P[t\langle 2, 1, 1\rangle + (1 - t)\langle 1, 3, 1\rangle]\} \cup \{P[1, -2, 0]\}$
    (d) $\{P[t\langle 3, 1, -2\rangle + (1 - t)\langle 1, 0, 0\rangle]\} \cup \{P[2, -1, 2]\}$

18. If $Q[\mathbf{q}]$ and $R[\mathbf{r}]$ are distinct points in the projective plane, how are the following sets of points related to each other?
    (a) $\{P[t\mathbf{q} + (1 - t)\mathbf{r}] \mid 0 \leq t \leq 1\}$ (b) $\{P[t\mathbf{q} - (1 - t)\mathbf{r}] \mid 0 \leq t \leq 1\}$

19. If $Q[\mathbf{q}]$ and $R[\mathbf{r}]$ are distinct points in the projective plane, show that
$$\{P[t\mathbf{q} + (1 - t)\mathbf{r}] \mid 0 \leq t \leq 1\}$$
intersects the ideal line if and only if q_3 and r_3 have opposite signs.

3

Projective Transformations

INTRODUCTION

In this chapter we describe projective transformations analytically. The study of projective transformations is central to the study of those properties of perspective discussed in Chapter 1. Applications of the theory presented in this chapter to computer graphics are given in Chapters 4, 5, 6, and 7.

In order to motivate the contents of this chapter, we first briefly discuss an important application: how to obtain analytically a two-dimensional perspective image of a three-dimensional object using an arbitrary viewing mechanism.

There are three fundamental steps involved in obtaining a two-dimensional perspective image of a three-dimensional object (see Figure 3.1): The first step is perspective projection of the three-dimensional object onto a two-dimensional image, which lies in a projective plane embedded in projective three-space. The second step is transformation of this two-dimensional embedded image into a two-dimensional image in the projective plane (this transformation is achieved through inverse parametrization of the embedded projective plane). The third step (not illustrated by Figure 3.1) is adjustment of the first two steps (that is, transformations) so that the desired image is obtained.

The first step—perspective projection—is described, for every point $P[p_1, p_2, p_3, p_4]$ of the object, as right multiplication by a 4×4 rank 3 matrix $M_1^{4 \times 4}$:

$$[p_1, p_2, p_3, p_4] \xrightarrow[\text{projection}]{\text{perspective}} [\langle p_1, p_2, p_3, p_4 \rangle M_1^{4 \times 4}]$$

The second step—inverse parametrization—is described, for every point of the resulting image, as right multiplication by a 4×3 rank 3 matrix $M_2^{4 \times 3}$; thus the first two steps are described (as a composite transformation) by right multiplication by the 4×3 rank 3 matrix $M_1^{4 \times 4} M_2^{4 \times 3}$:

$$[p_1, p_2, p_3, p_4] \xrightarrow[\text{projection}]{\text{perspective}} [\langle p_1, p_2, p_3, p_4 \rangle M_1^{4 \times 4}]$$

$$\xrightarrow[\text{parametrization}]{\text{inverse}} [\langle p_1, p_2, p_3, p_4 \rangle M_1^{4 \times 4} M_2^{4 \times 3}]$$

(Note that the product $\langle p_1, p_2, p_3, p_4 \rangle M_1^{4 \times 4} M_2^{4 \times 3}$ is a vector in R^3, so it has the correct number of components to describe an element of RP^2.)

There are several ways in which these first two steps can be adjusted to obtain the desired image. In this chapter we consider only two (in preparation for a more complete coverage of three-dimensional graphics to be presented in Chapter 5).

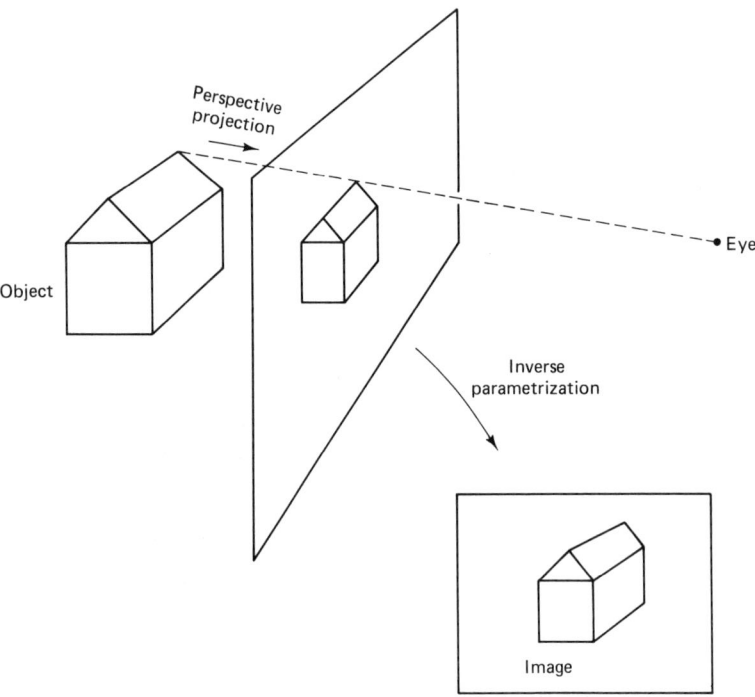

Figure 3.1 Obtaining a two-dimensional perspective image of a three-dimensional object.

The first type of adjustment is of the image in the projective plane. This is frequently the first adjustment of importance since, in practice, it is difficult to proceed with further adjustments if there is no (or poor) visual feedback. (For example—see Figure 3.2—the image might be rotated out of its natural position, it might not lie within the viewing window, or it might be out of scale.) Adjustment of the image in the projective plane (by rotation, translation, or scaling, for example) is accomplished by right multiplication by a 3×3 nonsingular matrix $M_3^{3 \times 3}$; combining this adjustment with the first two transformations (see Figure 3.3), we obtain

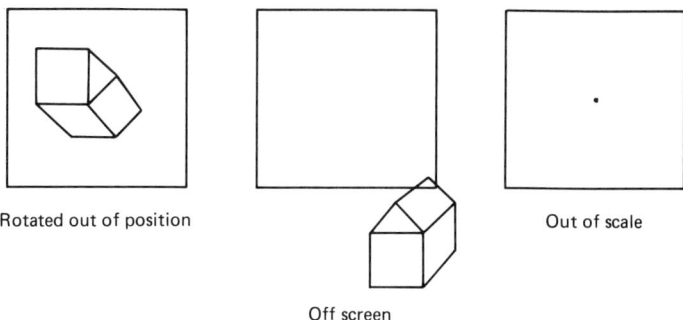

Figure 3.2 Poor perspective images.

$$[p_1, p_2, p_3, p_4] \xrightarrow[\text{projection}]{\text{perspective}} [\langle p_1, p_2, p_3, p_4 \rangle M_1^{4 \times 4}]$$

$$\xrightarrow[\text{parametrization}]{\text{inverse}} [\langle p_1, p_2, p_3, p_4 \rangle M_1^{4 \times 4} M_2^{4 \times 3}]$$

$$\xrightarrow[\text{adjustment}]{\text{image}} [\langle p_1, p_2, p_3, p_4 \rangle M_1^{4 \times 4} M_2^{4 \times 3} M_3^{3 \times 3}]$$

The second type of adjustment is of the original object itself. This type of adjustment (again including rotation, translation, or scaling, for example) is ac-

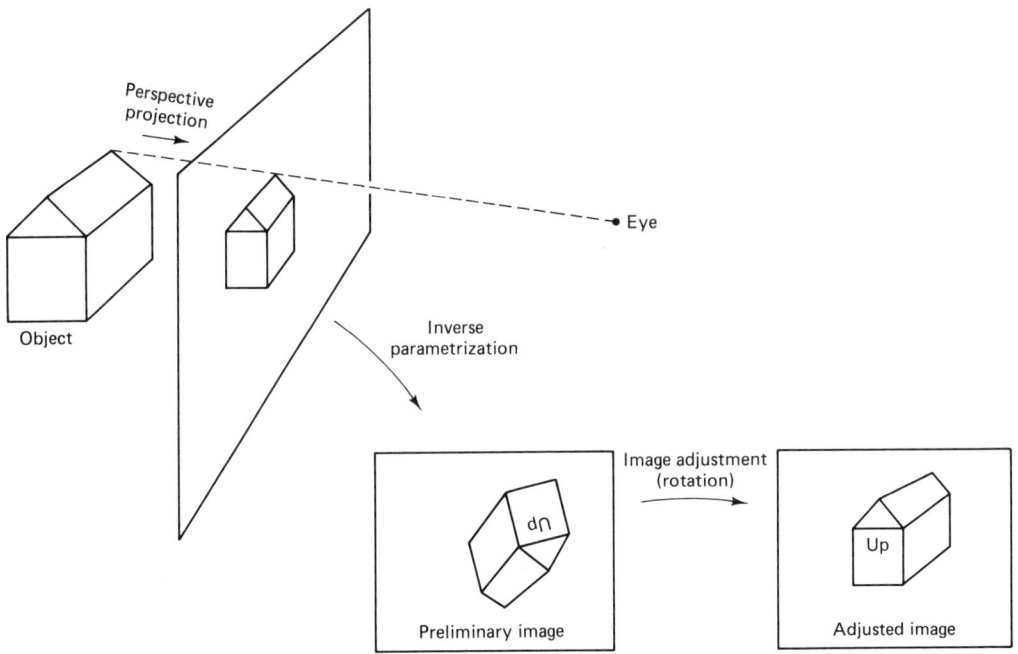

Figure 3.3 Obtaining a perspective image with image adjustment.

Introduction

complished by (initial) right multiplication by a 4 × 4 nonsingular matrix $M_0^{4\times 4}$; combining this transformation with the previous transformations (see Figure 3.4), we obtain

$$[p_1, p_2, p_3, p_4] \xrightarrow[\text{adjustment}]{\text{object}} [\langle p_1, p_2, p_3, p_4\rangle M_0^{4\times 4}]$$

$$\xrightarrow[\text{projection}]{\text{perspective}} [\langle p_1, p_2, p_3, p_4\rangle M_0^{4\times 4} M_1^{4\times 4}]$$

$$\xrightarrow[\text{parametrization}]{\text{inverse}} [\langle p_1, p_2, p_3, p_4\rangle M_0^{4\times 4} M_1^{4\times 4} M_2^{4\times 3}]$$

$$\xrightarrow[\text{adjustment}]{\text{image}} [\langle p_1, p_2, p_3, p_4\rangle M_0^{4\times 4} M_1^{4\times 4} M_2^{4\times 3} M_3^{3\times 3}]$$

(In Chapter 5 we also discuss adjustment of the viewing mechanism.)

The unifying concept here is that of projective transformation: Right multiplication by a 3 × 3 nonsingular matrix, such as $M_3^{3\times 3}$, is a projective transformation of the projective plane represented intrinsically, and right multiplication by a 4 × 4 nonsingular matrix, such as $M_0^{4\times 4}$, is a projective transformation of

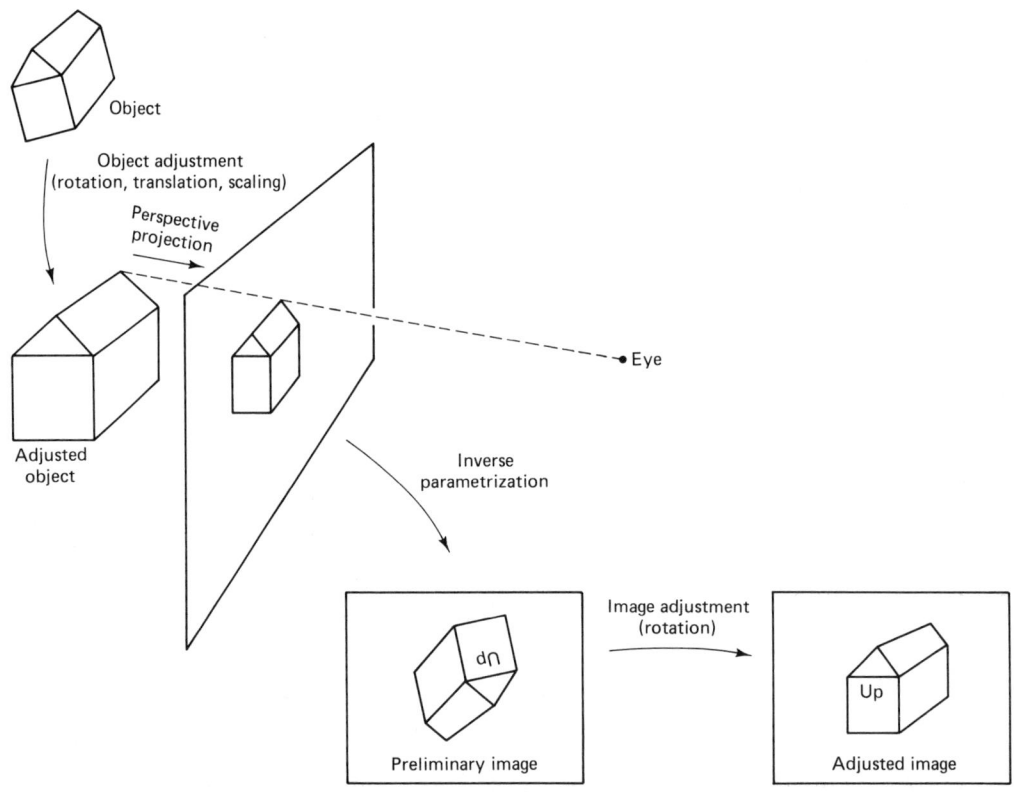

Figure 3.4 Obtaining a perspective image with image and object adjustments.

projective three-space represented intrinsically. Right multiplication by a 4 × 4 rank 3 matrix such as $M_1^{4 \times 4}$ is a projective transformation of projective planes represented extrinsically. Projective transformations, represented in two different ways (by singular and nonsingular matrix multiplication), are of central importance in obtaining a two-dimensional perspective image of a three-dimensional object.

In Section 1 we begin by considering projective transformations of lines, and in Section 2 we consider projective transformations of planes. The reason for considering lines first is not that they are a principle objective (actually planes are more frequently the principle objective of computer graphics); rather, the study of lines is analytically and geometrically simpler than the study of planes, and the study of planes follows immediately thereafter. (As indicated in Section 1.3, it is often convenient to address an issue by addressing the corresponding issue one dimension lower and then reasoning up a dimension by analogy.)

In Section 2 we see that projective transformations of projective planes can be represented in two ways. They can be represented extrinsically (as transformations of embedded projective planes) by right 4 × 4 rank 3 matrix multiplication, and they can be represented intrinsically (as transformations of the projective plane) by right 3 × 3 nonsingular matrix multiplication. These two representations are related by parametrizations (and inverse parametrizations) of projective planes (represented by 3 × 4 and 4 × 3 rank 3 matrix multiplication, respectively).

We discuss the geometry of projective transformations of the projective plane (represented intrinsically by right 3 × 3 nonsingular matrix multiplication), and the geometry of projective transformations of projective three-space (represented intrinsically by right 4 × 4 nonsingular matrix multiplication) in Section 3. One interesting feature of this section is that in it we explain analytically how vanishing points arise. (Later we analytically discuss the other concepts introduced in Chapter 1.)

In Section 4 we discuss general homogeneous coordinates. General homogeneous coordinates arise immediately from the study of projective transformations and are important because the use of general homogeneous coordinates can often make complicated geometric problems more tractable. In Section 5 we discuss homogeneous coordinate transformations.

In Section 6 we describe the cross ratio analytically. In Section 1.6 we described the cross ratio synthetically as a perspective scaling. In Section 6 we describe it quantitatively in terms of homogeneous coordinates.

Finally, in Section 7 we discuss a result (and an associated algorithm) that cannot be found in any one place in this chapter. It is a ubiquitous result, whose special cases appear continually throughout this chapter. (This being so, it is advisable to take a brief look at Section 7 before beginning this chapter.)

SECTION 1: TRANSFORMATIONS OF PROJECTIVE LINES

In this section we describe two ways to represent projective transformations of lines in the projective plane analytically: the first representation is by singular matrix multiplication, and the second is by nonsingular matrix multiplication. As indicated in the introduction of this chapter, lines are considered before planes because the study of lines is analytically and geometrically simpler than the study of planes, and the study of planes follows immediately thereafter.

Recall (see Section 1.6) that a projective transformation of lines, or a projectivity of lines, for short, is either a perspectivity (see Figure 3.5) or a composite of perspectivities of lines.

The following result describes the representation of projective transformations of lines in terms of singular matrix multiplication.

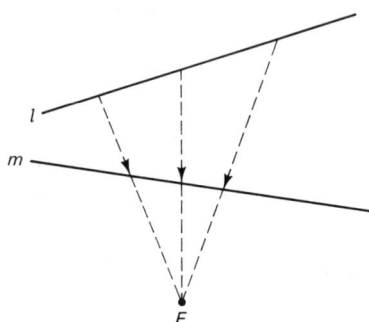

Figure 3.5 A perspectivity of lines.

1.1 Proposition. If $T : l \to m$ is a projective transformation of lines in the projective plane, then there is a 3×3 rank 2 matrix M for which $T([\mathbf{p}]) = [\mathbf{p}M]$ for every point $P[\mathbf{p}]$ on l. If T is a perspectivity whose center is $E[\mathbf{e}]$, then M is

$$\begin{pmatrix} m_2e_2 + m_3e_3 & -m_1e_2 & -m_1e_3 \\ -m_2e_1 & m_1e_1 + m_3e_3 & -m_2e_3 \\ -m_3e_1 & -m_3e_2 & m_1e_1 + m_2e_2 \end{pmatrix}$$

If T is a composite of perspectivities,

$$T = T_n \circ T_{n-1} \circ \cdots \circ T_2 \circ T_1$$

then M is the product—in reverse order—of the 3×3 matrices of rank 2 representing the component perspectivities of T:

$$M = M_1 M_2 \cdots M_{n-1} M_n$$

(Writing $T([\mathbf{p}]) = [\mathbf{p}M]$ is actually an abuse of notation: Since projective transformations are really transformations of points, rather than the coordinates of points, we should really be writing $T(P[\mathbf{p}]) = TP[\mathbf{p}M]$. For the sake of simplicity, however, we use the notation $T([\mathbf{p}]) = [\mathbf{p}M]$.)

Proof (of Proposition 1.1). Suppose T is a perspectivity. Given any point $P[p_1, p_2, p_3]$ on l (see Figure 3.6), first use Proposition 2.2.7(a) to compute the coordinates of the line PE determined by P and E. Then use Proposition 2.2.7(b) to find the point of intersection of PE and m. Simplifying the resulting expression, it follows that $T([\mathbf{p}]) = [\mathbf{p}M]$, where M is as above.

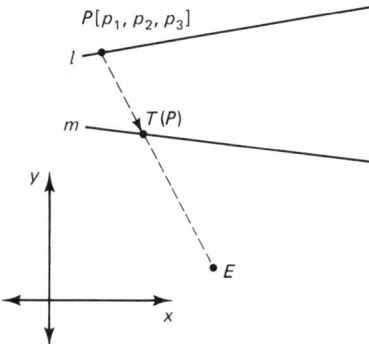

Figure 3.6 Computation of $T([p_1, p_2, p_3])$.

If the representative $k\mathbf{e}$ had been chosen for $[\mathbf{e}]$ instead of $\mathbf{e}$, the matrix representing T would be kM instead of M; and if the representative $k\mathbf{m}$ had been chosen for $[\mathbf{m}]$ instead of $\mathbf{m}$, the matrix representing T would be kM instead of M. Thus M is uniquely defined only up to a scalar multiple (see Exercise 5).

To see that M has rank 2, first observe that M cannot have rank 3 since the determinant of M is zero (verifying this is a straightforward computation). Thus the rank of M can only be 1 or 2. If the rank of M were 1, the determinants det M_{11}, det M_{22}, and det M_{33} of the 2×2 submatrices M_{11}, M_{22}, and M_{33} of M would all have to be zero. (Here M_{ij} is the submatrix of M obtained by omitting the ith row and jth column of M.) Hence det M_{11} + det M_{22} + det M_{33} would also have to be zero. But this is impossible because (expanding det M_{11}, det M_{22}, and det M_{33}, adding, and simplifying the resulting expression)

$$\det M_{11} + \det M_{22} + \det M_{33} = (e_1 m_1 + e_2 m_2 + e_3 m_3)^2$$

and $e_1 m_1 + e_2 m_2 + e_3 m_3$, and hence $(e_1 m_1 + e_2 m_2 + e_3 m_3)^2$, cannot be zero ($e_1 m_1 + e_2 m_2 + e_3 m_3$ cannot be zero since $E[\mathbf{e}]$ does not lie on $m[\mathbf{m}]$). Since the rank of M thus cannot be 1, it must be 2.

If T is a composite of perspectivities, the representation of T follows from repeated application of the fact that

$$(T_2 \circ T_1)([\mathbf{p}]) = T_2(T_1([\mathbf{p}])) = T_2([\mathbf{p}M_1]) = [\mathbf{p}M_1 M_2]$$

Since a product of 3×3 matrices of rank less than 3 is a 3×3 matrix of rank less than 3, the rank of $M = M_1 M_2 \cdots M_{n-1} M_n$ can again only be 1 or 2. But if the rank of M were 1, then there would be constants c_1, c_2, and c_3 for which

$$M = \begin{pmatrix} c_1 m_1 & c_2 m_1 & c_3 m_1 \\ c_1 m_2 & c_2 m_2 & c_3 m_2 \\ c_1 m_3 & c_2 m_3 & c_3 m_3 \end{pmatrix}$$

And if M were of this form then for every point $P[p_1, p_2, p_3]$ on l,

$$[\langle p_1, p_2, p_3 \rangle M] = [(p_1 m_1 + p_2 m_2 + p_3 m_3)\langle c_1, c_2, c_3 \rangle] = [c_1, c_2, c_3]$$

($p_1 m_1 + p_2 m_2 + p_3 m_3$ cannot be zero; see Exercise 7). But this contradicts the fact that T is one-to-one (all projective transformations are one-to-one). Since the rank of M cannot thus be 1, it must be 2. ∎

1.2 Example. If the center of the perspectivity T from $l[1, -1, -1]$ to $m[2, -1, -1]$ is $E[1, 2, 1]$ (see Figure 3.7), then

$$T([p_1, p_2, p_3]) = \left[\langle p_1, p_2, p_3 \rangle \begin{pmatrix} 3 & 4 & 2 \\ -1 & -1 & -1 \\ -1 & -2 & 0 \end{pmatrix} \right]$$

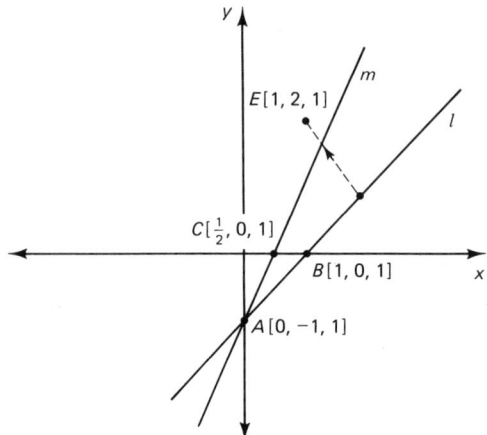

Figure 3.7 A perspectivity.

1.3 Example. If T_1 is the perspectivity from $l[1, -1, 5]$ to $m[2, -1, 3]$ whose center is $E_1[1, 2, 1]$ and T_2 is the perspectivity from $m[2, -1, 3]$ to $n[2, 0, 1]$ whose center is $E_2[1, -1, 0]$ (see Figure 3.8), then the matrices representing T_1 and T_2 are

$$M_1 = \begin{pmatrix} 1 & -4 & -2 \\ 1 & 5 & 1 \\ -3 & -6 & 0 \end{pmatrix} \quad \text{and} \quad M_2 = \begin{pmatrix} 0 & 2 & 0 \\ 0 & 2 & 0 \\ -1 & 1 & 2 \end{pmatrix}$$

respectively, and

$$T_2 \circ T_1([p_1, p_2, p_3]) = [\langle p_1, p_2, p_3 \rangle M_1 M_2] = \left[\langle p_1, p_2, p_3 \rangle \begin{pmatrix} -2 & -8 & -4 \\ 1 & 13 & 2 \\ 0 & -18 & 0 \end{pmatrix} \right]$$

In general, the only way in which the line $l[\mathbf{l}]$ enters into the representation of a perspectivity $T : l \to m$ is that the center E of perspectivity does not lie on l. In particular, the coordinates of l are not involved in M. Furthermore $\langle p_1, p_2, p_3 \rangle M = \langle 0, 0, 0 \rangle$ if and only if $[p_1, p_2, p_3] = [e_1, e_2, e_3]$ (see Exercise 6). We later see how this ambiguity in the specification of the domain of a perspectivity can be exploited.

Since a projective transformation T may, in general, be written as a composite of perspectivities in different ways (see Exercise 1.6.9), it is not obvious that the matrix M representing T is, or should be, unique (even up to scalar multiples). In general, it is not unique.

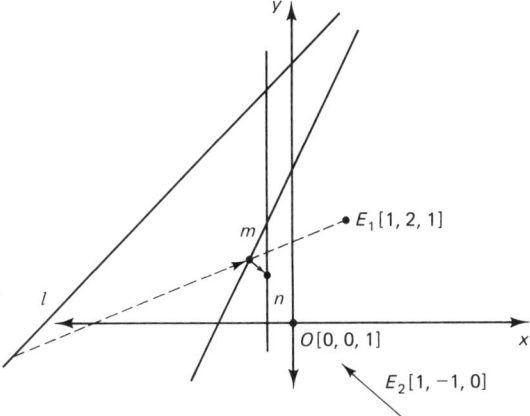

Figure 3.8 A projective transformation.

1.4 Example. For any real number k, let T_k be the perspectivity from $l[0, 1, 0]$ to $l[0, 1, 0]$ whose center is $E[1, -1, k]$. The matrix of T_k

$$M_k = \begin{pmatrix} 1 & 0 & 0 \\ 1 & 0 & k \\ 0 & 0 & 1 \end{pmatrix}$$

For every Euclidean point $P[t, 0, 1]$ on l,

$$T_k([t, 0, 1]) = [\langle t, 0, 1\rangle M_k] = [t, 0, 1]$$

and for the ideal point $P[1, 0, 0]$ on l

$$T_k([1, 0, 0]) = [\langle 1, 0, 0\rangle M_k] = [1, 0, 0]$$

Thus there is more than one way (in fact there are infinitely many ways) to represent the identity transformation on l.

Another way to represent projective transformations of projective lines is by nonsingular matrix multiplication. Before describing this representation, however, we need an important preliminary result.

1.5 Proposition. Let $Q[q_1, q_2, q_3]$, $R[r_1, r_2, r_3]$, and $S[s_1, s_2, s_3]$ be three distinct points on the line l in the projective plane. Then there is a 2×3 rank 2 matrix M, unique up to a scalar multiple, such that if the transformation $T : RP^1 \to l$ is given by

$$T([c_1, c_2]) = [\langle c_1, c_2\rangle M]$$

then $T([1, 0]) = [q_1, q_2, q_3]$, $T([0, 1]) = [r_1, r_2, r_3]$, $T([1, 1]) = [s_1, s_2, s_3]$, T is one-to-one and onto.

Note that the transformation T in this result is not a projective transformation in the sense of Definition 1.6.10 (a projective transformation of lines is a transformation of lines both of which are in the projective plane). Rather T is a parametrization of l (see Section 2.4) represented as matrix multiplication: If l

is the line passing through the points $Q[\mathbf{q}]$ and $R[\mathbf{r}]$, $\mathbf{q}$ and $\mathbf{r}$ are any two representatives of $[\mathbf{q}]$ and $[\mathbf{r}]$, respectively,

$$M = \begin{pmatrix} q_1 & q_2 & q_3 \\ r_1 & r_2 & r_3 \end{pmatrix}$$

and RP^1 is written

$$\{[t, 1 - t] \mid t \in R\} \cup \{[1, -1]\}$$

then the image of RP^1 under T (see Figure 3.9) is

$$\{P[\langle t, 1 - t \rangle M] \mid t \in R\} \cup \{P[\langle 1, -1 \rangle M]\}$$
$$= \{P[t\mathbf{q} + (1 - t)\mathbf{r}] \mid t \in R\} \cup \{P[\mathbf{q} - \mathbf{r}]\}$$

The nonuniqueness of parametrizations as discussed in Example 2.4.3 is due to the fact that according to Proposition 1.5, *three* points are necessary to determine a transformation, or parametrization, $T : RP^1 \to l$; two points are not enough.

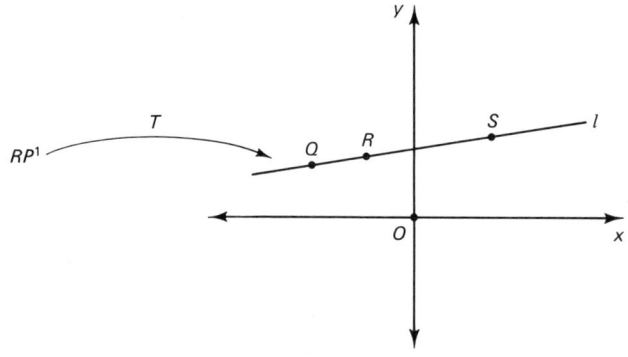

Figure 3.9 Parametrization of lines as matrix multiplication.

The existence of the matrix M of Proposition 1.5 is based on the fact that we can choose representatives $\langle q_1, q_2, q_3 \rangle$ of $[q_1, q_2, q_3]$, $\langle r_1, r_2, r_3 \rangle$ of $[r_1, r_2, r_3]$, and $\langle s_1, s_2, s_3 \rangle$ of $[s_1, s_2, s_3]$ for which

$$\langle q_1, q_2, q_3 \rangle + \langle r_1, r_2, r_3 \rangle + (-1)\langle s_1, s_2, s_3 \rangle = \langle 0, 0, 0 \rangle$$

or

$$\langle s_1, s_2, s_3 \rangle = \langle q_1, q_2, q_3 \rangle + \langle r_1, r_2, r_3 \rangle$$

Thus if

$$M = \begin{pmatrix} q_1 & q_2 & q_3 \\ r_1 & r_2 & r_3 \end{pmatrix}$$

then $T([1, 0]) = [q_1, q_2, q_3]$, $T([0, 1]) = [r_1, r_2, r_3]$, and $T([1, 1]) = [s_1, s_2, s_3]$.

Proof (of Proposition 1.5). Having constructed M above, it suffices to verify the remaining statements of Proposition 1.5.

First, since $Q[\mathbf{q}]$ and $R[\mathbf{r}]$ are distinct points, the rank of M is 2.

To see that M is unique up to a scalar multiple, observe that if

$$N = \begin{pmatrix} n_{11} & n_{12} & n_{13} \\ n_{21} & n_{22} & n_{23} \end{pmatrix}$$

is any other matrix satisfying the conditions of this result, then

$T([1, 0]) = [\langle 1, 0 \rangle N] = [n_{11}, n_{12}, n_{13}] = [q_1, q_2, q_3]$

$T([0, 1]) = [\langle 0, 1 \rangle N] = [n_{21}, n_{22}, n_{23}] = [r_1, r_2, r_3]$

$T([1, 1]) = [\langle 1, 1 \rangle N] = [\langle n_{11}, n_{12}, n_{13} \rangle + \langle n_{21}, n_{22}, n_{23} \rangle] = [s_1, s_2, s_3]$

Thus for some nonzero real constants c_1, c_2, and c_3,

$$\langle n_{11}, n_{12}, n_{13} \rangle = c_1 \langle q_1, q_2, q_3 \rangle$$

$$\langle n_{21}, n_{22}, n_{23} \rangle = c_2 \langle r_1, r_2, r_3 \rangle$$

$$\langle n_{11}, n_{12}, n_{13} \rangle + \langle n_{21}, n_{22}, n_{23} \rangle = c_3 \langle s_1, s_2, s_3 \rangle$$

Combining these equations, we find

$$c_3 \langle s_1, s_2, s_3 \rangle = c_1 \langle q_1, q_2, q_3 \rangle + c_2 \langle r_1, r_2, r_3 \rangle$$

Since Q, R, and S are distinct collinear points and

$$\langle s_1, s_2, s_3 \rangle = \langle q_1, q_2, q_3 \rangle + \langle r_1, r_2, r_3 \rangle,$$

it follows that $c_1 = c_2 = c_3$, and hence $M = cN$, where $c = c_1 = c_2 = c_3$.

To see that T is one-to-one, observe that since M has rank 2, M^+ exists. Thus if

$$T([s_1, s_2]) = T([t_1, t_2])$$

or

$$[\langle s_1, s_2 \rangle M] = [\langle t_1, t_2 \rangle M]$$

then

$$\langle s_1, s_2 \rangle M = c \langle t_1, t_2 \rangle M$$

for some nonzero real constant c. So

$$\langle s_1, s_2 \rangle = \langle s_1, s_2 \rangle M M^+ = c \langle t_1, t_2 \rangle M M^+ = c \langle t_1, t_2 \rangle$$

or $[s_1, s_2] = [t_1, t_2]$.

To see that T is onto, let $P[p_1, p_2, p_3]$ be any point of l and choose nonzero real constants c_1 and c_2 for which

$$c_1 \langle q_1, q_2, q_3 \rangle + c_2 \langle r_1, r_2, r_3 \rangle - \langle p_1, p_2, p_3 \rangle = \langle 0, 0, 0 \rangle$$

or

$$\langle p_1, p_2, p_3 \rangle = c_1 \langle q_1, q_2, q_3 \rangle + c_2 \langle r_1, r_2, r_3 \rangle$$

for some representative $\langle p_1, p_2, p_3 \rangle$ of $[p_1, p_2, p_3]$. Then

$$T([c_1, c_2]) = [\langle c_1, c_2 \rangle M]$$

$$= \left[\langle c_1, c_2 \rangle \begin{pmatrix} q_1 & q_2 & q_3 \\ r_1 & r_2 & r_3 \end{pmatrix} \right]$$

$$= [c_1 \langle q_1, q_2, q_3 \rangle + c_2 \langle r_1, r_2, r_3 \rangle]$$

$$= [p_1, p_2, p_3] \qquad \blacksquare$$

1.6 Example. For the three collinear points $A[2, 1, 3]$, $B[1, 2, -1]$, and $C[1, -4, 9]$,

$$\langle -4, -2, -6 \rangle + \langle 3, 6, -3 \rangle + \langle 1, -4, 9 \rangle = \langle 0, 0, 0 \rangle$$

(see Example 2.2.16) or

$$\langle 1, -4, 9 \rangle = \langle 4, 2, 6 \rangle + \langle -3, -6, 3 \rangle$$

Thus if $T : RP^1 \to l$ is given by

$$T([c_1, c_2]) = \left[\langle c_1, c_2 \rangle \begin{pmatrix} 4 & 2 & 6 \\ -3 & -6 & 3 \end{pmatrix} \right]$$

then $T([1, 0]) = [2, 1, 3]$, $T([0, 1]) = [1, 2, -1]$, and $T([1, 1]) = [1, -4, 9]$.

We now represent projective transformations of projective lines by nonsingular matrix multiplication.

1.7 Proposition. If $T_l : RP^1 \to l$ and $T_m : RP^1 \to m$ are parametrizations of the lines l and m in the projective plane, then for every projective transformation $T : l \to m$ there is a 2×2 nonsingular matrix M for which

$$T([\mathbf{p}]) = [\mathbf{p} M_l^+ M M_m]$$

where M_l and M_m are matrices representing T_l and T_m, respectively. In fact $M = M_l M_0 M_m^+$, where M_0 is a 3×3 rank 2 matrix representing T, and M is unique up to a scalar multiple.

In other words, given points Q, R, and S on l, Proposition 1.5 guarantees the existence of a parametrization $T_l : RP^1 \to l$, represented by a 2×3 rank 2 matrix M_l

$$T_l([c_1, c_2]) = [\langle c_1, c_2 \rangle M_l]$$

for which $T_l([1, 0]) = [\mathbf{q}]$, $T_l([0, 1]) = [\mathbf{r}]$, and $T_l([1, 1]) = [\mathbf{s}]$. Since T_l is one-to-one and onto, it has an inverse $T_l^{-1} : l \to RP^1$ (a *two-sided* inverse) represented by the 3×2 rank 2 matrix M_l^+:

$$T_l^{-1}([p_1, p_2, p_3]) = [\langle p_1, p_2, p_3 \rangle M_l^+]$$

(T_l^{-1} is a left inverse of T_l, since

$$T_l^{-1} \circ T_l([c_1, c_2]) = [\langle c_1, c_2 \rangle M_l M_l^+] = [c_1, c_2]$$

and it is a right inverse of T_l, that is,

$$T_l \circ T_l^{-1}([p_1, p_2, p_3]) = [p_1, p_2, p_3]$$

since $[p_1, p_2, p_3] = [\langle c_1, c_2 \rangle M_l]$ for some $[c_1, c_2]$, and $M_l M_l^+ = I$). Proposition 1.7 states (see Figure 3.10) that there is a 2×2 invertible matrix M for which T is the composite of T_l^{-1}, multiplication T_M by M,

$$T_M([c_1, c_2]) = [\langle c_1, c_2 \rangle M]$$

and T_m:

$$T([p_1, p_2, p_3]) = T_m \circ T_M \circ T_l^{-1}([p_1, p_2, p_3]) = [\langle p_1, p_2, p_3 \rangle M_l^+ M M_m]$$

The transformation $T_l^{-1} : RP^1 \to l$ is an *inverse parametrization* of l. (In Section 2 we discuss inverse parametrizations of planes; as indicated in the introduction

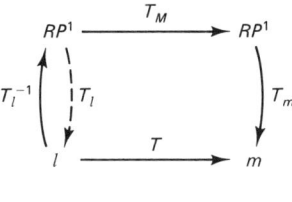

$$T = T_m \circ T_M \circ T_l^{-1}$$

Figure 3.10 A composite factoring of T.

of this chapter, inverse parametrizations of planes play an important role in obtaining a two-dimensional perspective image of a three-dimensional object.)

Proof (of Proposition 1.7). To verify that $T([\mathbf{p}]) = [\mathbf{p} M_l^+ M M_m]$, observe that $P[p_1, p_2, p_3]$ lies on l and $T([\mathbf{p}]) = [\langle p_1, p_2, p_3 \rangle M_0]$ lies on m, so

$$[\langle p_1, p_2, p_3 \rangle M_l^+ M M_m] = [\langle p_1, p_2, p_3 \rangle M_l^+ M_l M_0 M_m^+ M_m]$$
$$= [\langle p_1, p_2, p_3 \rangle M_0 M_m^+ M_m]$$
$$= [\langle p_1, p_2, p_3 \rangle M_0] = T([\mathbf{p}])$$

To see that

$$M = \begin{pmatrix} m_{11} & m_{12} \\ m_{21} & m_{22} \end{pmatrix}$$

is nonsingular, observe that if it were not, then there would be nonzero real constants c_1 and c_2 for which

$$c_1 \langle m_{11}, m_{12} \rangle + c_2 \langle m_{21}, m_{22} \rangle = \langle 0, 0 \rangle$$

but $P = P[\mathbf{p}] = P[c_1 \langle q_1, q_2, q_3 \rangle + c_2 \langle r_1, r_2, r_3 \rangle]$ is a point on l and

$$T([\mathbf{p}]) = T\left(\left[\langle c_1, c_2 \rangle \begin{pmatrix} q_1 & q_2 & q_3 \\ r_1 & r_2 & r_3 \end{pmatrix} \right] \right)$$
$$= T([\langle c_1, c_2 \rangle M_l])$$
$$= [\langle c_1, c_2 \rangle M_l M_l^+ M M_m]$$
$$= [\langle c_1, c_2 \rangle M M_m]$$
$$= \left[\langle c_1, c_2 \rangle \begin{pmatrix} m_{11} & m_{12} \\ m_{21} & m_{22} \end{pmatrix} M_m \right]$$
$$= [\langle 0, 0 \rangle M_m] = [0, 0, 0]$$

which is impossible.

Finally, to see that M is unique up to a scalar multiple, observe that if M_1 and M_2 are two nonsingular 2×2 matrices for which

$$[\langle p_1, p_2, p_3 \rangle M_l^+ M_1 M_m] = [\langle p_1, p_2, p_3 \rangle M_l^+ M_2 M_m]$$

for each point $P[p_1, p_2, p_3] = P[\langle c_1, c_2 \rangle M_l]$ on l, then

$$[\langle c_1, c_2 \rangle M_l M_l^+ M_1 M_m] = [\langle c_1, c_2 \rangle M_l M_l^+ M_2 M_m]$$

or

$$[\langle c_1, c_2 \rangle M_1 M_m] = [\langle c_1, c_2 \rangle M_2 M_m]$$

Sec. 1 Transformations of Projective Lines

The uniqueness of Proposition 1.5 now implies that $M_1 M_m$ and $M_2 M_m$ are the same up to a scalar multiple. Hence

$$M_1 = M_1(M_m M_m^+) = (M_1 M_m) M_m^+$$

and

$$M_2 = M_2(M_m M_m^+) = (M_2 M_m) M_m^+$$

are the same up to a scalar multiple. ∎

The converse of Proposition 1.7 is also true.

1.8 Proposition. If $T_l : RP^1 \to l$ and $T_m : RP^1 \to m$ are parametrizations of the lines l and m in the projective plane then for every 2×2 nonsingular matrix M the transformation $T : l \to m$ given by

$$T([\mathbf{p}]) = [\mathbf{p} M_l^+ M M_m]$$

where M_l and M_m are the matrices representing T_l and T_m, respectively, is a projective transformation.

Proof. Let $A = T_l([1, 0])$, $B = T_l([0, 1])$, and $C = T_l([1, 1])$. First we show the existence of a projective transformation $S : l \to m$ for which $S(A) = T(A)$, $S(B) = T(B)$, and $S(C) = T(C)$ (see Figure 3.11). Suppose l and m are distinct (this is the generic case), and let n be any other line passing through $T(A)$ other than m. Let E_1 be any point on the line determined by A and $T(A)$ other than A or $T(A)$, and let $S_1 : l \to n$ be the perspectivity whose center is E_1. Let E_2 be the point of intersection of the line determined by $T(B)$ and $S_1(B)$ and $T(C)$ and $S_1(C)$, and let $S_2 : n \to m$ be the perspectivity whose center

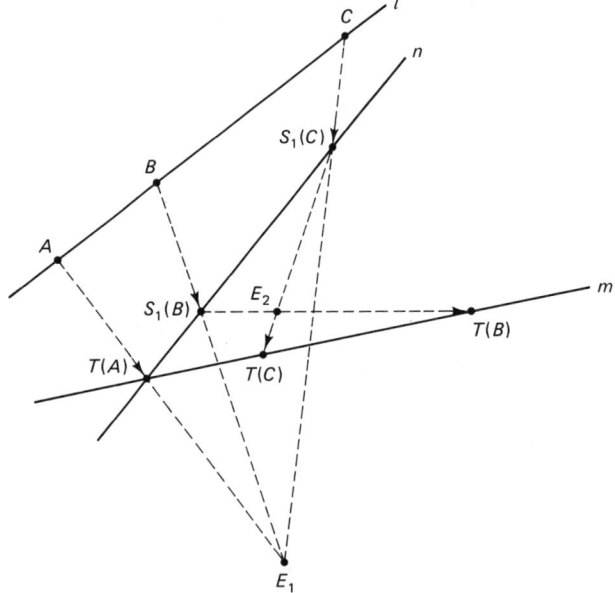

Figure 3.11 Constructing the projective transformation S.

is E_2. Then, by construction, $S = S_2 \circ S_1$ is a projective transformation, and $S(A) = T(A)$, $S(B) = T(B)$, and $S(C) = T(C)$.

According to Proposition 1.7, there is a unique 2×2 nonsingular matrix M_S for which
$$S([\mathbf{p}]) = [\mathbf{p}M_l^+ M_S M_m]$$
Since $S(A) = T(A)$, $S(B) = T(B)$, and $S(C) = T(C)$ it follows that if $\langle c_1, c_2 \rangle$ is $\langle 1, 0 \rangle$, $\langle 0, 1 \rangle$, or $\langle 1, 1 \rangle$ then
$$[\langle c_1, c_2 \rangle M_l M_l^+ M_S M_m] = [\langle c_1, c_2 \rangle M_l M_l^+ M M_m]$$
or
$$[\langle c_1, c_2 \rangle M_S M_m] = [\langle c_1, c_2 \rangle M M_m]$$
By the uniqueness of Proposition 1.5 this implies that $M_S M_m$ and $M M_m$ are the same up to a scalar multiple. This, in turn, implies that
$$M_S = M_S(M_m M_m^+) = (M_S M_m) M_m^+$$
and
$$M = M(M_m M_m^+) = (M M_m) M_m^+$$
are the same up to a scalar multiple. Consequently $S = T$, and since S is a projective transformation, T is also. ∎

As an immediate consequence of Proposition 1.8, we have the following important result. As its name indicates, this result is the major result of classical projective geometry.

1.9 Theorem. (The Fundamental Theorem of Projective Geometry for Lines) Let l and m be lines in the projective plane. If A, B, and C are three distinct points on l and P, Q, and R are three distinct points on m, then there is one and only one projective transformation T taking l to m and A to P, B to Q, and C to R.

Observe that this result ensures both the existence *and* the uniqueness of the projective transformation T.

Proof. If M_l is the 2×3 rank 2 matrix representing the transformation $T_l : RP^1 \to l$ for which $T_l([1, 0]) = A$, $T_l([0, 1]) = B$, and $T_l([1, 1]) = C$ and M_m is the 2×3 rank 2 matrix representing the transformation $T_m : RP^1 \to m$ for which $T_m([1, 0]) = P$, $T_m([0, 1]) = Q$, and $T_m([1, 1]) = R$, then the projective transformation $T : l \to m$ given by
$$T([\mathbf{p}]) = [\mathbf{p} M_l^+ M_m]$$
takes A to Q, B to R, and C to S (see Figure 3.12).

The uniqueness of T follows immediately from the proof of Proposition 1.8: there we started by constructing a projective transformation S taking A to P, B to Q, and C to R; but (and this is the key point) we then showed that if S is *any* projective transformation taking A to P, B to Q, and C to R, then $S = T$. It follows that the transformation T (constructed above) is unique. ∎

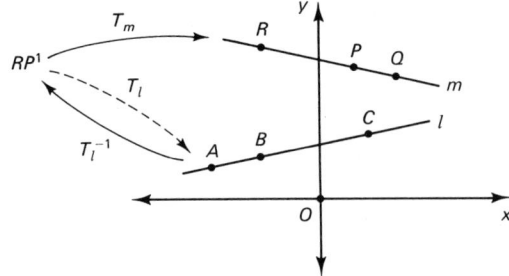

Figure 3.12 The Fundamental Theorem of Projective Geometry for Lines.

The Fundamental Theorem of Projective Geometry for Lines is of central importance throughout the rest of this book. The following example illustrates one way in which it will be used.

1.10 Example. Given two sets of three points on the projective line $A[1, 2]$, $B[-2, 1]$, $C[8, 1]$ and $P[1, 1]$, $Q[3, 1]$, $R[1, 0]$ the (intrinsically described) projective transformation T taking A to P, B to Q, and C to R is obtained as follows. Since

$$\mathbf{c} = \langle 8, 1 \rangle = \langle 2, 4 \rangle + \langle 6, -3 \rangle = \mathbf{a} + \mathbf{b}$$

right multiplication by the matrix

$$\begin{pmatrix} 2 & 4 \\ 6 & -3 \end{pmatrix}$$

defines a projective transformation T_1 of the projective line for which $T_1(I) = T_1([1, 0]) = A$, $T_1(O) = T_1([0, 1]) = B$, and $T_1(U) = T_1([1, 1]) = C$ (I denotes the ideal point, O the origin, and U the unit point). The inverse T_1^{-1} of T_1 is represented by

$$M_1 = \begin{pmatrix} -3 & -4 \\ -6 & 2 \end{pmatrix}$$

(see Exercise 5) and $T_1^{-1}(A) = I$, $T_1^{-1}(B) = O$, and $T_1^{-1}(C) = U$ (see Figure 3.13). Since

$$\mathbf{r} = \langle -2, 0 \rangle = \langle 1, 1 \rangle + \langle -3, -1 \rangle = \mathbf{p} + \mathbf{q}$$

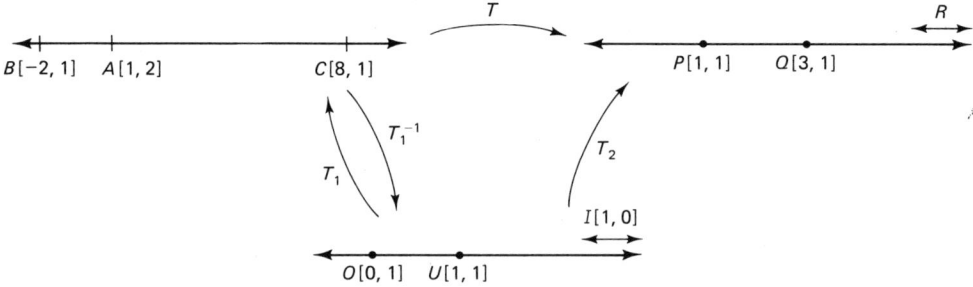

Figure 3.13 Application of the Fundamental Theorem of Projective Geometry for Lines.

right multiplication by the matrix

$$M_2 = \begin{pmatrix} 1 & 1 \\ -3 & -1 \end{pmatrix}$$

defines a projective transformation T_2 for which $T_2(I) = P$, $T_2(O) = Q$, and $T_2(U) = R$. Thus if T is the composite transformation $T = T_2 \circ T_1^{-1}$, which is represented by right multiplication by the matrix

$$M = M_1 M_2 = \begin{pmatrix} 9 & 1 \\ -12 & -8 \end{pmatrix}$$

then $T(A) = P$, $T(B) = Q$, and $T(C) = R$.

SECTION 1 EXERCISES

1. In each of the following cases, find the 3×3 rank 2 matrix representing the perspectivity $T : l \to m$ whose center is E. Draw a sketch illustrating the effect of T.
   (a) $l[3, 2, -6]$, $m[1, 0, -3]$, $E[0, 0, 1]$
   (b) $l[1, 2, -4]$, $m[1, 2, 2]$, $E[3, 3, 1]$
   (c) $l[2, 1, -2]$, $m[1, 0, 0]$, $E[1, 0, 0]$
   (d) $l[0, 1, 0]$, $m[1, -1, 0]$, $E[0, 1, 0]$
   (e) $l[1, 0, 0]$, $m[0, 1, 0]$, $E[1, -1, 0]$
   (f) $l[1, -4, 4]$, $m[2, 3, -6]$, $E[2, 3, -1]$

2. Verify the following technique for computing the matrix M representing a perspectivity given in Proposition 1.1: If representatives $\mathbf{m}$ and $\mathbf{e}$ of $[\mathbf{m}]$ and $[\mathbf{e}]$, respectively, are chosen so that $\mathbf{e}^{1 \times 3} \mathbf{m}^{3 \times 1} = 1$, then

$$M = I^{3 \times 3} - \mathbf{m}^{3 \times 1} \mathbf{e}^{1 \times 3}$$

3. Use Exercise 2 to show that if $P[\mathbf{p}]$ lies on $[\mathbf{m}]$, then $T([\mathbf{p}]) = [\mathbf{p}]$.

4. In each of the following cases, find the 3×3 rank 2 matrix representing the composite $T = T_2 \circ T_1 : l \to n$ of the perspectivities $T_1 : l \to m$, whose center is E_1, and $T_2 : m \to n$, whose center is E_2. Draw a sketch illustrating the effect of T.
   (a) $l[1, 0, 0]$, $m[1, 1, 0]$, $n[1, 0, -4]$, $E_1[2, 6, 1]$, $E_2[6, 0, 1]$
   (b) $l[3, 2, -12]$, $m[3, 2, -6]$, $n[1, -1, -6]$, $E_1[3, 5, 1]$, $E_2[10, 2, 1]$
   (c) $l[1, 1, -2]$, $m[0, 1, -2]$, $n[0, 1, -1]$, $E_1[0, 1, 0]$, $E_2[0, 0, 1]$
   (d) $l[1, 1, -3]$, $m[1, 0, -6]$, $n[1, -1, -9]$, $E_1[0, 0, 1]$, $E_2[2, 0, 1]$

†5. Show that if $[\mathbf{p}]$ is an element of RP^m and M is an $m \times n$ matrix for which $[\mathbf{p}M]$ is an element of RP^n (that is, $\mathbf{p}M$ is never $\mathbf{0}$), then $[\mathbf{p}M] = [\mathbf{p}kM]$ for every nonzero real constant k. (*Note:* As a consequence of this result, a matrix M representing a projective transformation can be replaced by a scalar multiple kM of M. This can be useful since, for an appropriate value of k, the entries of kM might be much nicer than the entries of M. For example, replacing M by kM might help reduce the number of minus signs or produce a matrix with integral—as opposed to rational—entries.)

6. Let M be the 3×3 rank 2 matrix representing the perspectivity $T : l \to m$ extrinsically.

Show that for any point $P[p_1, p_2, p_3]$ in the projective plane,
$$\langle p_1, p_2, p_3 \rangle M = \langle 0, 0, 0 \rangle$$
if and only if $P[p_1, p_2, p_3]$ is the center $E[e_1, e_2, e_3]$ of perspectivity of T.

7. (See the proof of Proposition 1.1.) Show that $p_1m_1 + p_2m_2 + p_3m_3$ cannot be zero. (*Hint:* Use the fact that if $p_1m_1 + p_2m_2 + p_3m_3 = 0$ then
$$\mathbf{p}M_1 \cdots M_n = \langle 0, 0, 0 \rangle$$
and repeated application of Exercise 6.)

8. In each of the following cases, three distinct collinear points Q, R, and S on a line l in the projective plane are given. Find a matrix representing the parametrization $T : RP^1 \to l$ of l for which $T([1, 0]) = Q$, $T([0, 1]) = R$, and $T([1, 1]) = S$.
   (a) $Q[-3, 2, 1]$, $R[1, -2, 1]$, $S[1, 2, -3]$
   (b) $Q[-3, 1, 1]$, $R[1, 0, -1]$, $S[1, -1, 1]$
   (c) $Q[2, 0, 1]$, $R[1, -1, 1]$, $S[1, 1, 0]$
   (d) $Q[1, 0, 0,]$, $R[0, 1, 1]$, $S[1, 1, 1]$
   (e) $Q[1, 0, 0]$, $R[1, 1, 1]$, $S[0, 1, 1]$
   (f) $Q[1, 0, 0]$, $R[0, 1, 0]$, $S[1, 1, 0]$

9. For each part of Exercise 1, matrices parametrizing the lines l and m are given below. With respect to these parametrizations, find the 2×2 nonsingular matrix representing each transformation intrinsically.

(a) $M_l = \begin{pmatrix} 2 & -3 & 0 \\ 2 & 0 & 1 \end{pmatrix}$, $M_m = \begin{pmatrix} 0 & 1 & 0 \\ 3 & 0 & 1 \end{pmatrix}$

(b) $M_l = \begin{pmatrix} 2 & 1 & 1 \\ 4 & 0 & 1 \end{pmatrix}$, $M_m = \begin{pmatrix} 0 & 1 & -1 \\ 2 & 0 & 1 \end{pmatrix}$

(c) $M_l = \begin{pmatrix} 1 & 0 & 1 \\ 0 & 2 & 1 \end{pmatrix}$, $M_m = \begin{pmatrix} 0 & 1 & 0 \\ 0 & 0 & 1 \end{pmatrix}$

(d) $M_l = \begin{pmatrix} 1 & 0 & 1 \\ 0 & 0 & 1 \end{pmatrix}$, $M_m = \begin{pmatrix} 1 & 1 & 0 \\ 0 & 0 & 1 \end{pmatrix}$

(e) $M_l = \begin{pmatrix} 0 & 1 & 2 \\ 0 & 2 & 1 \end{pmatrix}$, $M_m = \begin{pmatrix} 2 & 0 & 1 \\ 1 & 0 & 0 \end{pmatrix}$

(f) $M_l = \begin{pmatrix} 0 & 1 & 1 \\ 4 & 0 & -1 \end{pmatrix}$, $M_m = \begin{pmatrix} 3 & 0 & 1 \\ 0 & 2 & 1 \end{pmatrix}$

10. For each part of Exercise 9, a 2×2 nonsingular matrix is given below; this matrix represents a projective transformation intrinsically. Find the 3×3 rank 2 matrix representing this projective transformation extrinsically.

(a) $\begin{pmatrix} 1 & 2 \\ 1 & 1 \end{pmatrix}$ (b) $\begin{pmatrix} 0 & 1 \\ -1 & 0 \end{pmatrix}$ (c) $\begin{pmatrix} 1 & 0 \\ 1 & 1 \end{pmatrix}$

(d) $\begin{pmatrix} 2 & 1 \\ 1 & 3 \end{pmatrix}$ (e) $\begin{pmatrix} 1 & 2 \\ 0 & 1 \end{pmatrix}$ (f) $\begin{pmatrix} 1 & -1 \\ -1 & 1 \end{pmatrix}$

11. In each of the following cases, two sets of three distinct collinear points in the projective plane are given. Find the 3×3 rank 2 matrix M representing the projective transformation $T : l \to m$ taking A to P, B to Q, and C to R.

(a) $A[1, 0, 0]$, $B[0, 1, 1]$, $C[1, 1, 1]$ and $P[-3, 2, 1]$, $Q[1, -2, 1]$, $R[1, -2, 3]$
(b) $A[2, 0, 1]$, $B[1, -1, 1]$, $C[1, 1, 0]$ and $P[1, 0, 0]$, $Q[0, 1, 1]$, $R[1, 1, 1]$
(c) $A[1, 0, 0]$, $B[0, 1, 1]$, $C[1, 1, 1]$ and $P[1, 0, 0]$, $Q[1, 1, 1]$, $R[0, 1, 1]$
(d) $A[-3, 2, 1]$, $B[1, -2, 1]$, $C[1, 2, -3]$ and $P[-3, 1, 1]$, $Q[1, 0, -1]$, $R[1, -1, 1]$
(e) $A[-3, 1, 1]$, $B[1, 0, -1]$, $C[1, -1, 1]$ and $P[-3, 2, 1]$, $Q[1, -2, 1]$, $R[1, 2, -3]$
(f) $A[-3, 2, 1]$, $B[1, -2, 1]$, $C[1, 2, -3]$ and $P[2, 0, 1]$, $Q[1, -1, 1]$, $R[1, 1, 0]$

12. Let $Q[q_1, q_2, q_3, q_4]$, $R[r_1, r_2, r_3, r_4]$, and $S[s_1, s_2, s_3, s_4]$ be three distinct points on the line l in projective three-space. Show that there is a 2×4 rank 2 matrix M, unique up to a scalar multiple, such that if the transformation $T : RP^1 \to l$ is given by

$$T([c_1, c_2]) = [\langle c_1, c_2 \rangle M]$$

then $T([1, 0]) = [q_1, q_2, q_3, q_4]$, $T([0, 1]) = [r_1, r_2, r_3, r_4]$, and $T([1, 1]) = [s_1, s_2, s_3, s_4]$. (Again—see Section 2.4—T is a parametrization of l represented as matrix multiplication.)

13. In each of the following cases three distinct collinear points Q, R, and S on a line l in projective three-space are given. Find a matrix representing the parametrization $T : RP^1 \to l$ of l for which $T([1, 0]) = Q$, $T([0, 1]) = R$, and $T([1, 1]) = S$.
(a) $Q[1, -2, 3, 2]$, $R[5, 6, -1, 2]$, $S[3, 2, 1, 2]$
(b) $Q[1, -1, -6, -1]$, $R[1, 1, 2, 3]$, $S[2, 1, 0, 4]$
(c) $Q[3, 2, 1, 3]$, $R[0, 1, -1, 3]$, $S[2, 1, 1, 1]$
(d) $Q[5, 1, 2, -3]$, $R[2, 1, 1, 0]$, $S[1, 2, 1, 3]$

14. Apply the construction described in the proof of Proposition 1.8 to obtain projective transformations between the lines l and m in Figure 3.14 taking A to P, B to Q, and C to R.

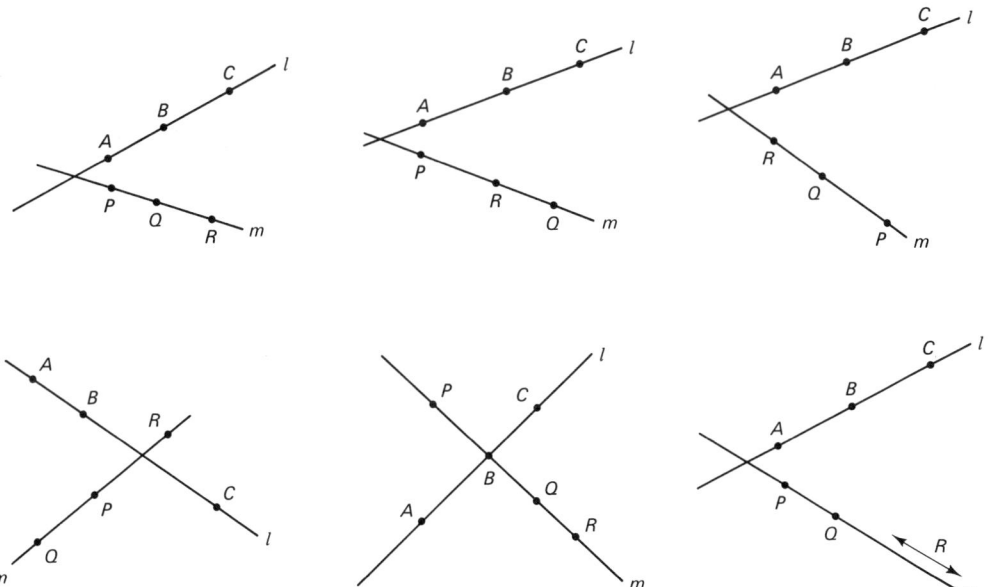

Figure 3.14 Constructing projective transformations.

Sec. 1   Transformations of Projective Lines

15. In each of the following cases, two sets of three points on the projective line are given. Find the projective transformation taking A to P, B to Q, and C to R.
   (a) A[1, 0], B[0, 1], C[1, 1] and P[2, 1], Q[1, 1], R[3, 1]
   (b) A[2, 1], B[0, 1], C[1, 3] and P[1, 1], Q[2, 3], R[−3, 2]
   (c) A[3, 2], B[1, 1], C[1, 4] and P[1, 2], Q[−1, 1], R[1, −2]
   (d) A[2, 1], B[3, 2], C[3, 1] and P[1, 2], Q[−1, 2], R[1, −3]
   (e) A[1, 0], B[0, 1], C[1, 1] and P[0, 1], Q[1, 0], R[1, 1]
   (f) A[1, 0], B[0, 1], C[1, 1] and P[1, 1], Q[1, 0], R[0, 1]

16. In Example 1.4 we showed that the identity transformation on the projective line can be extrinsically represented in an infinite number of ways. What does this imply about the extrinsic representation of projective transformations in general? (*Hint:* Any projective transformation can be written as itself composed with an identity transformation.)

†17. The rows of the matrix M representing a linear transformation T from R^m to R^n are the values of T on the standard basis vectors of R^m. Explain why the rows of a matrix $M^{2\times 2}$ representing a projective transformation T of the projective line intrinsically, are determined by the values of T on the points whose homogeneous coordinates are [1, 0], [0, 1], and [1, 1], although $T([1, 0])$ and $T([0, 1])$ are not the rows of M.

SECTION 2: TRANSFORMATIONS OF PROJECTIVE PLANES

In this section, projective transformations of planes in projective three-space are discussed in a manner parallel to the manner in which projective transformations of lines were discussed in Section 1. Since this parallelism is analytically so straightforward, proofs in this section are minimal.

We describe two ways to represent projective transformations of planes: The first representation is by singular matrix multiplication, and the second is by nonsingular matrix multiplication.

Recall (see Section 1.6) that a projective transformation of planes, or a projectivity of planes, for short, is either a perspectivity (see Figure 3.15) or a composite of perspectivities of planes.

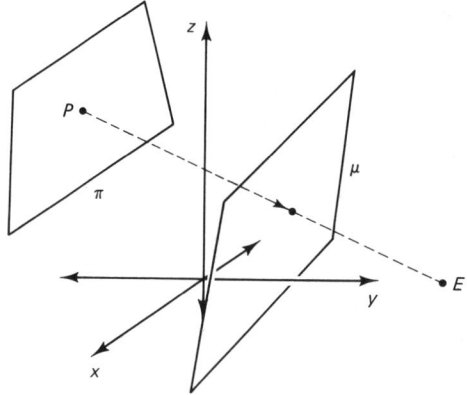

Figure 3.15 A perspectivity of planes.

The following result describes the representation of projective transformations of planes by singular matrix multiplication.

2.1 Proposition. If $T: \pi \to \mu$ is a projective transformation of planes in projective three-space, then there is a 4×4 rank 3 matrix M for which $T([\mathbf{p}]) = [\mathbf{p}M]$ for every point $P[\mathbf{p}]$ on π: If T is a perspectivity whose center is $E[\mathbf{e}]$ then M is

$$\begin{pmatrix} \mu_2 e_2 + \mu_3 e_3 + \mu_4 e_4 & -\mu_1 e_2 & -\mu_1 e_3 & -\mu_1 e_4 \\ -\mu_2 e_1 & \mu_1 e_1 + \mu_3 e_3 + \mu_4 e_4 & -\mu_2 e_3 & -\mu_2 e_4 \\ -\mu_3 e_1 & -\mu_3 e_2 & \mu_1 e_1 + \mu_2 e_2 + \mu_4 e_4 & -\mu_3 e_4 \\ -\mu_4 e_1 & -\mu_4 e_2 & -\mu_4 e_3 & \mu_1 e_1 + \mu_2 e_2 + \mu_3 e_3 \end{pmatrix}$$

If T is a composite of perspectivities,

$$T = T_n \circ T_{n-1} \circ \cdots \circ T_2 \circ T_1$$

then M is the product—in reverse order—of the 4×4 matrices of rank 3 representing the component perspectivities of T:

$$M = M_1 M_2 \cdots M_{n-1} M_n$$

2.2 Example. If the center of the perspectivity T from $\pi[1, 3, 2, 4]$ to $\mu[2, 1, -3, -2]$ is $E[2, 1, 0, -1]$, then

$$T([\mathbf{p}]) = \left[\mathbf{p}\begin{pmatrix} 3 & -2 & 0 & 2 \\ -2 & 6 & 0 & 1 \\ 6 & 3 & 7 & -3 \\ 4 & 2 & 0 & 5 \end{pmatrix}\right]$$

2.3 Example. If T_1 is the perspectivity from $\pi[1, 3, 2, 1]$ to $\mu[2, 1, 0, 1]$ whose center is $E_1[1, -2, 1, 1]$, and T_2 is the perspectivity from $\mu[2, 1, 0, 1]$ to $\nu[1, 1, 2, 0]$ whose center is $E_2[-3, 0, 1, -1]$, then the matrices representing T_1 and T_2 are

$$M_1 = \begin{pmatrix} -1 & 4 & -2 & -2 \\ -1 & 3 & -1 & -1 \\ 0 & 0 & 1 & 0 \\ -1 & 2 & -1 & 0 \end{pmatrix} \text{ and } M_2 = \begin{pmatrix} 2 & 0 & -1 & 1 \\ 3 & -1 & -1 & 1 \\ 6 & 0 & -3 & 2 \\ 0 & 0 & 0 & -1 \end{pmatrix}$$

respectively, and

$$T_2 \circ T_1([\mathbf{p}]) = [\mathbf{p}M_1 M_2] = \left[\mathbf{p}\begin{pmatrix} -2 & -4 & 3 & 1 \\ 1 & -3 & 1 & 1 \\ 6 & 0 & -3 & 2 \\ -2 & -2 & 2 & -1 \end{pmatrix}\right]$$

As before, the only way in which the plane $\pi[\boldsymbol{\pi}]$ enters into the representation of a perspectivity $T: \pi \to \mu$ is that the center E of perspectivity does not lie on π. In particular, the coordinates of π are not involved in M. Furthermore $\mathbf{p}M = 0$ if and only if $[\mathbf{p}] = [\mathbf{e}]$.

These facts have an important practical implication: Suppose we wish to obtain a two-dimensional perspective image of a three-dimensional object using μ as the image plane and E as the center of perspectivity. Even though the three-dimensional object may be bounded by two-dimensional faces which lie in different projective planes π_i (see Figure 3.16), the perspective projections $T_i : \pi_i \to \mu$ of the various faces onto μ are all represented by matrix multiplication by the *same* matrix. Consequently, we can unambiguously project all points of the object to μ via right multiplication by a single matrix. (It is natural to ask why, then, we describe a three-dimensional object by its two-dimensional faces. The answer to this question involves the consideration of hidden points, lines, and surfaces.) We return to this discussion in Chapter 5.

Since a projective transformation T may, in general, be written as a composite of perspectivities in different ways (see Exercise 1.6.9), it is again not obvious that the matrix M representing T is, or should be, unique (even up to scalar multiples). In general, it is not unique.

2.4 Example. For any real number k, let T be the perspectivity from $\pi[0, 0, 1, 0]$ to $\pi[0, 0, 1, 0]$ whose center is $E[1, 1, -1, k]$. The matrix of T_k

$$M_k = \begin{pmatrix} 1 & 0 & 0 & 0 \\ 0 & 1 & 0 & 0 \\ 1 & 1 & 0 & k \\ 0 & 0 & 0 & 1 \end{pmatrix}$$

For every Euclidean point $P[s, t, 0, 1]$ on π,

$$T_k([s, t, 0, 1]) = [\langle s, t, 0, 1 \rangle M_k] = [s, t, 0, 1]$$

and for every ideal point $P[s, t, 0, 0]$ on π,

$$T_k([s, t, 0, 0]) = [\langle s, t, 0, 0 \rangle M_k] = [s, t, 0, 0]$$

Thus there is more than one way (in fact there are infinitely many ways) to represent the identity transformation on π.

Another way to represent projective transformations of projective planes is by nonsingular matrix multiplication. Before describing this representation, however, we again need an important preliminary result.

2.5 Proposition. Let P, Q, R, and S be four distinct points, no three of which are collinear, on the plane π in projective three-space. Then there is a 3×4

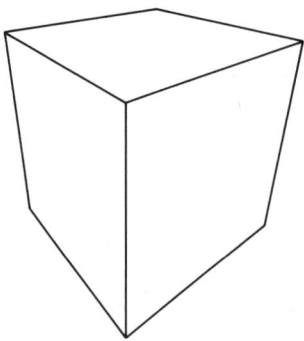

Figure 3.16 A two-dimensional perspective image of a three-dimensional object.

rank 3 matrix M, unique up to a scalar multiple, such that if the transformation $T: RP^2 \to \pi$ is given by
$$T([c_1, c_2, c_3]) = [\langle c_1, c_2, c_3 \rangle M]$$
then $T([1, 0, 0]) = P$, $T([0, 1, 0]) = Q$, $T([0, 0, 1]) = R$, $T([1, 1, 1]) = S$, T is one-to-one and onto.

There is a difference between the statements of Proposition 2.5 and the corresponding result, Proposition 1.5, for lines: in Proposition 2.5 no three of the points P, Q, R, and S can be collinear; in Proposition 1.5 there is no similar condition. (The key to the proof of Proposition 1.5 is that M has rank 2 and hence has a right inverse. The key to the proof of Proposition 2.5 is again that M has a right inverse; but now it is not adequate to assume that P, Q, R, and S are distinct—under this condition M could have rank less than 3 and hence have no right inverse. For example, the matrix associated to the points given in Example 2.3.11 would be
$$\begin{pmatrix} -6 & -3 & -9 & 0 \\ 0 & 0 & 0 & 0 \\ 1 & -4 & 9 & 3 \end{pmatrix}$$
a rank 2 matrix. Assuming that no three of the points $P, Q, R,$ and S are collinear does insure that M is of rank 3, and hence has a right inverse.)

The transformation T in Proposition 2.5 is, again, not a projective transformation in the sense of Definition 1.6.10 (a projective transformation of planes is a transformation of planes, both of which are in projective three-space). Rather, T is a parametrization of π (by RP^2) represented as matrix multiplication.

2.6 Example. The four points $A[2, 1, 3, 2]$, $B[1, 2, -1, -1]$, $C[1, 4, 1, 0]$, and $D[5, -3, 6, 5]$ in projective three-space are coplanar, but no three of these points are collinear. Since
$$\langle 6, 3, 9, 6 \rangle + \langle 1, 2, -1, -1 \rangle +$$
$$\langle -2, -8, -2, 0 \rangle + \langle -5, 3, -6, -5 \rangle = \langle 0, 0, 0, 0 \rangle$$
or
$$\langle 5, -3, 6, 5 \rangle = \langle 6, 3, 9, 6 \rangle + \langle 1, 2, -1, -1 \rangle + \langle -2, -8, -2, 0 \rangle$$
it follows that if $T: RP^2 \to \pi$ is given by
$$T([c_1, c_2, c_3]) = \left[\langle c_1, c_2, c_3 \rangle \begin{pmatrix} 6 & 3 & 9 & 6 \\ 1 & 2 & -1 & -1 \\ -2 & -8 & -2 & 0 \end{pmatrix} \right]$$
then $T([1, 0, 0]) = [2, 1, 3, 2]$, $T([0, 1, 0]) = [1, 2, -1, -1]$, $T([0, 0, 1]) = [1, 4, 1, 0]$ and $T([1, 1, 1]) = [5, -3, 6, 5]$.

We now represent projective transformations of projective planes by nonsingular matrix multiplication.

2.7 Proposition. If $T_\pi: RP^2 \to \pi$ and $T_\mu: RP^2 \to \mu$ are parametrizations of

the planes π and μ in projective three-space, then for every projective transformation $T : \pi \to \mu$ there is a 3×3 nonsingular matrix M for which
$$T([\mathbf{p}]) = [\mathbf{p} M_\pi^+ M M_\mu]$$
(see Figure 3.17), where M_π and M_μ are matrices representing T_π and T_μ, respectively. In fact $M = M_\pi M_0 M_\mu^+$, where M_0 is a 4×4 rank 3 matrix representing T, and M is unique up to scalar multiples.

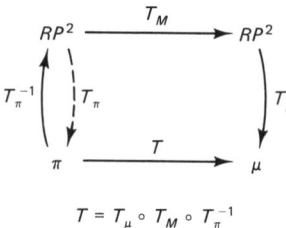

Figure 3.17 Description of a projective transformation in terms of nonsingular matrix multiplication.

In other words (as before), given points P, Q, R, and S on π, Proposition 2.5 guarantees the existence of a parametrization $T_\pi : RP^2 \to \pi$, represented by a 3×4 rank 3 matrix M_π
$$T_\pi([c_1, c_2, c_3]) = [\langle c_1, c_2, c_3 \rangle M_\pi]$$
for which $T_\pi([1, 0, 0]) = [\mathbf{p}]$, $T_\pi([0, 1, 0]) = [\mathbf{q}]$, $T_\pi([0, 0, 1]) = [\mathbf{r}]$, and $T_\pi([1, 1, 1]) = [\mathbf{s}]$. Since T_π is one-to-one and onto, it has an inverse $T_\pi^{-1} : \pi \to RP^2$ (a *two-sided* inverse) represented by the 4×3 rank 3 matrix M_π^+:
$$T_\pi^{-1}([p_1, p_2, p_3, p_4]) = [\langle p_1, p_2, p_3, p_4 \rangle M_\pi^+]$$
(T_π^{-1} is a left inverse of T_π since
$$T_\pi^{-1} \circ T_\pi([c_1, c_2, c_3]) = [\langle c_1, c_2, c_3 \rangle M_\pi M_\pi^+] = [c_1, c_2, c_3])$$
and it is a right inverse of T_π, that is,
$$T_\pi \circ T_\pi^{-1}([p_1, p_2, p_3, p_4]) = [p_1, p_2, p_3, p_4]$$
since $[p_1, p_2, p_3, p_4] = [\langle c_1, c_2, c_3 \rangle M_\pi]$ for some $[c_1, c_2, c_3]$, and $M_\pi M_\pi^+ = I$). Proposition 2.7 states (see Figure 3.17) that there is a 3×3 nonsingular matrix M for which T is the composite of T_π^{-1}, right multiplication T_M by M,
$$T_M([c_1, c_2, c_3]) = [\langle c_1, c_2, c_3 \rangle M]$$
and T_μ:
$$T([p_1, p_2, p_3, p_4]) = T_\mu \circ T_M \circ T_\pi^{-1}([p_1, p_2, p_3, p_4]) = [\langle p_1, p_2, p_3, p_4 \rangle M_\pi^+ M M_\mu]$$

The transformation T_π^{-1} is an *inverse parametrization* of π. Inverse parametrizations of planes in projective three-space play an important role in obtaining a two-dimensional perspective image of a three-dimensional object (see Figure 3.1): After perspectively projecting a three-dimensional object onto an image plane π in projective three-space, an inverse parametrization $T_\pi^{-1} : \pi \to RP^2$ (interpreted—via standard homogeneous coordinates—as a map onto the projective plane) transforms the image of the object—an image embedded in π—to an image of the object that can be displayed on a viewing device. This is pursued further in Chapter 5.

The converse of Proposition 2.7 is also true.

2.8 Proposition. If $T_\pi : RP^2 \to \pi$ and $T_\mu : RP^2 \to \mu$ are parametrizations of the planes π and μ in projective three-space, then for every 3×3 nonsingular matrix M, the transformation $T : \pi \to \mu$ given by

$$T([\mathbf{p}]) = [\mathbf{p} M_\pi^+ M M_\mu]$$

is a projective transformation.

Again, as an immediate consequence of Proposition 2.8, we have the following important result.

2.9 Theorem. (The Fundamental Theorem of Projective Geometry for Planes) Let π and μ be planes in projective three-space. If A, B, C, and D are four distinct points, no three of which are collinear, on π, and P, Q, R, and S are four distinct points, no three of which are collinear, on μ, then there is one and only one projective transformation taking π to μ and A to P, B to Q, C to R, and D to S.

The following example illustrates one way in which the Fundamental Theorem of Projective Geometry for Planes is used in the sequel.

2.10 Example. Given two sets of four points in the projective plane $A[1, 2, 1]$, $B[0, 1, 2]$, $C[-1, 4, 2]$, $D[1, 1, 2]$ and $P[0, 1, 1]$, $Q[1, 1, 0]$, $R[1, 1, 1]$, $S[0, 2, 1]$ such that no three in either set are collinear, the (intrinsically described) projective transformation T taking A to P, B to Q, C to R, and D to S is obtained as follows. Since

$$\mathbf{d} = \langle 3, 3, 6 \rangle = \langle 2, 4, 2 \rangle + \langle 0, 3, 6 \rangle + \langle 1, -4, -2 \rangle = \mathbf{a} + \mathbf{b} + \mathbf{c}$$

right multiplication by the matrix

$$\begin{pmatrix} 2 & 4 & 2 \\ 0 & 3 & 6 \\ 1 & -4 & -2 \end{pmatrix}$$

defines a projective transformation T_1 of the projective plane for which $T_1(I_x) = T_1([1, 0, 0]) = A$, $T_1(I_y) = T_1([0, 1, 0]) = B$, $T_1(O) = T_1([0, 0, 1]) = C$, and $T_1(U) = T_1([1, 1, 1]) = D$ (I_x denotes the ideal point in the x-direction, I_y the ideal point in the y-direction, O the origin, and U the unit point). The inverse T_1^{-1} of T_1 is represented by

$$M_1 = \begin{pmatrix} 6 & 0 & 6 \\ 2 & -2 & -4 \\ -1 & 4 & 2 \end{pmatrix}$$

and $T_1^{-1}(A) = I_x$, $T_1^{-1}(B) = I_y$, $T_1^{-1}(C) = O$, and $T_1^{-1}(D) = U$ (see Figure 3.18). Since

$$\mathbf{s} = \langle 0, 2, 1 \rangle = \langle 0, 2, 2 \rangle + \langle 1, 1, 0 \rangle + \langle -1, -1, -1 \rangle = \mathbf{p} + \mathbf{q} + \mathbf{r}$$

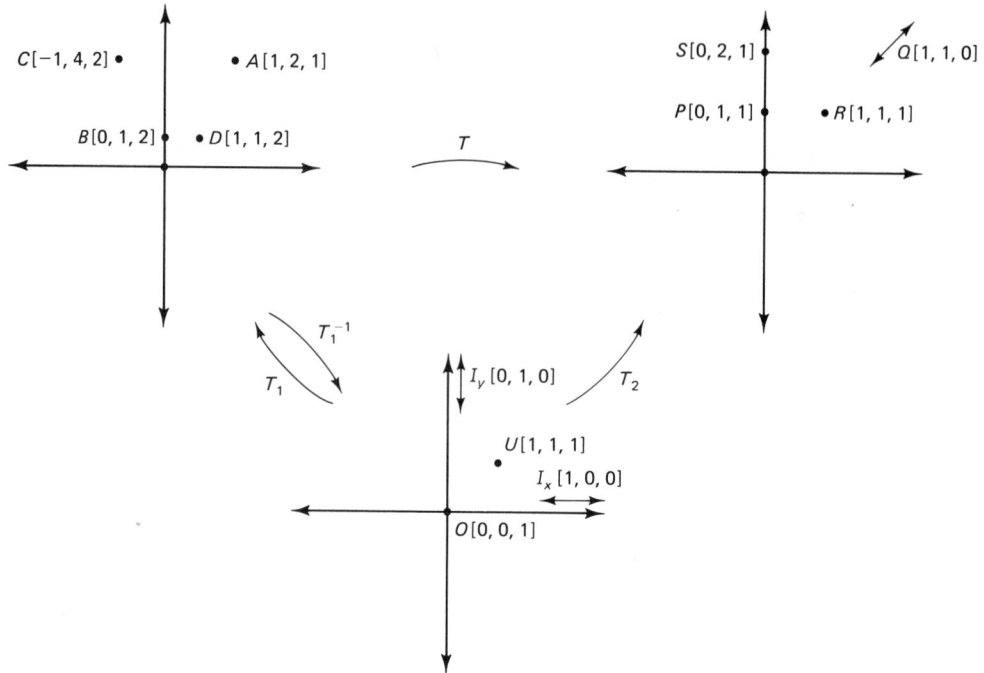

Figure 3.18 Application of the Fundamental Theorem of Projective Geometry for Planes.

right multiplication by

$$M_2 = \begin{pmatrix} 0 & 2 & 2 \\ 1 & 1 & 0 \\ -1 & -1 & -1 \end{pmatrix}$$

defines a projective transformation T_2 for which $T_2(I_x) = P$, $T_2(I_y) = Q$, $T_2(O) = R$, and $T_2(U) = S$. Thus if T is the composite transformation $T = T_2 \circ T_1$, which is represented by the matrix

$$M_1 M_2 = \begin{pmatrix} 6 & 0 & 6 \\ 2 & -2 & -4 \\ -1 & 4 & 2 \end{pmatrix} \begin{pmatrix} 0 & 2 & 2 \\ 1 & 1 & 0 \\ -1 & -1 & -1 \end{pmatrix} = \begin{pmatrix} -6 & 6 & 6 \\ 2 & 6 & 8 \\ 2 & 0 & -4 \end{pmatrix}$$

then $T(A) = P$, $T(B) = Q$, $T(C) = R$, and $T(D) = S$.

SECTION 2 EXERCISES:

1. In each of the following cases, find the 4×4 rank 3 matrix representing the perspectivity $T : \pi \to \mu$ whose center is E.
   (a) $\pi[1, 1, 1, -1]$, $\mu[0, 0, 1, -1]$, $E[0, 0, 0, 1]$
   (b) $\pi[6, -3, 2, -6]$, $\mu[0, 0, 1, 0]$, $E[0, 0, 10, 1]$

(c) $\pi[0, 0, 1, -1]$, $\mu[1, 1, 1, -4]$, $E[6, 6, 6, 1]$
(d) $\pi[0, 1, 2, -2]$, $\mu[0, 0, 1, -1]$, $E[0, 0, 1, 0]$

2. Verify the following technique for computing the matrix representing a perspectivity given in Proposition 2.1: If representatives μ and $\mathbf{e}$ of $[\mu]$ and $[\mathbf{e}]$, respectively, are chosen so that $\mathbf{e}^{1\times 4}\mu^{4\times 1} = 1$, then
$$M = I^{4\times 4} - \mu^{4\times 1}\mathbf{e}^{1\times 4}$$

3. In each of the following cases, find the 4×4 rank 3 matrix representing the composite $T = T_2 \circ T_1 : \pi \to \nu$ of the perspectivities $T_1 : \pi \to \mu$, whose center is E_1, and $T_2 : \mu \to \nu$, whose center is E_2.
    (a) $\pi[6, 3, 2, -1]$, $\mu[0, 1, 1, -4]$, $\nu[0, 0, 1, -1]$, $E_1[0, 2, 2, -1]$, $E_2[0, 0, 0, 1]$
    (b) $\pi[1, 1, -1, 1]$, $\mu[3, 2, 6, -6]$, $\nu[0, 0, 1, 0]$, $E_1[2, 3, -1, -1]$, $E_2[0, 0, 12, 1]$
    (c) $\pi[1, 1, -2, 4]$, $\mu[0, 0, 1, 0]$, $\nu[1, 1, 1, -1]$, $E_1[0, 0, 1, 0]$, $E_2[2, 2, 2, 1]$
    (d) $\pi[3, -2, -6, 6]$, $\mu[0, 0, 1, 0]$, $\nu[1, 0, 0, -2]$, $E_1[-4, 4, 4, 1]$, $E_2[3, 0, 4, 1]$

4. In each of the following cases, four coplanar points P, Q, R, and S in projective three-space are given. In each case verify that no three are collinear. If, in each case, π denotes the plane determined by P, Q, R, and S, find a matrix representing the parametrization $T : RP^2 \to \pi$ of π for which $T([1, 0, 0]) = P$, $T([0, 1, 0]) = Q$, $T([0, 0, 1]) = R$, and $T([1, 1, 1]) = S$.
    (a) $P[1, 2, 0, 1]$, $Q[3, 2, 1, 1]$, $R[0, 1, 2, 2]$, $S[1, -1, 0, -1]$
    (b) $P[1, 2, 3, 2]$, $Q[-2, 1, -1, 3]$, $R[0, 1, 1, -1]$, $S[0, 2, 2, 1]$
    (c) $P[2, 1, 3, 2]$, $Q[0, 0, 1, 0]$, $R[1, 0, -2, 1]$, $S[3, 1, 3, 3]$
    (d) $P[0, 2, 1, 2]$, $Q[1, 2, 3, 4]$, $R[0, 1, 1, 0]$, $S[1, -3, 0, 0]$

5. For each part of Exercise 1, matrices parametrizing the planes π and μ are given below. With respect to these parametrizations, find the 3×3 nonsingular matrix representing each transformation intrinsically.

(a) $M_\pi = \begin{pmatrix} 1 & 0 & 0 & 1 \\ 0 & 1 & 0 & 1 \\ 0 & 0 & 1 & 1 \end{pmatrix}$, $M_\mu = \begin{pmatrix} 1 & 0 & 0 & 0 \\ 0 & 1 & 0 & 0 \\ 0 & 0 & 1 & 1 \end{pmatrix}$

(b) $M_\pi = \begin{pmatrix} 1 & 0 & 0 & 1 \\ 1 & 2 & 0 & 0 \\ 0 & 0 & 3 & 1 \end{pmatrix}$, $M_\mu = \begin{pmatrix} 1 & 0 & 0 & 0 \\ 0 & 1 & 0 & 0 \\ 0 & 0 & 0 & 1 \end{pmatrix}$

(c) $M_\pi = \begin{pmatrix} 1 & 0 & 0 & 0 \\ 0 & 1 & 0 & 0 \\ 0 & 0 & 1 & 1 \end{pmatrix}$, $M_\mu = \begin{pmatrix} 1 & 0 & -1 & 0 \\ 0 & 1 & -1 & 0 \\ 2 & 1 & 1 & 1 \end{pmatrix}$

(d) $M_\pi = \begin{pmatrix} 0 & 0 & 1 & 1 \\ 0 & 2 & 0 & 1 \\ 1 & 0 & 0 & 0 \end{pmatrix}$, $M_\mu = \begin{pmatrix} 1 & 0 & 0 & 0 \\ 0 & 1 & 0 & 0 \\ 0 & 0 & 1 & 1 \end{pmatrix}$

6. For each part of Exercise 5, a 3×3 nonsingular matrix is given below; this matrix represents a projective transformation intrinsically. Find the 4×4 rank 3 matrix representing this projective transformation extrinsically.

(a) $\begin{pmatrix} 1 & -1 & 0 \\ 0 & 0 & 1 \\ 0 & 1 & 0 \end{pmatrix}$ (b) $\begin{pmatrix} 2 & 1 & 0 \\ 1 & 0 & 0 \\ 0 & 0 & 1 \end{pmatrix}$

(c) $\begin{pmatrix} 1 & 0 & 0 \\ 0 & -2 & 0 \\ 0 & 0 & 3 \end{pmatrix}$ (d) $\begin{pmatrix} 2 & 0 & 1 \\ 1 & 0 & 1 \\ 0 & 1 & 0 \end{pmatrix}$

7. In each of the following cases, two sets of four coplanar points in projective three-space are given. In each case show that no three in either set are collinear, and find the 4 × 4 rank 3 matrix M representing the projective transformation $T: \pi \to \mu$ from the plane π of $A, B, C,$ and D to the plane μ of $P, Q, R,$ and S, taking A to P, B to Q, C to R, and D to S.
   (a) $A[1, 2, 1, 0], B[0, 1, 0, 1], C[1, 0, 2, 0], D[2, 3, 3, 1]$ and $P[1, 0, 0, 0], Q[2, 1, 0, 1], R[0, 1, 2, 1], S[2, 1, 1, 1]$
   (b) $A[2, 0, 1, 0], B[1, 0, 0, 0], C[0, 2, 1, 1], D[3, 2, 2, 1]$ and $P[-2, 1, 2, 2], Q[3, 0, 1, 2], R[1, 1, 1, 0], S[1, 1, 2, 2]$
   (c) $A[1, 0, 0, 0], B[0, 1, 0, 0], C[0, 0, 1, 1], D[1, 1, 1, 1]$ and $P[0, 1, 0, 0], Q[1, 0, 0, 0], R[0, 0, 1, 1], S[1, 1, 1, 1]$
   (d) $A[1, 0, 0, 0], B[0, 1, 0, 0], C[0, 0, 1, 1], D[1, 1, 1, 1]$ and $P[3, 2, 1, 2], Q[-3, 0, 1, 0], R[0, 1, 0, 0], S[0, 0, 1, 1]$

8. In each of the following cases, two sets of four coplanar points in the projective plane are given. In each case show that no three in either set are collinear, and find the 3 × 3 nonsingular matrix M representing the projective transformation T taking A to P, B to Q, C to R, and D to S.
   (a) $A[1, 2, 1], B[0, 1, 0], C[1, 0, 2], D[2, 3, 3]$ and $P[1, 0, 0], Q[2, 1, 0], R[0, 1, 2], S[2, 1, 1]$
   (b) $A[2, 0, 1], B[1, 0, 0], C[0, 2, 1], D[3, 2, 2]$ and $P[-2, 1, 2], Q[3, 0, 1], R[1, 1, 1], S[1, 1, 2]$
   (c) $A[1, 0, 0], B[0, 1, 0], C[0, 0, 1], D[1, 1, 1]$ and $P[0, 1, 0], Q[1, 0, 0], R[0, 0, 1], S[1, 1, 1]$
   (d) $A[1, 0, 0], B[0, 1, 0], C[0, 0, 1], D[1, 1, 1]$ and $P[3, 2, 1], Q[-3, 0, 1], R[0, 1, 0], S[0, 0, 1]$

†9. Show that the projective transformation of the projective plane represented intrinsically by

$$\begin{pmatrix} 1 & 0 & -1 \\ 0 & 1 & -1 \\ 0 & 0 & 1 \end{pmatrix}$$

takes $A[1, 0, 1], B[0, 1, 1],$ and $C[0, 0, 1],$ and $D[1, 1, 1]$ to $P[1, 0, 0], Q[0, 1, 0], R[0, 0, 1],$ and $S[1, 1, 1]$, respectively. Generalize this result to projective three-space.

†10. Suppose T is a projective transformation of the projective plane represented intrinsically by the 3 × 3 nonsingular matrix M:

$$T([p_1, p_2, p_3]) = [\langle p_1, p_2, p_3 \rangle M^{3 \times 3}]$$

As a consequence of the definition of projective transformation, T takes lines to lines. Show that the coordinates of the image $T(l)$ of the line $l[\mathbf{l}]$ under T are $[M^{-1}\mathbf{l}]$. (Hint: Use Exercise 2.2.20 and the fact that $\mathbf{pl} = \mathbf{p}MM^{-1}\mathbf{l}$.)

11. Use Exercise 10 to find the coordinates of each of the following lines under the projective transformation T represented intrinsically by the corresponding 3 × 3 nonsingular matrix M.
   (a) $l[3, -2, 1], m[2, 0, -3], M = \begin{pmatrix} 1 & 0 & 1 \\ 0 & 1 & 1 \\ 1 & 0 & 0 \end{pmatrix}$

(b) $l[1, 5, 2]$, $m[1, 5, 3]$, $M = \begin{pmatrix} \cos 30° & \sin 30° & 0 \\ -\sin 30° & \cos 30° & 0 \\ 0 & 0 & 1 \end{pmatrix}$

(c) $l[2, 3, 1]$, $m[2, 3, 4]$, $M = \begin{pmatrix} 1 & 0 & 0 \\ 0 & 1 & 0 \\ 3 & 5 & 1 \end{pmatrix}$

(d) $l[1, 3, 1]$, $m[1, 3, 3]$, $M = \begin{pmatrix} 2 & 0 & 0 \\ 0 & 3 & 0 \\ 0 & 0 & 1 \end{pmatrix}$

(e) $l[2, 1, 1]$, $m[2, 1, 0]$, $M = \begin{pmatrix} 1 & 2 & 0 \\ 0 & 1 & 0 \\ 0 & 0 & 1 \end{pmatrix}$

(f) $l[1, 2, 3]$, $m[1, 2, 1]$, $M = \begin{pmatrix} 1 & 0 & 2 \\ 0 & 1 & 3 \\ 0 & 0 & 1 \end{pmatrix}$

12. Do the following for each part of Exercise 11.
    (a) Find the point of intersection of l and m.
    (b) Find the image under T of the point of intersection of l and m.
    (c) Find the point of intersection of $T(l)$ and $T(m)$.
    (*Note:* In general the image under T of the intersection of l and m is the intersection of $T(l)$ and $T(m)$.)

13. Show that the point of intersection of the Euclidean lines $m_i[a_i, b_i, c_i]$, $i = 1, 2$, in the projective plane is given by $[\langle 0, 0, 1\rangle M^{-1}]$, where

$$M = \begin{pmatrix} a_1 & a_2 & 0 \\ b_1 & b_2 & 0 \\ c_1 & c_2 & 1 \end{pmatrix}$$

(*Hint:* Use Exercise 10 to show that right multiplication by M^{-1} is a projective transformation of the projective plane which takes the coordinate lines $l_1[1, 0, 0]$, $l_2[0, 1, 0]$, and $l_3[0, 0, 1]$ to $m_1[a_1, b_1, c_1]$, $m_2[a_2, b_2, c_2]$, and $m_3[0, 0, 1]$, respectively. The image of what point under this transformation is the point of intersection of m_1 and m_2?)

14. Use Exercise 13 to find the coordinates of the point of intersection of the Euclidean lines given in Exercise 11.

15. State and verify the dual of the result described in Exercise 13.

16. In each of the following cases, use Exercise 15 to find the coordinates of the line determined by the given Euclidean points.
    (a) $A[1, 2, 1]$, $B[2, 3, 1]$
    (b) $A[2, 1, 3]$, $B[1, 2, 1]$
    (c) $A[5, 2, 1]$, $B[-1, 1, 3]$
    (d) $A[2, 1, 2]$, $B[1, 3, 5]$

17. Right multiplication by a 4×4 invertible matrix M represents (intrinsically) a projective transformation of projective three-space:

$$T([p_1, p_2, p_3, p_4]) = [\langle p_1, p_2, p_3, p_4\rangle M^{4\times 4}]$$

Show that the coordinates of the image $T(\pi)$ of the plane $\pi[\boldsymbol{\pi}]$ under T are $[M^{-1}\boldsymbol{\pi}]$.

18. Show that the point of intersection of the Euclidean planes $\mu_i[a_i, b_i, c_i, d_i]$, $i = 1$, 2, 3, in the projective plane is given by $[\langle 0, 0, 0, 1\rangle M^{-1}]$, where

Sec. 2   Transformations of Projective Planes

$$M = \begin{pmatrix} a_1 & a_2 & a_3 & 0 \\ b_1 & b_2 & b_3 & 0 \\ c_1 & c_2 & c_3 & 0 \\ d_1 & d_2 & d_3 & 1 \end{pmatrix}$$

19. In each of the following cases, use Exercise 18 to find the point of intersection of the Euclidean planes.
    (a) $x + y + z = 0$, $2x - y + z = -3$, $x - 2y + z = -6$
    (b) $y + z = 0$, $x - y + 2z = 3$, $2x + 3y - 4z = 3$
    (c) $2x + y - z = 1$, $x - y + z = 1$, $x + 2y - 3z = 1$
    (d) $x + y - 3z = 1$, $2x + y + z = 4$, $3x - y + 4z = -4$

20. State and verify the dual of the result described in Exercise 18.

21. In each of the following cases, use Exercise 20 to find the equation of the plane determined by the Euclidean points.
    (a) $A(0, 1, 1)$, $B(4, 0, -1)$, $C(-3, 1, 2)$
    (b) $A(1, 1, 2)$, $B(0, 3, -1)$, $C(2, -1, 2)$
    (c) $A(2, 3, 1)$, $B(0, 1, 1)$, $C(-1, 1, 2)$
    (d) $A(0, 0, 1)$, $B(1, 1, 0)$, $C(2, -2, -3)$

†22. (The Fundamental Theorem of Projective Geometry for Three-Space) Show that if A, B, C, D, and E are five points in projective three-space, no four of which are coplanar, and P, Q, R, S, and T are five points in projective three-space, no four of which are coplanar, then there is one and only one projective transformation taking A to P, B to Q, C to R, D to S, and E to T.

23. In each of the following cases, two sets of five points in projective three-space are given. In each case show that no four points in either set are coplanar, and find the invertible 4×4 matrix M representing the projective transformation taking A to P, B to Q, C to R, D to S, and E to T.
    (a) $A[0, 1, 1, 1]$, $B[1, 0, 1, 1]$, $C[1, 1, 0, 1]$, $D[1, 1, 1, 0]$, $E[1, 1, 1, 1]$ and $P[1, 0, 0, 0]$, $Q[1, 1, 0, 0]$, $R[1, 0, 1, 0]$, $S[1, 0, 0, 1]$, $T[1, 2, 3, 4]$
    (b) $A[1, 0, 0, 0]$, $B[1, 1, 0, 0]$, $C[0, 1, 1, 0]$, $D[0, 0, 1, 1]$, $E[0, 0, 0, 1]$ and $P[2, -1, 0, 0]$, $Q[-1, 2, -1, 0]$, $R[0, -1, 2, -1]$, $S[0, 0, -1, 2]$, $T[0, 0, 0, 1]$
    (c) $A[1, 0, 0, 0]$, $B[1, 1, 0, 0]$, $C[1, 1, 1, 0]$, $D[1, 1, 1, 1]$, $E[2, 5, 3, 1]$ and $P[2, 1, 0, 0]$, $Q[1, 2, 1, 0]$, $R[0, 1, 2, 1]$, $S[0, 0, 1, 2]$, $T[3, 2, -3, -1]$
    (d) $A[2, 1, 0, 0]$, $B[1, 2, 1, 0]$, $C[0, 1, 2, 1]$, $D[0, 0, 1, 2]$, $E[1, 2, 3, 4]$ and $P[1, -1, -3, -1]$, $Q[1, 2, 0, -1]$, $R[3, 4, -2, 0]$, $S[1, 0, -2, 3]$, $T[1, 1, -1, -6]$

SECTION 3: THE GEOMETRY OF PROJECTIVE TRANSFORMATIONS

The goal of this section is to obtain a better understanding of the geometry of projective transformations of the projective plane and projective three-space. First, however, we consider affine coordinates.

3.1 Definition. If P is a point in the Euclidean plane whose Cartesian coordinates are (x, y), then the *affine coordinates* of P are $(x, y, 1)$, the triple whose first two entries are the Cartesian coordinates of P and whose third entry is 1.

Affine coordinates arose earlier in the discussion of homogeneous coordinates (see Proposition 2.1.4 and the remark following Definition 2.1.5), although they were not then identified in any special way. Once they are identified, however, their importance can also be identified: The projective plane is an extension of the Euclidean plane in the same way homogeneous coordinates are an extension of affine coordinates. (Affine coordinates can be defined on the Euclidean line and on Euclidean three-space as well as on the Euclidean plane.)

Affine coordinates are important for another reason that has nothing to do with projective geometry: The Euclidean transformations (rotations, translations, and reflections) of the Euclidean plane and Euclidean three-space can all be represented by matrix multiplication if affine coordinates are used. This is not true if Cartesian coordinates are used. Thus affine coordinates lend themselves to the development of transformation formulas more readily than Cartesian coordinates do.

In this section we see that all projective transformations of the projective plane (and projective three-space) are composites of two types of transformations: extensions to the projective plane (and projective three-space) of transformations of the Euclidean plane (and Euclidean three-space) represented in affine coordinates, and transformations known as perspective projections. We study both types of transformations in this section. In our study of the first type of transformation we use Euclidean concepts such as the measure of a segment, the measure of an angle, parallelism, and direction. Although these are *not* projective concepts (see Section 1.1) they become projective concepts once a decomposition of RP^2 (and RP^3) of the form given in Proposition 2.1.4 has been identified; that is, once a Euclidean subspace of the projective plane (and projective three-space) has been identified. (Different decompositions determine different geometric interpretations of these transformations.)

We begin by discussing projective transformations of the projective plane.

Test patterns, such as the one illustrated in Figure 3.19, play an important role in our analysis. A (two-dimensional) *test pattern* consists of a finite set of points in the projective plane together with a set of line segments determined

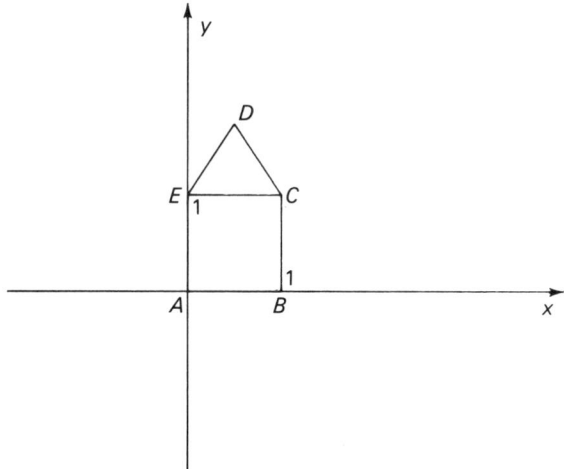

Figure 3.19 A two-dimensional test pattern.

by these points. (Even though segment is not a projective concept, it is, in a restricted sense, still a graphics concept. More about this in Section 4.3.)

A test pattern with n vertices can be stored and reconstructed via two matrices: an $n \times 3$ *point matrix* and an $n \times n$ *incidence matrix*. Each row of the point matrix consists of the homogeneous coordinates of a vertex of the test pattern, and the incidence matrix is the matrix whose (i, j)th entry is 1 if there is a segment in the test pattern joining the ith and jth vertices and 0 otherwise. (This is not necessarily the best data structure with which to do graphics. It is used in this section just for simplicity.) The point and incidence matrices for the test pattern of Figure 3.19 are given in Figure 3.20.

$$
\begin{array}{c}
A \\ B \\ C \\ D \\ E
\end{array}
\begin{pmatrix}
0 & 0 & 1 \\
1 & 0 & 1 \\
1 & 1 & 1 \\
\frac{1}{2} & \frac{1+\sqrt{3}}{2} & 1 \\
0 & 1 & 1
\end{pmatrix}
\qquad
\begin{array}{c} \\ A \\ B \\ C \\ D \\ E \end{array}
\begin{array}{c} A \quad B \quad C \quad D \quad E \end{array}
\begin{pmatrix}
1 & 1 & 0 & 0 & 1 \\
1 & 1 & 1 & 0 & 0 \\
0 & 1 & 1 & 1 & 1 \\
0 & 0 & 1 & 1 & 1 \\
1 & 0 & 1 & 1 & 1
\end{pmatrix}
$$

(a) \qquad\qquad (b)

Figure 3.20 (a) The point and (b) the incidence matrices for the test pattern.

The reason test patterns are important is that we can illustrate the geometry of a transformation by comparing the image of the test pattern under the transformation to the original test pattern.

3.2 Example. Let T be the transformation

$$T([p_1, p_2, p_3]) = \left[\langle p_1, p_2, p_3\rangle \begin{pmatrix} \cos\theta & \sin\theta & 0 \\ -\sin\theta & \cos\theta & 0 \\ 0 & 0 & 1 \end{pmatrix}\right]$$

To demonstrate the geometry of T, we apply T to the vertices of the test pattern and then use the original incidence matrix to construct the test pattern image (see Figure 3.21). The transformation T represents counterclockwise *rotation* about the origin through the angle θ. In Cartesian coordinates, rotation is represented by right multiplication by the matrix

$$N = \begin{pmatrix} \cos\theta & \sin\theta \\ -\sin\theta & \cos\theta \end{pmatrix}$$

and N is precisely the upper left-hand 2×2 submatrix of the matrix that represents T. This is typical: Any linear transformation of the Euclidean plane represented in Cartesian coordinates as right multiplication by a 2×2 nonsingular matrix N can be extended to a projective transformation of the projective plane. This projective transformation is represented intrinsically by the 3×3 nonsingular matrix obtained by replacing the upper left hand 2×2 submatrix of the 3×3 identity matrix by N:

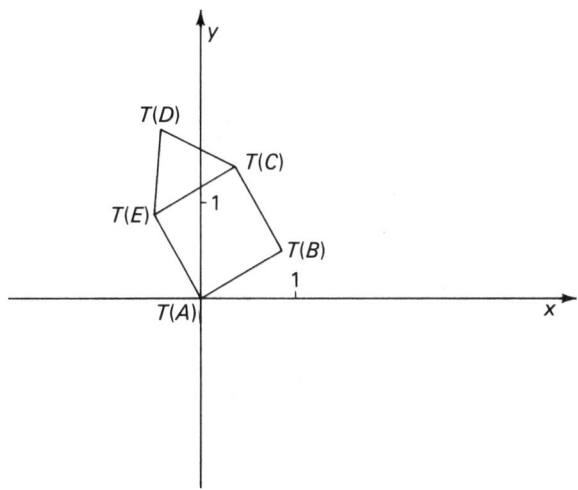

Figure 3.21 Rotation ($\theta = 30°$).

$$\left(\begin{array}{c|c} N & \begin{array}{c} 0 \\ 0 \end{array} \\ \hline 00 & 1 \end{array}\right)$$

3.3 Example. The transformation

$$T([p_1, p_2, p_3]) = \left[\langle p_1, p_2, p_3\rangle \begin{pmatrix} 1 & 0 & 0 \\ 0 & 1 & 0 \\ h & k & 1 \end{pmatrix}\right]$$

is the *translation* that takes the origin $O[0, 0, 1]$ to the point $TO = TO[h, k, 1]$ (see Figure 3.22). (Translation is a Euclidean transformation of the Euclidean

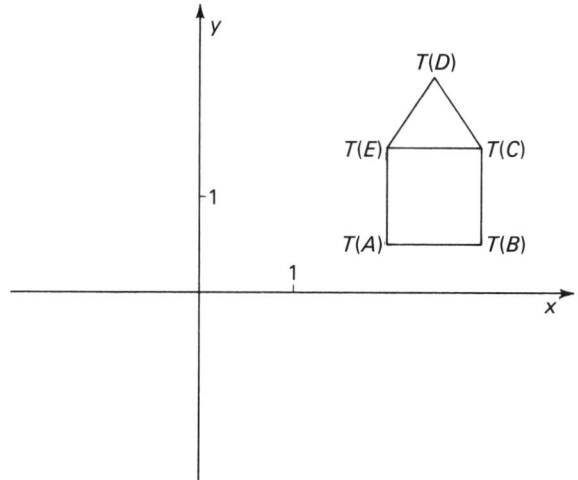

Figure 3.22 Translation ($h = 2$, $k = 1/2$).

Sec. 3   The Geometry of Projective Transformations

plane. Translation cannot be represented by matrix multiplication in Cartesian coordinates, but it can be represented by matrix multiplication in affine coordinates. See Exercise 1.)

3.4 Example. The transformation

$$T([p_1, p_2, p_3]) = \left[\langle p_1, p_2, p_3 \rangle \begin{pmatrix} m_x & 0 & 0 \\ 0 & m_y & 0 \\ 0 & 0 & m_z \end{pmatrix} \right]$$

where m_x, m_y and m_z are nonzero real constants, is *scaling* by a factor of m_x/m_z in the x-direction and by a factor of m_y/m_z in the y-direction (see Figure 3.23); T is a positive scaling in the x- (y-) direction if m_x/m_z (m_y/m_z) is positive, a positive scaling with a reflection in the x- (y-) direction if m_x/m_z (m_y/m_z) is negative, and a positive scaling with reflection in the origin if both m_x/m_z and m_y/m_z are negative. Scaling by m_x and m_y is *local scaling*, and scaling by $1/m_z$ is *global scaling*.

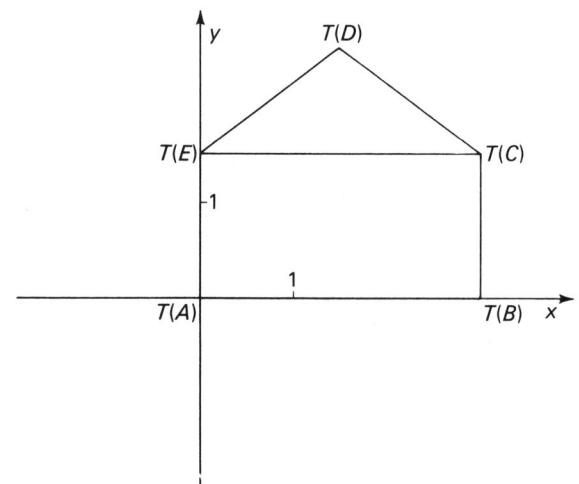

Figure 3.23 Composite scaling ($m_x = 6$, $m_y = 3$, $m_z = 2$).

3.5 Example. The transformation

$$T([p_1, p_2, p_3]) = \left[\langle p_1, p_2, p_3 \rangle \begin{pmatrix} 1 & a & 0 \\ 0 & 1 & 0 \\ 0 & 0 & 1 \end{pmatrix} \right]$$

is *shearing* by a in the y direction (see Figure 3.24(a)). If the projective plane is thought of as an infinite number of thin and rigid bands parallel to the y-axis, then shearing by a in the y direction displaces the band x units from the y-axis along itself through the distance ax. The transformation

$$T([p_1, p_2, p_3]) = \left[\langle p_1, p_2, p_3 \rangle \begin{pmatrix} 1 & 0 & 0 \\ b & 1 & 0 \\ 0 & 0 & 1 \end{pmatrix} \right]$$

is shearing by b in the x direction (see Figure 3.24(b)).

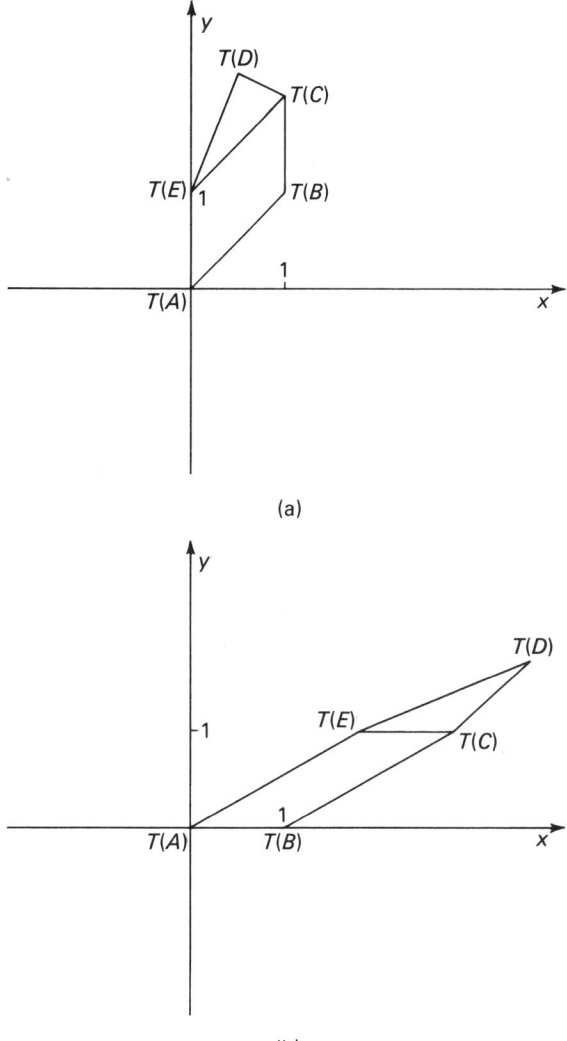

Figure 3.24 Shearing. (a) $a = 1$. (b) $b = 1.8$.

The preceding transformations are familiar from Euclidean geometry. It is natural to ask if there are any other projective transformations. There is just one more type. This type of projective transformation is important since a projective transformation of the type illustrated in Figure 1.12 (see also Figures 1.57 and 1.58), when described intrinsically, will be of this type.

3.6 Example. The transformation

$$T([p_1, p_2, p_3]) = \left[\langle p_1, p_2, p_3 \rangle \begin{pmatrix} 1 & 0 & a \\ 0 & 1 & b \\ 0 & 0 & 1 \end{pmatrix} \right]$$

where a and b are real constants, is a *perspective projection* (see Figure 3.25; the test pattern has been translated for purposes of illustration.)

Sec. 3   The Geometry of Projective Transformations

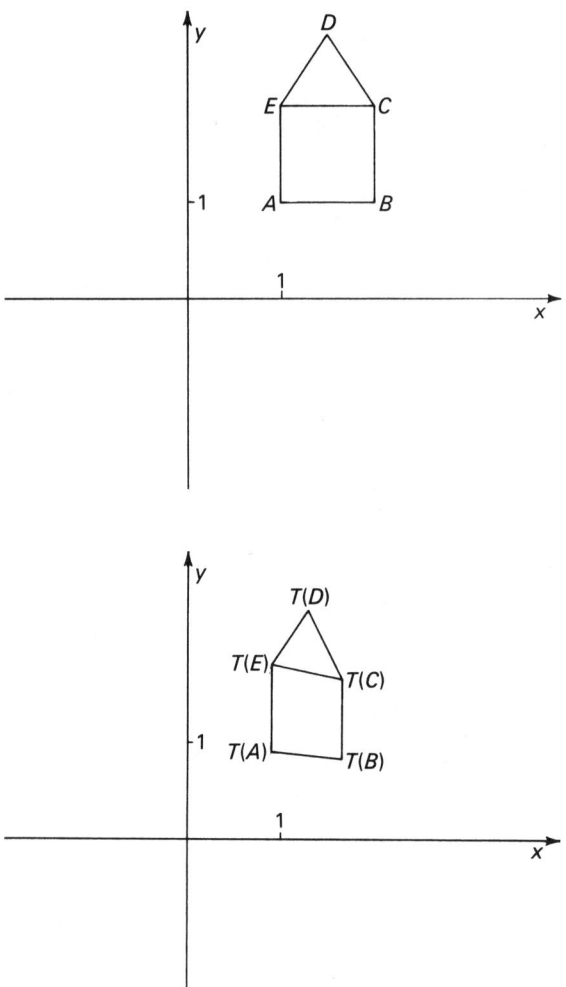

Figure 3.25 Perspective transformation ($a = 0.1$, $b = 0$) after an initial translation ($h = 1$, $k = 1$).

Geometrically this transformation is the composite (see Figure 3.26) of the projection $T_1 : \pi \to \mu$ parallel to the z-axis from the standard embedded projective plane $\pi[0, 0, 1, -1]$ to the projective plane $\mu[a, b, -1, 1]$ (a perspectivity whose center is $E[0, 0, 1, 0]$), followed by the perspectivity $T_2 : \mu \to \pi$ from μ to π, whose center is the origin $O[0, 0, 0, 1]$. This can be seen, on one hand, by observing that since

$$T([p_1, p_2, 1]) = [p_1, p_2, ap_1 + bp_2 + 1]$$

T is an extension to the standard embedded projective plane of the transformation from the standard embedded Euclidean plane to the standard embedded projective plane given by

$$(p_1, p_2, 1) \xrightarrow[\text{projection}]{\text{parallel}} (p_1, p_2, ap_1 + bp_2 + 1)$$

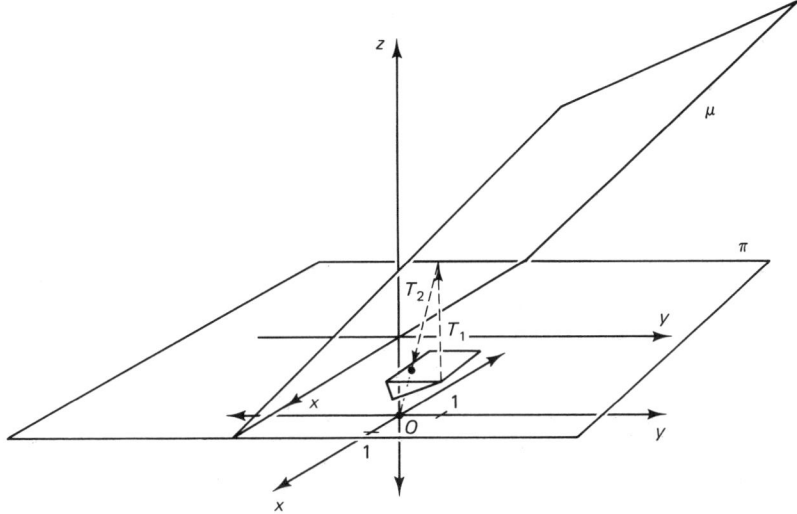

Figure 3.26 The geometry of perspective transformations.

$$\xrightarrow[\text{projection}]{\text{perspective}} [p_1, p_2, ap_1 + bp_2 + 1]$$

Alternately, if π is parametrized (see Proposition 2.5) by

$$M_\pi = \begin{pmatrix} 1 & 0 & 0 & 0 \\ 0 & 1 & 0 & 0 \\ 0 & 0 & 1 & 1 \end{pmatrix}$$

so that

$$M_\pi^+ = \begin{pmatrix} 1 & 0 & 0 \\ 0 & 1 & 0 \\ 0 & 0 & \frac{1}{2} \\ 0 & 0 & \frac{1}{2} \end{pmatrix}$$

then (see Proposition 2.1) T_1 and T_2 are represented by

$$M_1 = \begin{pmatrix} 1 & 0 & a & 0 \\ 0 & 1 & b & 0 \\ 0 & 0 & 0 & 0 \\ 0 & 0 & 1 & 1 \end{pmatrix} \quad \text{and} \quad M_2 = \begin{pmatrix} 1 & 0 & 0 & 0 \\ 0 & 1 & 0 & 0 \\ 0 & 0 & 1 & 1 \\ 0 & 0 & 0 & 0 \end{pmatrix}$$

respectively, and

$$T([p_1, p_2, p_3]) = [\langle p_1, p_2, p_3 \rangle M_\pi M_1 M_2 M_\pi^+]$$

$$= \left[\langle p_1, p_2, p_3 \rangle \begin{pmatrix} 1 & 0 & a \\ 0 & 1 & b \\ 0 & 0 & 1 \end{pmatrix} \right]$$

With perspective projection, as with global scaling, it is sometimes necessary to normalize the third component of the coordinates of each point given by the modified point matrix to 1 in order to plot the test pattern image. As the next example illustrates, it is thus possible for Euclidean points to be mapped to ideal points.

3.7 Example. If T is the perspective projection for which $a = 1$ and $b = 0$ (see Example 3.6) then the image of the test pattern illustrated in Figure 3.27(a) is illustrated in Figure 3.27(b). The image of the left-hand side of the test pattern is off at infinity. To explain this geometrically requires an extrinsic argument based on a somewhat complicated three-dimensional figure. To simplify the explanation, we drop our discussion down a dimension and reason by analogy.

Consider the perspective projection of the projective line l

$$T([p_1, p_2]) = \left[\langle p_1, p_2\rangle \begin{pmatrix} 1 & a \\ 0 & 1 \end{pmatrix}\right]$$

for some nonzero real constant a. Geometrically, this transformation is the composite (see Figure 3.28) of the projection $T_1 : l \to m$ parallel to the z-axis from the standard embedded projective line $l[0, 1, -1]$ to the projective line $m[a, -1, 1]$ (a perspectivity whose center is $E[0, 1, 0]$), followed by the perspectivity $T_2 : m \to l$ from m to l, whose center is the origin $O[0, 0, 1]$. Again, this can be seen by observing that since

$$T([p_1, 1]) = [p_1, ap_1 + 1]$$

T is an extension to the standard embedded projective line of the transformation from the standard embedded Euclidean line to the standard embedded projective line given by

$$(p_1, 1) \xrightarrow[\text{projection}]{\text{parallel}} (p_1, ap_1 + 1)$$
$$\xrightarrow[\text{projection}]{\text{perspective}} [p_1, ap_1 + 1]$$

Take as a test pattern the line segment illustrated in Figure 3.29(a), where the tic marks are evenly spaced and $a = 1$. The geometry of T is extrinsically illustrated in Figure 3.29(b). (The object and image have been displaced slightly from l for the purpose of artistic clarity.) Observe that since $T_1(A)$ lies on the x-axis, $T_2(T_1(A)) = T(A)$ is ideal.

It is also possible for perspective projections of the projective plane to map ideal points to Euclidean points; the result is that we obtain vanishing points.

3.8 Example. If T is the perspective projection for which $a = 0.5$ and $b = 0$ (see Example 3.6) then the image of the test pattern illustrated in Figure 3.27(a) is illustrated in Figure 3.30. Observe that the image of the ideal point whose homogeneous coordinates are $[1, 0, 0]$ under this transformation is the vanishing point whose homogeneous coordinates are $[2, 0, 1]$ and whose Cartesian coordinates are $(2, 0)$.

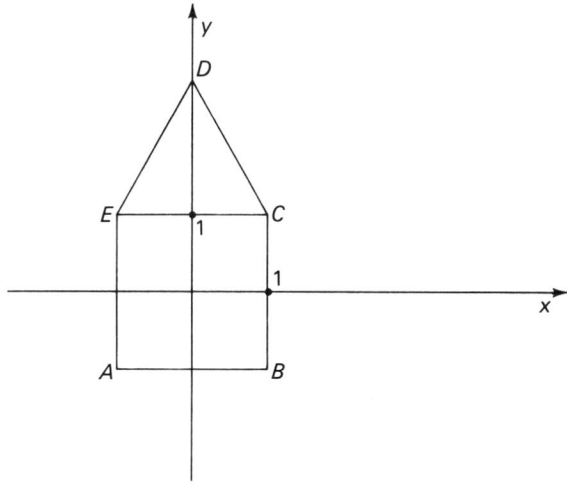

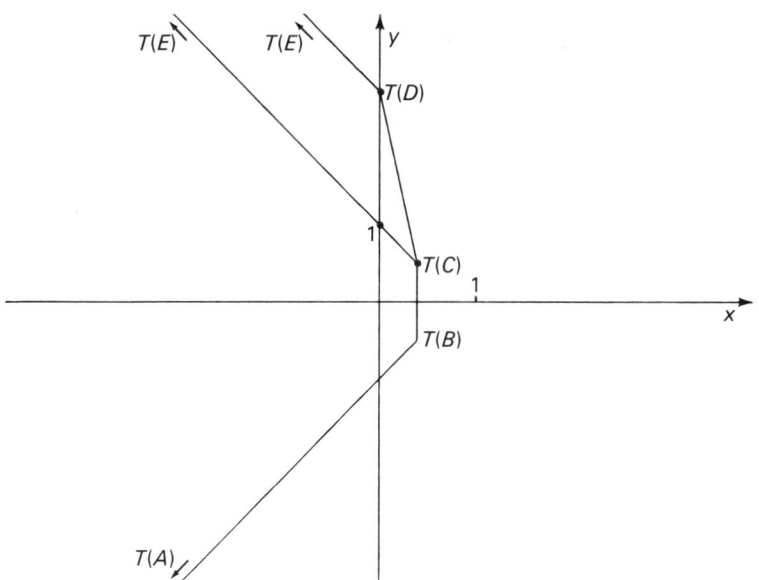

Figure 3.27 A perspective projection. (a) The test pattern. (b) The image.

Again, to explain this geometrically requires an extrinsic argument based on a somewhat complicated three-dimensional figure. Thus we again drop the discussion down a dimension and reason by analogy: Consider the perspective projection of the projective line for which $a = 0.5$. The geometry of this trans-

Sec. 3   The Geometry of Projective Transformations   **135**

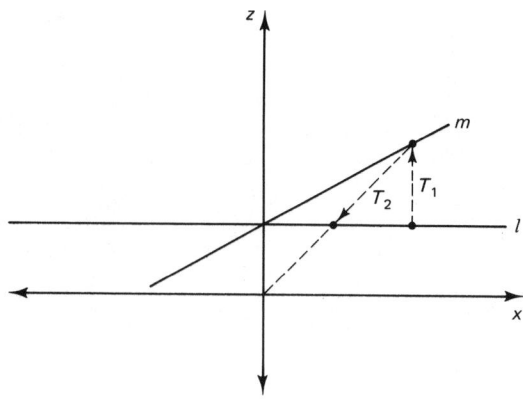

Figure 3.28 The extrinsic geometry of perspective transformations of lines.

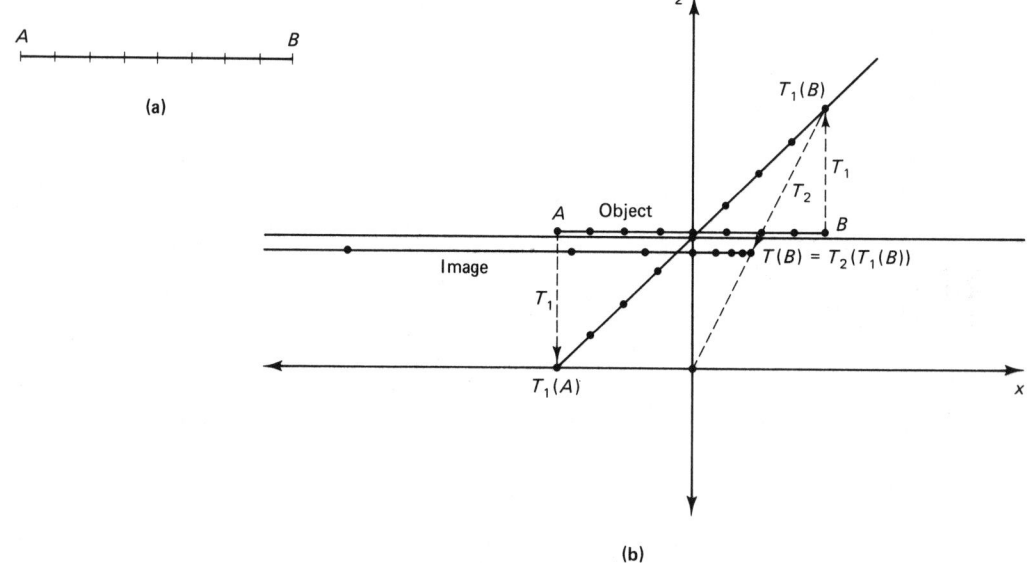

Figure 3.29 How Euclidean points become ideal.

formation is illustrated in Figure 3.31. Observe that the image $T_2([1, 0])$ of the ideal point of m is visible from the origin (that is, it is a Euclidean point on l).

To summarize, the projective transformations T of the projective plane represented by 3×3 nonsingular matrix multiplication

$$T([\mathbf{p}]) = [\mathbf{p}M^{3 \times 3}]$$

include rotation, translation, local scaling, global scaling, reflections, shearing,

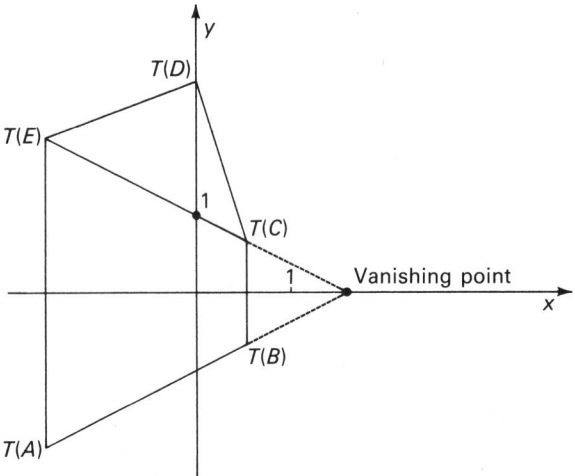

Figure 3.30 Another perspective projection.

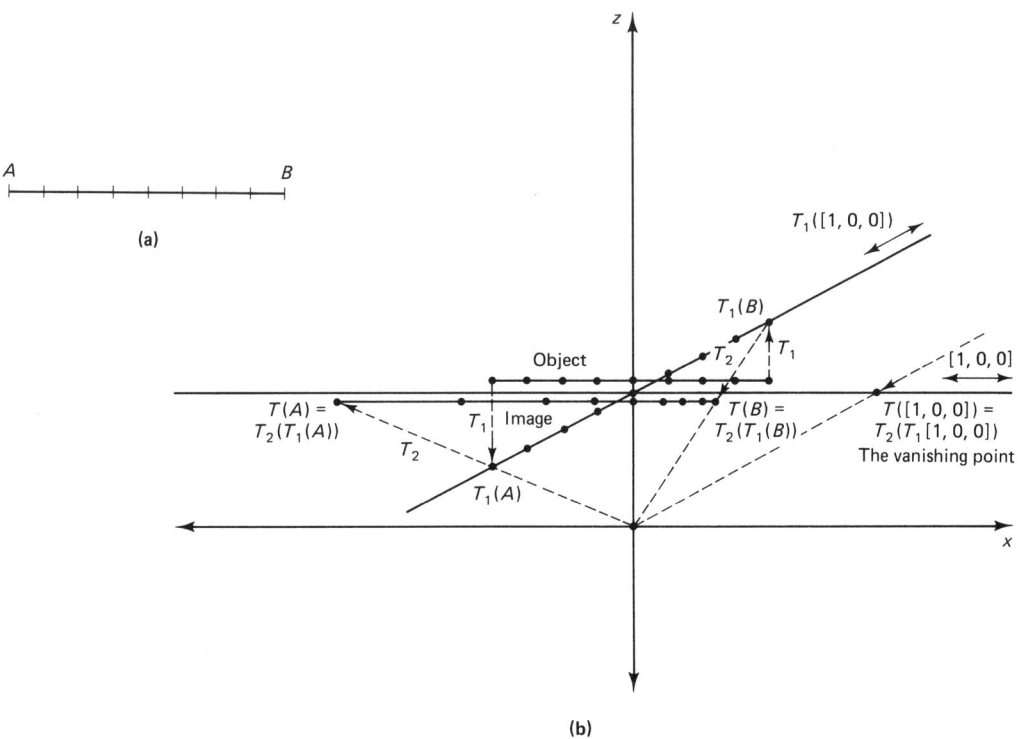

Figure 3.31 How vanishing points arise. (a) The test pattern. (b) The geometry of T extrinsically.

Sec. 3 The Geometry of Projective Transformations

perspective projection, and combinations (composites) of such transformations. These are all the possible projective transformations of the projective plane.

3.9 Proposition. All projective transformations T of the projective plane are, or can be written as composites of, rotations, translations, local scalings, global scalings, reflections, shearings, and perspective projections.

Proof. Consider the set of all nonsingular 3×3 matrices that are obtained by interchanging rows, interchanging columns, and replacing any single entry of the 3×3 identity matrix with a nonzero real constant (see Figure 3.32). Since this is the set of elementary row and column matrices, every nonsingular 3×3 matrix can be written as a product of matrices in this set. On the other hand, a matrix obtained by interchanging the rows or interchanging the columns of the 3×3 identity matrix represents a reflection, and if a nonzero entry is placed in

1. The (3, 1) or (3, 2) position of the 3×3 identity matrix, T is translation
2. The (1, 1) or (2, 2) position, T is local scaling
3. The (3, 3) position, T is global scaling
4. The (2, 1) or (1, 2) position, T is shearing
5. The (1, 3) or (2, 3) position, T is a perspective projection ∎

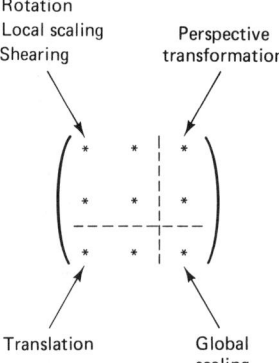

Figure 3.32 The components of nonsingular 3×3 matrix multiplication.

Observe that right multiplication by the matrix

$$M = \begin{pmatrix} 0 & 0 & 1 \\ 0 & 1 & 0 \\ 1 & 0 & 0 \end{pmatrix}$$

is treated, in this result, as a reflection. In particular, it is reflection in the ideal line, *not* a Euclidean reflection!

We now consider transformations of projective three-space. To illustrate the geometry of these transformations, we apply each to a three-dimensional test pattern (such as that in Figure 3.33) and reconstruct the image the same

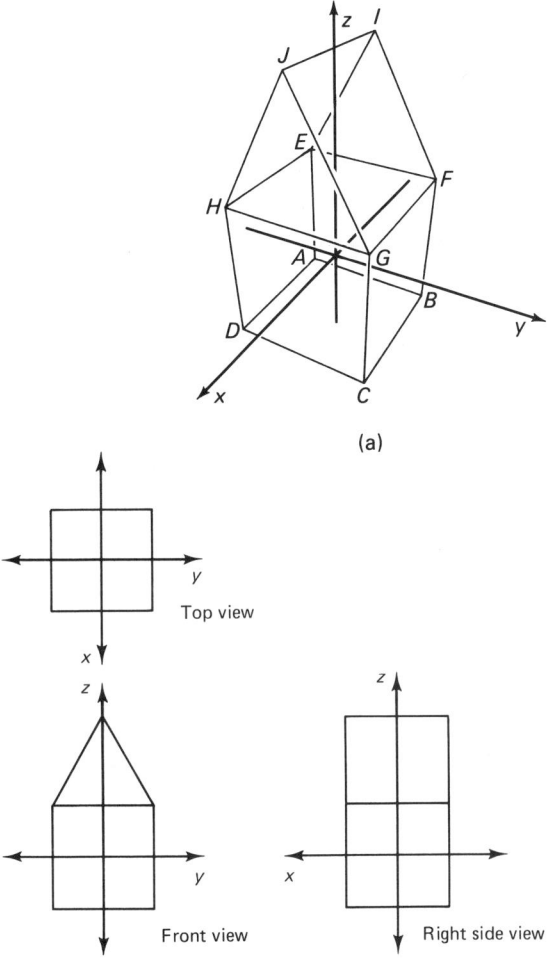

Figure 3.33 A three-dimensional test pattern. (a) A perspective view. (b) Multiview orthographic projection.

way we did when considering transformations of the projective plane. The point and incidence matrices for the test pattern of Figure 3.33 are given in Figure 3.34.

The test pattern and its image are three-dimensional, but illustrations are, of necessity, two-dimensional. Thus we must use parallel and perspective projections in the display of our illustrations. We do so here without further comment, postponing the discussion of how these figures are obtained until Section 5.4. Also, to improve the clarity of our illustrations, line segments have been hidden as necessary. Again, we postpone the discussion of how this is done until Section 5.4.

The general description of projective transformations of projective three-space is similar to that of projective transformations of the projective plane and

Sec. 3   The Geometry of Projective Transformations   139

$$
\begin{array}{c}
A \\ B \\ C \\ D \\ E \\ F \\ G \\ H \\ I \\ J
\end{array}
\begin{pmatrix}
-\frac{1}{2} & -\frac{1}{2} & -\frac{1}{2} & 1 \\
-\frac{1}{2} & \frac{1}{2} & -\frac{1}{2} & 1 \\
\frac{1}{2} & \frac{1}{2} & -\frac{1}{2} & 1 \\
\frac{1}{2} & -\frac{1}{2} & -\frac{1}{2} & 1 \\
-\frac{1}{2} & -\frac{1}{2} & \frac{1}{2} & 1 \\
-\frac{1}{2} & \frac{1}{2} & \frac{1}{2} & 1 \\
\frac{1}{2} & \frac{1}{2} & \frac{1}{2} & 1 \\
\frac{1}{2} & -\frac{1}{2} & \frac{1}{2} & 1 \\
-\frac{1}{2} & 0 & \frac{1+\sqrt{3}}{2} & 1 \\
-\frac{1}{2} & 0 & \frac{1+\sqrt{3}}{2} & 1
\end{pmatrix}
\qquad
\begin{array}{c} \\ A \\ B \\ C \\ D \\ E \\ F \\ G \\ H \\ I \\ J \end{array}
\begin{pmatrix}
A & B & C & D & E & F & G & H & I & J \\
1 & 1 & 0 & 1 & 1 & 0 & 0 & 0 & 0 & 0 \\
1 & 1 & 1 & 0 & 0 & 1 & 0 & 0 & 0 & 0 \\
0 & 1 & 1 & 1 & 0 & 0 & 1 & 0 & 0 & 0 \\
1 & 0 & 1 & 1 & 0 & 0 & 0 & 1 & 0 & 0 \\
1 & 0 & 0 & 0 & 1 & 1 & 0 & 1 & 1 & 0 \\
0 & 1 & 0 & 0 & 1 & 1 & 1 & 0 & 1 & 0 \\
0 & 0 & 1 & 0 & 0 & 1 & 1 & 1 & 0 & 1 \\
0 & 0 & 0 & 1 & 1 & 0 & 1 & 1 & 0 & 1 \\
0 & 0 & 0 & 0 & 1 & 1 & 0 & 0 & 1 & 1 \\
0 & 0 & 0 & 0 & 0 & 0 & 1 & 1 & 1 & 1
\end{pmatrix}
$$

(a) \hspace{5cm} (b)

Figure 3.34 (a) The point and (b) the incidence matrices for the three-dimensional test pattern.

is illustrated in Figure 3.35. In particular, any linear transformation of Euclidean three-space representable in Cartesian coordinates as right multiplication by a 3×3 nonsingular matrix N can be extended to a projective transformation of projective three-space by replacing the upper-left-hand 3×3 submatrix of the 4×4 identity matrix by N:

$$\left(\begin{array}{c|c} N & \begin{array}{c} 0 \\ 0 \\ 0 \end{array} \\ \hline 000 & 1 \end{array} \right)$$

In fact, we have the following analog of Proposition 3.9.

3.10 Proposition. All projective transformations T of projective three-space are, or can be written as composites of, rotations, translations, local scalings, global scalings, reflections, shearings, and perspective projections.

We consider rotations first. In this example, an *axis* is a directed line and *counterclockwise rotation* about such an axis through the angle θ obeys the right hand rule: If you grasp the axis with your right hand in such a manner that your thumb points in the positive direction, then your fingers will curl in the direction of positive rotation.

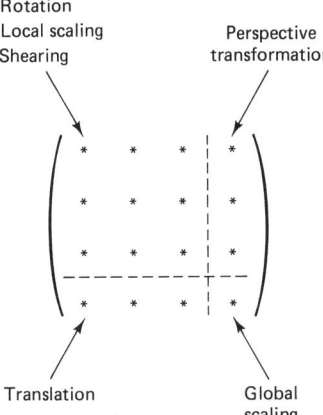

Figure 3.35 The component effects of invertible 4 × 4 matrix multiplication.

3.11 Example. The transformation

$$T([p_1, p_2, p_3, p_4]) = \left[\langle p_1, p_2, p_3, p_4 \rangle \begin{pmatrix} \cos\theta & \sin\theta & 0 & 0 \\ -\sin\theta & \cos\theta & 0 & 0 \\ 0 & 0 & 1 & 0 \\ 0 & 0 & 0 & 1 \end{pmatrix} \right]$$

is counterclockwise *rotation about the z-axis* through the angle θ (see Figure 3.36(a)). The transformation

$$T([p_1, p_2, p_3, p_4]) = \left[\langle p_1, p_2, p_3, p_4 \rangle \begin{pmatrix} \cos\theta & 0 & -\sin\theta & 0 \\ 0 & 1 & 0 & 0 \\ \sin\theta & 0 & \cos\theta & 0 \\ 0 & 0 & 0 & 1 \end{pmatrix} \right]$$

is counterclockwise *rotation about the y-axis* through the angle θ (see Figure 3.36(b)). The transformation

$$T([p_1, p_2, p_3, p_4]) = \left[\langle p_1, p_2, p_3, p_4 \rangle \begin{pmatrix} 1 & 0 & 0 & 0 \\ 0 & \cos\theta & \sin\theta & 0 \\ 0 & -\sin\theta & \cos\theta & 0 \\ 0 & 0 & 0 & 1 \end{pmatrix} \right]$$

is counterclockwise *rotation about the x-axis* through the angle θ (see Figure 3.36(c)).

3.12 Example. The transformation

$$T([p_1, p_2, p_3, p_4]) = \left[\langle p_1, p_2, p_3, p_4 \rangle \begin{pmatrix} 1 & 0 & 0 & 0 \\ 0 & 1 & 0 & 0 \\ 0 & 0 & 1 & 0 \\ h & k & l & 1 \end{pmatrix} \right]$$

is the *translation* that takes the point $O[0, 0, 0, 1]$ to the point $TO = TO[h, k, l, 1]$ (see Figure 3.37).

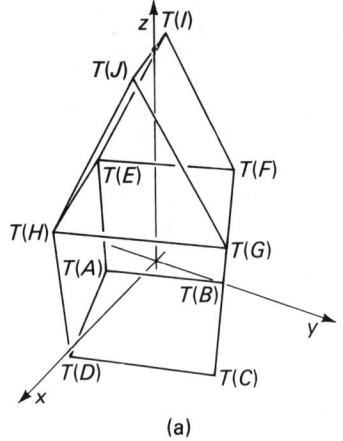

(a)

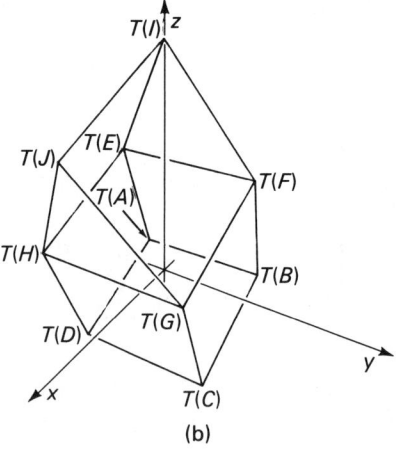

(b)

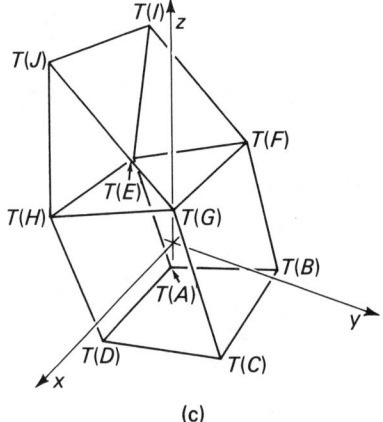

(c)

Figure 3.36 Rotations about the coordinate axes. (a) The z-axis. (b) The y-axis. (c) The x-axis.

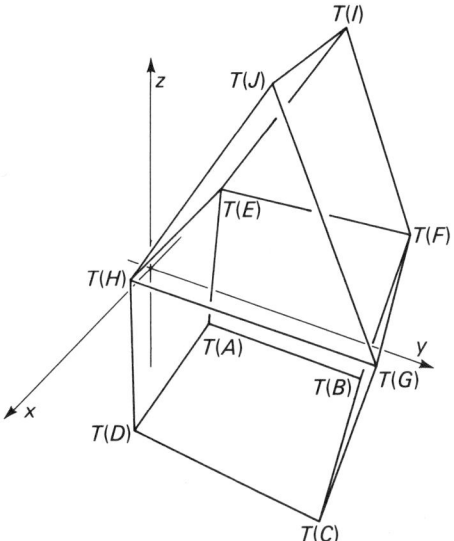

Figure 3.37 Translation ($h = 1$, $k = 2$, $l = 0.5$).

3.13 Example. The transformation

$$T([p_1, p_2, p_3, p_4]) = \left[\langle p_1, p_2, p_3, p_4 \rangle \begin{pmatrix} m_x & 0 & 0 & 0 \\ 0 & m_y & 0 & 0 \\ 0 & 0 & m_z & 0 \\ 0 & 0 & 0 & m_w \end{pmatrix}\right]$$

is *scaling* by a factor of m_x/m_w in the x-direction, by m_y/m_w in the y-direction, and by m_z/m_w in the z-direction (see Figure 3.38); this is a positive scaling in the x- (y- or z-) direction if m_x/m_w (m_y/m_w or m_z/m_w) is positive and a positive scaling with a reflection in the x- (y- or z-) direction if m_x/m_w (m_y/m_w or m_z/m_w)

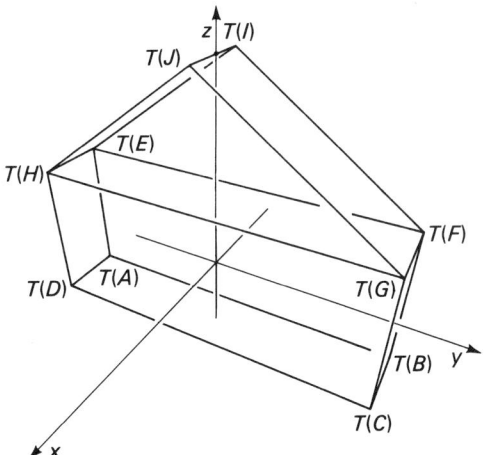

Figure 3.38 Scaling ($m_x = 1$, $m_y = 5$, $m_x = 2$, $m_w = 2$).

is negative. The scaling associated with m_x, m_y, and m_z is *local scaling*, and the scaling associated with $1/m_z$ is *global scaling*.

3.14 Example. The transformation

$$T([p_1, p_2, p_3, p_4]) = \langle p_1, p_2, p_3, p_4 \rangle \begin{pmatrix} 1 & a & 0 & 0 \\ 0 & 1 & 0 & 0 \\ 0 & 0 & 1 & 0 \\ 0 & 0 & 0 & 1 \end{pmatrix}$$

is *shearing* by a parallel to the *yz*-plane in the *y* direction (see Figure 3.39). If projective three-space is thought of as an infinite number of thin and rigid plates parallel to the *yz*-plane, then shearing by a parallel to the *yz*-plane in the *y*-direction displaces the plate *x* units from the *yz*-plane along itself in the *y* direction through the distance ax. (See Exercise 24.)

3.15 Example. The transformation

$$T([p_1, p_2, p_3, p_4]) = \left[\langle p_1, p_2, p_3, p_4 \rangle \begin{pmatrix} 1 & 0 & 0 & a \\ 0 & 1 & 0 & b \\ 0 & 0 & 1 & c \\ 0 & 0 & 0 & 1 \end{pmatrix} \right]$$

is a *perspective projection* of projective three-space (see Figure 3.40).

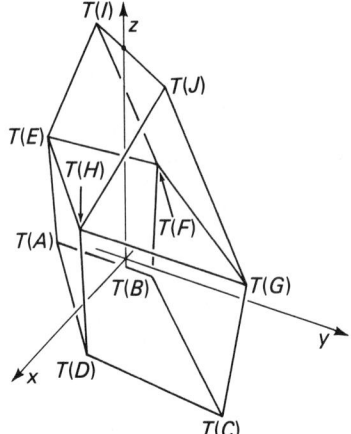

Figure 3.39 Shearing by a parallel to the *yz*-plane in the *y*-direction ($a = 1$).

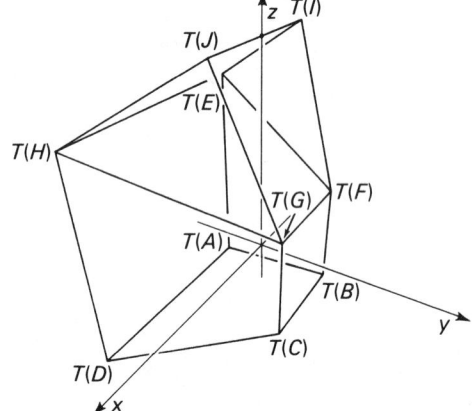

Figure 3.40 A perspective transformation of projective three-space ($a = 0$, $b = 0.4$, $c = 0$).

SECTION 3 EXERCISES:

1. (a) Show that translation cannot be represented by matrix multiplication in Cartesian coordinates, but that it can be represented by matrix multiplication in affine coordinates.
   (b) Why is 1 used as the third entry in affine coordinates? Why not use some other number? Why not use 0?

2. In what way is the relationship between an extrinsic Euclidean plane and the intrinsic Euclidean plane the same as the relationship between affine coordinates and Cartesian coordinates?

3. The *affine coordinates* **l** of the line l in the Euclidean plane whose equation is $ax + by + c = 0$ are

$$\begin{pmatrix} a \\ b \\ c \end{pmatrix}$$

We indicate the correspondence between l and its affine coordinates by $l(\mathbf{l})$.
   (a) What triples (a, b, c) do *not* correspond to the affine coordinates of a line?
   (b) Show that $l(\mathbf{l})$ is perpendicular to $m(\mathbf{m})$ if and only if

$$\mathbf{l}^T \begin{pmatrix} 1 & 0 & 0 \\ 0 & 1 & 0 \\ 0 & 0 & 0 \end{pmatrix} \mathbf{m} = 0$$

   (c) Show that $l(\mathbf{l})$ is parallel to $m(\mathbf{m})$ if and only if

$$\begin{pmatrix} 1 & 0 & 0 \\ 0 & 1 & 0 \\ 0 & 0 & 0 \end{pmatrix} \mathbf{l} = k \begin{pmatrix} 1 & 0 & 0 \\ 0 & 1 & 0 \\ 0 & 0 & 0 \end{pmatrix} \mathbf{m}$$

   for some nonzero constant k.
   (d) Show that the line l perpendicular to $m(\mathbf{m})$ and passing through the point $P(x_0, y_0)$ has the affine coordinates

$$\begin{pmatrix} 1 & 0 & 0 \\ 0 & 1 & 0 \\ -x_0 & -y_0 & 1 \end{pmatrix} \begin{pmatrix} 0 & 1 & 0 \\ -1 & 0 & 0 \\ 0 & 0 & 0 \end{pmatrix} \mathbf{m}$$

   (e) Show that the line l parallel to $m(\mathbf{m})$ and passing through the point $P(x_0, y_0)$ has the affine coordinates

$$\begin{pmatrix} 1 & 0 & 0 \\ 0 & 1 & 0 \\ -x_0 & -y_0 & 1 \end{pmatrix} \begin{pmatrix} 1 & 0 & 0 \\ 0 & 1 & 0 \\ 0 & 0 & 0 \end{pmatrix} \mathbf{m}$$

4. Give the explicit form of the 3×3 matrix representing each of the following projective transformations of the projective plane.
   (a) Rotation through 30°; rotation through 135°
   (b) Translation of the origin to the point $A[2, -3, 4]$; translation of the point $A[2, -3, 4]$ to the origin
   (c) Scaling by a factor of 2 in the x-direction and 3 in the y-direction
   (d) Global scaling by a factor of 5
   (e) Reflection in the x-axis; reflection in the line $y = x$; reflection in the origin
   (f) Shearing by 2 parallel to the x-axis; shearing by 3 parallel to the y-axis

5. Derive formulas for reflection in the following lines.
   (a) $l[1, 0, -a]$
   (b) $l[0, 1, -b]$
   (c) $l[0, 0, 1]$

6. Show that reflection of the projective plane in the line l that passes through the origin and whose inclination with respect to the x-axis is θ is represented by

$$\begin{pmatrix} \cos 2\theta & \sin 2\theta & 0 \\ \sin 2\theta & -\cos 2\theta & 0 \\ 0 & 0 & 1 \end{pmatrix}$$

(*Hint:* Think of this reflection as rotation of l to the x-axis followed by reflection in the x-axis followed by rotation of the x-axis back to l.)

7. Show that shearing of the projective plane by a parallel to the line l that passes through the origin and whose inclination with respect to the x-axis is θ is represented by

$$\begin{pmatrix} 1 - a\sin\theta\cos\theta & -a\sin^2\theta & 0 \\ a\cos^2\theta & 1 + a\sin\theta\cos\theta & 0 \\ 0 & 0 & 1 \end{pmatrix}$$

Show that this matrix can be written as $I + a\mathbf{n}^T\mathbf{v}$ where $\mathbf{n} = [-\sin\theta, \cos\theta, 0]$ is normal to l, and $\mathbf{v} = [\cos\theta, \sin\theta, 0]$ is parallel to l. (*Hint:* Use the same technique you used to do Exercise 6.)

8. Show that the composition of reflection of the projective plane in the line that passes through the origin and whose inclination is θ and reflection in the line that passes through the origin and whose inclination is ϕ is rotation about the origin through the angle $2(\phi - \theta)$.

The next two exercises show that the value in representing transformations by matrix multiplication is in derivation of formulas, not necessarily in implementation of them.

9. Using projective transformations, show that counterclockwise rotation through an angle θ about the point $C(h, k)$ in the Euclidean plane is given by

$$u = x\cos\theta - y\sin\theta - h\cos\theta + k\sin\theta + h$$
$$v = x\sin\theta + y\cos\theta - h\sin\theta - k\cos\theta + k$$

where (x, y) are the before-rotation coordinates and (u, v) are the after-rotation coordinates. (*Hint:* This transformation is the composite of translation of C to the origin, counterclockwise rotation through an angle θ about the origin, and translation of the origin back to C.)

10. (a) How many additions and multiplications must be performed in computing the following matrix product?

$$\langle p_1, p_2, 1 \rangle \begin{pmatrix} 1 & 0 & 0 \\ 0 & 1 & 0 \\ h & k & 1 \end{pmatrix}$$

(b) How many additions and multiplications must be performed in computing the vector $\langle p_1 + h, p_2 + k, 1 \rangle$?

11. Verify analytically what happens to midpoints of segments and parallelism of lines under each of the following projective transformations.
    (a) Rotation
    (b) Translation
    (c) Scaling
    (d) Shearing
    (e) Perspective projection

12. (Do this exercise using a specific example.) Consider a test pattern consisting of two intersecting Euclidean lines. Geometrically describe the image of this test pattern

under a transformation taking the point of intersection of the two lines to infinity. Verify your conclusion analytically.

†13. Explain in what sense the geometry of the transformations discussed in this section is relative to the choice of coordinate system in the following sense: Given a test pattern (whose vertices are fixed as *points* in the projective plane) and a matrix M representing a transformation of the projective plane, what effects does moving the coordinate system have on the image of the test pattern under the transformation described by M?

14. Show that the group of projective transformations of the projective plane is not commutative (in general that $T_2 \circ T_1 \neq T_1 \circ T_2$). What effect does this have on the development of transformation formulas?

15. Show that the composition of two of each of the following types of transformations of the projective plane is a transformation of the same type.
    (a) Rotations
    (b) Translations
    (c) Scalings
    (d) Perspective projections

16. Show that the composition of two shearings of the projective plane is not necessarily a shearing.

†17. Suppose M is the matrix repesenting the perspective projection for which $a \neq 0$ and $b = 0$ (see Example 3.6). Show that a point is *fixed* under the transformation represented by M,
$$[\langle p_1, p_2, p_3 \rangle M] = [p_1, p_2, p_3]$$
if and only if it lies on the line $l[1, 0, 0]$. Show that the line l is *fixed* under the transformation represented by M,
$$\left[M^{-1} \begin{pmatrix} l_1 \\ l_2 \\ l_3 \end{pmatrix} \right] = \left[\begin{pmatrix} l_1 \\ l_2 \\ l_3 \end{pmatrix} \right]$$
if and only if l passes through the origin $O[0, 0, 1]$.

18. Find the set of points and the set of lines fixed by the transformations represented by the following matrices.

    (a) $M = \begin{pmatrix} 1 & 0 & a \\ 0 & 1 & b \\ 0 & 0 & 1 \end{pmatrix}$ (b) $M = \begin{pmatrix} 1 & 0 & 0 \\ 0 & 1 & 0 \\ h & k & 1 \end{pmatrix}$

19. Investigate the geometry of reflection in the ideal line. (*Hint:* Consider some simple test patterns. What is the image of each under reflection in the ideal line?)

20. Give the explicit form of the 4 × 4 matrix representing each of the following projective transformations of projective three-space.
    (a) Rotation about the *x*-axis through 45°; rotation about the *z*-axis through 30°
    (b) Translation of the origin to the point $P[1, -2, 3, 2]$; translation of the point $P[1, -2, 3, 2]$ to the origin
    (c) Scaling by a factor of 2 in the *x*-direction, 3 in the *y*-direction, and 4 in the *z*-direction
    (d) Overall scaling by a factor of 3
    (e) Reflection in the plane $z = 0$; reflection in the plane $y = x$; reflection in the plane $x + 2y - z = 0$; reflection in the origin

Sec. 3   The Geometry of Projective Transformations

(f) Shearing by 2 parallel to the *yz*-coordinate plane in the *y*-direction; shearing by 3 parallel to the *xy*-coordinate plane in the direction of the vector $\langle 2, 1, 0 \rangle$

21. Let $\mathbf{v} = [a, b, c, 0]$ be a unit vector in projective three-space other than $[0, 0, 1, 0]$, and let $\mathbf{n} = [-b/\sqrt{a^2 + b^2}, a/\sqrt{a^2 + b^2}, 0, 0]$ be the unit normal vector to $\mathbf{v}$ in the plane $p_3 = 0$. Show that the matrix

$$M = \begin{pmatrix} a & -\dfrac{b}{\sqrt{a^2+b^2}} & -\dfrac{ac}{\sqrt{a^2+b^2}} & 0 \\ b & \dfrac{a}{\sqrt{a^2+b^2}} & -\dfrac{bc}{\sqrt{a^2+b^2}} & 0 \\ c & 0 & \sqrt{a^2+b^2} & 0 \\ 0 & 0 & 0 & 1 \end{pmatrix}$$

represents the projective transformation that takes the vectors $\mathbf{v}$, $\mathbf{n}$, and $\mathbf{v} \times \mathbf{n}$ to the vectors $[1, 0, 0, 0]$, $[0, 1, 0, 0]$, and $[0, 0, 1, 0]$, respectively. Show that $M^{-1} = M^T$. (Since $M^{-1} = M^T$, the projective transformation represented by M^T takes the vectors $[1, 0, 0, 0]$, $[0, 1, 0, 0]$, and $[0, 0, 1, 0]$ to $\mathbf{v}$, $\mathbf{n}$, and $\mathbf{v} \times \mathbf{n}$, respectively.)

22. Show that counterclockwise rotation of projective three-space through the angle θ about the axis whose positive direction is that of the unit vector $\mathbf{v} = [a, b, c, 0]$ is represented by the matrix

$$\begin{pmatrix} a^2 + (1 - a^2)\cos\theta & ab(1 - \cos\theta) + c\sin\theta & ac(1 - \cos\theta) - b\sin\theta & 0 \\ ab(1 - \cos\theta) - c\sin\theta & b^2 + (1 - b^2)\cos\theta & bc(1 - \cos\theta) + a\sin\theta & 0 \\ ac(1 - \cos\theta) + b\sin\theta & bc(1 - \cos\theta) - a\sin\theta & c^2 + (1 - c^2)\cos\theta & 0 \\ 0 & 0 & 0 & 1 \end{pmatrix}$$

(*Hint:* Let $\mathbf{n} = [-b/\sqrt{a^2 + b^2}, a/\sqrt{a^2 + b^2}, 0, 0]$ be the unit normal vector to $\mathbf{v}$ in the plane $p_3 = 0$. The transformation we wish to represent is the composite of rotation of the vectors $\mathbf{v}$, $\mathbf{n}$, and $\mathbf{v} \times \mathbf{n}$ to the vectors $[1, 0, 0, 0]$, $[0, 1, 0, 0]$, and $[0, 0, 1, 0]$, respectively, followed by counterclockwise rotation through the angle θ about the *x*-axis, followed by rotation of the vectors $[1, 0, 0, 0]$, $[0, 1, 0, 0]$, and $[0, 0, 1, 0]$ back to $\mathbf{v}$, $\mathbf{n}$, and $\mathbf{v} \times \mathbf{n}$, respectively. See Exercise 21.)

23. Show that reflection of projective three-space in the plane which passes through the origin and whose unit normal vector is $\mathbf{n} = [a, b, c, 0]$ is represented by the matrix

$$\begin{pmatrix} 1 - 2a^2 & -2ab & -2ac & 0 \\ -2ab & 1 - 2b^2 & -2bc & 0 \\ -2ac & -2bc & 1 - 2c^2 & 0 \\ 0 & 0 & 0 & 1 \end{pmatrix}$$

(*Hint:* Use the same technique you used to do Exercise 22.) Show that this matrix can be written as $I - 2\,\mathbf{n}^T\mathbf{n}$.

24. Give a geometric interpretation of the shearing of projective three-space represented by the 4×4 matrix obtained by replacing the 0 in the (1, 2), (1, 3), (2, 1), (2, 3), (3, 1), or (3, 2) position of the 4×4 identity matrix with a nonzero constant a.

25. Show that shearing of projective three-space by a parallel to the *yz*-plane in the direction of the unit vector $[0, b, c, 0]$ is represented by the matrix

$$\begin{pmatrix} 1 & ab & ac & 0 \\ 0 & 1 & 0 & 0 \\ 0 & 0 & 1 & 0 \\ 0 & 0 & 0 & 1 \end{pmatrix}$$

26. Show that shearing of projective three-space by a parallel to the plane π whose unit normal vector is $\mathbf{n}$ in the direction of the unit vector $\mathbf{v}$ that lies in π is $I + a\mathbf{n}^T\mathbf{v}$. (*Hint:* Use the same technique you used to do Exercise 22.)

SECTION 4: GENERAL HOMOGENEOUS COORDINATES

To simplify complicated problems in Euclidean geometry, we can either transform geometric objects or transform coordinates. The same thing is true in projective geometry. Until now we have discussed transformations of objects. In this section we discuss transformations of coordinates.

As before, we begin by considering the projective line because it is analytically and geometrically easier to study than the projective plane and projective three-space, and because the study of the projective plane and projective three-space follow immediately thereafter.

Recall (see Proposition 1.5) that if $Q[\mathbf{q}]$, $R[\mathbf{r}]$, and $S[\mathbf{s}]$ are three distinct points on a line l in the projective plane (or projective three-space) then there is a 2×3 (or 2×4) rank 2 matrix M, unique up to a scalar multiple, such that if $T: RP^1 \to l$ is (the parametrization) given by

$$T([c_1, c_2]) = [\langle c_1, c_2 \rangle M]$$

then $T([1, 0]) = [\mathbf{q}]$, $T([0, 1]) = [\mathbf{r}]$, and $T([1, 1]) = [\mathbf{s}]$. The following result is proved similarly.

4.1 Proposition. If $Q[\mathbf{q}]$, $R[\mathbf{r}]$, and $S[\mathbf{s}]$ are three distinct points on the projective line l, then there is a 2×2 nonsingular matrix M, unique up to a scalar multiple, such that if the parametrization $T: RP^1 \to l$ is given by

$$T([c_1, c_2]) = [\langle c_1, c_2 \rangle M]$$

then $T([1, 0]) = [\mathbf{q}]$, $T([0, 1]) = [\mathbf{r}]$, and $T([1, 1]) = [\mathbf{s}]$.

The proof of this result is based on the fact that if $\mathbf{q}$, $\mathbf{r}$, and $\mathbf{s}$ are three distinct vectors in R^2, then there are nonzero real constants k_1 and k_2 for which

$$\mathbf{s} = k_1 \mathbf{q} + k_2 \mathbf{r}$$

Absorbing the constants k_1 and k_2 into the representatives of $[\mathbf{q}]$ and $[\mathbf{r}]$, respectively, we have

$$\langle s_1, s_2 \rangle = \langle q_1, q_2 \rangle + \langle r_1, r_2 \rangle$$

We may then take M to be the matrix

$$M = \begin{pmatrix} q_1 & q_2 \\ r_1 & r_2 \end{pmatrix}$$

4.2 Example. The points $A[2, 5]$, $B[1, 3]$, and $C[4, 7]$ are distinct points on the projective line. Since
$$\langle 4, 7 \rangle = 5\langle 2, 5 \rangle - 6\langle 1, 3 \rangle = \langle 10, 25 \rangle + \langle -6, -18 \rangle$$
it follows that if
$$M = \begin{pmatrix} 10 & 25 \\ -6 & -18 \end{pmatrix}$$
and $T : RP^1 \to l$ is given by
$$T([c_1, c_2]) = [\langle c_1, c_2 \rangle M]$$
then $T([1, 0]) = [2, 5]$, $T([0, 1]) = [1, 3]$, and $T([1, 1]) = [4, 7]$.

There is a great deal of similarity between Proposition 4.1 and the Fundamental Theorem of Projective Geometry for Lines (Theorem 1.9). In fact, the only difference between these results is point of view: According to the Fundamental Theorem of Projective Geometry for Lines, given three points $Q[\mathbf{q}]$, $R[\mathbf{r}]$, and $S[\mathbf{s}]$ on the projective line, there is a *transformation* of the projective line taking $I[1, 0]$ to Q, $O[0, 1]$ to R, and $U[1, 1]$ to S. According to Proposition 4.1 there is a *parametrization* of the projective line taking $[1, 0]$ to Q, $[0, 1]$ to R, and $[1, 1]$ to S. (Transformations and parametrizations are really quite closely related to each other.)

When we affix coordinates to the Euclidean line (see Figure 3.41), we choose a positive direction, an origin O, and a unit point U, and then we associate numbers to points on the Euclidean line by using an equal-spacing algorithm. What we are really doing is parametrizing the Euclidean line by the set of real numbers, the specific parametrization being defined by the choice of positive direction, origin, and unit point. A different parametrization would yield different coordinates.

Figure 3.41 The coordinate Euclidean line.

We obtain standard homogeneous coordinates on the projective line by completing the coordinate Euclidean line with its ideal point I (see Figure 3.42). Topologically, a projective line is a circle, and even though the points I, O, and U play an important role in assigning coordinates to the projective line, they are still just three points on a circle used to define a parametrization of the projective line by RP^1. The idea behind general homogeneous coordinates is that we can choose *any* three points I', O', and U' on the projective line to define a *new* parametrization of the projective line by RP^1. This new parametrization can be thought of as a new association of homogeneous coordinates to points, and in this new coordinate system I' is the ideal point, O' is the origin, and U' is the unit point.

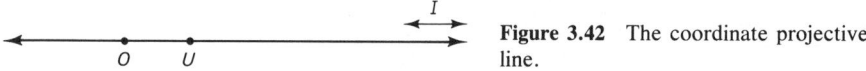

Figure 3.42 The coordinate projective line.

The coordinates of a point with respect to the new coordinate system are related to the coordinates with respect to the old coordinate system as follows: According to Proposition 4.1, there is a transformation $T : RP^1 \to l$ for which $T([1, 0]) = I'$, $T([0, 1]) = O'$, and $T([1, 1]) = U'$; T transforms the new coordinates of a point into the old coordinates. For example,

$T([1, 0])$ = $[i_1, i_2]$
T(new coordinates of I')   old coordinates of I'
$T([0, 1])$ = $[o_1, o_2]$
T(new coordinates of O')   old coordinates of O'
$T([1, 1])$ = $[u_1, u_2]$
T(new coordinates of U')   old coordinates of U'

Conversely, T^{-1} (which is represented by M^{-1}) transforms the old coordinates of a point into the new ones.

The reason that homogeneous coordinates transform in this manner is discussed in Section 5. Essentially the reason is this (see Figure 3.43): To determine the coordinates of a point P on the projective line l with respect to the coordinate system whose ideal point is I', whose origin is O', and whose unit point is U', first assume that the projective line is embedded in the projective plane. Let m be any other ("standard") projective line passing through O', whose "standard" homogeneous coordinates have O' as the origin. Let E be the point of intersection of the line determined by I' and the ideal point of m and the line determined by U' and the unit point of m. Each point P on l corresponds to a point Q on m under the perspectivity whose center is E, and the general homogeneous coordinates of P are the standard homogeneous coordinates of Q.

4.3 Definition. Three distinct points I', O', and U' on the projective line l, in a specified order, form a *general homogeneous coordinate system* on l. The *general homogeneous coordinates* of the point P on l with respect to this coordinate system are given by

$$T^{-1}(P) = T^{-1}([p_1, p_2]) = [\langle p_1, p_2 \rangle M^{-1}]$$

where $[p_1, p_2]$ are the standard homogeneous coordinates of P and M^{-1} is the

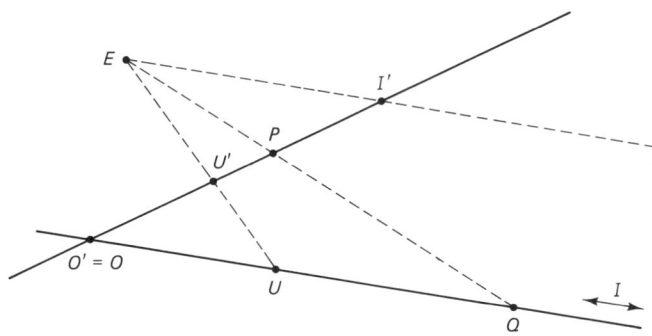

Figure 3.43 Transformation of general homogeneous coordinates on the projective line.

inverse of the matrix M described above. With respect to the homogeneous coordinate system determined by I', O', and U', I' is the *ideal point*, O' is the *origin*, and U' is the *unit point*.

4.4 Example. Suppose the points A, B, and C of Example 4.2 are taken to be the ideal point, the origin, and the unit point, respectively, of a coordinate system on the projective line. Then the new coordinates of any point $P[p_1, p_2]$ are

$$[\langle p_1, p_2 \rangle M^{-1}] = \left[\langle p_1, p_2 \rangle \begin{pmatrix} \frac{3}{5} & \frac{5}{6} \\ -\frac{1}{5} & -\frac{1}{3} \end{pmatrix} \right]$$

For example, the new coordinates of $A[2, 5]$ are $[1, 0]$, the new coordinates of $B[1, 3]$ are $[0, 1]$, the new coordinates of $C[4, 7]$ are $[1, 1]$, and the new coordinates of the point $P[5, -1]$ are $[\frac{16}{5}, \frac{27}{6}]$.

Before proceeding, let us again consider the relationship between general homogeneous coordinates and projective transformations: A homogeneous coordinate system on the projective line l can best be thought of as a parametrization $T : RP^1 \to l$. The comparison between standard homogeneous coordinates (thought of as a parametrization $T_s : RP^1 \to l$) and a set of general homogeneous coordinates $T_l : RP^1 \to l$ on l is made through a transformation T_M of RP^1 represented by a 2×2 nonsingular matrix M (See Figure 3.44). A projective transformation $T : l \to l$ of l is a transformation that can be represented (with respect to standard homogeneous coordinates $T_s : RP^1 \to l$) by a transformation T_M of RP^1, which is represented by a 2×2 nonsingular matrix M (see Figure 3.45). In both discussions, a transformation T_M of RP^1 represented by a 2×2 nonsingular matrix M occurs, but within different contexts this transformation has different interpretations: In one case the coordinates of points on l are being changed, and in the other, points on l are being moved.

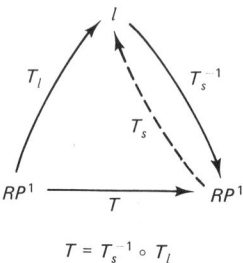

Figure 3.44 Transformation of homogeneous coordinates.

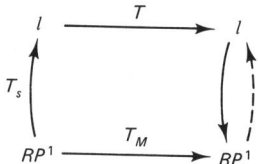

Figure 3.45 Projective transformation of a line.

Again, the reason for introducing general homogeneous coordinates is that they can often make complicated geometric problems more tractable. We illustrate this shortly in proving another fundamental result of projective geometry—Pappus' Theorem. First, however, we observe that what we have just done for the projective line can easily be generalized to the projective plane and projective three-space.

4.5 Proposition. If $P[\mathbf{p}]$, $Q[\mathbf{q}]$, $R[\mathbf{r}]$, and $S[\mathbf{s}]$ are four distinct points in the projective plane π, no three of which are collinear, then there is a 3×3 nonsingular matrix M, unique up to a scalar multiple, such that if the transformation $T: RP^2 \to \pi$ is given by
$$T([c_1, c_2, c_3]) = [\langle c_1, c_2, c_3 \rangle M]$$
then $T([1, 0, 0]) = [\mathbf{p}]$, $T([0, 1, 0]) = [\mathbf{q}]$, $T([0, 0, 1]) = [\mathbf{r}]$, and $T([1, 1, 1]) = [\mathbf{s}]$.

This result is analagous to Proposition 4.1.

4.6 Example. The points $A[1, -1, 0]$, $B[-3, 0, 2]$, $C[1, -2, -1]$, and $D[4, -4, 5]$ are distinct points in the projective plane, no three of which are collinear. Since
$$\langle 4, -4, 5 \rangle = -2\langle 1, -1, 0 \rangle - 1\langle -3, 0, 2 \rangle + 3\langle 1, -2, -1 \rangle$$
$$= \langle -2, 2, 0 \rangle + \langle 3, 0, -2 \rangle + \langle 3, -6, -3 \rangle$$
it follows that if
$$M = \begin{pmatrix} -2 & 2 & 0 \\ 3 & 0 & -2 \\ 3 & -6 & -3 \end{pmatrix}$$
and $T: RP^2 \to \pi$ is given by
$$T([c_1, c_2, c_3]) = [\langle c_1, c_2, c_3 \rangle M]$$
then $T([1, 0, 0]) = [1, -1, 0]$, $T([0, 1, 0]) = [-3, 0, 2]$, $T([0, 0, 1]) = [1, -2, -1]$, and $T([1, 1, 1]) = [4, -4, 5]$.

Four points are required to define a coordinate system on the projective plane.

4.7 Definition. Four distinct points I'_x, I'_y, O', and U' on the projective plane π in a specified order, no three of which are collinear, form a *general homogeneous coordinate system* on π. The *general homogeneous coordinates* of the point P on π with respect to this coordinate system are given by
$$T^{-1}(P) = T^{-1}([p_1, p_2, p_3]) = [\langle p_1, p_2, p_3 \rangle M^{-1}]$$
where $[p_1, p_2, p_3]$ are the standard homogeneous coordinates of P and M^{-1} is the inverse of the matrix M described in Proposition 4.5. With respect to the homogeneous coordinate system determined by I'_x, I'_y, O', and U', I'_x and I'_y are the *ideal points* (in the x- and y-directions, respectively), O' is the *origin*, and U' is the *unit point*.

4.8 Example. Suppose the points A, B, C, and D of Example 4.6 are taken to be the ideal points, the origin, and the unit point, respectively, of a coordinate system on the projective plane. Then the new coordinates of any point $P[p_1, p_2, p_3]$ are

$$[\langle p_1, p_2, p_3 \rangle M^{-1}] = \left[\langle p_1, p_2, p_3 \rangle \begin{pmatrix} -\frac{2}{5} & \frac{1}{5} & -\frac{2}{15} \\ \frac{1}{10} & \frac{1}{5} & -\frac{2}{15} \\ -\frac{3}{5} & -\frac{1}{5} & -\frac{1}{5} \end{pmatrix} \right]$$

For example, the new coordinates of $A[1, -1, 0]$ are $[1, 0, 0]$, the new coordinates of $B[-3, 0, 2]$ are $[0, 1, 0]$, the new coordinates of $C[1, -2, -1]$ are $[0, 0, 1]$, the new coordinates of $D[4, -4, 5]$ are $[1, 1, 1]$, and the new coordinates of the point $P[5, 1, 2]$ are $[-\frac{31}{10}, \frac{4}{5}, -\frac{12}{15}]$.

In Section 6 we use the cross ratio to give a description of general homogeneous coordinates in the projective plane similar to the classical description of Cartesian coordinates in the Euclidean plane. The cross ratio is a perspective scaling. Given a general homogeneous coordinate system in the projective plane (see Figure 3.54), every Euclidean point P has general homogeneous coordinates $[x, y, 1]$, where x is the perspective scaling of P_x along the (general) x-axis and y is the perspective scaling of P_y along the (general) y-axis. Figure 3.46 is the analog of Figure 3.43 in terms of explaining the association of general homogeneous coordinates to points in the projective plane.

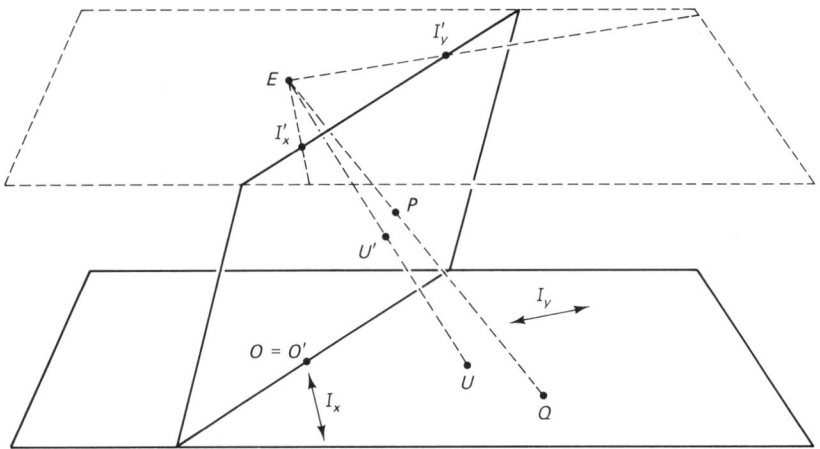

Figure 3.46 Transformation of general homogeneous coordinates on the projective plane.

Finally, general homogeneous coordinates can be defined in projective three-space.

4.9 Proposition. If $P[p]$, $Q[q]$, $R[r]$, $S[s]$, and $U[u]$ are five distinct points in projective three-space, no four of which are coplanar, then there is a 4×4 nonsingular matrix M, unique up to a scalar multiple, such that if the transformation T from RP^3 to projective three-space is given by

$$T([c_1, c_2, c_3, c_4]) = [\langle c_1, c_2, c_3, c_4 \rangle M]$$

then $T([1, 0, 0, 0]) = [\mathbf{p}]$, $T([0, 1, 0, 0]) = [\mathbf{q}]$, $T([0, 0, 1, 0]) = [\mathbf{r}]$, $T([0, 0, 0, 1]) = [\mathbf{s}]$, and $T([1, 1, 1, 1]) = [\mathbf{u}]$.

This result is again analagous to Proposition 4.1.

4.10 Example. The points $A[2, 0, 0, 3]$, $B[1, 0, 2, 1]$, $C[1, -2, -1, 1]$, $D[0, 2, 1, 3]$ and $E[2, 8, 6, 7]$ are distinct points in projective three-space, no four of which are coplanar. Since

$$\langle 2, 8, 6, 7 \rangle = 2\langle 2, 0, 0, 3 \rangle + \langle 1, 0, 2, 1 \rangle - 3\langle 1, -2, -1, 1 \rangle + \langle 0, 2, 1, 3 \rangle$$
$$= \langle 4, 0, 0, 6 \rangle + \langle 1, 0, 2, 1 \rangle + \langle -3, 6, 3, -3 \rangle + \langle 0, 2, 1, 3 \rangle$$

it follows that if

$$M = \begin{pmatrix} 4 & 0 & 0 & 6 \\ 1 & 0 & 2 & 1 \\ -3 & 6 & 3 & -3 \\ 0 & 2 & 1 & 3 \end{pmatrix}$$

and the transformation T from RP^3 to projective three-space is given by

$$T([p_1, p_2, p_3, p_4]) = [\langle p_1, p_2, p_3, p_4 \rangle M]$$

then $T([1, 0, 0, 0]) = [2, 0, 0, 3]$, $T([0, 1, 0, 0]) = [1, 0, 2, 1]$, $T([0, 0, 1, 0]) = [1, -2, -1, 1]$, $T([0, 0, 0, 1]) = [0, 2, 1, 3]$ and $T([1, 1, 1, 1]) = [2, 8, 6, 7]$.

Five points are required to define a coordinate system on projective three-space.

4.11 Definition. Five distinct points I'_x, I'_y, I'_z, O', and U' in projective three-space in a specified order, no four of which are coplanar, form a *general homogeneous coordinate system* on projective three-space. The *general homogeneous coordinates* of the point P in projective three-space with respect to this coordinate system are given by

$$T^{-1}(P) = T^{-1}([p_1, p_2, p_3, p_4]) = [\langle p_1, p_2, p_3, p_4 \rangle M^{-1}]$$

where $[p_1, p_2, p_3, p_4]$ are the standard homogeneous coordinates of P and M^{-1} is the inverse of the matrix M described in Proposition 4.9. With respect to the homogeneous coordinate system determined by I'_x, I'_y, I'_z, O', and U', I'_x, I'_y, and I'_z are the *ideal points* (in the x-, y-, and z-directions, respectively), O' is the *origin*, and U' is the *unit point*.

4.12 Example. Suppose the points A, B, C, D, and E of Example 4.10 are taken to be the ideal points, the origin, and the unit point, respectively, of a coordinate system on projective three-space. Then the new coordinates of any point $P[p_1, p_2, p_3, p_4]$ in projective three-space are

$$[\langle p_1, p_2, p_3, p_4 \rangle M] = \left[\langle p_1, p_2, p_3, p_4 \rangle \begin{pmatrix} \frac{2}{5} & 0 & \frac{1}{5} & -\frac{3}{5} \\ \frac{9}{40} & -\frac{1}{4} & \frac{13}{60} & -\frac{3}{20} \\ -\frac{3}{20} & \frac{1}{2} & -\frac{1}{30} & \frac{1}{10} \\ -\frac{1}{10} & 0 & -\frac{2}{15} & \frac{2}{5} \end{pmatrix} \right]$$

For example, the new coordinates of $A[2, 0, 0, 3]$ are $[1, 0, 0, 0]$, the new coordinates of $B[1, 0, 2, 1]$ are $[0, 1, 0, 0]$, the new coordinates of $C[1, -2, -1, 1]$ are $[0, 0, 1, 0]$, the new coordinates of $D[0, 2, 1, 3]$ are $[0, 0, 0, 1]$, the new coordinates of $E[2, 8, 6, 7]$ are $[1, 1, 1, 1]$, and the new coordinates of the point $P[1, 4, 3, 2]$ are $[12, 0, 11, -3]$.

As an application of general homogeneous coordinates, we now prove Pappus' Theorem.

4.13 Theorem (Pappus' Theorem). If A, B, C, and A', B', C' are two triples of distinct points on two distinct lines in the projective plane, then the intersections of AB' and $A'B$, of AC' and $A'C$, and of BC' and $B'C$ are collinear.

Proof. Denote the point of intersection of AB' and $A'B$ by P, of AC' and $A'C$ by Q, and of BC' and $B'C$ by R (see Figure 3.47). We consider the case in which A, B', C, and Q are distinct and noncollinear, all other cases being immediate.

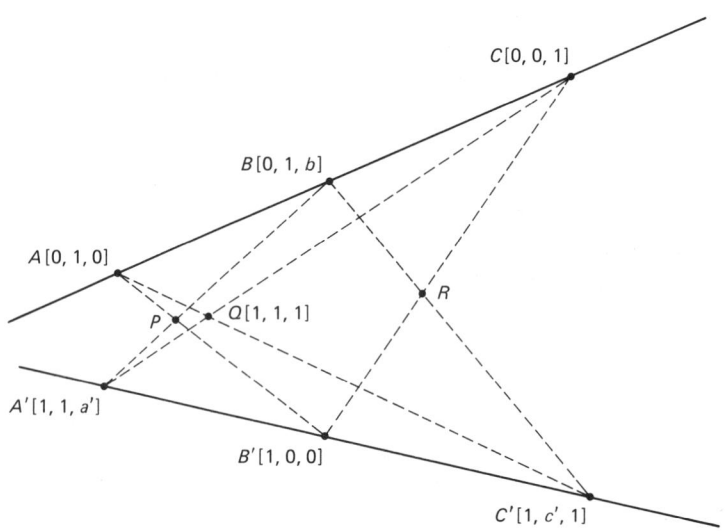

Figure 3.47 Pappus' Theorem.

Let $T : RP^2 \to \pi$ denote the transformation (or parametrization) for which $T([1, 0, 0]) = B'$, $T([0, 1, 0]) = A$, $T([0, 0, 1]) = C$, and $T([1, 1, 1]) = Q$; that is, choose (general homogeneous) coordinates on π so that the coordinates of B' are $[1, 0, 0]$, those of A are $[0, 1, 0]$, those of C are $[0, 0, 1]$, and those of Q are $[1, 1, 1]$. Since T and its inverse (considered as projective transformations of the projective plane) take lines to lines, the result follows from computing the (general homogeneous) coordinates of P, Q, and R and verifying the collinearity condition of Proposition 2.2.12.

To this end, observe that the (general homogeneous) coordinates of B are of the form $[0, 1, b]$ for some nonzero real number b (since B lies on the line determined by A and C), the (general homogeneous) coordinates of A' are of the form $[1, 1, a']$ for some real number a' (since A' lies on the line determined by Q and C), and the (general homogeneous) coordinates of C' are of the form $[1, c', 1]$ for some real number c' (since C' lies on the line determined by Q and A). Furthermore, since A', B', and C' are collinear, it follows that

$$1 - a'c' = \begin{vmatrix} 1 & 0 & 0 \\ 1 & 1 & a' \\ 1 & c' & 1 \end{vmatrix} = 0$$

The general homogeneous coordinates of the line BC' are

$$\left[\begin{vmatrix} 1 & b \\ c' & 1 \end{vmatrix}, -\begin{vmatrix} 0 & b \\ 1 & 1 \end{vmatrix}, \begin{vmatrix} 0 & 1 \\ 1 & c' \end{vmatrix} \right] = [1 - bc', b, -1]$$

and the general homogeneous coordinates of $B'C$ are

$$\left[\begin{vmatrix} 0 & 1 \\ 0 & 0 \end{vmatrix}, -\begin{vmatrix} 0 & 1 \\ 1 & 0 \end{vmatrix}, \begin{vmatrix} 0 & 0 \\ 1 & 0 \end{vmatrix} \right] = [0, 1, 0]$$

so the general homogeneous coordinates of R are

$$\left[\begin{vmatrix} b & -1 \\ 1 & 0 \end{vmatrix}, -\begin{vmatrix} 1-bc' & -1 \\ 0 & 0 \end{vmatrix}, \begin{vmatrix} 1-bc' & b \\ 0 & 1 \end{vmatrix} \right] = [1, 0, 1 - bc']$$

A similar calculation shows that the coordinates of P are $[b, b - a', 0]$.

Thus to show that P, Q, and R are collinear, it suffices to observe that

$$\begin{vmatrix} b & b-a' & 0 \\ 1 & 1 & 1 \\ 1 & 0 & 1-bc' \end{vmatrix} = b(1 - a'c') = 0 \qquad \blacksquare$$

SECTION 4 EXERCISES

1. In each of the following cases, find the 2×2 matrix M representing the parametrization $T : RP^1 \to l$ of the projective line for which $T([1, 0]) = A$, $T([0, 1]) = B$, and $T([1, 1]) = C$ (that is, the matrix representing change of coordinates from general homogeneous coordinates to standard homogeneous coordinates). In each case, also find the matrix M^{-1} representing change of coordinates from standard homogeneous coordinates to general homogeneous coordinates.
   (a) $A[1, 3]$, $B[1, 0]$, $C[2, 1]$
   (b) $A[1, 1]$, $B[2, 3]$, $C[1, 0]$
   (c) $A[-2, 3]$, $B[0, -1]$, $C[1, -2]$
   (d) $A[1, -2]$, $B[1, 2]$, $C[3, 2]$

2. For each part of Exercise 1, a general homogeneous coordinate system on the projective line with ideal point A, origin B, and unit point C is determined. With respect to each coordinate system, find the standard homogeneous coordinates of

the point P whose general homogeneous coordinates are given and the general homogeneous coordinates of the point Q whose standard homogeneous coordinates are given.

(a) $P[2, 2]$, $Q[3, 2]$ (b) $P[-1, 1]$, $Q[5, 1]$
(c) $P[2, 3]$, $Q[1, 1]$ (d) $P[1, 0]$, $Q[-1, 3]$

†3. In each of the following cases, find the 3×3 matrix M representing the parametrization $T : RP^2 \to \pi$ of the projective plane for which $T([1, 0, 0]) = A$, $T([0, 1, 0]) = B$, $T([0, 0, 1]) = C$, and $T([1, 1, 1]) = D$ (that is, the matrix representing change of coordinates from general homogeneous coordinates to standard homogeneous coordinates). In each case, also find the matrix M^{-1} representing change of coordinates from standard homogeneous coordinates to general homogeneous coordinates.

(a) $A[1, 2, 1]$, $B[2, 1, 0]$, $C[1, 0, 2]$, $D[-2, 1, 3]$
(b) $A[1, 1, 0]$, $B[1, 0, 1]$, $C[0, 1, 1]$, $D[1, 1, 1]$
(c) $A[1, 0, 0]$, $B[1, 1, 0]$, $C[1, 1, 1]$, $D[2, 3, 4]$
(d) $A[3, -1, -1]$, $B[2, -1, 3]$, $C[1, 0, 2]$, $D[-1, 1, 0]$

4. For each part of Exercise 3, a general homogeneous coordinate system on the projective plane with the x-direction ideal point A, the y-direction ideal point B, origin C, and unit point D is determined. With respect to each coordinate system, find the standard homogeneous coordinates of the point P whose general homogeneous coordinates are given and the general homogeneous coordinates of the point Q whose standard homogeneous coordinates are given.

(a) $P[3, 1, 2]$, $Q[1, 2, -1]$ (b) $P[1, -2, 3]$, $Q[2, 4, 1]$
(c) $P[2, 1, 0]$, $Q[1, -2, 1]$ (d) $P[1, 2, 3]$, $Q[2, -1, -1]$

†5. In each of the following cases, find the 4×4 matrix M representing the parametrization T of projective three-space by RP^3 for which $T([1, 0, 0, 0]) = A$, $T([0, 1, 0, 0]) = B$, $T([0, 0, 1, 0]) = C$, $T([0, 0, 0, 1]) = D$, and $T([1, 1, 1, 1]) = E$ (that is, the matrix representing change of coordinates from general homogeneous coordinates to standard homogeneous coordinates). In each case, also find the matrix M^{-1} representing change of coordinates from standard homogeneous coordinates to general homogeneous coordinates.

(a) $A[1, 0, 0, 0]$, $B[1, 1, 0, 0]$, $C[1, 1, 1, 0]$, $D[1, 1, 1, 1]$, $E[2, 5, 3, -1]$
(b) $A[1, 1, 1, 0]$, $B[1, 1, 0, 1]$, $C[1, 0, 1, 1]$, $D[0, 1, 1, 1]$, $E[1, 1, 1, 1]$
(c) $A[1, -1, -3, -1]$, $B[1, 2, 0, 1]$, $C[3, 4, -2, 0]$, $D[1, 0, -2, 3]$, $E[1, 1, -1, 6]$
(d) $A[2, 1, 0, 0]$, $B[1, 2, 1, 0]$, $C[0, 1, 2, 1]$, $D[0, 0, 1, 2]$, $E[3, 2, -3, -1]$

6. For each part of Exercise 5, a general homogeneous coordinate system on projective three-space with the x-direction ideal point A, the y-direction ideal point B, the z-direction ideal point C, origin D, and unit point E is determined. With respect to each coordinate system, find the standard homogeneous coordinates of the point P whose general homogeneous coordinates are given and the general homogeneous coordinates of the point Q whose standard homogeneous coordinates are given.

(a) $P[1, 1, 2, 1]$, $Q[2, 0, 1, 3]$ (b) $P[-1, 0, 2, 1]$, $Q[2, 1, 0, 2]$
(c) $P[2, 1, 3, 2]$, $Q[1, 2, -2, 1]$ (d) $P[2, 1, 2, 2]$, $Q[1, 3, 0, 1]$

7. Given a general homogeneous coordinate system in the projective plane, how are the standard homogeneous coordinates of a line related to the general homogeneous coordinates of the line?

8. For each part of Exercise 3, find the standard homogeneous coordinates of the line l whose general homogeneous coordinates are given and the general homogeneous coordinates of the line m whose standard homogeneous coordinates are given.

(a) $l[1, 2, 1]$, $m[2, 1, 3]$ (b) $l[1, 0, -1]$, $m[0, 2, 1]$
(c) $l[1, 0, 0]$, $m[0, 1, 0]$ (d) $l[0, 0, 1]$, $m[1, 0, -1]$

9. Given a general homogeneous coordinate system in projective three-space, how are the standard homogeneous coordinates of a plane related to the general homogeneous coordinates of the plane?

10. For each part of Exercise 5, find the standard homogeneous coordinates of the plane π whose general homogeneous coordinates are given and the general homogeneous coordinates of the plane μ whose standard homogeneous coordinates are given.
    (a) $\pi[1, 2, 1, 0]$, $\mu[2, 0, 1, 3]$ (b) $\pi[1, 0, -1, 0]$, $\mu[0, 2, 0, 1]$
    (c) $\pi[1, 0, 0, 0]$, $\mu[0, 0, 1, 0]$ (d) $\pi[0, 0, 0, 1]$, $\mu[1, 0, 0, -1]$

†11. In each of the following cases, four coplanar points A, B, C, and D in projective three-space, no three of which are collinear, are given. In each case, find the parametrization T of the plane determined by these points for which $T([0, 0, 1]) = A$, $T([1, 0, 0]) = B$, $T([1, 1, 1]) = C$, and $T([0, 1, 0]) = D$.
    (a) $A[3, 3, 3, 1]$, $B[1, -1, 0, 0]$, $C[1, -1, 2, 0]$, $D[0, 1, -1, 0]$
    (b) $A[0, 0, 1, 1]$, $B[1, 0, 0, 0]$, $C[0, 1, 0, 0]$, $D[1, 1, 1, 1]$
    (c) $A[0, 0, 0, 1]$, $B[1, 0, 0, 0]$, $C[0, 1, 0, 0]$, $D[1, 1, 0, 1]$
    (d) $A[2, 1, -3, 2]$, $B[1, 0, 0, 0]$, $C[2, 0, -1, 0]$, $D[3, 0, -1, 0]$

†12. In each of the following cases, four coplanar points A, B, C, and D in projective three-space, no three of which are collinear, are given. In each case, find the parametrization T of the plane determined by these points for which $T([0, 0, 1]) = A$, $T([1, 0, 1]) = B$, $T([1, 1, 1]) = C$, and $T([0, 1, 1]) = D$.
    (a) $A[3, 3, 3, 1]$, $B[1, -1, 0, 0]$, $C[1, -1, 2, 0]$, $D[0, 1, -1, 0]$
    (b) $A[0, 0, 1, 1]$, $B[1, 0, 0, 0]$, $C[0, 1, 0, 0]$, $D[1, 1, 1, 1]$
    (c) $A[2, 1, -3, 2]$, $B[1, 0, 0, 0]$, $C[2, 0, -1, 0]$, $D[3, 0, -1, 0]$
    (d) $A[2, 1, 3, 3]$, $B[1, 0, 1, 1]$, $C[0, 1, 1, 1]$, $D[1, 1, 1, 1]$

13. In Section 2.1 we discussed a decomposition of RP^2 that arose in associating standard homogeneous coordinates to points in the projective plane. In what way does the assignment of general homogeneous coordinates to points in the projective plane give rise to other decompositions of RP^2?

SECTION 5: COORDINATE TRANSFORMATIONS

The purpose of this section is to explain why homogeneous coordinates transform the way they do.

We begin by considering the projective line l. Recall (see Section 4) that three distinct points $I'[i_1, i_2]$, $O'[o_1, o_2]$, and $U'[u_1, u_2]$ on l, in a specified order, form a general homogeneous coordinate system on l. The general homogeneous coordinates of the point P on l with respect to this coordinate system are given by

$$T^{-1}(P) = T^{-1}([p_1, p_2]) = [\langle p_1, p_2 \rangle M^{-1}]$$

where $[p_1, p_2]$ are the standard homogeneous coordinates of P and M^{-1} is the inverse of the 2×2 matrix M given as follows: If $\langle i_1, i_2 \rangle$, $\langle o_1, o_2 \rangle$, and $\langle u_1, u_2 \rangle$ are representatives of the coordinates of I', O', and U', respectively, for which

$$\langle u_1, u_2 \rangle = \langle i_1, i_2 \rangle + \langle o_1, o_2 \rangle$$

then

$$M = \begin{pmatrix} i_1 & i_2 \\ o_1 & o_2 \end{pmatrix}$$

Suppose (see Figure 3.48) a two-dimensional observer is located at the origin of the projective plane. As this observer looks at the plane, using the line $l[0, 1, -1]$ as the image line, there is a natural candidate for a coordinate system on l: The origin has planar coordinates [0, 1, 1] (and linear coordinates on l of [0, 1]), the unit point has planar coordinates [1, 1, 1] (and linear coordinates on l of [1, 1]), and the ideal point (which appears to the observer to be at infinity) has planar coordinates [1, 0, 0] (and linear coordinates on l of [1, 0]).

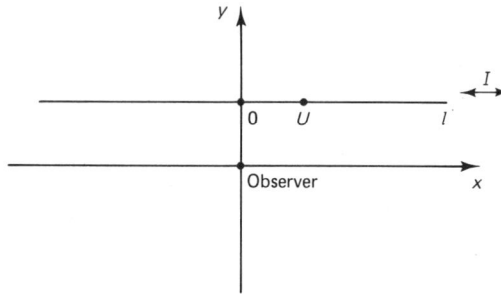

Figure 3.48 Standard homogeneous coordinates.

Now let us suppose (see Figure 3.49) that there is a line m in the projective plane coordinatized by an origin O' (whose coordinates on m are [0, 1]), a unit point U' (whose Euclidean distance from O' is 1 and whose coordinates on m are [1, 0]), and an ideal point I' (at infinity and whose coordinates on m are [1, 0]). If the observer looks at this line (using l as the image line), there is now another candidate for a coordinate system on l: In this new coordinate system, the origin O is the projection of O' onto l, the unit point U is the projection of U' onto l, and the ideal point I is the vanishing point on l of m. With respect to this new coordinate system on l, the coordinates of each point on l are the coordinates of the corresponding point on m.

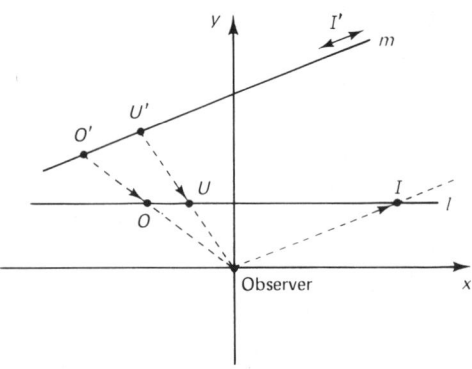

Figure 3.49 General homogeneous coordinates.

160   Projective Transformations   Chap. 3

More precisely (see Figure 3.50), let $T_l : RP^1 \to l$ be the parametrization of l for which $T_l([1, 0]) = [1, 0, 0]$, $T_l([0, 1]) = [0, 1, 1]$, and $T_l([1, 1]) = [1, 1, 1]$. Then the old coordinates of a point P on l are given by $T_l^{-1}(P)$. If $T_m : RP^1 \to m$ is the parametrization of m for which $T_m([1, 0]) = I'$, $T_m([0, 1]) = O'$, and $T_m([1, 1]) = U'$, then the linear coordinates of a point Q on m are given by $T_m^{-1}(Q)$. Thus if $T_0 : m \to l$ is the perspectivity whose center is $[0, 0, 1]$, the new coordinates of a point P on l are given by $T_m^{-1} \circ T_0^{-1}(P)$. Here is how these two sets of coordinates are related: Since

$$T_l^{-1}(P) = T_l^{-1} \circ T_0 \circ T_m \circ T_m^{-1} \circ T_0^{-1}(P) = T(T_m^{-1} \circ T_0^{-1}(P))$$

the image of the new coordinates of P under the transformation

$$T = T_l^{-1} \circ T_0 \circ T_m : RP^1 \to RP^1$$

are the old coordinates of P. To determine the matrix M representing T (which is effectively an intrinsically represented projective transformation of the projective line), we need only to know how T maps three points. Since $T([1, 0]) = [i_1, i_2]$, $T([0, 1]) = [o_1, o_2]$, and $T([1, 1]) = [u_1, u_2]$, it follows that if representatives $\langle i_1, i_2 \rangle$, $\langle o_1, o_2 \rangle$, and $\langle u_1, u_2 \rangle$ are representatives of the coordinates of I, O, and U, respectively, for which

$$\langle u_1, u_2 \rangle = \langle i_1, i_2 \rangle + \langle o_1, o_2 \rangle$$

then

$$M = \begin{pmatrix} i_1 & i_2 \\ o_1 & o_2 \end{pmatrix}$$

Similar arguments can be made for the projective plane and projective three-space.

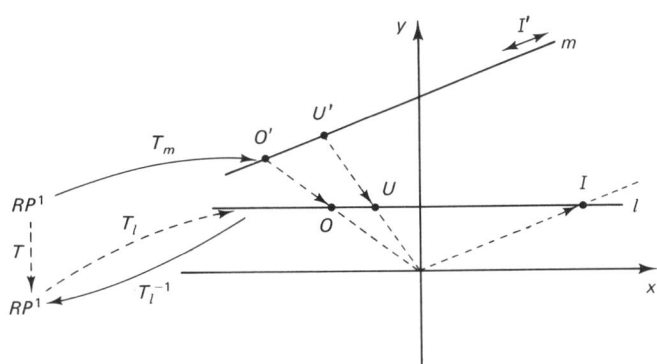

Figure 3.50 An analysis of coordinate transformations.

SECTION 5 EXERCISES

1. In what way does the assignment of general homogeneous coordinates relate to the checkerboard pattern discussed in Section 1.1?
2. Describe why homogeneous coordinates on the projective plane transform the way they do.

SECTION 6: THE CROSS RATIO

We introduced the cross ratio in Section 1.6. We can think of the cross ratio as a perspective scaling. In this section we continue the discussion of the cross ratio in the context of homogeneous coordinates.

Recall (see Section 1.6) that the cross ratio is a Euclidean concept (it is defined in terms of distances), which extends to projective geometry (see Figure 3.51): If A, B, C, and D are any four distinct points on a coordinate Euclidean line (either the Euclidean line or a Euclidean line in the Euclidean plane or Euclidean three-space) then

$$R(A, B, C, D) = \frac{\text{dist}(A, C) \times \text{dist}(B, D)}{\text{dist}(B, C) \times \text{dist}(A, D)}$$

where dist(A, B) denotes the directed Euclidean distance from the point A to the point B. This expression is a well-defined extended real number even if only three of the points A, B, C, and D are distinct. (An extended real number is a finite real number or $\pm\infty$. The inclusion of $\pm\infty$ is necessitated by the event that either $B = C$ or $A = D$.) Indeed, this expression is also a naturally well-defined extended real number if any of A, B, C, and D are allowed to be ideal points. In other words, the extended real number $R(A, B, C, D)$ can be defined for any four points A, B, C, and D, at least three of which are distinct, that lie on a projective line (either the projective line or a projective line in the projective plane or in projective three-space).

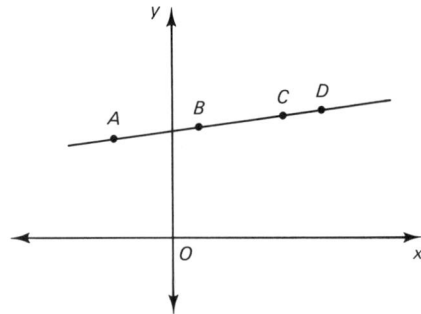

Figure 3.51 The cross ratio on a line in the projective plane.

6.1 Definition. For any four points A, B, C, and D, at least three of which are distinct, on a projective line, the extended real number $R(A, B, C, D)$ is the *cross ratio* associated to A, B, C, and D.

The following result describes the cross ratio for points on the projective line in terms of homogeneous coordinates.

6.2 Proposition. If $A[a_1, a_2]$, $B[b_1, b_2]$, $C[c_1, c_2]$, and $D[d_1, d_2]$ are points on the projective line (see Figure 3.52), at least three of which are distinct, then

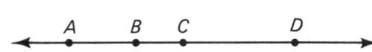

Figure 3.52 The cross ratio on the projective line.

$$R(A, B, C, D) = \frac{\begin{vmatrix} a_1 & a_2 \\ c_1 & c_2 \end{vmatrix} \begin{vmatrix} b_1 & b_2 \\ d_1 & d_2 \end{vmatrix}}{\begin{vmatrix} b_1 & b_2 \\ c_1 & c_2 \end{vmatrix} \begin{vmatrix} a_1 & a_2 \\ d_1 & d_2 \end{vmatrix}}$$

if the denominator is nonzero and $R(A, B, C, D)$ is $\pm\infty$ if the denominator is zero.

Proof. This result follows immediately from the definition of the cross ratio, from the sact that if $A[a_1, a_2] = A[a_1/a_2, 1]$ and $B[b_1, b_2] = B[b_1/b_2, 1]$ are two Euclidean points on the projective line, then the directed distance from A to B

$$\text{dist}(A, B) = \frac{b_1}{b_2} - \frac{a_1}{a_2} = \frac{b_1 a_2 - a_1 b_2}{a_2 b_2} = -\frac{1}{a_2 b_2}\begin{vmatrix} a_1 & a_2 \\ b_1 & b_2 \end{vmatrix}$$

and, finally, from the fact that if $A[a_1, a_2]$, $B[b_1, b_2]$, and $C[c_1, c_2]$ are three Euclidean points on the projective line, then

$$\lim_{C \to \text{ideal point}} \frac{\text{dist}(A, C)}{\text{dist}(B, C)} = \lim_{c_2 \to 0} \frac{(c_1 a_2 - a_1 c_2)/a_2 c_2}{(c_1 b_2 - b_1 c_2)/b_2 c_2}$$

$$= \lim_{c_2 \to 0} \frac{a_2 b_2 c_1 - a_1 b_2 c_2}{a_2 b_2 c_1 - a_2 b_1 c_2}$$

$$= \frac{a_2 b_2 c_1}{a_2 b_2 c_1} = 1$$

Limits are necessary in proving results about the cross ratio. This is because the cross ratio is a Euclidean concept (it is defined in terms of distances) that extends to projective geometry, and passing from Euclidean geometry to projective geometry inherently involves limits. (In particular, ideal points are "limits" of Euclidean points.) ■

6.3 Example. Given points $A[12, 1]$, $B[2, 1]$, $C[6, 1]$, and $D[4, 1]$ on the projective line,

$$R(A, B, C, D) = \frac{\begin{vmatrix} 12 & 1 \\ 6 & 1 \end{vmatrix} \begin{vmatrix} 2 & 1 \\ 4 & 1 \end{vmatrix}}{\begin{vmatrix} 2 & 1 \\ 6 & 1 \end{vmatrix} \begin{vmatrix} 12 & 1 \\ 4 & 1 \end{vmatrix}} = \frac{(6)(-2)}{(-4)(8)} = \frac{3}{8}$$

And, given $E[1, 0]$,

$$R(A, B, C, E) = \frac{\begin{vmatrix} 12 & 1 \\ 6 & 1 \end{vmatrix} \begin{vmatrix} 2 & 1 \\ 1 & 0 \end{vmatrix}}{\begin{vmatrix} 2 & 1 \\ 6 & 1 \end{vmatrix} \begin{vmatrix} 12 & 1 \\ 1 & 0 \end{vmatrix}} = \frac{(6)(-1)}{(-4)(-1)} = -\frac{3}{2}$$

Sec. 6   The Cross Ratio

As the next result illustrates, the cross ratio may be viewed as a parametrization (or, more precisely, as an inverse parametrization) of the projective line (see Section 2.4).

6.4 Proposition. Given distinct collinear points A, B, and C on the projective line, $R(A, B, C, \text{---})$ as a function of one variable (from the set of points D on the projective line to the set of extended real numbers) is one-to-one, onto, and continuous.

Proof. To verify that $R(A, B, C, \text{---})$ is one-to-one, assume that $R(A, B, C, D) = R(A, B, C, E)$ for two points D and E. Using the definition of the cross ratio to rewrite this equation, expanding, and simplifying, it follows that $d_1 e_2 = e_1 d_2$ and hence that $D = E$. (Either e_1 or e_2 must be nonzero. If e_1 is nonzero, for example, then
$$[d_1, d_2] = [e_1 d_1, e_1 d_2] = [e_1 d_1, d_1 e_2] = [e_1, e_2]$$
A similar argument can be used if e_2 is nonzero.)

To show that $R(A, B, C, \text{---})$ is onto, for any extended real number t we must show there is a point D for which $R(A, B, C, D) = t$. If $t = \pm\infty$, then we may take $D[d_1, d_2] = A[a_1, a_2]: R(A, B, C, D) = \pm\infty$. Otherwise, by using the definition of cross ratio, expanding, and rearranging terms, it follows that $R(A, B, C, D) = t$ if and only if

$$\left(\begin{vmatrix} a_1 & a_2 \\ c_1 & c_2 \end{vmatrix} b_1 - \begin{vmatrix} b_1 & b_2 \\ c_1 & c_2 \end{vmatrix} a_1 t \right) d_2 = \left(\begin{vmatrix} a_1 & a_2 \\ c_1 & c_2 \end{vmatrix} b_2 - \begin{vmatrix} b_1 & b_2 \\ c_1 & c_2 \end{vmatrix} a_2 t \right) d_1$$

Thus if t is a finite real number and $D = D[d_1, d_2]$, where

$$d_1 = \begin{vmatrix} a_1 & a_2 \\ c_1 & c_2 \end{vmatrix} b_1 - \begin{vmatrix} b_1 & b_2 \\ c_1 & c_2 \end{vmatrix} a_1 t$$

$$d_2 = \begin{vmatrix} a_1 & a_2 \\ c_1 & c_2 \end{vmatrix} b_2 - \begin{vmatrix} b_1 & b_2 \\ c_1 & c_2 \end{vmatrix} a_2 t$$

then $R(A, B, C, D) = t$.

Verification that $R(A, B, C, \text{---})$ is continuous is a topic not within the scope of this book, and will not be pursued. ∎

As a consequence of the following result, the cross ratio is independent of the choice of coordinate system (on the Euclidean line) with respect to which it is defined.

6.5 Proposition. If T is a transformation of the projective line represented by a 2×2 nonsingular matrix M,
$$T([p_1, p_2]) = [\langle p_1, p_2 \rangle M]$$
then, for points A, B, C, and D, at least three of which are distinct,
$$R(A, B, C, D) = R(T(A), T(B), T(C), T(D))$$

Proof. If $\mathbf{p}M = \mathbf{p}'$ and $\mathbf{q}M = \mathbf{q}'$, then

$$\begin{vmatrix} p_1' & p_2' \\ q_1' & q_2' \end{vmatrix} = \left| \begin{pmatrix} p_1 & p_2 \\ q_1 & q_2 \end{pmatrix} M \right| = \begin{vmatrix} p_1 & p_2 \\ q_1 & q_2 \end{vmatrix} \det M$$

Thus, canceling factors of $\det M$ in numerator and denominator,

$$R(T(A), T(B), T(C), T(D)) = \frac{\begin{vmatrix} a_1' & a_2' \\ c_1' & c_2' \end{vmatrix} \begin{vmatrix} b_1' & b_2' \\ d_1' & d_2' \end{vmatrix}}{\begin{vmatrix} b_1' & b_2' \\ c_1' & c_2' \end{vmatrix} \begin{vmatrix} a_1' & a_2' \\ d_1' & d_2' \end{vmatrix}}$$

$$= \frac{\begin{vmatrix} a_1 & a_2 \\ c_1 & c_2 \end{vmatrix} \begin{vmatrix} b_1 & b_2 \\ d_1 & d_2 \end{vmatrix}}{\begin{vmatrix} b_1 & b_2 \\ c_1 & c_2 \end{vmatrix} \begin{vmatrix} a_1 & a_2 \\ d_1 & d_2 \end{vmatrix}} = R(A, B, C, D) \quad \blacksquare$$

We now show that the cross ratio on any line in the projective plane or in projective three-space is independent of parametrization. Combining this result with Proposition 6.5 and Proposition 1.7, it follows that the cross ratio is invariant under projective transformations of lines: If T is a projective transformation of projective lines in the projective plane or in projective three-space, then

$$R(A, B, C, D) = R(T(A), T(B), T(C), T(D))$$

Furthermore, the next result will be the key to a simple algorithm for computing the cross ratio for points on an arbitrary projective line in the projective plane or in projective three-space.

6.6 Proposition. Let $T : RP^1 \to l$ be a parametrization of a projective line l in the projective plane or in projective three-space. If elements of RP^1 are identified with points of the projective line via standard homogeneous coordinates, then for any four points A, B, C, and D, at least three of which are distinct,

$$R(A, B, C, D) = R(T(A), T(B), T(C), T(D))$$

Proof. We verify this result in the case that $A = A[a, 1]$, $B = B[b, 1]$, $C = C[c, 1]$, and $D = D[d, 1]$ are distinct Euclidean points and l is a projective line in the projective plane (this is the generic case).

If T is represented by the matrix

$$M = \begin{pmatrix} q_1 & q_2 & q_3 \\ r_1 & r_2 & r_3 \end{pmatrix}$$

$\mathbf{q} = \langle q_1, q_2, q_3 \rangle$ and $\mathbf{r} = \langle r_1, r_2, r_3 \rangle$, then the undirected (Euclidean) distance from $T(A) = a\mathbf{q} + \mathbf{r}$ to $T(C) = c\mathbf{q} + \mathbf{r}$ is

$$|(a\mathbf{q} + \mathbf{r}) - (c\mathbf{q} + \mathbf{r})| = |c - a||\mathbf{q}|$$

Using the positive direction of the real numbers

$$R = \{[r, 1] \in RP^1 \mid r \in R\} \subseteq RP^1$$

to determine the positive direction for the set of Euclidean points on l, the directed distance $\text{dist}(T(A), T(C))$ from $T(A)$ to $T(C)$ is

$$\text{dist}(T(A), T(C)) = (c - a)|\mathbf{q}|$$

Similarly,
$$\text{dist}(T(B), T(D)) = (d - b)|\mathbf{q}|$$
$$\text{dist}(T(B), T(C)) = (c - b)|\mathbf{q}|$$
$$\text{dist}(T(A), T(D)) = (d - a)|\mathbf{q}|$$

Thus, canceling factors of $|\mathbf{q}|$ in numerator and denominator,
$$R(T(A), T(B), T(C), T(D)) = \frac{\text{dist}(T(A), T(C)) \times \text{dist}(T(B), T(D))}{\text{dist}(T(B), T(C)) \times \text{dist}(T(A), T(D))}$$
$$= \frac{(a-c)(b-d)}{(b-c)(a-d)} = R(A, B, C, D) \qquad \blacksquare$$

We now use Propositions 6.5 and 6.6 to establish a simple algorithm for computing the cross ratio for collinear points in the projective plane and in projective three-space.

6.7 Proposition. If A, B, C, and D are collinear points in the projective plane, at least three of which are distinct, and N is a 3×2 rank 2 matrix such that the transformation T_N from the line determined by A, B, C, and D to the projective line given by
$$T_N([p_1, p_2, p_3]) = [\langle p_1, p_2, p_3 \rangle N^{3 \times 2}]$$
takes A, B, C, and D to four points, at least three of which are distinct, then
$$R(A, B, C, D) = R(T_N(A), T_N(B), T_N(C), T_N(D))$$
Similarly, if A, B, C, and D are collinear points in projective three-space, at least three of which are distinct, and N is a 4×2 rank 2 matrix such that the transformation T_N from the line determined by A, B, C, and D to the projective line given by
$$T_N([p_1, p_2, p_3, p_4]) = [\langle p_1, p_2, p_3, p_4 \rangle N^{4 \times 2}]$$
takes A, B, C, and D to four points, at least three of which are distinct, then
$$R(A, B, C, D) = R(T_N(A), T_N(B), T_N(C), T_N(D))$$

Proof. If l is the line determined by A, B, C, and D and $T : RP^1 \to l$ is any parametrization of l represented by the 2×3 rank 2 matrix M,
$$T([\mathbf{p}]) = [\mathbf{p}M]$$
then $A = A[\mathbf{a}M]$, $B = B[\mathbf{b}M]$, $C = C[\mathbf{c}M]$, and $D = D[\mathbf{d}M]$ for some $[\mathbf{a}]$, $[\mathbf{b}]$, $[\mathbf{c}]$, and $[\mathbf{d}]$ in RP^1. Thus
$$R(T_N(A), T_N(B), T_N(C), T_N(D)) = R([\mathbf{a}MN], [\mathbf{b}MN], [\mathbf{c}MN], [\mathbf{d}MN])$$
$$= R([\mathbf{a}], [\mathbf{b}], [\mathbf{c}], [\mathbf{d}])$$
$$= R([\mathbf{a}M], [\mathbf{b}M], [\mathbf{c}M], [\mathbf{d}M])$$
$$= R(A, B, C, D)$$
the first equality following from the definition of T_N, the second following from Proposition 6.5 (since at least three of $[\mathbf{a}MN]$, $[\mathbf{b}MN]$, $[\mathbf{c}MN]$, and $[\mathbf{d}MN]$ are distinct, the vectors $\mathbf{a}MN$, $\mathbf{b}MN$, $\mathbf{c}MN$, and $\mathbf{d}MN$ in R^3 cannot be coplanar; it follows that MN is invertible), and the third following from Proposition 6.6. $\blacksquare$

6.8 Example. The points $A[1, 1, -1]$, $B[-3, 0, 1]$, $C[1, -2, 1]$, and $D[-1, -4, 3]$ are distinct collinear points in the projective plane (see Figure 3.53). If

$$N = \begin{pmatrix} 1 & 0 \\ 0 & 1 \\ 0 & 0 \end{pmatrix}$$

then $T_N(A) = [1, 1]$, $T_N(B) = [-3, 0]$, $T_N(C) = [1, -2]$, and $T_N(D) = [-1, -4]$, and at least three of these points are distinct. Thus

$$R(A, B, C, D) = \frac{\begin{vmatrix} 1 & 1 \\ 1 & -2 \end{vmatrix} \begin{vmatrix} -3 & 0 \\ -1 & -4 \end{vmatrix}}{\begin{vmatrix} -3 & 0 \\ 1 & -2 \end{vmatrix} \begin{vmatrix} 1 & 1 \\ -1 & -4 \end{vmatrix}} = \frac{(-3)(12)}{(6)(-3)} = 2$$

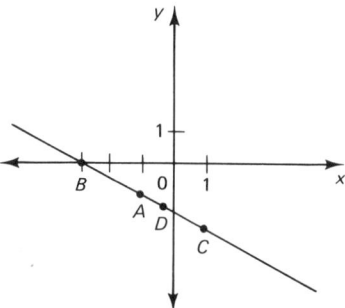

Figure 3.53 The cross ratio.

In this example, we in effect projected A, B, C, and D onto their first two coordinates and then computed the cross ratio of the resulting points. This is, in general, the algorithm for computing the cross ratio of four (collinear) points, at least three of which are distinct, in the projective plane or projective three-space: Determine two coordinates onto which the projections of any two of A, B, C, and D are different. At least three of the projections A', B', C', and D' of A, B, C, and D onto these coordinates will automatically be distinct, and

$$R(A, B, C, D) = R(A', B', C', D')$$

6.9 Example. The points $A[1, 2, 1]$, $B[3, 2, 0]$, $C[0, 4, 3]$, and $D[2, 0, -1]$ are distinct collinear points in the projective plane. Since the projections of A and B onto their first two coordinates are different, we may project A, B, C, and D onto their respective first two coordinates, obtaining $A'[1, 2]$, $B'[3, 2]$, $C'[0, 4]$, and $D'[2, 0]$, and

$$R(A, B, C, D) = R(A', B', C', D') = \frac{\begin{vmatrix} 1 & 2 \\ 0 & 4 \end{vmatrix} \begin{vmatrix} 3 & 2 \\ 2 & 0 \end{vmatrix}}{\begin{vmatrix} 3 & 2 \\ 0 & 4 \end{vmatrix} \begin{vmatrix} 1 & 2 \\ 2 & 0 \end{vmatrix}} = \frac{(4)(-4)}{(12)(-4)} = \frac{1}{3}$$

We could also have projected onto the first and third coordinates or onto the second and third coordinates and obtained the same result.

In general, however, not all possible combinations of projections will work.

6.10 Example. The points $A[1, 1, 1, 0]$, $B[2, 1, 1, 2]$, $C[1, 2, 2, 2]$, and $D[4, 3, 3, 2]$ are distinct collinear points in projective three-space. Since the projections of A and B onto their first two coordinates are different, we may project A, B, C, and D onto their respective first two coordinates, obtaining $A'[1, 1]$, $B'[2, 1]$, $C'[1, 2]$, and $D'[4, 3]$, and

$$R(A, B, C, D) = R(A', B', C', D') = -\tfrac{2}{3}$$

In this case we could also have projected onto the first and third coordinates, the first and fourth coordinates, the second and fourth coordinates, or the third and fourth coordinates. However, we could not project onto the second and third coordinates, since the projections of A, B, C, and D onto these coordinates only give one point: $P[1, 1]$.

One further note: the projections of A, B, C, and D must all be points.

6.11 Example. The points $A[1, 0, 0]$, $B[1, 1, 1]$, $C[2, 1, 1]$, and $D[3, 2, 2]$ are distinct collinear points in the projective plane. Since the second and third coordinates of A are both 0 and $[0, 0]$ is meaningless, we cannot project onto the second and third coordinates. We can, however, project onto the first two coordinates, and doing this we find $R(A, B, C, D) = \tfrac{1}{2}$.

We close this section by relating the cross ratio and general homogeneous coordinates of a point in the projective plane. The context of this relationship is a description of general homogeneous coordinates in the projective plane similar to the classical description of Cartesian coordinates in the Euclidean plane. A similar description is valid for projective three-space.

Suppose (see Figure 3.54) that four distinct points I_x, I_y, O, and U, no three of which are collinear, are given in the projective plane. (Recall that I_x and I_y represent the ideal points in the x- and y-directions, respectively, O represents the origin, and U represents the unit point.) We assign coordinates to these points as follows: $I_x = I_x[1, 0, 0]$, $I_y = I_y[0, 1, 0]$, $O = O[0, 0, 1]$, and $U = U[1, 1, 1]$. Using Proposition 2.2.7(a) it follows that the line OI_x passing through O and I_x has coordinates $[0, 1, 0]$, the line OI_y passing through O and I_y has coordinates $[1, 0, 0]$, and the line $I_x I_y$ passing through I_x and I_y (the ideal line of this general homogeneous coordinate system) has coordinates $[0, 0, 1]$. From Proposition 2.2.7(b) it follows that the point U_x of intersection of the lines UI_y and OI_x has coordinates $[1, 0, 1]$ and the point U_y of intersection of the lines UI_x and OI_y has coordinates $[0, 1, 1]$.

Let P be an arbitrary (Euclidean) point in the projective plane with general homogeneous coordinates $[x, y, 1]$. From Proposition 2.2.7 it follows that the point P_x of intersection of the lines PI_y and OI_x has coordinates $[x, 0, 1]$ and the point P_y of intersection of the lines PI_x and OI_y has coordinates $[0, y, 1]$. Thus $x = R(I_x, O, U_x, P_x)$ and $y = R(I_y, O, U_y, P_y)$. For example, to verify that

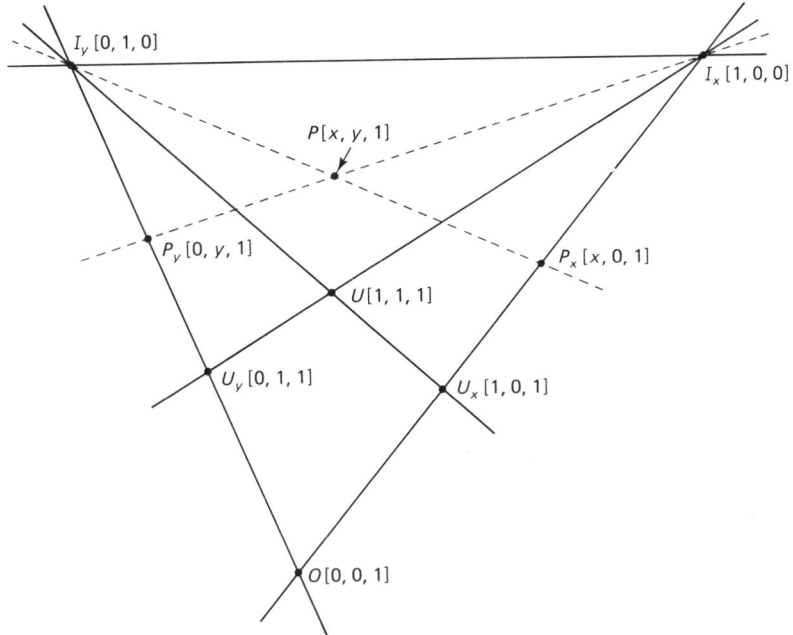

Figure 3.54 The cross ratio and general homogeneous coordinates.

$x = R(I_x, O, U_x, P_x)$, we can project I_x, I_y, O, and U_x onto the first and third coordinates and compute

$$R(I_x, I_y, O, U_x) = \frac{\begin{vmatrix} 1 & 0 \\ 1 & 1 \end{vmatrix} \begin{vmatrix} 0 & 1 \\ x & 1 \end{vmatrix}}{\begin{vmatrix} 0 & 1 \\ 1 & 1 \end{vmatrix} \begin{vmatrix} 1 & 0 \\ x & 1 \end{vmatrix}} = \frac{(1)(-x)}{(-1)(1)} = x$$

Conversely, to locate the point P whose general homogeneous coordinates are $[x, y, 1]$, first locate the point P_x along the "x-axis" for which $x = R(I_x, O, U_x, P_x)$ ($P_x = P_x[x, 0, 1]$) and the point P_y along the "y-axis" for which $y = R(I_y, O, U_y, P_y)$ ($P_y = P_y[0, y, 1]$). The point $P = P[x, y, 1]$ is the intersection of the lines $P_x I_y$ and $P_y I_x$.

In other words, the cross ratio is a perspective scaling. The general homogeneous coordinates of a (Euclidean) point are determined by this scaling along the x- and y-axes of the general homogeneous coordinate system. (Note, in particular, the similarity between Figure 3.54 and Figure 3.55 as it would be viewed in perspective.)

SECTION 6 EXERCISES:

1. In each of the following cases, four points A, B, C, and D on the projective line are given. In each case find $R(A, B, C, D)$.

Sec. 6   The Cross Ratio

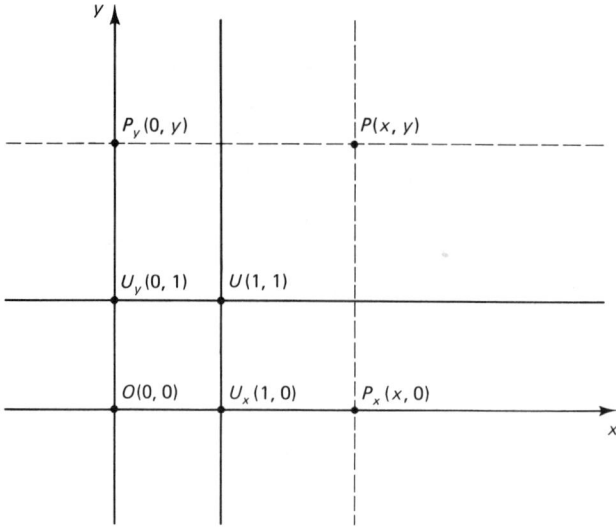

Figure 3.55 Coordinates in the Euclidean plane.

    (a) $A[1, 2]$, $B[1, 5]$, $C[2, 9]$, $D[1, 3]$
    (b) $A[1, 1]$, $B[3, 1]$, $C[0, 1]$, $D[5, 2]$
    (c) $A[1, 2]$, $B[1, 0]$, $C[3, -2]$, $D[1, 1]$
    (d) $A[0, 1]$, $B[1, 1]$, $C[1, 0]$, $D[2, 3]$

2. In each of the following cases, four collinear points A, B, C, and D in the projective plane are given. In each case find $R(A, B, C, D)$.
    (a) $A[1, 2, -3]$, $B[2, 1, 1]$, $C[3, 3, -2]$, $D[3, 0, 5]$
    (b) $A[2, 1, -1]$, $B[1, 2, 1]$, $C[1, 1, 0]$, $D[0, 1, 1]$
    (c) $A[1, -2, 1]$, $B[2, 1, 1]$, $C[0, 5, -1]$, $D[1, 3, 0]$
    (d) $A[1, 1, 1]$, $B[1, 2, 3]$, $C[0, 1, 2]$, $D[1, 0, -1]$

3. In each of the following cases, four collinear points A, B, C, and D in projective three-space are given. In each case, find $R(A, B, C, D)$.
    (a) $A[1, 2, 1, 0]$, $B[2, 0, 0, 1]$, $C[3, 2, 1, 1]$, $D[0, 4, 2, -1]$
    (b) $A[3, -2, 1, 2]$, $B[1, 1, 2, 1]$, $C[2, -3, -1, 1]$, $D[2, 2, 4, 2]$
    (c) $A[3, 2, 1, 0]$, $B[0, 1, 2, 3]$, $C[1, 1, 1, 1]$, $D[1, 0, -1, -2]$
    (d) $A[1, -2, 0, 3]$, $B[2, 0, 1, 3]$, $C[3, -2, 1, 6]$, $D[5, 6, 4, 3]$

4. Verify that if $R(A, B, C, D) = t$, $t \neq 0$, then
    (a) $R(B, A, C, D) = 1/t$
    (b) $R(C, B, A, D) = t/(t - 1)$
    (c) $R(D, B, C, A) = 1 - t$
    (d) $R(A, C, B, D) = 1 - t$
    (e) $R(A, D, C, B) = t/(t - 1)$
    (f) $R(A, B, D, C) = 1/t$

5. Suppose (see Figure 3.56) A, B, C, and D are four points on a line l in the projective plane, and O is a point in the projective plane not on l. Verify that

$$R(A, B, C, D) = \frac{(\sin \angle AOC)(\sin \angle BOD)}{(\sin \angle BOC)(\sin \angle AOD)}$$

6. Show that the points A, B, C, and D form a harmonic quadrad (see Exercise 1.3.17) if and only if $R(A, B, C, D) = -1$. (*Hint:* Use general homogeneous coordinates.)

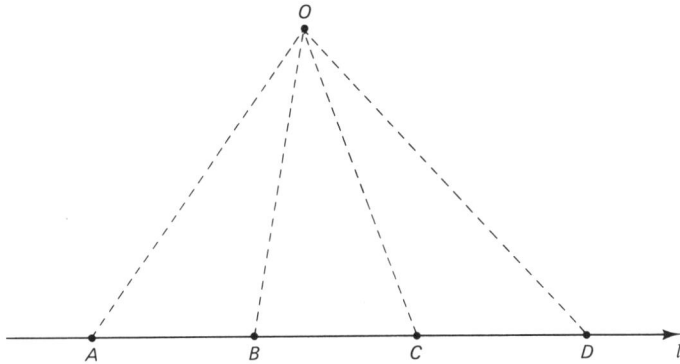

Figure 3.56 A version of the cross ratio.

SECTION 7: AN IMPORTANT RESULT

One important result cannot be found in any one place in this chapter. It is a ubiquitous result whose special cases have appeared continually throughout the chapter.

We let e_i denote the vector whose components are all 0 except for the ith component, which is 1. We let (m, n) denote any one of the following six pairs of integers:

$$(2, 1) \quad (2, 2) \quad (3, 2) \quad (2, 3) \quad (3, 3) \quad (4, 3)$$

(in general, m and n are positive integers for which $m - 1 \leq n$).

7.1 Proposition. If $[\mathbf{p}_1], \ldots, [\mathbf{p}_{m+1}]$ are $m + 1$ elements of RP^n for which

$$\mathbf{p}_{m+1} = c_1 \mathbf{p}_1 + \cdots + c_m \mathbf{p}_m$$

for some constants $c_1, \ldots, c_m$, none of which are zero, and if M is the $m \times n$ matrix whose ith row is $c_i \mathbf{p}_i$, then right multiplication by M represents a transformation $T : RP^{m-1} \to RP^n$,

$$T([\mathbf{p}]) = [\mathbf{p} M^{m \times n}]$$

for which $T([\mathbf{e}_i]) = [\mathbf{p}_i]$, for $i = 1, \ldots, m$, and $T([1, \ldots, 1]) = [\mathbf{p}_{m+1}]$.

Proof. For $i = 1, \ldots, m$, $\mathbf{e}_i M$ is the ith row of M. And $\langle 1, \ldots, 1 \rangle M$ is the sum of the rows of M, which is $c_1 \mathbf{p}_1 + \cdots + c_m \mathbf{p}_m = \mathbf{p}_{m+1}$. ∎

Special cases of Proposition 7.1 arise when discussing the following:

1. Projective transformations and, in particular, the Fundamental Theorem of Projective Geometry for Lines, Planes, and Projective Three-Space.
2. The parametric description of projective lines in the projective plane and in projective three-space and the parametric description of planes in projective three-space.
3. General homogeneous coordinates on the projective line, on the projective plane, and on projective three-space.

Indeed, it is this result that relates the concepts of projective transformation, parametrization, and general homogeneous coordinates.

What makes Proposition 7.1 useful is that there is a simple algorithm that determines whether the equation of Proposition 7.1 holds and, if so, what the constants $c_1, \ldots, c_m$ are.

7.2 Algorithm. Let $[\mathbf{p}_1], \ldots, [\mathbf{p}_{m+1}]$ be $m + 1$ elements of RP^n. There exist constants $c_1, \ldots, c_m$ for which

$$\mathbf{p}_{m+1} = c_1 \mathbf{p}_1 + \cdots + c_m \mathbf{p}_m$$

if and only if the matrix NN^T, where N is the $m \times n$ matrix whose ith row is $\mathbf{p}_i$, is invertible (that is, $\det(NN^T) \neq 0$). In this case,

$$\langle c_1, \ldots, c_m \rangle = \mathbf{p}_{m+1} N^+ = \mathbf{p}_{m+1} N^T (NN^T)^{-1}$$

Proof. If

$$c_1 \mathbf{p}_1 + \cdots + c_m \mathbf{p}_m = \mathbf{p}_{m+1}$$

then

$$\langle c_1, \ldots, c_m \rangle N = \mathbf{p}_{m+1}$$

and

$$\langle c_1, \ldots, c_m \rangle NN^T = \mathbf{p}_{m+1} N^T$$

This equation has a solution $\langle c_1, \ldots, c_m \rangle$ if and only if NN^T is invertible. And if NN^T is invertible,

$$\langle c_1, \ldots, c_m \rangle = \mathbf{p}_{m+1} N^+ = \mathbf{p}_{m+1} N^T (NN^T)^{-1} \quad \blacksquare$$

7.3 Example. Given $[\mathbf{p}_1] = [2, 1, -3, -1]$, $[\mathbf{p}_2] = [1, -2, 1, 2]$, and $[\mathbf{p}_3] = [2, -1, -1, 1]$ (so $m = 2$ and $n = 3$),

$$N = \begin{pmatrix} 2 & 1 & -3 & -1 \\ 1 & -2 & 1 & 2 \end{pmatrix}$$

so

$$NN^T = \begin{pmatrix} 15 & -5 \\ -5 & 10 \end{pmatrix}$$

Since $\det(NN^T) = 125 \neq 0$, NN^T is invertible and, in fact,

$$(NN^T)^{-1} = \frac{1}{125} \begin{pmatrix} 10 & 5 \\ 5 & 15 \end{pmatrix} = \frac{1}{25} \begin{pmatrix} 2 & 1 \\ 1 & 3 \end{pmatrix}$$

Thus

$$N^+ = N^T (NN^T)^{-1} = \tfrac{1}{5} \begin{pmatrix} 1 & 1 \\ 0 & -1 \\ -1 & 0 \\ 0 & 1 \end{pmatrix}$$

and
$$\langle c_1, c_2 \rangle = \langle 2, -1, -1, 1 \rangle N^+ = \langle \tfrac{3}{5}, \tfrac{4}{5} \rangle$$
This checks since
$$c_1 \langle 2, 1, -3, -1 \rangle + c_2 \langle 1, -2, 1, 2 \rangle$$
$$= \langle \tfrac{6}{5}, \tfrac{3}{5}, -\tfrac{9}{5}, -\tfrac{3}{5} \rangle + \langle \tfrac{4}{5}, -\tfrac{8}{5}, \tfrac{4}{5}, \tfrac{8}{5} \rangle = \langle 2, -1, -1, 1 \rangle$$

Since we are working with representatives of elements of RP^3, we can simplify our arithmetic by ignoring multiplicative constants (see Exercise 1.5): That is, we can let

$$NN^T = \begin{pmatrix} 3 & -1 \\ -1 & 2 \end{pmatrix}$$

in which case

$$(NN^T)^{-1} = \begin{pmatrix} 2 & 1 \\ 1 & 3 \end{pmatrix}$$

so

$$N^+ = \begin{pmatrix} 1 & 1 \\ 0 & -1 \\ -1 & 0 \\ 0 & 1 \end{pmatrix}$$

and
$$\langle c_1, c_2 \rangle = \langle 2, -1, -1, 1 \rangle N^+ = \langle 3, 4 \rangle$$
This checks since
$$c_1 \langle 2, 1, -3, -1 \rangle + c_2 \langle 1, -2, 1, 2 \rangle$$
$$= \langle 6, 3, -9, -3 \rangle + \langle 4, -8, 4, 8 \rangle = \langle 10, -5, -5, 5 \rangle$$
and, although $\langle 2, -1, -1, 1 \rangle \neq \langle 10, -5, -5, 5 \rangle$, both represent the same element of RP^3. We can also ignore multiplicative constants when working with RP^1 and RP^2.

Since $c_1 = 3$ and $c_2 = 4$ are both nonzero, it follows that the matrix M described in Proposition 7.1 exists in this case: If

$$M = \begin{pmatrix} 6 & 3 & -9 & -3 \\ 4 & -8 & 4 & 8 \end{pmatrix}$$

then right multiplication by M represents a transformation $T : RP^1 \to RP^3$, for which $T([1, 0]) = [\mathbf{p}_1] = [2, 1, -3, -1]$, $T([0, 1]) = [\mathbf{p}_2] = [1, -2, 1, 2]$, and $T([1, 1]) = [\mathbf{p}_3] = [2, -1, -1, 1]$.

7.4 Example. Given $[\mathbf{p}_1] = [1, 0, 0]$, $[\mathbf{p}_2] = [0, 1, 0]$, $[\mathbf{p}_3] = [0, 0, 1]$, and $[\mathbf{p}_4] = [1, 1, 0]$ (so $m = 3$ and $n = 2$), it follows that $c_1 = 1$, $c_2 = 1$, and $c_3 = 0$. Despite the fact that there are constants c_1, c_2, and c_3 for which $\mathbf{p}_4 =$

$c_1\mathbf{p}_1 + c_2\mathbf{p}_2 + c_3\mathbf{p}_3$, since $c_3 = 0$, there is no matrix M representing a transformation for which $T([0, 0, 1]) = [\mathbf{p}_3]$. In particular, if

$$M = \begin{pmatrix} 1 & 0 & 0 \\ 0 & 1 & 0 \\ 0 & 0 & 0 \end{pmatrix}$$

it would follow that $T([0, 0, 1]) = [0, 0, 0]$, which is meaningless.

One final observation: The matrix M has rank m, and there is a simple technique for finding M^+.

7.5 Algorithm. The right inverse M^+ of M is the $n \times m$ matrix whose ith column is $\mathbf{q}_i/c_i$ where $\mathbf{q}_i$ is the ith column of N^+.

This algorithm is useful because we must inevitably find N^+ in the computation of M; thus, having computed M, M^+ can be written down without any further computations. (Otherwise, it would be necessary to compute the matrices MM^T, $(MM^T)^{-1}$, and $M^T(MM^T)^{-1} = M^+$.)

Proof (of Algorithm 7.5). If $N = (p_{ij})$ and $N^+ = (q_{ij})$, then

$$NN^+ = \begin{pmatrix} p_{11} & \cdots & p_{1n} \\ \vdots & & \vdots \\ p_{m1} & \cdots & p_{mn} \end{pmatrix} \begin{pmatrix} q_{11} & \cdots & q_{1m} \\ \vdots & & \vdots \\ q_{n1} & \cdots & q_{nm} \end{pmatrix} = Id^{m \times m}$$

Thus, for each i and j, $\Sigma_k\, p_{ik}q_{kj}$ is 1 if $i = j$, and 0 otherwise.

Now consider

$$MM^+ = \begin{pmatrix} c_1p_{11} & \cdots & c_1p_{1n} \\ \vdots & & \vdots \\ c_mp_{m1} & \cdots & c_mp_{mn} \end{pmatrix} \begin{pmatrix} \frac{q_{11}}{c_1} & \cdots & \frac{q_{1m}}{c_m} \\ \vdots & & \vdots \\ \frac{q_{n1}}{c_1} & \cdots & \frac{q_{nm}}{c_m} \end{pmatrix}$$

The (i, i)th entry of this product,

$$(MM^+)_{ii} = \sum_k (c_i p_{ik})\left(\frac{q_{ki}}{c_i}\right) = \sum_k p_{ik}q_{ki} = 1$$

and if $i \neq j$, the (i, j)th entry of this product is

$$(MM^+)_{ij} = \sum_k (c_i p_{ik})\left(\frac{q_{kj}}{c_j}\right) = \left(\frac{c_i}{c_j}\right)\sum_k p_{ik}q_{kj} = \left(\frac{c_i}{c_j}\right)0 = 0 \qquad \blacksquare$$

7.6 Example. (Continuation of Example 7.3): Since $c_1 = 3$, $c_2 = 4$, and

$$N^+ = \begin{pmatrix} 1 & 1 \\ 0 & -1 \\ -1 & 0 \\ 0 & 1 \end{pmatrix}$$

it follows that

$$M^+ = \begin{pmatrix} \frac{1}{3} & \frac{1}{4} \\ 0 & -\frac{1}{4} \\ -\frac{1}{3} & 0 \\ 0 & \frac{1}{4} \end{pmatrix}$$

Observe that

$$MM^+ = \begin{pmatrix} 6 & 3 & -9 & -3 \\ 4 & -8 & 4 & 8 \end{pmatrix} \begin{pmatrix} \frac{1}{3} & \frac{1}{4} \\ 0 & -\frac{1}{4} \\ -\frac{1}{3} & 0 \\ 0 & \frac{1}{4} \end{pmatrix}$$

$$= \begin{pmatrix} 2 & 1 & -3 & -1 \\ 1 & -2 & 1 & 2 \end{pmatrix} \begin{pmatrix} 1 & 1 \\ 0 & -1 \\ -1 & 0 \\ 0 & 1 \end{pmatrix} = NN^+ = \begin{pmatrix} 1 & 0 \\ 0 & 1 \end{pmatrix}$$

SECTION 7 EXERCISES:

1. For each choice of (m, n) for which Proposition 7.1 applies, describe the geometric interpretation of the transformation $T : RP^{m-1} \to RP^n$.
2. In each of the following cases, $m + 1$ elements of RP^n are given. Find the matrix representing the transformation $T : RP^{m-1} \to RP^n$ for which $T([e_i]) = [p_i]$ for $i = 1, \ldots, m$, and $T([u]) = p_{m+1}$.
   (a) [2, 5], [1, 3], [4, 7]
   (b) [2, 1, 3], [1, 2, −1], [1, −4, 9]
   (c) [1, −1, 0], [−3, 0, 2], [1, −2, −1], [4, −4, 5]
   (d) [2, −1, 3, 1], [2, 3, −5, −7], [4, 0, 2, −2]
   (e) [2, 1, 3, 2], [1, 2, −1, −1], [1, 4, 1, 0], [5, −3, 6, 5]
   (f) [2, −1, 0, 0], [−1, 2, −1, 0], [0, −1, 2, −1], [0, 0, −1, 2], [0, 0, 0, 1]

4

Two-Dimensional Graphics

INTRODUCTION

In this chapter we investigate the use of projective geometry in the study of two-dimensional graphics. Applications of two-dimensional graphics include floor plans of buildings (of interest to architects, engineers, builders, interior designers, and interior decorators), the design of building exteriors, machine parts and other manufactured objects (multiview orthographic plans), the design and manufacture of electronic microcircuitry, graphs, road maps, and star charts, to name just a few.

In Section 1 we discuss some basic concepts and terminology, and then analyze the complete viewing operation of two-dimensional imaging. In Section 2 we describe a simple two-dimensional graphics library. In Section 3 we consider a general purpose two-dimensional display program. This program not only demonstrates how the library is used, but also shows several ways in which projective geometry arises in two-dimensional graphics.

The library developed in this chapter is a pedagogical tool. There is no intention that it should be as complete and general as possible. In particular, to make this library more useful you might want to add, for example, error messages and commands useful to a given family of applications. Despite this, it is strongly recommended that you implement the library (in its simplest form, at least): doing so not only reinforces what has been said, but the resulting software also provides a useful tool for later analysis.

SECTION 1: THE COMPLETE VIEWING OPERATION IN TWO DIMENSIONS

Several different coordinate spaces can arise in the solution of a single graphics problem. The two most basic such coordinate spaces are world coordinate space and device coordinate space.

World coordinate two-space is a coordinate Euclidean two-space within which the object, or objects, of interest naturally exist. The coordinates of world coordinate two-space are *world coordinates*. Suppose, for example, a floor plan of an office building is being created for use in an interactive environment by an interior decorator. The world coordinate space might appear as illustrated in Figure 4.1. Observe that in establishing this world coordinate system, choices of origin, axes, and units are made. Choices of world coordinate systems are not unique. In particular, although in many problems there may be a natural choice of units, there is not always a natural choice for the origin and coordinate axes of a rectangular coordinate system (or the pole and polar axis of a polar coordinate system).

A *window* is that part of world coordinate space to be displayed at a given time. For example, if the floor plan of the office building described above is too large to fit on the designated display device (and, at the same time, maintain reasonable resolution), it is necessary to restrict that part of the floor plan that is displayed at any given time (see Figure 4.2).

Device coordinate space is the natural coordinate space of a graphics device. Some examples of device coordinate spaces are illustrated in Figure 4.3. Observe that device coordinates can vary in many ways from device to device: Location of the origin, position of coordinate axes, orientation, and units all can vary.

A *viewport* is that part of device coordinate space in which the window of world coordinate space is displayed. For example, if we are working on the device illustrated in Figure 4.3(a), we might wish to display the window in world coordinates in the region illustrated in Figure 4.4, leaving room at the bottom of the display for a menu or room around the edges for a border.

Viewports are of limited size. The process of screening geometric objects so that only those parts of objects within the viewport are displayed is known as *clipping*. The implementation of clipping is largely device- and application-dependent. We discuss clipping only to the extent that techniques of projective geometry are relevant to it. (For further reference, see [Newman and Sproull] and [Foley and van Dam].)

The world coordinate system is natural for the user; the device coordinate system is natural for the device. The process of mapping the window to the viewport (world coordinates to device coordinates) is the *complete viewing operation* (see Figure 4.5). There are intrinsic differences between world coordinate space and device coordinate space. For one, world coordinate space is infinite and unbounded, while device coordinate space is discrete and bounded. For another, points in world coordinate space do not have length or width, while points in device coordinate space do have length and width. Consequently, the complete

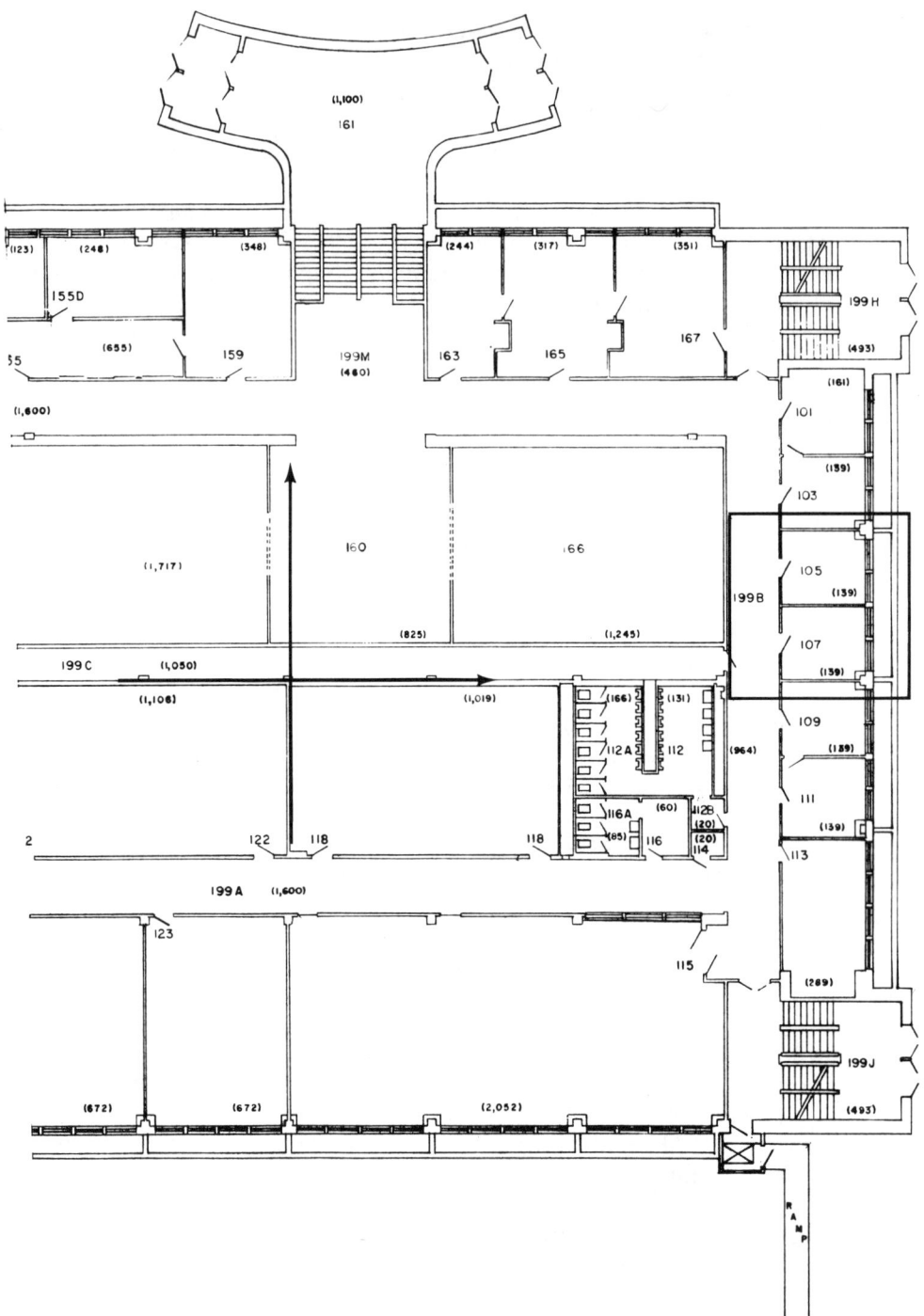

Figure 4.1 World coordinate space for an office building floor plan.

178 Two-Dimensional Graphics Chap. 4

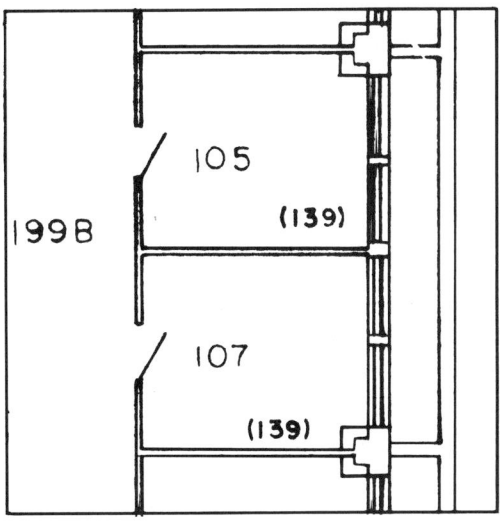

Figure 4.2 A window in the office building floor plan.

viewing operation is many-to-one, and implementing it involves more than simply implementing a mathematical transformation between two Euclidean spaces. It involves, for example, making decisions when drawing a line (or curve) about which points are plotted so that the appearance of the line (or curve) is acceptable. In this book we study the complete viewing operation only as a mathematical transformation and do not pursue these device-dependent issues.

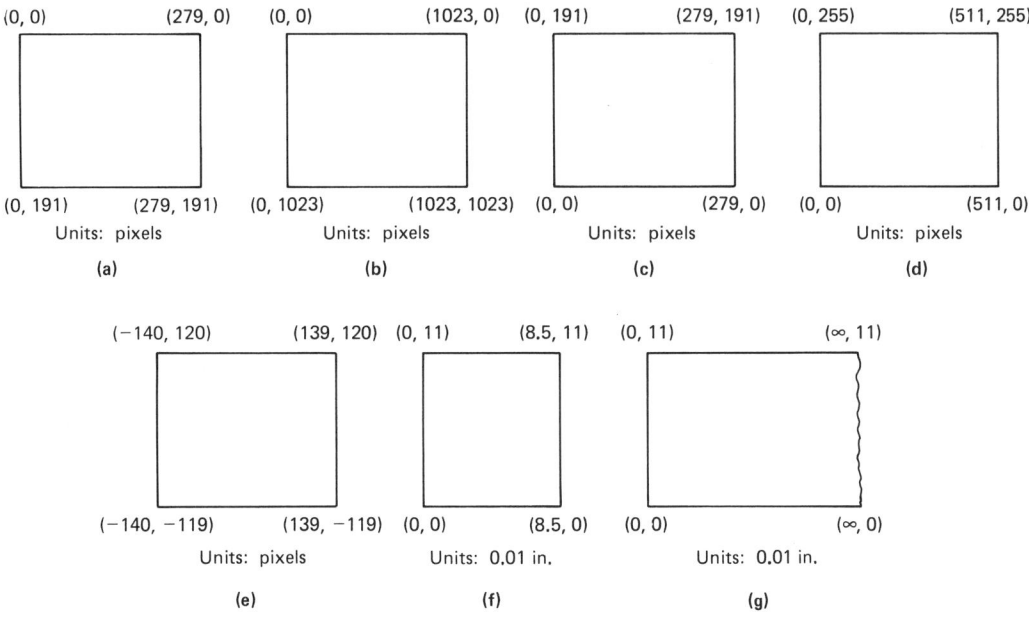

Figure 4.3 Examples of device coordinates.

Sec. 1   The Complete Viewing Operation in Two Dimensions

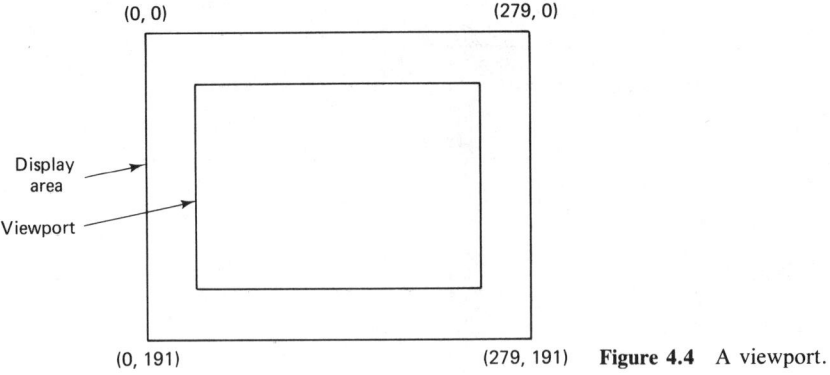

Figure 4.4 A viewport.

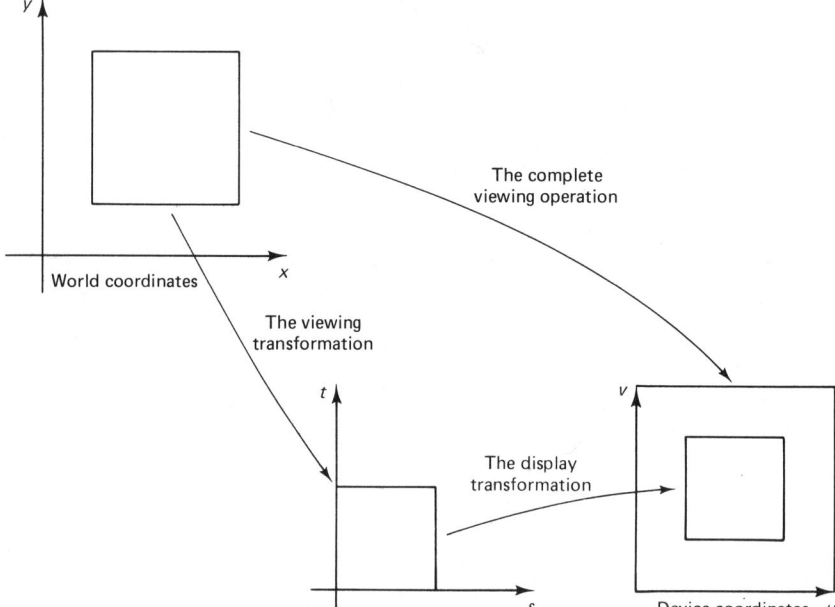

Figure 4.5 The complete viewing operation.

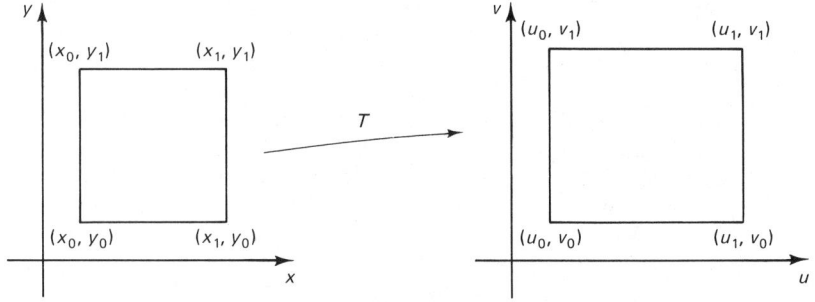

Figure 4.6 Transformation of rectangles.

Assuming the window and viewport are rectangles whose sides are parallel to the world and device coordinate axes, respectively, the complete viewing operation is described by the transformation given in Proposition 1.1: This transformation takes a rectangle in one coordinate plane to a rectangle in another coordinate plane. To describe the complete viewing operation we choose the window in world coordinate space for the first rectangle and the viewport in device coordinate space for the second rectangle.

1.1 Proposition. The transformation taking the rectangle whose vertices have coordinates (x_0, y_0), (x_1, y_0), (x_1, y_1) and (x_0, y_1) to the rectangle whose vertices have coordinates (u_0, v_0), (u_1, v_0), (u_1, v_1) and (u_0, v_1), respectively (see Figure 4.6) is given by

$$u = \frac{u_1 - u_0}{x_1 - x_0}x + \frac{u_0 x_1 - x_0 u_1}{x_1 - x_0}$$

$$v = \frac{v_1 - v_0}{y_1 - y_0}y + \frac{v_0 y_1 - y_0 v_1}{y_1 - y_0}$$

where (u, v) are the coordinates of the image of the point whose coordinates are (x, y).

1.2 Example. To describe the complete viewing operation given the window and viewport illustrated in Figure 4.7, let $(x_0, y_0) = (10, 16)$, $(x_1, y_1) = (58, 52)$, $(u_0, v_0) = (16, 50)$, and $(u_1, v_1) = (216, 210)$. Then

$$u = \frac{216 - 16}{58 - 10}x + \frac{16 \times 58 - 10 \times 216}{58 - 10} = 4.1667 x - 25.6667$$

$$v = \frac{210 - 50}{52 - 16}y + \frac{50 \times 52 - 16 \times 210}{52 - 16} = 4.4444 y - 21.1111$$

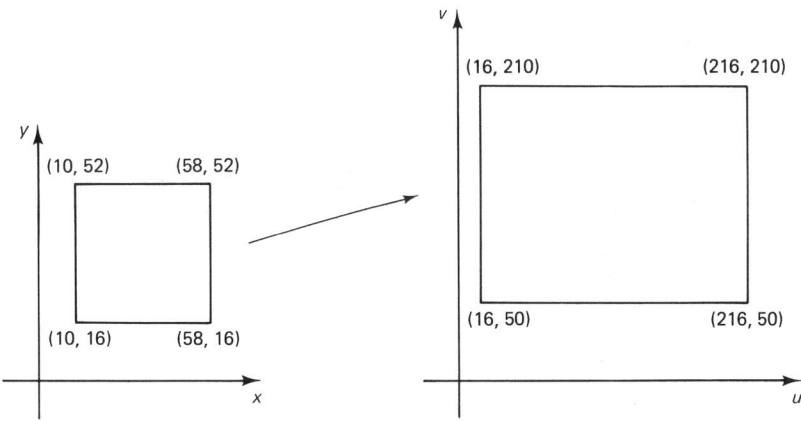

Figure 4.7 A complete viewing operation.

To verify that this is the correct transformation, observe that if you substitute (x_0, y_0) and (x_1, y_1) into these equations for (x, y), the resulting (u, v) values are (u_0, v_0) and (u_1, v_1).

Proof (of Proposition 1.1). We derive this result in two ways—first, using techniques of Euclidean geometry, and second, using techniques of projective geometry. The Euclidean derivation is simple and straightforward. The projective derivation introduces some important concepts, and it generalizes to useful constructions in both two and three dimensions.

1.3 Lemma. Given two pairs of coordinates (s_0, t_0) and (s_1, t_1), the linear function $t = f(s)$ for which $t_0 = f(s_0)$ and $t_1 = f(s_1)$ (see Figure 4.8) is given by

$$f(s) = \frac{t_1 - t_0}{s_1 - s_0} s + \frac{t_0 s_1 - s_0 t_1}{s_1 - s_0}$$

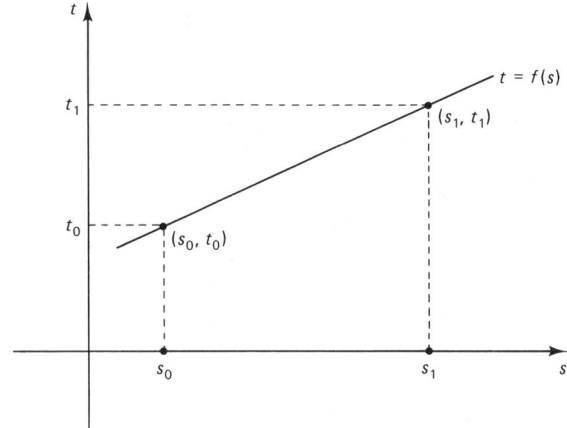

Figure 4.8 Linear interpolation.

Proposition 1.1 follows from two applications of this result: once in the x-direction (in which case we use Lemma 1.3 to map x_0 and x_1 to u_0 and u_1, respectively), and once in the y-direction (in which case we use Lemma 1.3 to map y_0 and y_1 to v_0 and v_1, respectively).

The projective derivation of Proposition 1.1 is based on the construction given in Example 3.2.10. However, instead of using the intermediate points $I_s[1, 0, 0]$, $I_t[0, 1, 0]$, $O[0, 0, 1]$, and $U[1, 1, 1]$, we use a different set of intermediate points. The transformation described in Proposition 1.1 can be regarded (see Figure 4.9) as the composite of two projective transformations—first, the inverse T_{object}^{-1} of the transformation T_{object} that takes $O[0, 0, 1]$, $U_s[1, 0, 1]$, $U_t[0, 1, 1]$, and $U[1, 1, 1]$ to $A_{\text{object}}[x_0, y_0, 1]$, $B_{\text{object}}[x_1, y_0, 1]$, $C_{\text{object}}[x_1, y_1, 1]$, and $D_{\text{object}}[x_0, y_1, 1]$, respectively; and second, the transformation T_{image} that takes $O[0, 0, 1]$, $U_s[1, 0, 1]$, $U_t[0, 1, 1]$, and $U[1, 1, 1]$ to $A_{\text{image}}[u_0, v_0, 1]$, $B_{\text{image}}[u_1, v_0, 1]$, $C_{\text{image}}[u_1, v_1, 1]$, and $D_{\text{image}}[u_0, v_1, 1]$, respectively. Now T_{object} is represented by

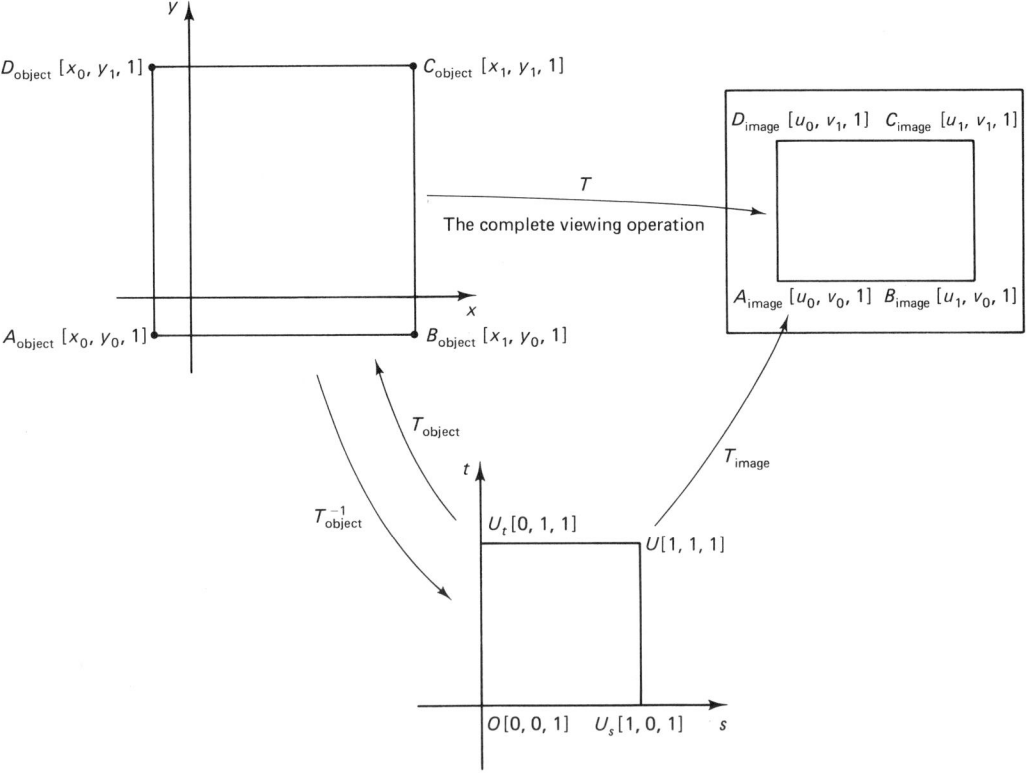

Figure 4.9 The complete viewing operation as a composition.

$$M_{\text{object}} = \begin{pmatrix} x_1 - x_0 & 0 & 0 \\ 0 & y_1 - y_0 & 0 \\ x_0 & y_0 & 1 \end{pmatrix}$$

(see Exercise 4), so T_{object}^{-1} is represented by

$$M_{\text{object}}^{-1} = \begin{pmatrix} \dfrac{1}{x_1 - x_0} & 0 & 0 \\ 0 & \dfrac{1}{y_1 - y_0} & 0 \\ \dfrac{-x_0}{x_1 - x_0} & \dfrac{-y_0}{y_1 - y_0} & 1 \end{pmatrix}$$

and T_{image} is represented by

$$M_{\text{image}} = \begin{pmatrix} u_1 - u_0 & 0 & 0 \\ 0 & v_1 - v_0 & 0 \\ u_0 & v_0 & 1 \end{pmatrix}$$

The conclusion follows since $M = M_{\text{object}}^{-1} M_{\text{image}}$ represents T (see Exercise 5). ■

There is a natural division of the complete viewing operation into two parts—T_{object}^{-1}, the *viewing transformation,* and T_{image}, the *display transformation.* (We refer to the viewing operation as the *complete* viewing operation to avoid any potential confusion between it and the viewing transformation.) The choice of the intermediate rectangle does not affect the product matrix $M = M_{\text{object}}^{-1} M_{\text{image}}$ which represents the complete viewing operation (see Exercise 8). When we select the rectangle with vertices O, U_s, U_t, and U in the intermediate space, the coordinates of the intermediate space are *normalized device coordinates.*

Normalized device coordinates assist in the development of portable graphics software since they act as a standard communication link between hardware and software: for a different display device we only need to change the display transformation (see Section 2). We can also implement a simple clipping algorithm using normalized device coordinates: a point is plotted if both its normalized device coordinates are between 0 and 1 (see Exercise 2.23). Since this algorithm operates in normalized device coordinate space, it is independent of the size and shape of the window and viewport.

By using the full generality of Example 3.2.10, we can allow the window to be rotated with respect to the world coordinate axes, the viewport to be rotated with respect to the device coordinate axes, or both (see Figure 4.10). We can, in fact, allow the window or viewport to be an arbitrary quadrilateral (see Exercise 10). If the window is a rectangle and the viewport is an arbitrary quadrilateral, the resulting image appears to be in perspective (see Figure 4.11). This method of achieving a three-dimensional effect with two-dimensional projective geometry is known as *billboarding.*

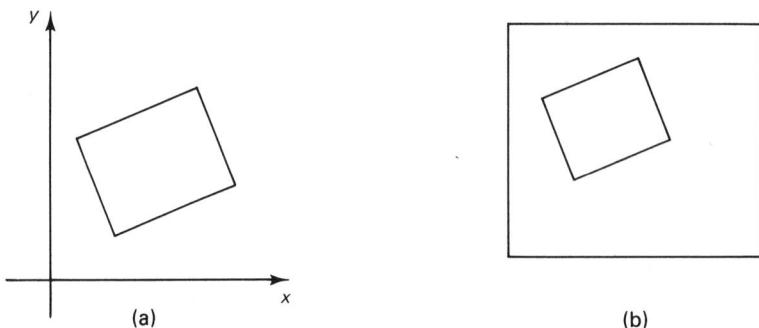

Figure 4.10 (a) Rotated window. (b) Rotated viewport.

To avoid distortion in the image, it is important to choose the window and viewport so that the height-to-width ratio of the window equals the height-to-width ratio of the viewport. When measuring the viewport for this purpose, some common unit, such as centimeters, must be used in both directions. Device units may not be appropriate, since the actual length of a vertical device unit may be different than the actual length of a horizontal device unit.

1.4 Example. Let H and W be the height and width of the window in world

Figure 4.11 Billboarding.

coordinate space and let H' and W' be the actual height and width of the viewport. Figure 4.12(a) and (b) show a grid and a 30° rotation of the grid with $H/W = H'/W' = 0.75$. Figure 4.12(c) shows the result of a 30° rotation with $H/W = 0.75$ but $H'/W' = 1.00$.

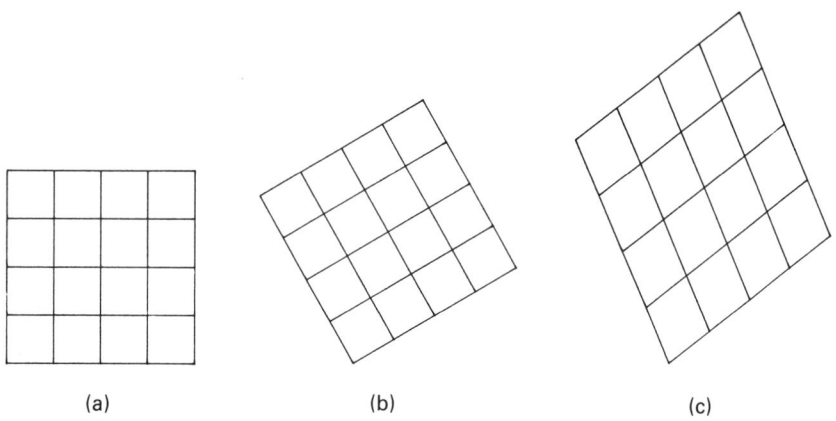

Figure 4.12 Distortion caused by different height-to-width ratios of the window and viewport.

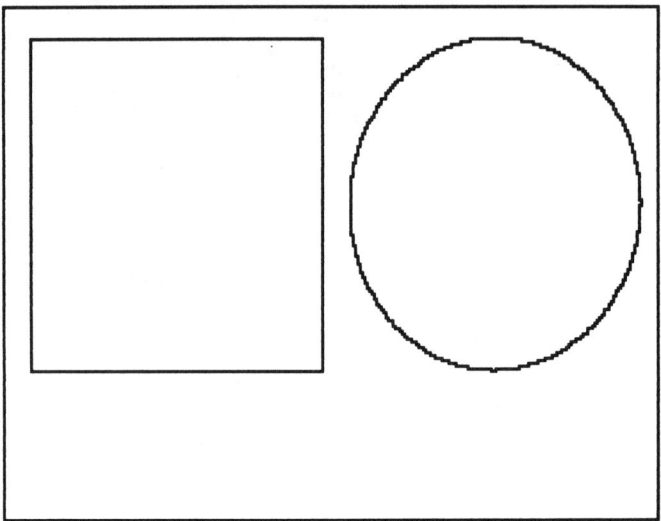

Figure 4.13 A "square" and a "circle" in device units.

Since we use device units to specify the size of the viewport, it is often necessary to compare length measurements of the viewport in device units to length measurements in some common unit such as centimeters. To do this we use the aspect ratio. The *aspect ratio*, AspectRatio, of a display device is the ratio of the length of a vertical device unit to the length of a horizontal device unit. If this ratio is not 1, then a "square" 100 device units by 100 device units on the display device will not, in fact, be square, and a "circle" in device units will not be circular (see Figure 4.13).

To compute the aspect ratio of a display device, we draw a rectangle v device units high by u device units wide on the display device. With a ruler, we measure the height y in centimeters and the width x in centimeters of this rectangle. The height of one device unit is y/v centimeters, the width of one device unit is x/u centimeters, and the

$$\text{AspectRatio} = \frac{y/v}{x/u}$$

(The quantities y/v and x/u are the *resolution* of the display device in the vertical and horizontal directions, respectively. The smaller y/v and x/u, the more device units per centimeter and the finer the resolution of the display device.)

1.5 Example. The device coordinate space for a microcomputer is 192 pixels high by 280 pixels wide. The display area of a monitor attached to the microcomputer is measured to be approximately 12.3 cm high by 15.1 cm wide, which gives the monitor an aspect ratio of

$$\frac{12.3/192}{15.1/280} = 1.2$$

Since exact measurements are impossible to make, this value of the aspect ratio might not give the best possible results. Through trial-and-error experimentation (drawing squares with varying values of AspectRatio), we can compute the aspect ratio more accurately. In this case, for example, an aspect ratio of 1.19 might give better results. (We can also compute the aspect ratio of a plotter or a printer attached to the microcomputer. If these aspect ratios are different, we must take care to avoid differences between what we see on the monitor and the hard copy we get from the plotter or printer. We return to this subject in Sections 2 and 3.)

If a rectangle drawn on a display device is the same number of device units wide as it is high ($u = v$), then the formula for the aspect ratio becomes AspectRatio = y/x, or

$$x \text{AspectRatio} = y$$

This is illustrated in Figure 4.14(a) in which $u = v = 2$ and AspectRatio = $3/2$. This result can be stated as follows.

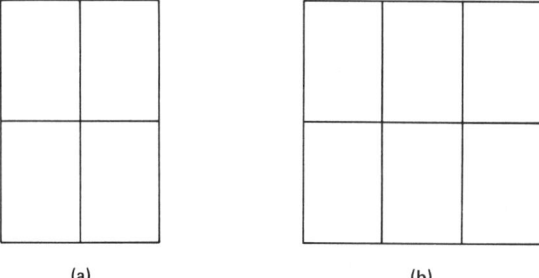

(a)   (b)   **Figure 4.14** The use of the aspect ratio.

1.6 Proposition. The actual length of a given number of horizontal device units times the aspect ratio is the actual length of that number of vertical device units.

If a true square is drawn on the display device ($x = y$), the formula for the aspect ratio becomes AspectRatio = u/v, or

$$v \text{AspectRatio} = u$$

This is illustrated in Figure 4.14(b) in which AspectRatio = $3/2$ again. This result can be stated as follows.

1.7 Proposition. The number of vertical device units in a given length times the aspect ratio is the number of horizontal device units in that length.

SECTION 1 EXERCISES

1. Make a survey of the graphics devices available to you. Include a description of device type, device units, device coordinate systems (where the origin is located and

what the directions of the positive coordinate axes are), the maximum viewport size, and whether clipping is automatic. Compute the aspect ratio for each device and the resolution in the horizontal and vertical directions. What special features (if any) are available in each case?

2. Verify that Proposition 1.1 holds in the case $(x_0, y_0) = (10, 16)$, $(x_1, y_1) = (58, 52)$, $(u_0, v_0) = (16, 176)$, and $(u_1, v_1) = (216, 16)$. How do the location of the origin and the orientation differ from those in Example 1.2?

3. In each of the following cases, the coordinates of the antipodal vertices of two rectangles are given: (x_0, y_0), (x_1, y_1) and (u_0, v_0), (u_1, v_1). Find the transformation taking the rectangle whose vertices have coordinates (x_0, y_0), (x_1, y_0), (x_1, y_1), and (x_0, y_1) to the rectangle whose vertices have coordinates (u_0, v_0), (u_1, v_0), (u_1, v_1), and (u_0, v_1), respectively.
   (a) (1, 2), (3, 4); (2, 1), (5, 6) (b) (−1, −1), (1, 1); (2, 3), (4, 5)
   (c) (−1, 2), (1, 3); (−1, −1), (1, 1) (d) (4, 2), (5, 3); (6, 2), (8, 4)

4. Show that the following technique can be used to compute the matrix M_{object} in the proof of Proposition 1.1:
$$\begin{pmatrix} 1 & 0 & 1 \\ 0 & 1 & 1 \\ 0 & 0 & -1 \end{pmatrix} M_{object} = \begin{pmatrix} x_1 & y_0 & 1 \\ x_0 & y_1 & 1 \\ -x_0 & -y_0 & -1 \end{pmatrix}$$
so that
$$M_{object} = \begin{pmatrix} 1 & 0 & 1 \\ 0 & 1 & 1 \\ 0 & 0 & -1 \end{pmatrix} \begin{pmatrix} x_1 & y_0 & 1 \\ x_0 & y_1 & 1 \\ -x_0 & -y_0 & -1 \end{pmatrix}$$

5. Compute the product $M_{object}^{-1} M_{image}$ used in the projective derivation of Proposition 1.1 and verify the conclusion of Proposition 1.1 by computing T in projective notation and then rewriting T in Euclidean notation.

6. Use the data of Exercise 3 and normalized device coordinates to find the matrices M_{object}^{-1} and M_{image}.

7. Use the data of Exercise 3 and the intermediate rectangle whose antipodal vertices have coordinates $(s_0, t_0) = (-1, -1)$ and $(s_1, t_1) = (1, 1)$ to find the matrices M_{object}^{-1} and M_{image}.

8. Show that the product $M_{object}^{-1} M_{image}$, that represents the complete viewing operation, is the same whether normalized device coordinates are used or some other intermediate coordinates are used. Illustrate this by comparing your answers to Exercises 3, 6, and 7.

9. Show that the transformation T_{object} in the proof of Proposition 1.1 also transforms I_s to I_x and I_t to I_y, so that M_{object} could also be defined as the matrix representing the transformation that takes I_s, I_t, O, and U to I_x, I_y, A_{object}, and C_{object}, respectively.

10. In each of the following cases, two sets of four points are given. In each case verify that no three points in either set are collinear, and find the transformation taking A to P, B to Q, C to R, and D to S.
    (a) $A(0, 0)$, $B(1, 0)$, $C(1, 1)$, $D(0, 1)$; $P(2, 1)$, $Q(6, 3)$, $R(5, 5)$, $S(1, 3)$
    (b) $A(-1, -1)$, $B(1, -1)$, $C(1, 1)$, $D(-1, 1)$; $P(2, -2)$, $Q(4, -4)$, $R(8, 0)$, $S(6, 2)$
    (c) $A(1, 0)$, $B(2, 1)$, $C(1, 2)$, $D(0, 1)$; $P(2, 1)$, $Q(4, 1)$, $R(4, 4)$, $S(2, 4)$
    (d) $A(0, -1)$, $B(1, 0)$, $C(0, 1)$, $D(-1, 0)$; $P(-1, -1)$, $Q(2, -1)$, $R(2, 3)$, $S(-1, 3)$

11. Give a geometric interpretation of the alternate formula for the aspect ratio

$$\text{AspectRatio} = \frac{y/x}{v/u}$$

12. If k is the height-to-width ratio of the viewport, explain why selecting window vertices whose coordinates are (± 1, $\pm k$) avoids distortion caused by aspect ratio problems.
13. Verify the linear interpolation formula used in Lemma 1.3.
14. What issues are faced in the development of device-independent graphics software? Consider in particular the use of normalized device coordinates.

SECTION 2: A SIMPLE TWO-DIMENSIONAL GRAPHICS LIBRARY

In this section we define a small library of graphics commands with which we demonstrate applications of projective geometry to two-dimensional computer graphics. These commands are of two types: primitive and procedural. A *primitive command* (or *primitive*, for short) is a device-dependent command whose implementation is not discussed here. A *procedural command* is a command defined by a procedure using primitives and other procedural commands. (The primitives in our library should be present in one form or another in most graphics packages. If not, they have to be implemented as procedures. See Exercise 1.)

The first commands in our library are the primitives

GraphicsMode

which puts the display device into its graphics mode and

TextMode

which returns the display device to text mode. These commands merely toggle the state of the graphics display device: Neither performs any initialization.

Points are plotted in device coordinate space by the primitive

PlotInDC(U,V)

where (U, V) are the device coordinates of the point P to be plotted. (As usual, points are plotted, not coordinates. To be consistent, this command should be PlotInDCP(U,V). We do not use this notation, however, since doing so would create later notation problems.) In the same way that points are the most basic geometric object, this is the most basic plotting command: Without it, creating graphics is impossible, and with it the only limitations are those of the display device.

A line segment is drawn in device coordinate space by the primitive

DrawInDCFrom(U0,V0)To(U1,V1)

where (U0, V0) and (U1, V1) are the device coordinates of the endpoints P and Q of the segment to be drawn.

The primitive command for adding text to the display is

WriteCharacter("Character";U0,V0;Theta;Scale)

This command draws the character Character at the point whose device coordinates are (U0, V0), rotated through the angle Theta, and magnified by the scale factor Scale.

The remaining library commands are all procedural.

After the display device has been put into graphics mode, we call the procedure

$$\text{Initialize2DimensionalGraphics}$$

to initialize the graphics environment. We use this procedure to declare and initialize variables used by the other library procedures. For example, we declare three 3×3 matrices M_0, M_1 and M. We use M_0 to represent object adjustments, and initialize it as the 3×3 identity matrix. (We discuss M_0 later in this section.) We use M_1 to represent the initial complete viewing operation, and $M = M_0 M_1$ to represent the complete viewing operation with object adjustments. The matrices M_1 and M are computed automatically after we call procedures that define the window and viewport.

The procedure

$$\text{DefineWindow(MinX,MinY;MaxX,MaxY)}$$

defines as the window that rectangle whose lower left vertex has world coordinates (MinX,MinY) and whose upper right vertex has world coordinates (MaxX,MaxY). The procedure

$$\text{DefineViewport(MinU,MinV;MaxU,MaxV)}$$

defines as the viewport that rectangle whose lower left vertex has device coordinates (MinU,MinV) and whose upper right vertex has device coordinates (MaxU,MaxV). As indicated in Section 1, distortion can be avoided by choosing the vertices of the window and viewport so that the height-to-width ratio of the viewport (measured in some common unit, such as centimeters) is equal to that of the window. (See also Exercise 22, in which we discuss modifications that make our library device-independent.) Despite the use of the mnemonics Max and Min in DefineWindow and DefineViewport, it is *not* assumed, for example, that MinU < MaxU and MinV < MaxV. This flexibility is allowed so that we may take into account rotation and orientation of the viewport in the viewing transformation.

Once DefineWindow and DefineViewport have been called, the matrix M_1 is automatically computed and the matrix $M = M_0 M_1$ of the complete viewing operation is initialized to M_1 (see Figure 4.15). If DefineWindow or DefineViewport is called again later, both M_1 and M are automatically recomputed.

Since the complete viewing operation is represented by $M = (M[I,J])$, the complete viewing operation is implemented by

```
PROCEDURE Transform(X,Y)To(U,V)
  BEGIN
    W := M[1,3]*X + M[2,3]*Y + M[3,3]
    U := (M[1,1]*X + M[2,1]*Y + M[3,1])/W
    V := (M[1,2]*X + M[2,2]*Y + M[3,2])/W
  END
```

and points with world coordinates (X, Y) are plotted by

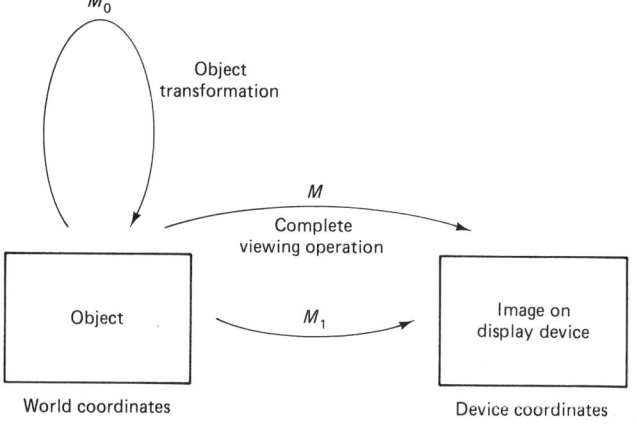

Figure 4.15 The complete viewing operation.

```
PROCEDURE Plot(X,Y)
  BEGIN
    Transform(X,Y)To(U,V)
    PlotInDC(U,V)
  END
```

2.1 Example. Given the window and viewport of Example 1.2, the point P whose world coordinates are (X, Y) is plotted by the following sequence:

```
    ⋮
DefineWindow(10,16;58,52)
DefineViewport(16,50;216,210)
Plot(X,Y)
    ⋮
```

The command for drawing a line segment in world coordinate space is implemented by the following procedure. This command draws the line segment whose endpoints have world coordinates (X0,Y0) and (X1,Y1).

```
PROCEDURE DrawFrom(X0,Y0)To(X1,Y1)
  BEGIN
    Transform(X0,Y0)To(U0,V0)
    Transform(X1,Y1)To(U1,V1)
    DrawInDCFrom(U0,V0)To(U1,V1)
  END
```

Similarly (see Exercise 4) we define the procedure

WriteString("CharacterString";X0,Y0;Theta;Scale)

that writes the character string CharacterString from left to right, beginning at the point with world coordinates (X0,Y0), rotated through the angle Theta, and magnified by a scale factor Scale.

Sec. 2   A Simple Two-Dimensional Graphics Library

The next addition to our library is CurrentPosition. CurrentPosition is the location of a graphics cursor in world coordinates, and is defined by the variables CurrentPositionX and CurrentPositionY. The advantage to using CurrentPosition is that doing so allows us to introduce and use incremental techniques; under certain circumstances, using such techniques reduces the data which must be processed. The coordinates of CurrentPosition are declared and initialized by Initialize2DimensionalGraphics and are available to the user.

With CurrentPosition defined, the procedures

$$\text{MoveAbsolute}(X,Y) \quad \text{and} \quad \text{MoveRelative}(X,Y)$$
$$\text{PlotAbsolute}(X,Y) \quad \text{and} \quad \text{PlotRelative}(X,Y)$$
$$\text{DrawAbsolute}(X,Y) \quad \text{and} \quad \text{DrawRelative}(X,Y)$$

can be implemented. In the case of each Absolute command, (X,Y) are the world coordinates of a point with respect to which an action is performed. In the case of each Relative command, (X,Y) are the coordinates of a point relative to CurrentPosition and with respect to which an action is performed.

For example (see Figure 4.16), MoveAbsolute(X,Y) redefines CurrentPosition as the point whose world coordinates are (X,Y):

```
PROCEDURE MoveAbsolute(X,Y)
   BEGIN
      CurrentPositionX := X
      CurrentPositionY := Y
   END
```

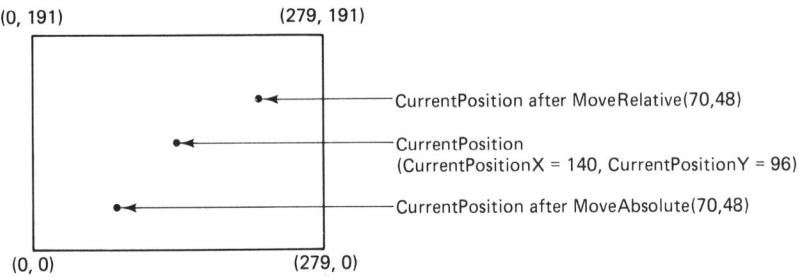

Figure 4.16 MoveAbsolute and MoveRelative.

MoveRelative(X,Y) redefines CurrentPosition as the point whose world coordinates are (X,Y) relative to CurrentPosition:

```
PROCEDURE MoveRelative(X,Y)
   BEGIN
      CurrentPositionX := X + CurrentPositionX
      CurrentPositionY := Y + CurrentPositionY
   END
```

PlotAbsolute(X,Y) moves CurrentPosition to the point whose world coordinates are (X,Y) and plots this point; PlotRelative(X,Y) moves CurrentPosition to the point whose world coordinates are (X,Y) relative to CurrentPosition and plots this point (see Figure 4.17). DrawAbsolute(X,Y) draws a line segment from CurrentPosition to the point whose world coordinates are (X,Y) and then moves CurrentPosition to (X,Y); DrawRelative(X,Y) draws a line segment from CurrentPosition to the point whose world coordinates are (X,Y) relative to CurrentPosition and then moves CurrentPosition to this point (see Figure 4.18).

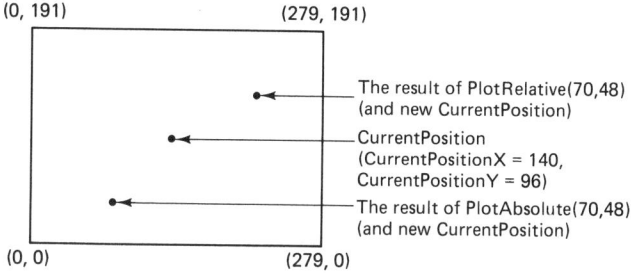

Figure 4.17 PlotAbsolute and PlotRelative.

The last commands in our library are object adjustments. They include rotation, translation, scaling, shearing, perspective projection, and general (projective) transformation. All these transformations can be represented by right 3×3 matrix multiplication (see Section 3.3). The matrix M_0 representing object adjustment is initially the identity matrix. If an object adjustment, represented by the matrix M_0', is to be performed, M_0 is replaced by the product $M_0 M_0'$, and the matrix M is automatically computed. The facts that $M = M_0 M_0' M_1$ (that is, that the factor M_0' is between M_0 and M_1) and matrix multiplication is not commutative explain why M_0 and M_1 are stored separately. Recomputing M has the effect of changing Transform, Plot, and all other commands defined in terms of them.

2.2 Example. The geometric effect of the command

$$\text{Rotate(Theta)}$$

is to rotate the object through the angle Theta about the origin of the world

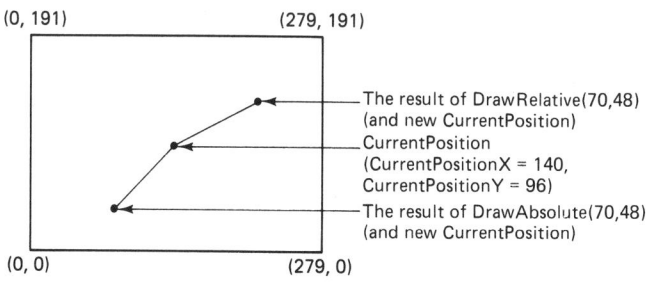

Figure 4.18 DrawAbsolute and DrawRelative.

coordinate system. Calling Rotate(Theta) causes the following operations to be performed automatically:
$$M_0 := M_0 M_0'$$
$$M := M_0 M_1$$
where (see Section 3.3) M_0' is
$$\begin{pmatrix} \cos(\text{Theta}) & \sin(\text{Theta}) & 0 \\ -\sin(\text{Theta}) & \cos(\text{Theta}) & 0 \\ 0 & 0 & 1 \end{pmatrix}$$

Similarly we add the following commands to our library:
$$\text{Translate}(H,K)$$
that translates world coordinates so that the point whose world coordinates are (H,K) becomes the origin;
$$\text{Scale}(MX,MY;MZ)$$
that scales locally by MX in the x-direction and by MY in the y-direction, and that scales globally by $1/MZ$;
$$\text{ShearY}(A), \text{ShearX}(B)$$
that shear by A in the y-direction and by B in the x-direction, respectively;
$$\text{PerspectivelyProject}(A,B)$$
that performs a perspective projection by A in the x-direction and B in the y-direction; and
$$\text{GeneralTransformation}(A,B,C;D,E,F;G,H,I)$$
that transforms by the matrix
$$\begin{pmatrix} A & B & C \\ D & E & F \\ G & H & I \end{pmatrix}$$

It is important to understand that each of these transformations transforms the object as it is seen on the display device as opposed to the data that represents the object. This is best illustrated by an example.

2.3 Example. The complete viewing operation illustrated in Figure 4.19 is represented by the matrix
$$M = M_0 M_1 = M_0 M_{\text{object}}^{-1} M_{\text{image}}$$
$$= \begin{pmatrix} 1 & 0 & 0 \\ 0 & 1 & 0 \\ 0 & 0 & 1 \end{pmatrix} \begin{pmatrix} 1/3.8 & 0 & 0 \\ 0 & 1/3 & 0 \\ 1/2 & 1/2 & 1 \end{pmatrix} \begin{pmatrix} 279 & 0 & 0 \\ 0 & 191 & 0 \\ 0 & 0 & 1 \end{pmatrix}$$
$$= \begin{pmatrix} 1/3.8 & 0 & 0 \\ 0 & 1/3 & 0 \\ 1/2 & 1/2 & 1 \end{pmatrix} \begin{pmatrix} 279 & 0 & 0 \\ 0 & 191 & 0 \\ 0 & 0 & 1 \end{pmatrix}$$

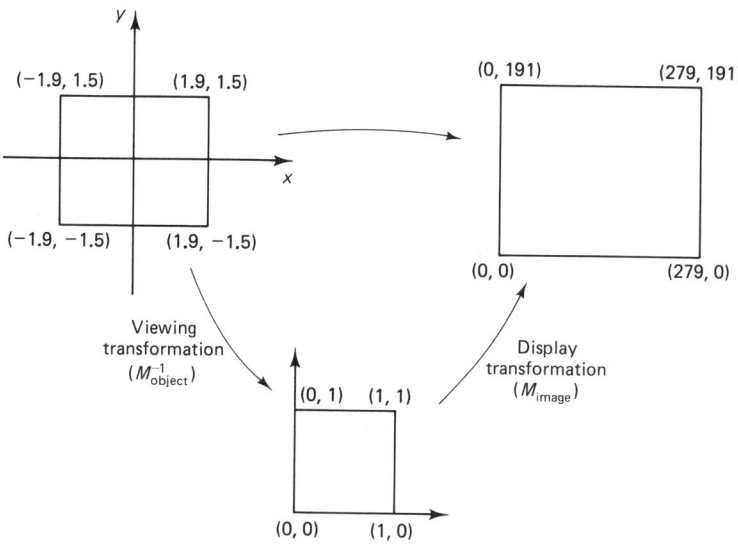

Figure 4.19 A complete viewing operation.

The image under this complete viewing operation of the world coordinate axes is shown in Figure 4.20. Having established this complete viewing operation, Plot(1,0) has the effect of plotting the point illustrated in Figure 4.20(a). The device coordinates of the point to be plotted are the first two coordinates of the product

$$\langle 1, 0, 1 \rangle \begin{pmatrix} 1/3.8 & 0 & 0 \\ 0 & 1/3 & 0 \\ 1/2 & 1/2 & 1 \end{pmatrix} \begin{pmatrix} 279 & 0 & 0 \\ 0 & 191 & 0 \\ 0 & 0 & 1 \end{pmatrix}$$

On the other hand,

```
    ⋮
Rotate(30°)
Plot(1,0)
    ⋮
```

has the effect of plotting the point illustrated in Figure 4.20(b). The effect of Rotate(30°) is to update M to

$$\begin{pmatrix} \cos 30° & \sin 30° & 0 \\ -\sin 30° & \cos 30° & 0 \\ 0 & 0 & 1 \end{pmatrix} \begin{pmatrix} 1/3.8 & 0 & 0 \\ 0 & 1/3 & 0 \\ 1/2 & 1/2 & 1 \end{pmatrix} \begin{pmatrix} 279 & 0 & 0 \\ 0 & 191 & 0 \\ 0 & 0 & 1 \end{pmatrix}$$

Thus the device coordinates of the point plotted by Plot(1,0) are given by

$$\langle 1, 0, 1 \rangle \begin{pmatrix} \cos 30° & \sin 30° & 0 \\ -\sin 30° & \cos 30° & 0 \\ 0 & 0 & 1 \end{pmatrix} \begin{pmatrix} 1/3.8 & 0 & 0 \\ 0 & 1/3 & 0 \\ 1/2 & 1/2 & 1 \end{pmatrix} \begin{pmatrix} 279 & 0 & 0 \\ 0 & 191 & 0 \\ 0 & 0 & 1 \end{pmatrix}$$

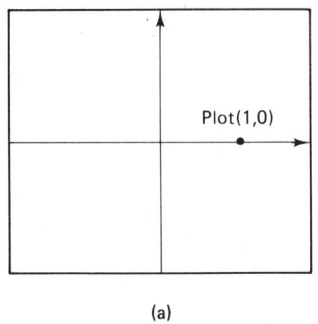

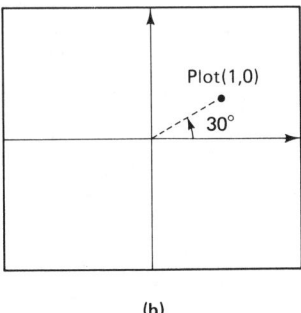
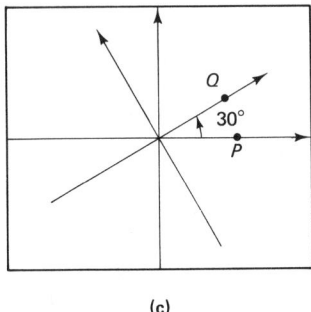

(b)   (c)

Figure 4.20 The point with coordinates (1, 0).

The effect of Rotate(30°) is to rotate P about the origin of the world coordinate system: The world coordinates of the point Q obtained by rotating P through 30° about the origin (see Figure 4.20(c)) are given by

$$\langle 1, 0, 1 \rangle \begin{pmatrix} \cos 30° & \sin 30° & 0 \\ -\sin 30° & \cos 30° & 0 \\ 0 & 0 & 1 \end{pmatrix}$$

Thus the device coordinates of the image of Q under the original complete viewing operation are given by

$$\langle 1, 0, 1 \rangle \begin{pmatrix} \cos 30° & \sin 30° & 0 \\ -\sin 30° & \cos 30° & 0 \\ 0 & 0 & 1 \end{pmatrix} \begin{pmatrix} 1/3.8 & 0 & 0 \\ 0 & 1/3 & 0 \\ 1/2 & 1/2 & 1 \end{pmatrix} \begin{pmatrix} 279 & 0 & 0 \\ 0 & 191 & 0 \\ 0 & 0 & 1 \end{pmatrix}$$

Since the rotation matrix

$$\begin{pmatrix} \cos 30° & \sin 30° & 0 \\ -\sin 30° & \cos 30° & 0 \\ 0 & 0 & 1 \end{pmatrix}$$

is automatically absorbed into M (and hence Transform), calling Plot(1, 0) after Rotate(30°) has the effect of plotting P in a rotated coordinate system, leaving unchanged the data ((1, 0) in Cartesian coordinates or [1, 0, 1] in homogeneous coordinates) that represents P.

SECTION 2 EXERCISES

1. Examine the two-dimensional graphics libraries available to you and determine which primitive commands are available. Which commands correspond to the primitive commands GraphicsMode, TextMode, PlotInDC, DrawInDCFromTo and WriteCharacter? If one of these commands is not present as a primitive, implement it as a procedure.
2. Implement the library presented in this section on a specific graphics device.
3. If the lowest level library commands available include CurrentPosition, MoveRelative, PlotRelative, and DrawRelative, how would you implement MoveAbsolute, PlotAbsolute, and DrawAbsolute?
4. Write the procedure WriteString described in this section, using the primitive WriteCharacter and the procedure Transform.
5. The following commands might be useful library additions. Give the meaning of each one and show how they are implemented.
   (a) SaveEnvironment and RecallEnvironment
   (b) ClearScreen
   (c) InverseTransform(U,V)To(X,Y)
   (d) Polar (changes future world coordinate references to polar coordinates) and Cartesian
   (e) PlotPoints(P) (P is an $n \times 2$ array whose rows are the coordinates of the points to be plotted)
   (f) PlotPolygonalLine(P) (P is an $n \times 2$ array)
6. Add to the library presented in this section the image transformations RotateInDC, TranslateInDC, and ScaleInDC that operate on device coordinate space. What distortion problems arise when these transformations are used if the aspect ratio of the display device is not 1?
7. What commands (in addition to those of Exercise 5 above) would make useful library additions? Consider ColorOf(U0,V0), which stores in an array the color of the point whose device coordinates are (U0,V0).
8. Suppose the angle CurrentHeading is attached to the graphics cursor at CurrentPosition. Give the meaning of each of the following commands and show how they are implemented.
   (a) Move(Distance)
   (b) TurnRelative(Theta)
   (c) TurnAbsolute(Theta)
9. Implement a procedure Circle(H,K;R) to draw the circle in world coordinate space whose center has coordinates (H,K) and whose radius is R.
10. Implement a procedure Arc(H,K;R;Alpha;Omega) that draws that part of the circle in world coordinate space whose center has coordinates (H,K) and whose radius is R, from the initial angle Alpha to the terminal angle Omega.
11. Write programs that graph the ellipse $x^2/a^2 + y^2/b^2 = 1$ in two ways.
    (a) Use $y = \pm(b/a)\sqrt{a^2 - x^2}$ and select values for x.
    (b) Use $x = a\cos\theta$, $y = b\sin\theta$ and select values for θ.
    Which program is easier to write? Which program yields a better spacing of the points on the ellipse?
12. Write a program to graph each of the following.
    (a) Quadratic polynomials $y = ax^2 + bx + c$
    (b) Arbitrary polynomials

13. Write a program to graph the polar coordinate equation $r = a + b \cos((n/m)\theta)$ (see Figure 4.21). (*Hint:* Use the parametrization $x = r \cos\theta$, $y = r \sin\theta$. The domain for the parameter θ must be $[0, 2\pi m]$ to obtain a closed curve.

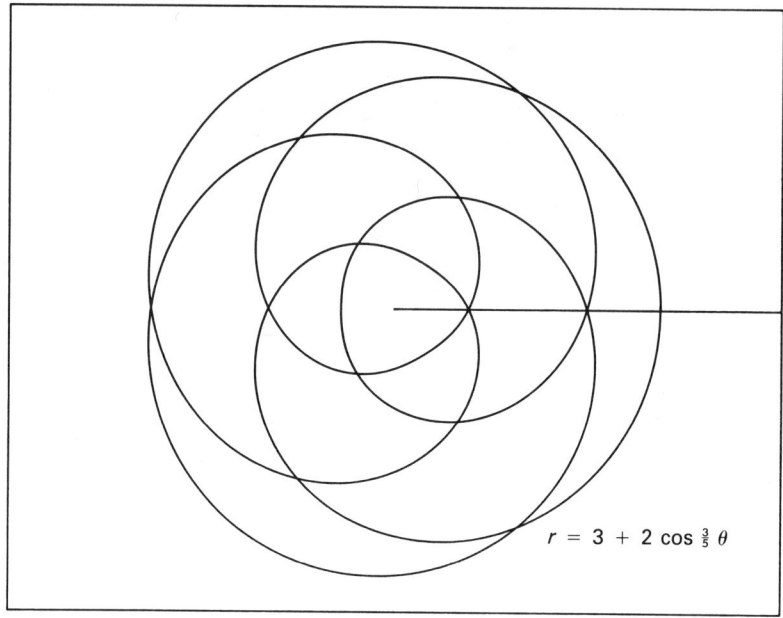

Figure 4.21 The graph of $r = a + b \cos((n/m)\theta)$.

14. (a) Write a program to graph $y = a \sin(bx) + c \cos(dx)$.
    (b) Write a program to simulate an oscilloscope.
15. Write a program to draw the floor plans for your house (see Figure 4.22).
16. Write a program to draw bar charts (see Figure 4.23).
17. Write a program to draw pie charts (see Figure 4.24; this figure is distorted because of an uncorrected aspect ratio problem in the program that produced it).
18. Write a program to draw line graphs (see Figure 4.25).
19. Why are the viewing and display transformations concatenated into the complete viewing operation before any points are transformed? Support your argument with an example.
20. Describe in your own words what happens when Rotate(Theta) is called.
21. If the vector **p** is a representative of the (homogeneous) world coordinates of an object point P, and a complete viewing operation is represented by the matrix M, then the (homogeneous) device coordinates of the image of P (under the complete viewing operation) are given by $[\mathbf{p}M]$. If a rotation of the object is represented by the matrix M_0, then the (homogeneous) device coordinates of the image of P under the modified complete viewing operation are given by $[\mathbf{p}M_0 M]$. In this section, rotation was interpreted in two ways as moving points (see Figure 4.20(b)) or rotating

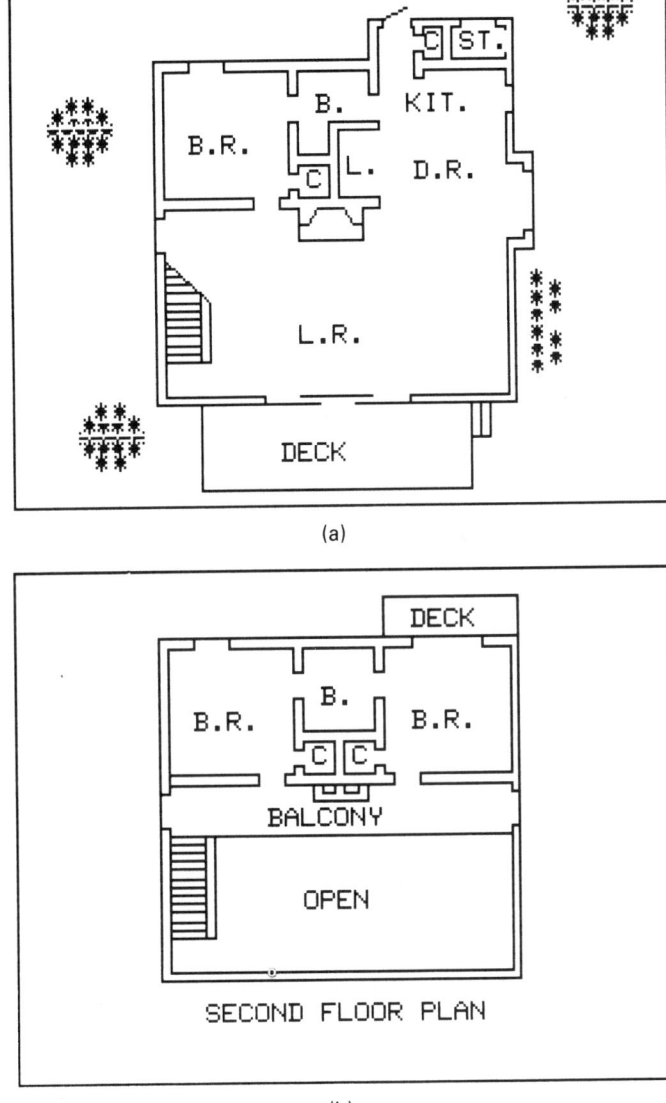

Figure 4.22 The floor plans for a vacation home.

the coordinate system (see Figure 4.20(c)). In what ways do the groupings $[(\mathbf{p}M_0)M]$ and $[\mathbf{p}(M_0 M)]$ correspond to these two interpretations?

22. Modify the library presented in this section so that applications programs can be made device-independent. (*Hints:*

   (a) Use normalized device coordinates as an intermediate step in the transformation process. Have Initialize2DimensionalGraphics declare M_0, M_{object}^{-1}, M_{image}, and M.
   (b) Introduce a command OutputDevice(DeviceName). Store a matrix M_{image} for each

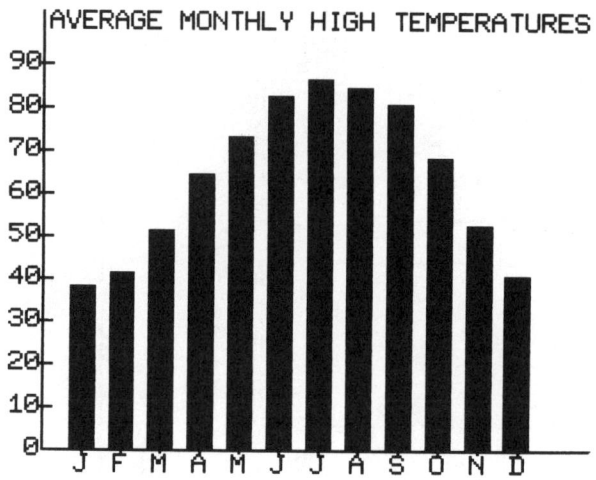

Figure 4.23 A bar chart.

display device. Let M_{image} represent the transformation of the entire square of normalized device space to a large square region of the display area. This corrects any aspect ratio problems without the attention of the user, but may restrict the user to only part of the available display area.

(c) Have DefineWindow trigger the calculation of M_{object}^{-1}.

(d) Have DefineViewport operate in normalized device coordinate space.

(e) Calculate M automatically after the procedures OutputDevice, DefineWindow, and DefineViewport have been called.)

23. Implement the clipping algorithm in normalized device coordinate space mentioned in Section 1 and incorporate it into your modified library from Exercise 22. Recall that a point is plotted if both of its normalized device coordinates are between 0 and 1. Extend your algorithm to one that clips line segments. (*Hint:* This latter algorithm can use a binary search.)

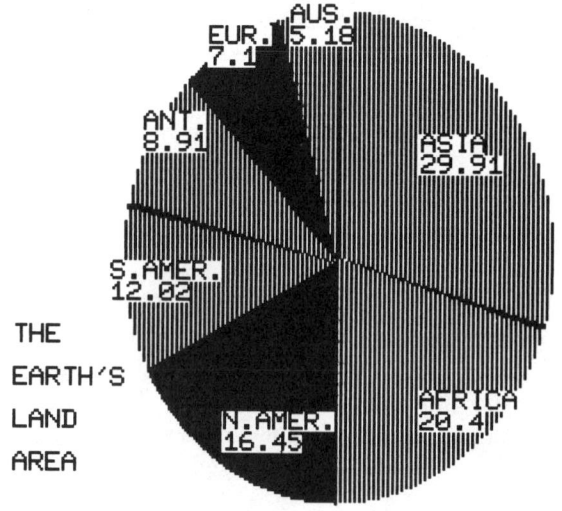

Figure 4.24 A pie chart.

Figure 4.25 A line graph.

SECTION 3: A TWO-DIMENSIONAL GRAPHICS DISPLAY PROGRAM

In this section we outline a general purpose interactive two-dimensional graphics display program. This program allows us to display, manipulate with object transformations, and redisplay any two-dimensional object described by vertices and edges. (This program, in fact, was used to produce figures for Section 3.3.) In Section 5.4 we outline a corresponding program for three-dimensional graphics.

We assume throughout this section that clipping has been implemented. Here are the commands that will be on the menu of this program.

(a) Window: This prompts the user for the corners of the window and then calls DefineWindow.

(b) Viewport: This prompts the user for the corners of the viewport and then calls DefineViewport.

(c) GetPictureFile: This prompts the user for the name of a data file. The data file consists of an integer (the number of vertices), a list of pairs (the Euclidean coordinates of the vertices), an integer (the number of edges), and a list of those pairs of vertices that define edges. This information is read into NumberOfVertices, an array Vertex, NumberOfEdges, and an array Edge.

(d) DisplayPicture: This causes the current complete viewing transformation to be applied to the data and the picture to be displayed on the monitor.

(e) Rotate: This prompts the user for an angle Theta and calls Rotate(Theta).

(f) Translate: This prompts the user for the translation distances H and K in the *x*- and *y*-directions and calls Translate(H,K).

(g) Scale: This prompts the user for the local scale factors MX and MY in the *x*- and *y*-directions and the global scale factor MZ and calls Scale(MX,MY;MZ).

(h) Shear: This prompts the user for the direction of shear (X or Y) and the shear factor (A or B, respectively) and calls ShearY(A) or ShearX(B).
(i) PerspectivelyProject: This prompts the user for the components A and B of perspective projection in the x- and y-directions and calls PerspectivelyProject(A,B).
(j) Matrix: This prompts the user for the nine entries of an arbitrary 3×3 matrix and calls GeneralTransformation.
(k) Reinitialize: This resets the initial object transformation matrix M_0 to the identity matrix to allow for a fresh start.
(l) PrintPicture (optional): This dumps the current picture to a plotter or printer.

The main program is:

```
PROGRAM Graph2D
  BEGIN
    Initialize2DimensionalGraphics
    DefineWindow
    DefineViewport
    GetPictureFile
    REPEAT
      WRITE(Menu)
      READ(Command)
      CASE Command OF
        A: Window
        B: Viewport
        C: GetPictureFile
        D: DisplayPicture
        E: Rotate
        F: Translate
        G: Scale
        H: Shear
        I: PerspectivelyProject
        J: Matrix
        K: Reinitialize
        L: PrintPicture
        Q: Finished
      ENDCASE
    UNTIL Finished
  END
```

For a distortion-free image, the ratios of actual height to actual width of the window and the viewport should be the same.

We consider several versions of the procedure DisplayPicture. The first and simplest is as follows.

```
PROCEDURE DisplayPicture
  BEGIN
```

```
    GraphicsMode
    FOR I: = 1 TO NumberOfEdges DO
      DrawFrom(Vertex[Edge[I,1],1],Vertex[Edge[I,1],2])
        To(Vertex[Edge[I,2],1],Vertex[Edge[I,2],2])
    Pause for viewing picture
    TextMode
  END
```

To demonstrate the capabilities and limitations of this version of DisplayPicture, we experiment with perspective projection. Let T be the transformation represented by the matrix

$$\begin{pmatrix} 1 & 0 & a \\ 0 & 1 & 0 \\ 0 & 0 & 1 \end{pmatrix}$$

and consider the test pattern of Figure 4.26(a); it has vertices $A(-1, -1)$, $B(1, -1)$, $C(1, 1)$, $D(0, 1.7321)$ and $E(-1, 1)$. Figure 4.26(a)–(d) illustrate the results of applying T with increasing values of a to the test pattern. Recall (see Exercise 3.3.17) that lines through the origin are fixed (as lines) by T, with the vertical line through the origin fixed pointwise. If $a = 0.2$, the point A is moved away from the origin on line AB while point B is moved nearer; the same is true for points E and C. This makes segment AE lengthen and segment BC shorten. Point D, which is on the vertical line through the origin, remains fixed. These geometric changes continue as a is increased to 0.8.

The first limitation of this version of DisplayPicture appears with this particular test pattern when $a = 1$. In this case T transforms two vertices, A and E, to ideal points. The procedure Transform attempts to divide by zero, and the program crashes.

To correct this problem, we modify the procedure Transform so that DisplayPicture can trap for points whose third coordinate is zero. (Recall $M = (M[I,J])$ is the matrix of the complete viewing transformation.)

```
    PROCEDURE DisplayPictureWithTrap
      BEGIN
        GraphicsMode
        FOR I := 1 TO NumberOfEdges DO
          BEGIN
            Transform(Vertex[Edge[I,1],1],Vertex[Edge[I,1],2])
              To(U1,V1,W1)
            Transform(Vertex[Edge[I,2],1],Vertex[Edge[I,2],2])
              To(U2,V2,W2)
            IF W1 ≠ 0 AND W2 ≠ 0 THEN
              DrawInDCFrom(U1/W1,V1/W1)To(U2/W2,V2/W2)
          END
        Pause for viewing picture
        TextMode
      END
```

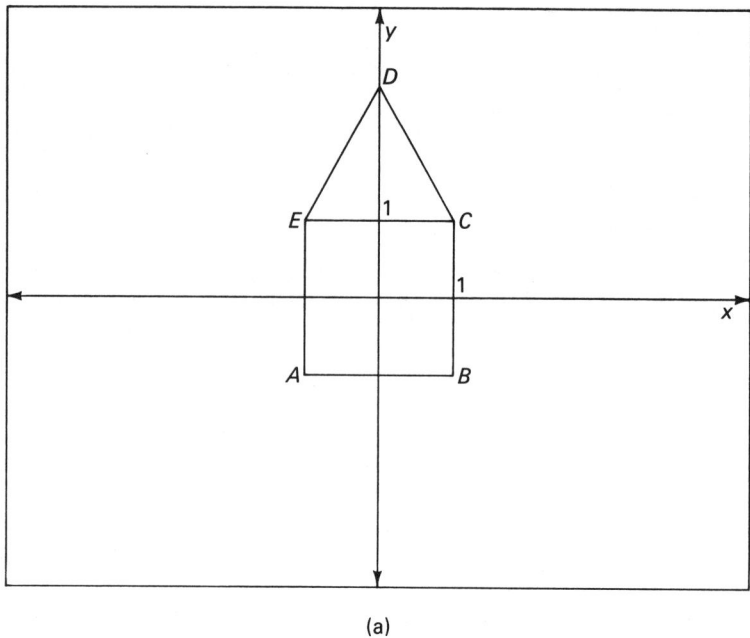

(a)

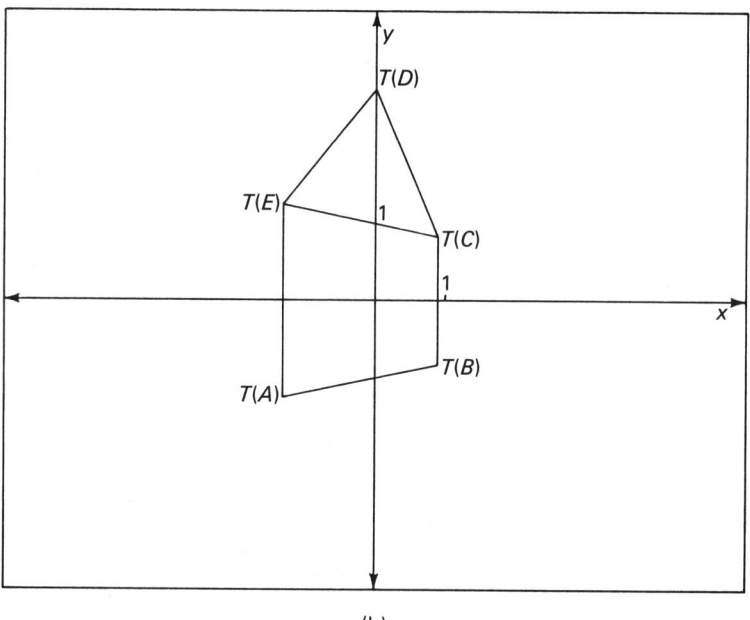

(b)

Figure 4.26 Perspective projection (a) $a = 0$, (b) $a = 0.2$, (c) $a = 0.8$, (d) $a = 1.5$.

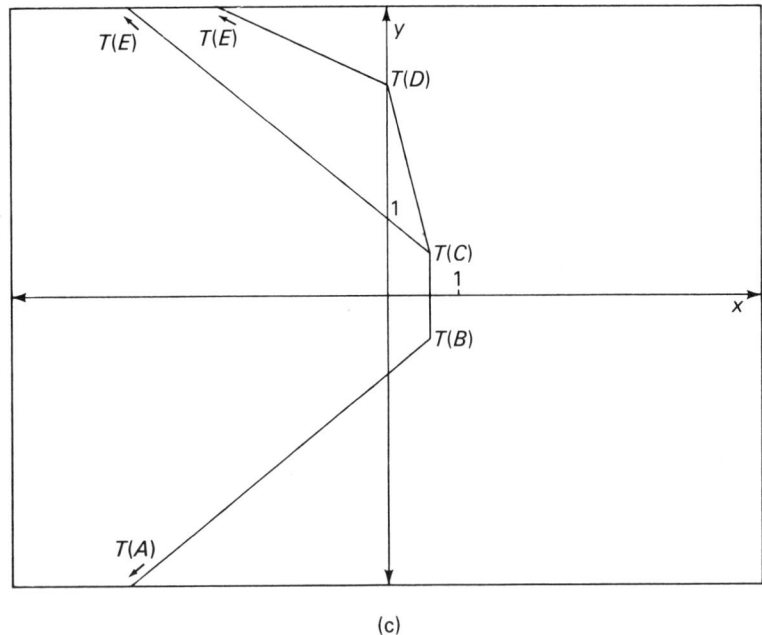

(c)

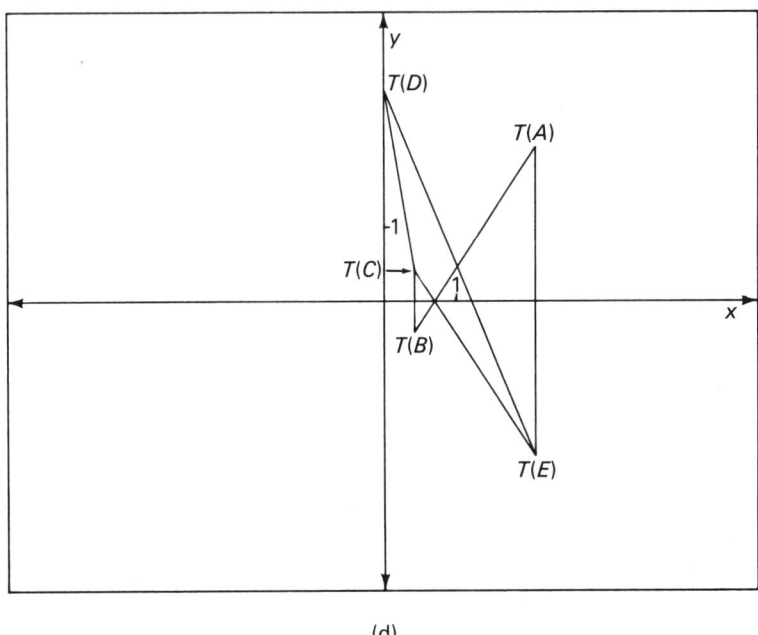

(d)

Figure 4.26 Continued

```
PROCEDURE Transform(X,Y)To(U,V,W)
  BEGIN
    U := M[1,1]*X + M[2,1]*Y + M[3,1]
    V := M[1,2]*X + M[2,2]*Y + M[3,2]
    W := M[1,3]*X + M[2,3]*Y + M[3,3]
  END
```

Using DisplayPictureWithTrap, Graph2D now creates Figure 4.27 when $a = 1$. Although the problem of division by zero has been eliminated, another problem has arisen: The picture is incomplete.

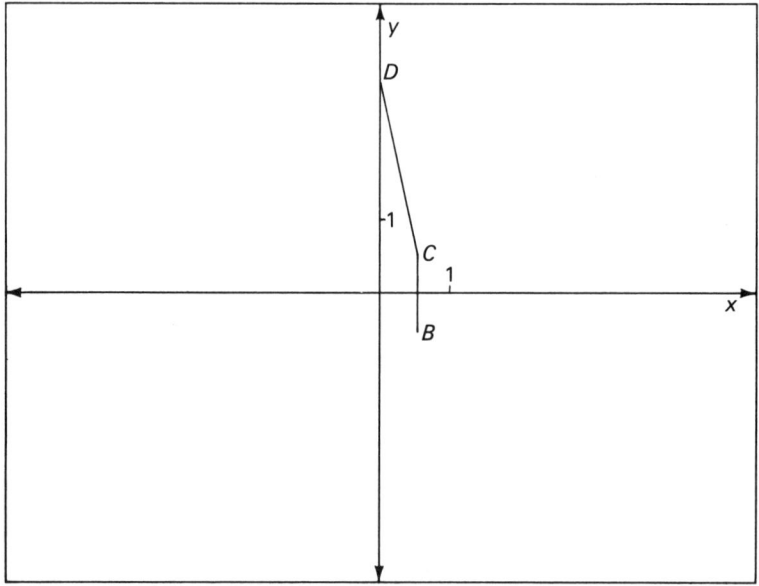

Figure 4.27 Incomplete perspective projection ($a = 1$).

Furthermore, this version and the original version of DisplayPicture have a second limitation, that appears when $a > 1$ (see Figure 4.26(d)). The segments *AB*, *CE* and *DE* have all been drawn incorrectly.

To understand what should be drawn both when $a = 1$ and when $a > 1$, recall (see Section 3.3) that *T* is the composite of projection $T_1 : \pi \to \mu$ parallel to the *z*-axis from the standard embedded projective plane $\pi[0, 0, -1, 1]$ to the projective plane $\mu[a, 0, -1, 1]$, followed by the perspectivity $T_2 : \mu \to \pi$ from μ to π whose center is the origin $O[0, 0, 0, 1]$. Figure 4.28 illustrates the corresponding transformation, down one dimension, of the segment *AB* for varying values of *a*. When $a = 1$, the image of *AB* should be a ray that disappears off to the left. When $a > 1$, the image of *AB* should be a ray that disappears off to the left and reappears on the right. Figure 4.26(d) was drawn incorrectly because the points A and B were connected, ignoring the fact that we are working in the projective plane: Since a line in the projective plane is topologically a circle, two points in the projective plane determine two different segments (see Figure 1.43) and in this case the wrong one was drawn.

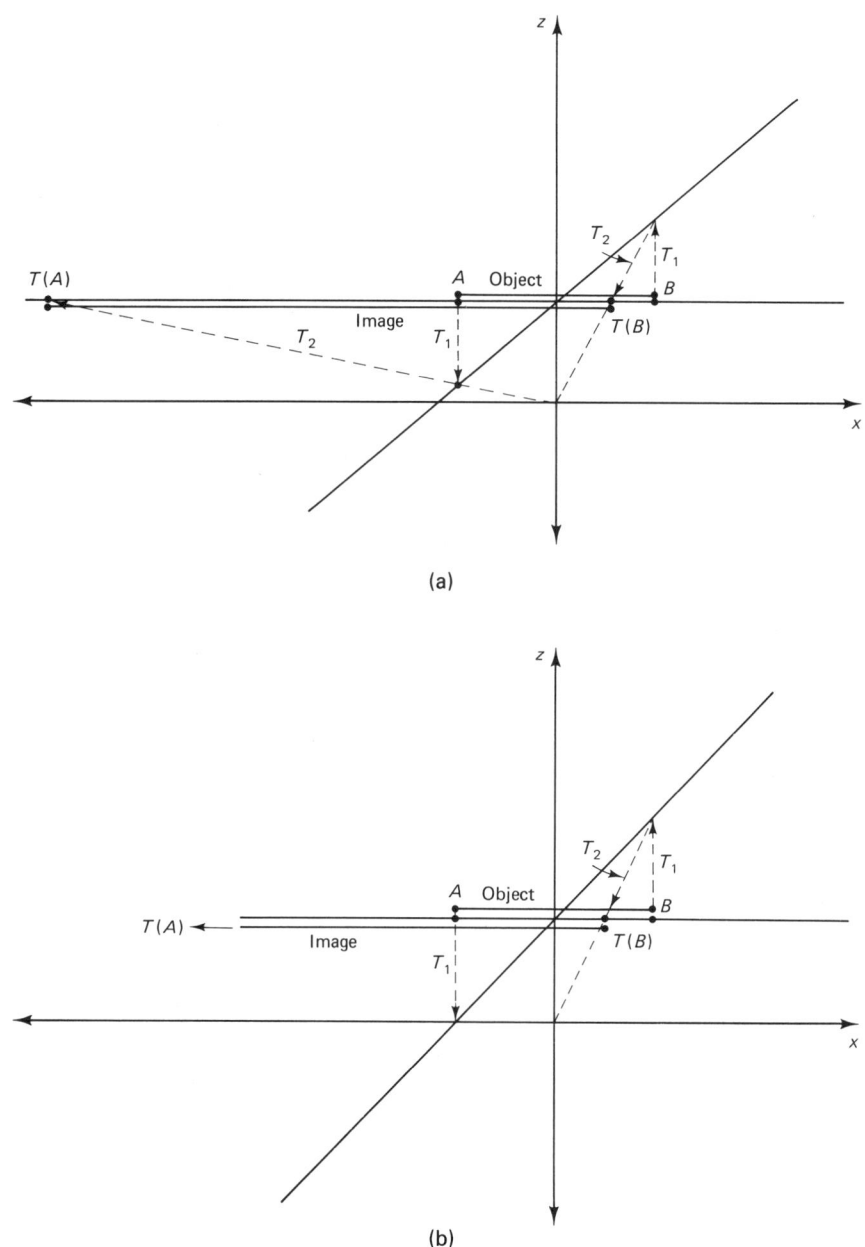

Figure 4.28 The transformation T with (a) $a = 0.8$, (b) $a = 1$, and (c) $a = 1.5$.

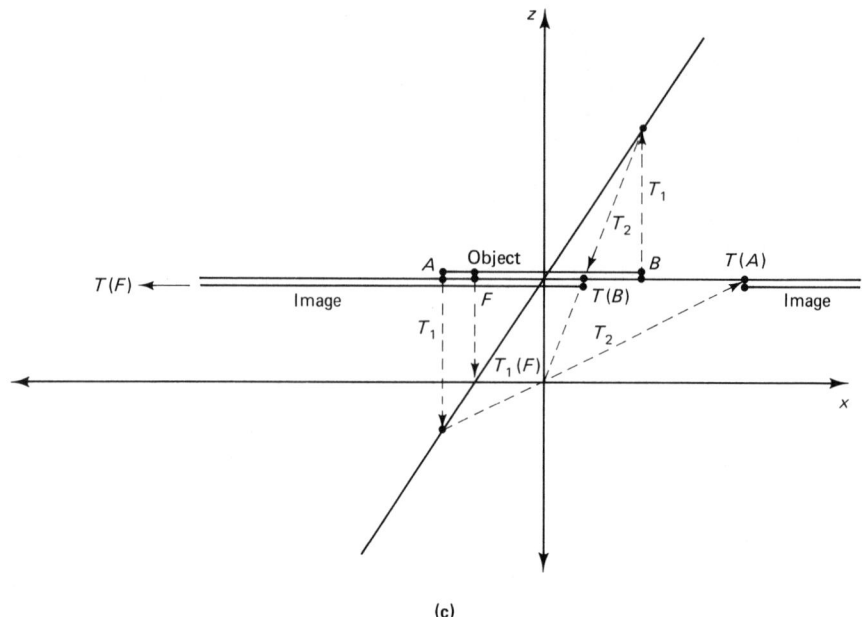

(c)

The limitations of DisplayPicture (and DisplayPictureWithTrap) arise when T maps points of the test pattern to infinity: the first limitation arises at vertices and the second at interior points of edges. This may appear to demonstrate only peculiar properties of perspective projection, but we see in Section 5.4 that translation, scaling, and rotation in three dimensions can also cause points to move to infinity.

To correct the second limitation and the problem of incompleteness that arises with DisplayPictureWithTrap, we plot points on each segment. The idea is roughly this: Since the concept of segment is not a projective concept, the image of a Euclidean segment under a projective transformation is not well defined. The image of a (parametrized) set *is*, however, well defined. We parametrize the Ith segment of the object by

$$\{[(1-T)*\langle \text{Vertex}[\text{Edge}[I,1],1], \text{Vertex}[\text{Edge}[I,1],2],1\rangle$$
$$+ T*\langle \text{Vertex}[\text{Edge}[I,2],1], \text{Vertex}[\text{Edge}[I,2],2],1\rangle]\}$$

where $0 \leq T \leq 1$. This segment is transformed, under right multiplication by M, into the parametrized set

$$\{[(1-T)*\langle \text{Vertex}[\text{Edge}[I,1],1], \text{Vertex}[\text{Edge}[I,1],2],1\rangle *M$$
$$+ T*\langle \text{Vertex}[\text{Edge}[I,2],1], \text{Vertex}[\text{Edge}[I,2],2],1\rangle *M]\}$$

If we let $P_1[U_1, V_1, W_1]$ and $P_2[U_2, V_2, W_2]$ be the transformed endpoints, this set can be written

$$\{[(1-T)*\langle U_1, V_1, W_1\rangle + T*\langle U_2, V_2, W_2\rangle]\} =$$
$$\{[(1-T)*U_1 + T*U_2, (1-T)*V_1 + T*V_2, (1-T)*W_1 + T*W_2]\}$$

This image is one of the two subsets of the projective line that passes through P_1 and P_2 and is determined by P_1 and P_2. In general, only one of the subsets of this line will contain an ideal point. The ideal point occurs for that value of T for which

$$(1 - T)*W_1 + T*W_2 = 0$$

or

$$T = \frac{W_1}{W_1 - W_2}$$

If this critical value of T is between 0 and 1, the image of the segment contains an ideal point. We should see the line disappear off one edge of the display area and reappear on another. If the critical value of T is not between 0 and 1, then the image of the original segment is the Euclidean segment determined by P_1 and P_2.

To find a more workable criterion for whether the critical value of T is between 0 and 1 we observe that

1. If $W_1 - W_2$ is positive, then $0 < T < 1$ if and only if W_1 is positive and W_2 is negative.
2. If $W_1 - W_2$ is negative, then $0 < T < 1$ if and only if W_1 is negative and W_2 is positive.

In other words, the critical value of T is between 0 and 1 if and only if W_1 and W_2 have opposite signs.

The procedure DisplayPictureByParametrizing uses the procedure

DrawByParametrizingFrom(U1,V1,W1)To(U2,V2,W2)

DrawByParametrizing detects whether an ideal point lies on the image of the segment from P_1[U1, V1, W1] to P_2[U2, V2, W2] by checking the signs of W1 and W2: if W1 and W2 are both nonzero and have the same sign (W1*W2 > 0), there is no ideal point on the segment; otherwise (W1*W2 ≤ 0) there is an ideal point (see Exercises 2.4.16 and 2.4.19).

DrawByParametrizing uses a value NumberOfSteps to control the plotting of each segment. The image of each segment is subdivided into NumberOfSteps short segments. If a short segment does not contain an ideal point it is drawn; otherwise it is not.

Several somewhat subtle issues raised by DrawByParametrizing are discussed in Exercises 11 and 12.

```
PROCEDURE DisplayPictureByParametrizing
  BEGIN
    GraphicsMode
    FOR I := 1 TO NumberOfEdges DO
      BEGIN
        Transform(Vertex[Edge[I,1],1],Vertex[Edge[I,1],2])
          To(U1,V1,W1)
```

```
            Transform(Vertex[Edge[I,2],1],Vertex[Edge[I,2],2])
               To(U2,V2,W2)
            DrawByParametrizingFrom(U1,V1,W1)To(U2,V2,W2)
         END
      Pause for viewing picture
      TextMode
   END


   PROCEDURE DrawByParametrizingFrom(U1,V1,W1)To(U2,V2,W2)
      BEGIN
         DeltaT := 1/NumberOfSteps
         (PreviousU,PreviousV,PreviousW) := (U1,V1,W1)
         (U,V,W) := (U2 - U1,V2 - V1,W2 - W1)
         IF W1 = 0 AND W2 = 0 THEN      (* Do nothing - line is
                                                    ideal *)
         ELSE IF W1*W2>0 THEN            (* No ideal point *)
            FOR I := 1 TO NumberOfSteps DO
               BEGIN
                  (CurrentU,CurrentV,CurrentW) :=
                     (PreviousU,PreviousV,PreviousW) + DeltaT*(U,V,W)
                  DrawInDCFrom
                     (PreviousU/PreviousW,PreviousV/Previous W)
                     To(CurrentU/CurrentW,CurrentV/CurrentW)
                  (PreviousU,PreviousV,PreviousW) :=
                     (CurrentU,CurrentV,CurrentW)
               END
         ELSE                             (* There is an ideal point*)
            BEGIN
               CriticalT := W1/(W1 - W2)
               FOR I := 1 TO NumberOfSteps DO
                  BEGIN
                     (CurrentU,CurrentV,CurrentW) :=
                        (PreviousU,PreviousV,PreviousW)
                     IF I*DeltaT < CriticalT OR
                        (I - 1)*DeltaT > CriticalT THEN
                           DrawInDCFrom
                              (PreviousU/PreviousW,PreviousV/PreviousW)
                              To(CurrentU/CurrentW,CurrentV/CurrentW)
                     (PreviousU,PreviousV,PreviousW) :=
                        (CurrentU,CurrentV,CurrentW)
                  END
            END
      END
```

Since different figures may need different values for NumberOfSteps, an option

⋮
M: ChangeNumberOfSteps
⋮

might be added to Graph2D. A large NumberOfSteps has the advantage of slowing down the graphing so that the order and direction in which the lines are drawn can be seen. Using Graph2D with DisplayPictureByParametrizing and the parameter values $a = 1$ and $a = 1.5$, we produce the accurate Figures 4.29(a) and (b). Note that the positions of A and E in Figure 4.29(b) have been interchanged. This is because the display area is really a window in the projective plane, and the projective plane contains a Moebius band (see Figure 4.30 and Section 1.4).

There are thus different possible versions of Graph2D. Two comments apply to the versions discussed above.

First, in the program Graph2D the coordinates of data points are stored in the matrix Vertex. When an object transformation, represented by a matrix M_0', is performed, M_0' is automatically concatenated with M_0 and the complete viewing transformation is automatically recomputed. In particular, the effect of performing an object transformation is to change M_0 and M but not to change the original data stored in Vertex.

It is possible to produce a program with equivalent output in which the effect of performing an object transformation is to change the data stored in Vertex (by multiplying Vertex by M_0') but not change the viewing transformation matrix M_1. The method used in Graph2D and in the library has several advantages over this technique. For one, it achieves the same results while performing fewer calculations (see Exercise 9). For another, it permits recovery of the original data without creating a duplicate data file in memory. Finally, it permits interactive graphics—transformations can be computed before the points to be transformed are specified.

A great deal of geometry can be illustrated with Graph2D. For example, in viewing a picture we often want to zoom in to view a magnified portion of the picture. Graph2D can do this in either of two ways. One way is to scale up and (if necessary) translate the picture. The other is to define a new window or sequence of windows that zero in on the desired region. Figure 4.31 illustrates the result of zooming in on a second-floor bedroom of the house plan of Figure 4.22. The window has lower left corner (20, 21) and upper right corner (56.6, 51). Panning can be accomplished similarly by translating the window across the picture. (Note that both zooming and panning as described here require that clipping be implemented.)

In both two and three dimensions, line segments and curves in world coordinate space are normally idealized conceptually—they contain an infinite

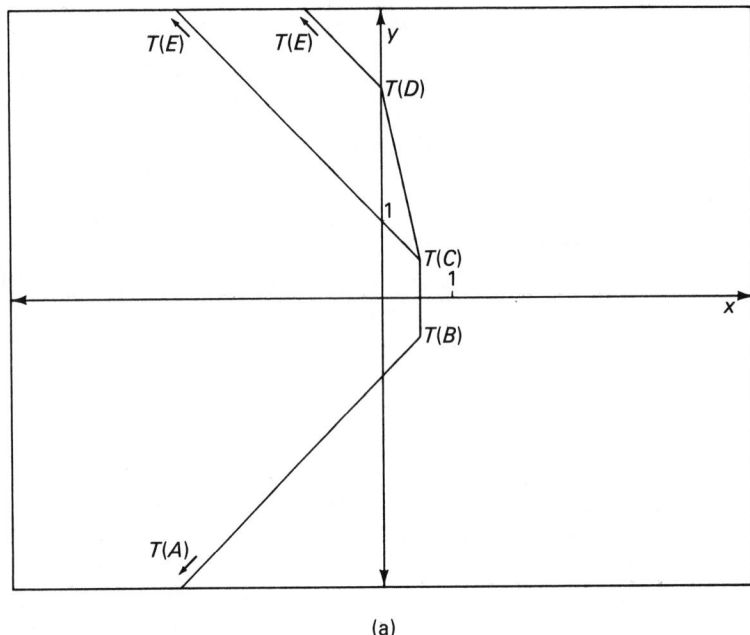

(a)

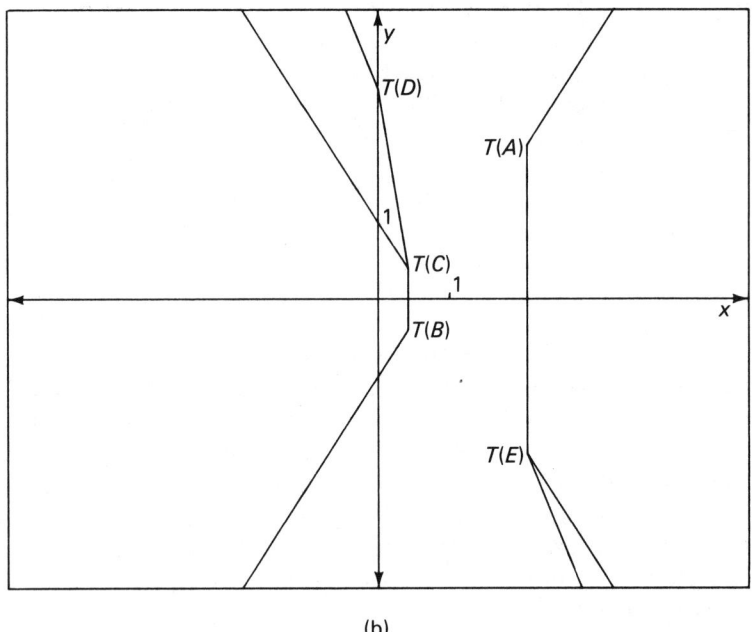

(b)

Figure 4.29 Correct perspective projection. (a) $a = 1$. (b) $a = 1.5$.

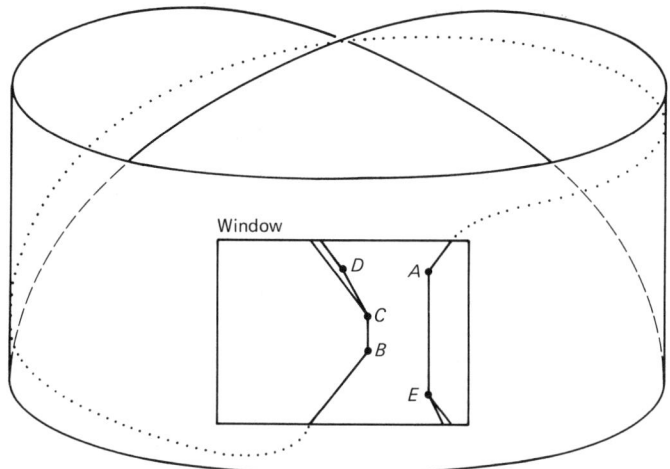

Figure 4.30 The display area as a window in the projective plane.

number of points and are without width. Line segments and curves in device coordinate space contain a finite number of points and do have width. The generation of line segments and curves in device coordinate space requires a mathematical (world coordinate) description in order that their appearances are not distorted by transformations. In other words, when we zoom in or scale up, for example, points will still be points and lines will not become wider. If the figure 2, for example, is scaled up in world coordinates (software scaling)

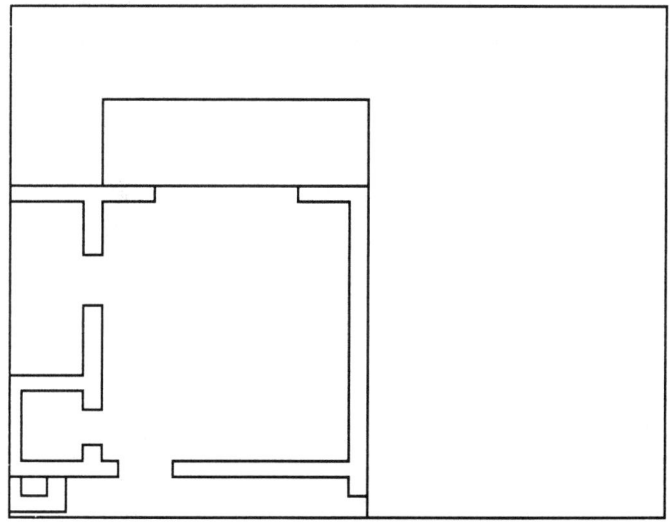

Figure 4.31 An enlarged view of a bedroom of the house plan in Fig. 4.22.

and then drawn, it will appear as in Figure 4.32(a). If it were converted to device coordinates first and then scaled up (hardware scaling), it would appear as in Figure 4.32(b).

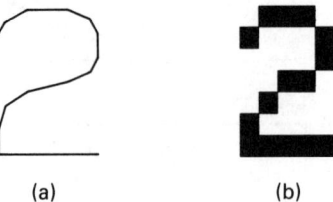

Figure 4.32 The figure 2 scaled up in (a) world coordinates and (b) device coordinates.

(a)   (b)

SECTION 3 EXERCISES

1. Implement Graph2D on a specific graphics device.
2. Use Graph2D to experiment with various test patterns. In particular, use a circle as a test pattern and note the results of perspective projection. What curves arise and why?
3. Verify that perspective projection with $a = 1$ of the test pattern of Figure 4.26 does cause vertices A and E to become ideal. Are there any other critical values of a?
4. Find matrices to be used with the command Matrix of Graph2D, which will convert a square test pattern into each of the figures in Figure 4.33.

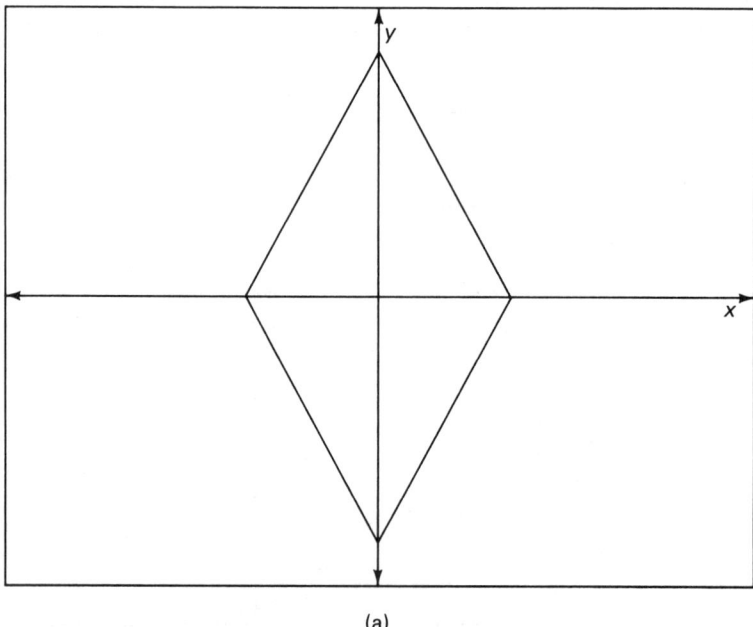

(a)

Figure 4.33 Projective transformations of a square.

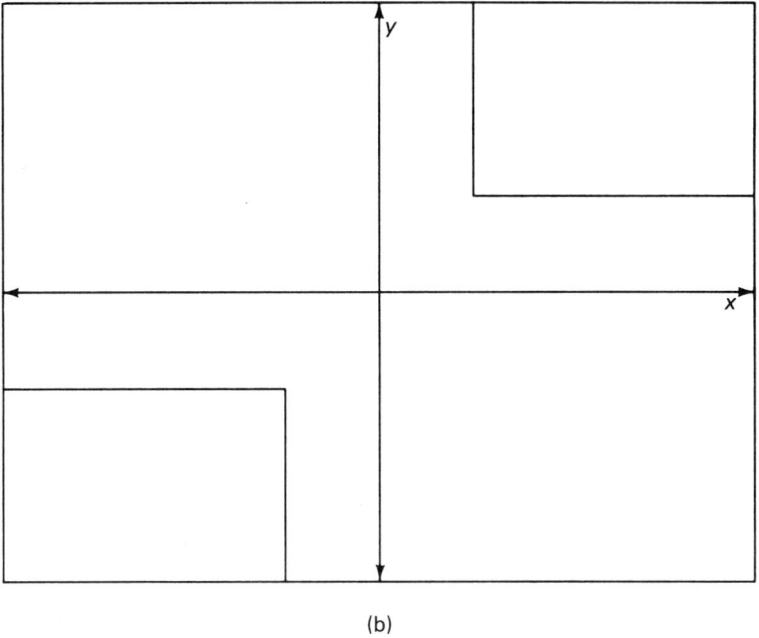

(b)

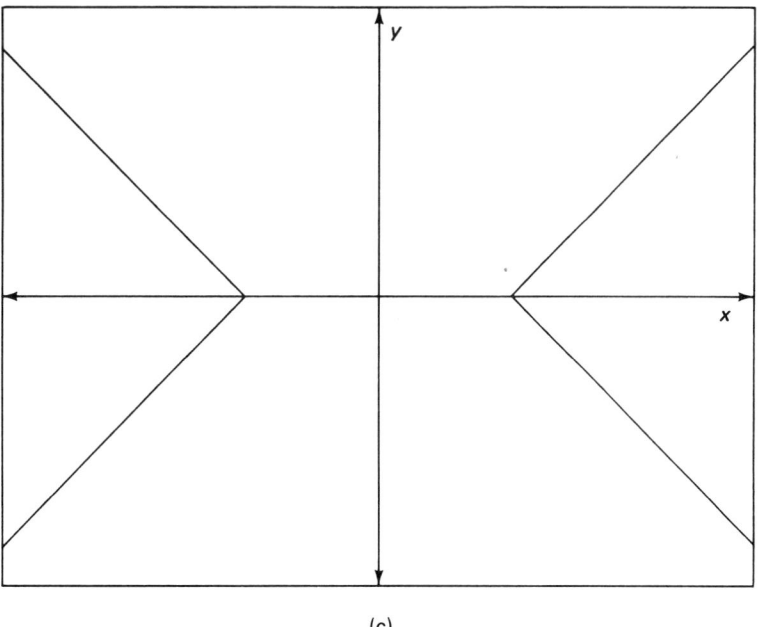

(c)

Figure 4.33 Continued.

5. Transformation by right multiplication by the matrix

$$\begin{pmatrix} \cos\theta & \sin\theta & 0 \\ -\sin\theta & \cos\theta & 0 \\ 0 & 0 & 1 \end{pmatrix}$$

is rotation with the origin as a fixed point and the ideal line mapped into itself. In this language describe the transformation represented by the matrix

$$\begin{pmatrix} \cos\theta & 0 & \sin\theta \\ 0 & 1 & 0 \\ -\sin\theta & 0 & \cos\theta \end{pmatrix}$$

Experiment with Graph2D and various value of θ.

6. When the perspective projection represented by the matrix

$$\begin{pmatrix} 1 & 0 & a \\ 0 & 1 & b \\ 0 & 0 & 1 \end{pmatrix}$$

is applied to a square test pattern (vertices $A(-1, -1)$, $B(1, -1)$, $C(1, 1)$, $D(-1, 1)$), the resulting quadrilateral has two vanishing points. Compute their coordinates.

7. Perspective projection of the test pattern of Figure 4.26(a) for small values of a may appear to be rotation of the two-dimensional test pattern in Euclidean three-space (see Figure 4.26(b)). However, this is not the case. Verify this by comparing Figure 4.34 to Figure 4.28(a). What qualitative differences are there between these two transformations? (Another way to see the differences between these transformations

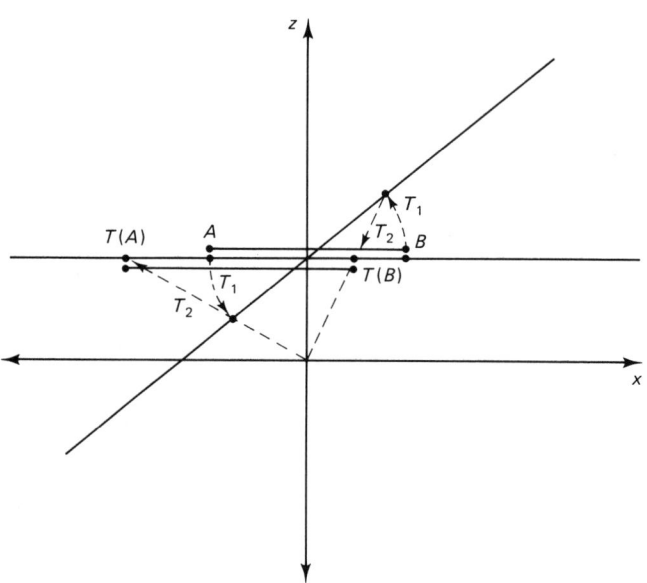

Figure 4.34 Rotation of a two-dimensional object in three-space.

is to compare the images (see Figure 4.35(b) and (c)) of the test pattern illustrated in Figure 4.35(a) under these transformations.)

8. How realistic is Figure 4.29(b)? Could you ever actually see this view? Would it help if you had an eye in the back of your head?

9. Suppose a test pattern has n vertices and we perform m consecutive object transformations and then display the test pattern. Compare the number of additions and multiplications required in each case.
   (a) The matrices of the object transformations are concatenated before being applied to the vertex data.
   (b) The matrices modify the vertex data one at a time.
   If n is large, which is more efficient?

10. How would you implement a command Zoom in Graph2D?

11. Would it be possible to define absolute and relative Move and Draw commands, such as MoveAbsolute[X,Y,Z], whose arguments are homogeneous coordinates? (Such commands might provide a more elegant version of DrawByParametrizing.)

12. The line of code in DrawByParametrizing

    (CurrentU,CurrentV,CurrentW) :=
            (PreviousU,PreviousV,PreviousW) + DeltaT*(U,V,W)

    appears to contradict the third remark after Proposition 2.1.3 that homogeneous coordinates cannot be added in a consistent way. Why is this *not* the case?

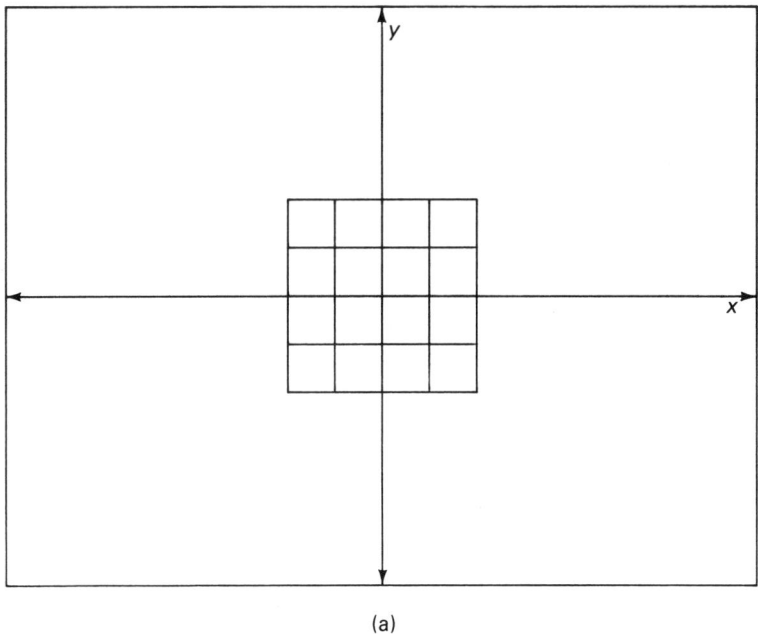

(a)

Figure 4.35 Test pattern (a) and its images after (b) 35° rotation and (c) perspective projection ($a = \tan 35° = 0.7002$, $b = 0$).

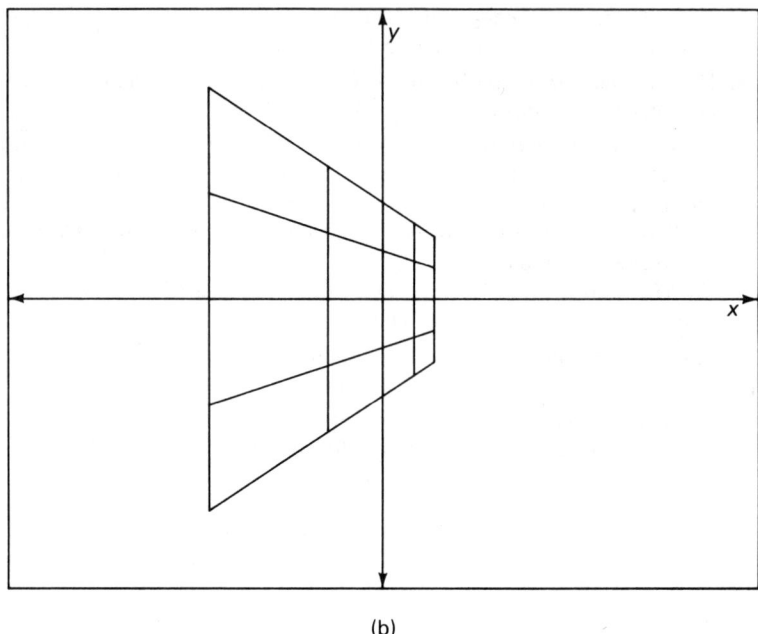

(b)

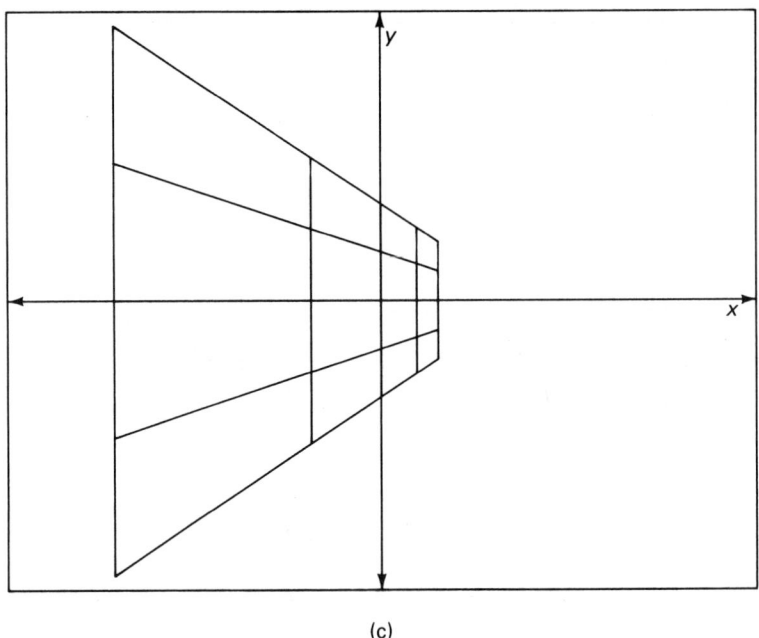

(c)

Figure 4.35 Continued.

13. Use Graph2D to experiment with windows and viewports that have different height-to-width ratios. What distortion is created by object transformations when these ratios are different? How might such distortion be used to advantage?
14. Suppose Graph2D is running on a computer that has a plotter or printer with a different aspect ratio than the monitor. As Graph2D is written, the window or the viewport needs to be adjusted first in order to plot or print exactly what the monitor shows. Write Graph2D using the modified library of Exercise 2.22 to achieve device-independence and avoid this difficulty.

5

Three-Dimensional Graphics

INTRODUCTION

This chapter is devoted to the study of three-dimensional graphics. With the advances in computer graphics systems that have been made in recent years, three-dimensional graphics has become an ever present part of our daily lives. The uses of three-dimensional computer graphics (and the related field of image reconstruction, to be considered in the next chapter) are so widespread that any enumeration of them would surely omit significant applications. We therefore attempt no such enumeration here.

In Section 1 we discuss some concepts and terminology related to simple three-dimensional imaging. In Section 2 we analyze the complete viewing operation of three-dimensional imaging. In Section 3 we present a simple three-dimensional graphics library similar to the two-dimensional library presented in Section 4.2. Again, we strongly recommend that (some form of) this library be implemented. Section 4 presents a general purpose three-dimensional imaging program. In Section 5 we apply a specialized three-dimensional imaging program to drawing the graphs of functions of two real variables. In Section 6, with these programs in hand, we discuss the effects resulting from variations in the viewing mechanism. Section 7 presents an alternate approach to three-dimensional imaging which is particularly useful in depth clipping. In Section 8 we present a program that simulates an enhanced attitude indicator for an aircraft.

SECTION 1: SOME CONCEPTS AND TERMINOLOGY

We begin our study of three-dimensional graphics with a brief discussion of some fundamental three-dimensional graphics concepts and terminology. The quantitative analysis of these elements is addressed in the next section.

The geometry of simple perspective imaging is illustrated in Figure 5.1. The elements of this geometry include the object, the center of perspectivity, the viewplane, a parametrizing plane, a parametrization of the viewplane, and a display transformation.

The object, center of perspectivity, and viewplane all exist in *world coordinate three-space*. The coordinates of world coordinate three-space are *world coordinates*. The process of taking the object to the viewport, which is the composite of perspective projection, inverse parametrization, and the display transformation, is the *complete viewing operation*. The composite of perspective projection and inverse parametrization is the *viewing transformation;* thus (as in two dimensions) the complete viewing operation is the composite of a viewing transformation and a display transformation.

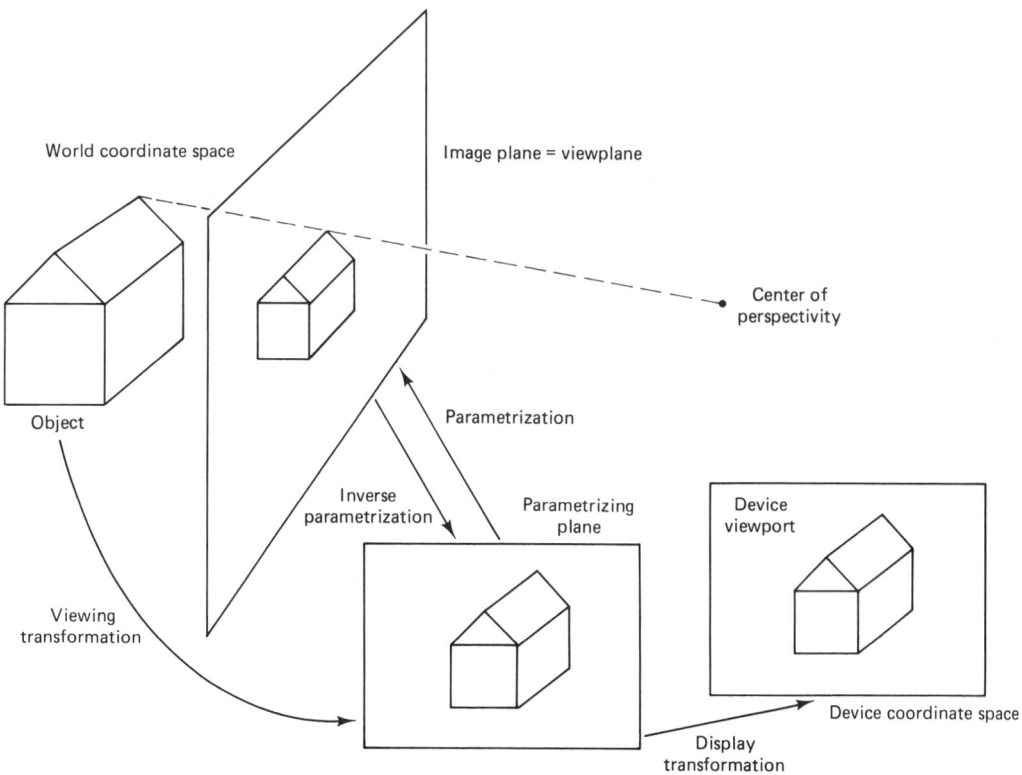

Figure 5.1 The geometry of simple perspective imaging.

The center of perspectivity is specified by its coordinates. As we will see in Section 5, spherical coordinates can be particularly useful in specifying the center of perspectivity.

The viewplane can be specified in several ways. One way (see Figure 5.2(a)) is by a point on the viewplane, the *viewplane reference point,* and a vector normal to the viewplane, the *viewplane normal.* Another way to specify the viewplane (see Figure 5.2(b)) is by a point that is central to the object, the *center of the object,* and a distance, the *viewplane distance:* the viewplane is the plane perpendicular to the line determined by the center of the object and the center of perspectivity, the *principal line of sight,* and at the viewplane distance from the center of the object. (The center of the object is a user-defined point that is "central" to the object in the sense that it will be located—when the imaging process is complete—at the "center" of the object. It is not assumed, unless the user wishes, that the center of the object is the centroid—or any other "natural" point—associated to the object.) Yet another way to specify the viewplane is by its homogeneous coordinates.

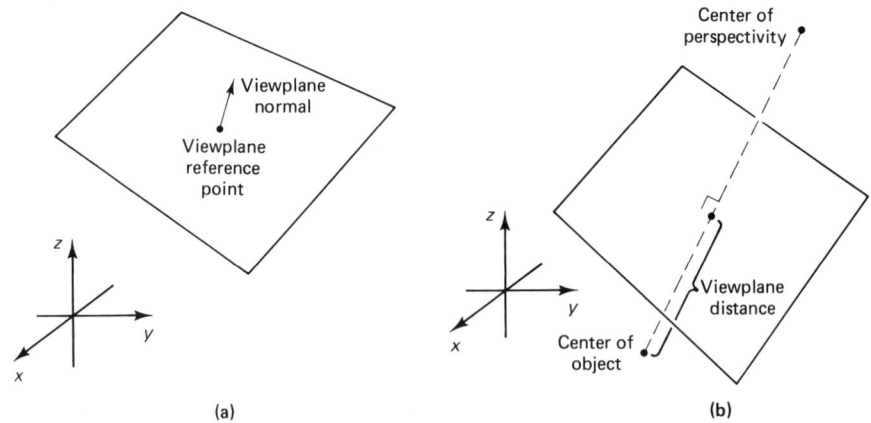

Figure 5.2 Specification of the viewplane.

Various factors determine the particular viewplane specification to be used. For example, specification of the viewplane by a center of the object and a viewplane distance is useful if it is desired to move the entire viewing mechanism about the object with as little respecification on each move as possible: given the center of the object and the viewplane distance, the entire viewing mechanism is specified by the center of perspectivity, and it can be respecified by redefining (just) the center of perspectivity. The price paid for the utility of this viewing mechanism is a certain amount of inflexibility: the principal line of sight is perpendicular to the viewplane. This inflexibility *could* be circumvented by positioning the center of the object away from the object, but this defeats the purpose for using this viewing mechanism.

To complete specification of the viewing transformation, we must specify the viewplane parametrization and the display transformation. This can be done in several ways. One way is as follows:

Let the display transformation be the transformation T_{image} taking $A(-h, -k)$, $B(h, -k)$, $C(h, k)$, and $D(-h, k)$ to $A_{\text{image}}(u_0, v_0)$, $B_{\text{image}}(u_1, v_0)$, $C_{\text{image}}(u_1, v_1)$, and $D_{\text{image}}(u_0, v_1)$, respectively (see Figure 5.3). (This is a special case of the display transformation discussed in Section 4.1.) Parametrization of the viewplane is then specified by the following type of geometric data: a viewplane reference point, a viewplane up-direction vector, and a viewplane window (see Figure 5.4).

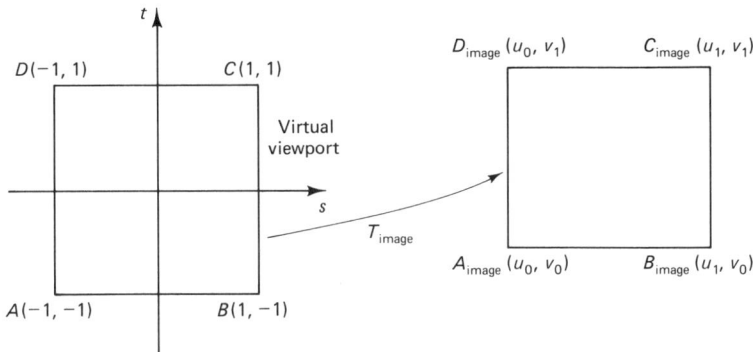

Figure 5.3 The display transformation.

The purpose of the viewplane reference point and the viewplane up-direction vector is to determine a coordinate system on the viewplane: The viewplane reference point is a point on the viewplane to be taken as the origin of the coordinate system. The viewplane up-direction vector is a vector emanating from the viewplane reference point that is not perpendicular to the viewplane and whose projection onto the viewplane determines the positive direction of the "vertical" axis of our coordinate system. The positive direction of the "horizontal" axis is chosen so that the resulting coordinate system is right-handed (as viewed from the center of perspectivity) and orthogonal (the angle between the coordinate axes is a right angle). The distance between two points in this coordinate system on the viewplane is the same as the distance between the two points as measured in world coordinate three-space.

The viewplane window is that (rectangular) part of the viewplane to be displayed. The parametrization T_{object} of the viewplane is determined by requiring that it take $A(-h, -k)$, $B(h, -k)$, $C(h, k)$, and $D(-h, k)$ to $A_{\text{object}}(-h, -k)$, $B_{\text{object}}(h, -k)$, $C_{\text{object}}(h, k)$, and $D_{\text{object}}(-h, k)$, respectively. (The coordinates of the vertices of the viewplane window are coordinates in the viewplane coordinate system constructed above.)

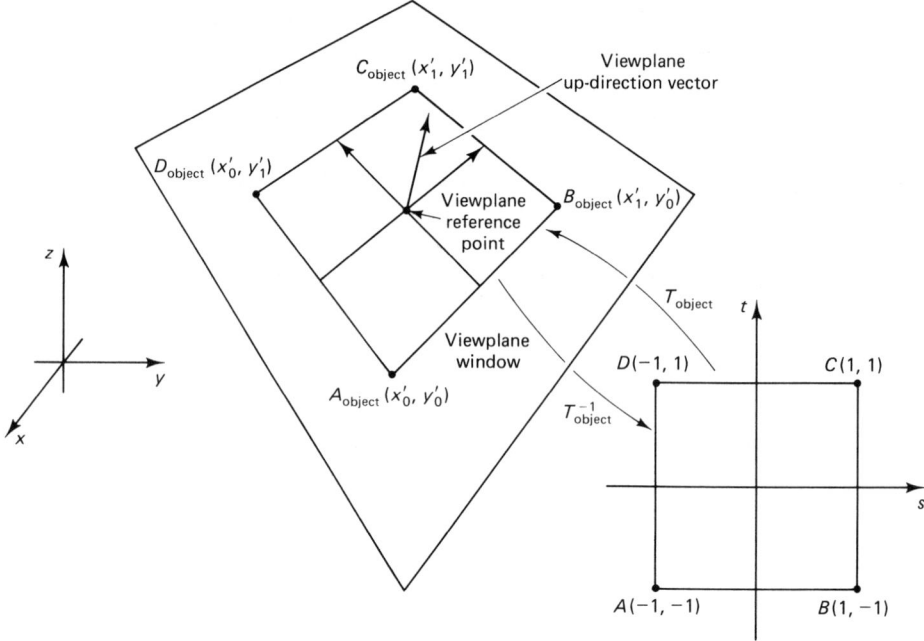

Figure 5.4 Parametrization of the viewplane.

The analytic description of the viewing transformation will be pursued in the next section. We close this section with some general comments.

First, since the analytic description of the viewing transformation used in simple perspective imaging is derived using homogeneous coordinates, the same description applies to the viewing transformation used in parallel projection imaging. This is because points and directions in Euclidean geometry are both treated as points in projective geometry. Thus the only difference between parallel projection and perspective projection is that a *direction of projection* replaces a center of perspectivity.

Second, the way the viewing transformation is implemented here, clipping outside the cone of vision (see Figure 5.5) is simple and straightforward. However it is sometimes also desirable to clip the cone of vision to a finite view volume (see Figure 5.6) with a *front* (or *hither*) *clipping plane* and a *back* (or *yon*) *clipping plane*. This type of clipping can be accomplished by modifying the viewing procedure presented in this section (to take depth into account), or it can be accomplished via a completely different geometry; this latter approach is discussed in Section 7.

SECTION 1 EXERCISES

1. What is required to specify a complete viewing operation? How many parameters are necessary? What assumptions can reduce the number of parameters?

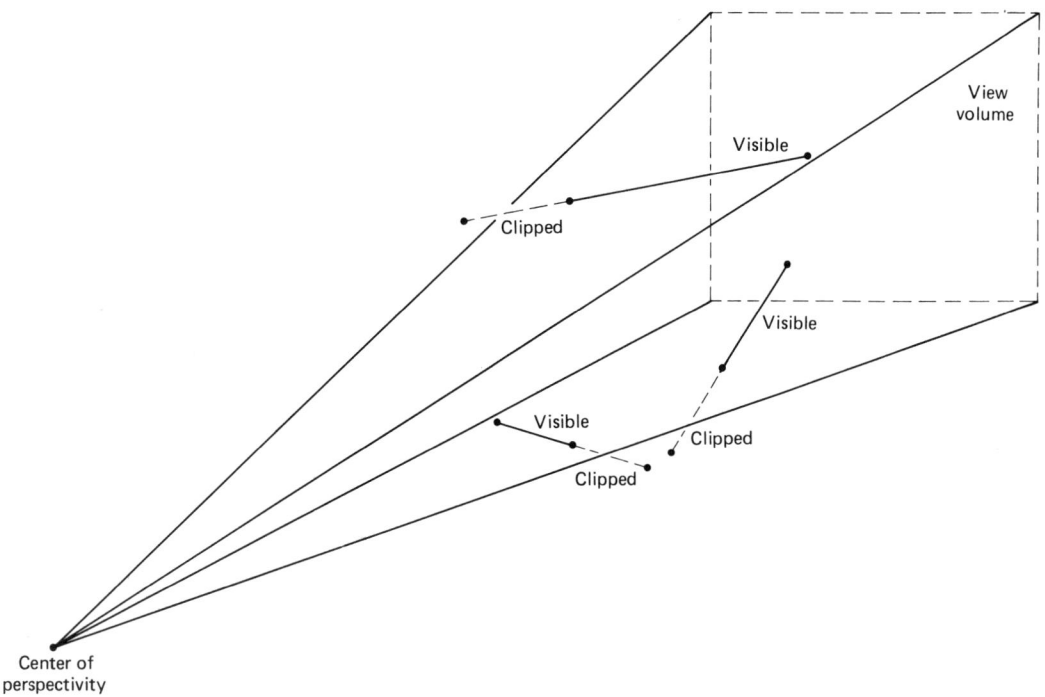

Figure 5.5 Lateral clipping.

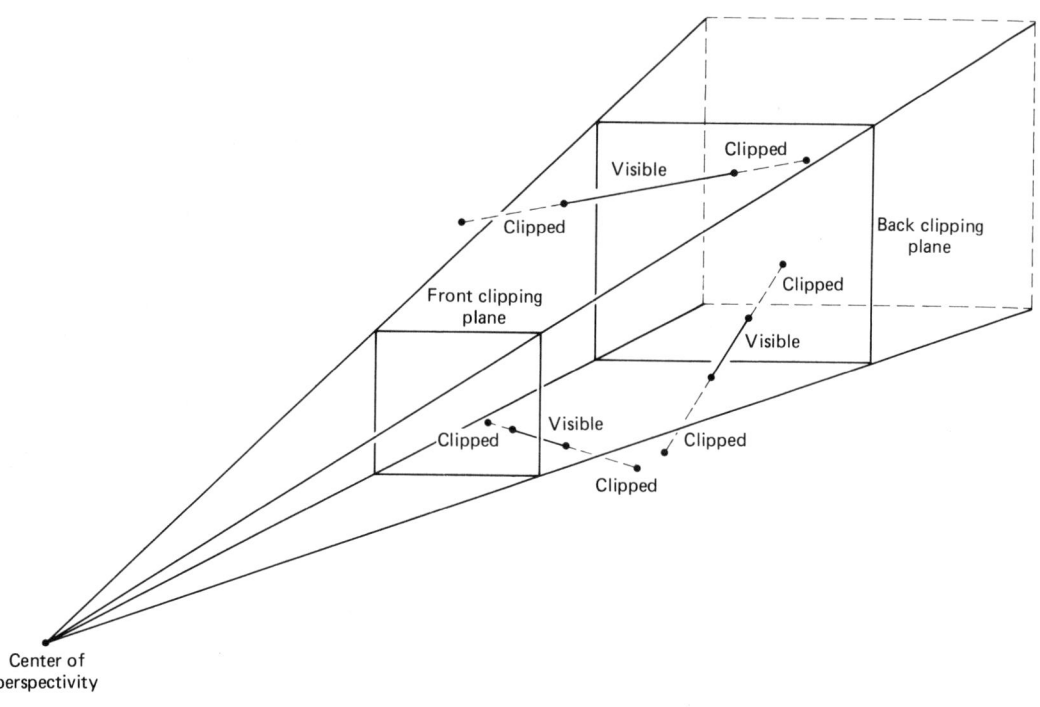

Figure 5.6 Lateral and depth clipping.

Sec. 1   Some Concepts and Terminology

The following two exercises are pursued in Section 6.

2. If the center of perspectivity is specified by spherical coordinates (ρ, θ, ϕ), and the viewplane is specified by the center of the object being the origin and a viewplane distance d (as described in this section), what effects will changing each of the parameters ρ, θ, ϕ, and d individually have on the image?

3. Suppose a complete viewing operation that will produce a perspective image has been defined.
   (a) How will the image be changed if the object and the viewplane are fixed and the center of perspectivity is moved? What if the center of perspectivity is moved to infinity?
   (b) How will the image be changed if the object and center of perspectivity are fixed and the viewplane is moved?

4. Why should one be concerned about depth clipping?

SECTION 2: THE COMPLETE VIEWING OPERATION IN THREE DIMENSIONS

In this section we derive the formulas necessary to implement the complete viewing operation. We do this in several ways: We begin by considering the case of General Perspective Projection, in which the elements determining the complete viewing operation for perspective projection are arbitrary. In certain special cases, these formulas simplify greatly. (The price to be paid for generality is the complexity of the resulting formulas.) We investigate four Special Cases: three are special cases of General Perspective Projection, and the fourth case deals with parallel projection.

For simplicity, we assume throughout this section that the vertices of the viewport in the parametrizing plane are $A(-h, -k)$, $B(h, -k)$, $C(h, k)$, and $D(-h, k)$ and that the vertices of the viewport in device coordinate space are $A_{\text{image}}(u_0, v_0)$, $B_{\text{image}}(u_1, v_0)$, $C_{\text{image}}(u_1, v_1)$, and $D_{\text{image}}(u_0, v_1)$. Furthermore, we assume that the viewplane reference point is the center (that is, the intersection of the diagonals) of the window.

Throughout this section the coordinates that specify the elements determining the complete viewing operation are Cartesian, as opposed to homogeneous. The reason for this is that when (in the next section) we extend the simple graphics library presented in Section 4.2 (to support simple three-dimensional imaging), we will use Cartesian coordinates. We do this not because homogeneous coordinates (and projective geometry) are dispensable in the end; on the contrary, homogeneous coordinates (and projective geometry) play an important role in this section. Rather, we do this because Cartesian coordinates (and Euclidean geometry) are more familiar than homogeneous coordinates (and projective geometry) to many of those who would use the library.

GENERAL PERSPECTIVE PROJECTION

We begin by assuming that the complete viewing operation is specified (see Figure 5.7) by

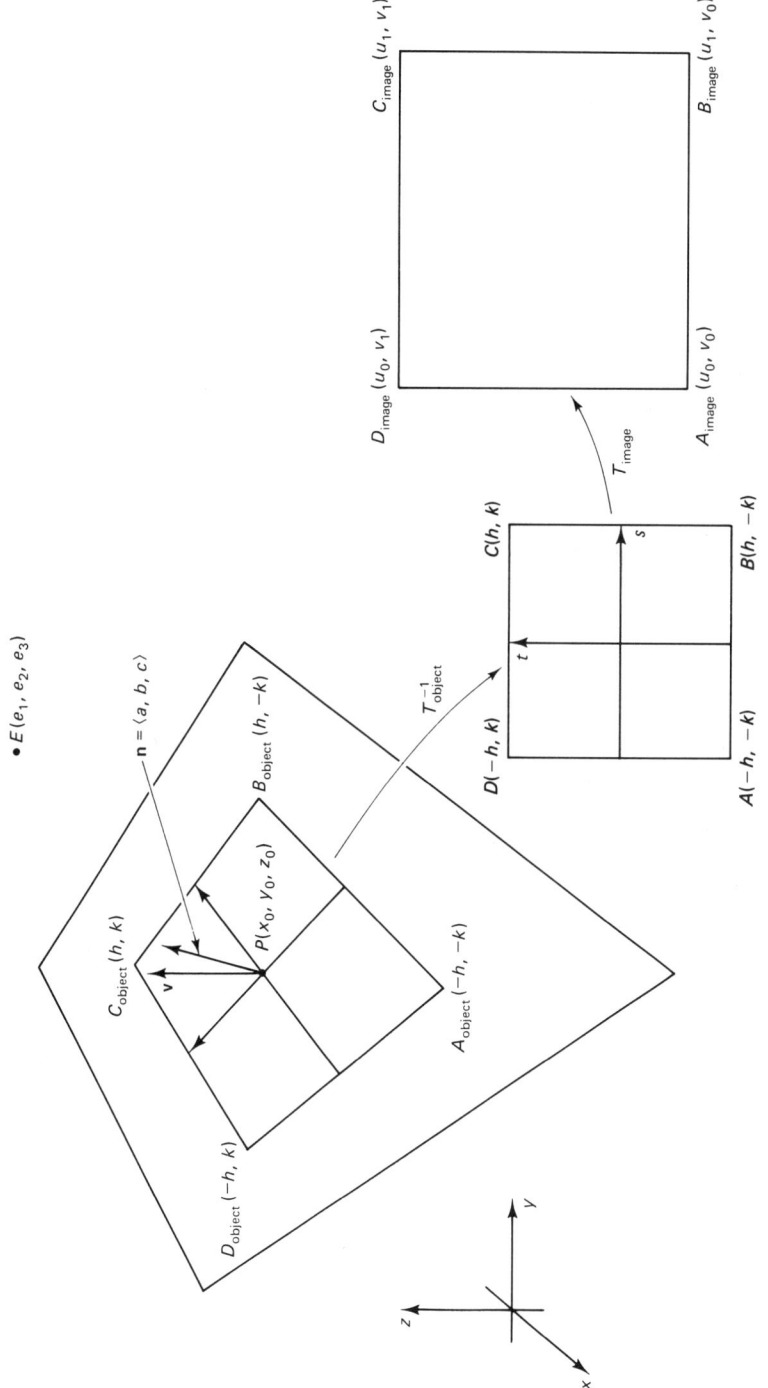

Figure 5.7 The complete viewing operation for General Perspective Projection.

1. A center of perspectivity, or eye, $E(e_1, e_2, e_3)$
2. A viewplane reference point $P(x_0, y_0, z_0)$
3. A viewplane normal vector $\mathbf{n} = \langle a, b, c \rangle$
4. A viewplane up-direction vector $\mathbf{v}$
5. A window whose vertices are $A_{\text{object}}(-h, -k)$, $B_{\text{object}}(h, -k)$, $C_{\text{object}}(h, k)$, and $D_{\text{object}}(-h, k)$

The complete viewing operation is the composite of three transformations: perspective projection, inverse parametrization, and display transformation. When described in homogeneous coordinates, each of these transformations can be represented by right matrix multiplication: Projection is represented by a 4 × 4 rank 3 matrix M_1, inverse parametrization is represented by a 4 × 3 rank 3 matrix M_2, and the display tranformation is represented by a 3 × 3 rank 3 matrix M_4. Consequently the complete viewing operation is represented by right multiplication by the 4 × 3 rank 3 matrix.

$$M = M_1 M_2 M_4$$

We compute M_1 as follows: Given the viewplane reference point $P(x_0, y_0, z_0)$ and the viewplane normal $\mathbf{n} = \langle a, b, c \rangle$, the equation of the viewplane is

$$a(x - x_0) + b(y - y_0) + c(z - z_0) = 0$$

Thus the homogeneous coordinates of the viewplane μ are

$$[\mu_1, \mu_2, \mu_3, \mu_4] = [a, b, c, -(ax_0 + by_0 + cz_0)]$$

The homogeneous coordinates for the center of perspectivity are $[e_1, e_2, e_3, 1]$, and Proposition 3.2.1 gives the matrix M_1 representing projection to be

$$\begin{pmatrix} \mu_2 e_2 + \mu_3 e_3 + \mu_4 & -\mu_1 e_2 & -\mu_1 e_3 & -\mu_1 \\ -\mu_2 e_1 & \mu_1 e_1 + \mu_3 e_3 + \mu_4 & -\mu_2 e_3 & -\mu_2 \\ -\mu_3 e_1 & -\mu_3 e_2 & \mu_1 e_1 + \mu_2 e_2 + \mu_4 & -\mu_3 \\ -\mu_4 e_1 & -\mu_4 e_2 & -\mu_4 e_3 & \mu_1 e_1 + \mu_2 e_2 + \mu_3 e_3 \end{pmatrix}$$

Since the display transformation is defined on the rectangle whose vertices are $A(-h, -k)$, $B(h, -k)$, $C(h, k)$, and $D(-h, k)$, the matrix M_4 representing the display transformation (see Section 4.1) is

$$M_4 = \begin{pmatrix} \dfrac{u_1 - u_0}{2h} & 0 & 0 \\ 0 & \dfrac{v_1 - v_0}{2k} & 0 \\ \dfrac{u_1 + u_0}{2} & \dfrac{v_1 + v_0}{2} & 1 \end{pmatrix}$$

As in Chapter 4, if normalized device coordinates are used, the matrix M_4 can be used in the factored form $M_4 = M_{\text{object}}^{-1} M_{\text{image}}$.

To compute the matrix M_2 representing inverse parametrization, we first examine the coordinate system defined in the viewplane by the viewplane reference point P and the viewplane up-direction vector $\mathbf{v}$ (see Figure 5.8): With the correct choice of $\pm$ sign, the vector

$$\mathbf{w}_1 = \langle w_{11}, w_{12}, w_{13} \rangle = \pm \frac{\mathbf{v} \times \mathbf{n}}{|\mathbf{v} \times \mathbf{n}|}$$

is a unit vector in the direction of the positive "horizontal" axis, and

$$\mathbf{w}_2 = \langle w_{21}, w_{22}, w_{23} \rangle = \frac{\mathbf{n} \times \mathbf{w}_1}{|\mathbf{n} \times \mathbf{w}_1|} = \frac{\mathbf{n} \times \mathbf{w}_1}{|\mathbf{n}|}$$

is a unit vector in the direction of the positive "vertical" axis. Thus the Cartesian coordinates of C_{object} are the components of the position vector

$$\mathbf{OP} + h\mathbf{w}_1 + k\mathbf{w}_2$$

Given the Cartesian coordinates of $C_{\text{object}}(c_1, c_2, c_3)$, the homogeneous coordinates of C_{object} are $[c_1, c_2, c_3, 1]$.

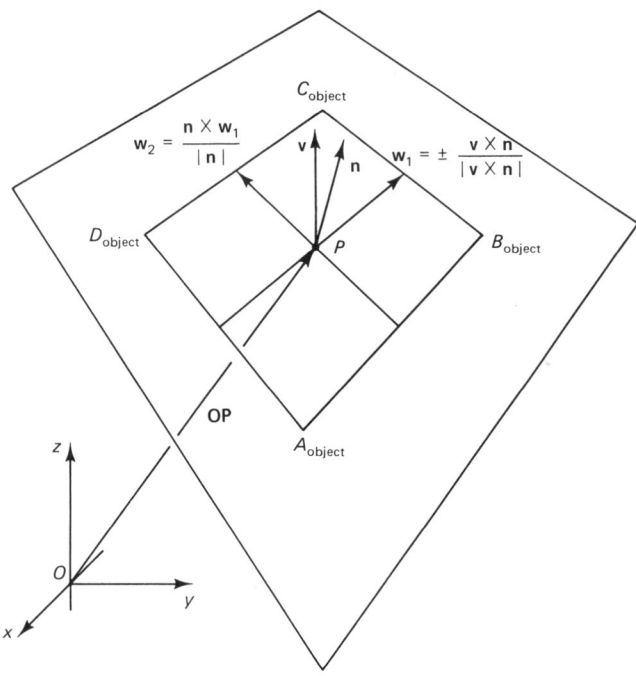

Figure 5.8 Determination of the inverse parametrization.

The matrix M_2 is the right inverse M_μ^+ of the 3×4 rank 3 matrix M_μ representing the parametrization $T_\mu : RP^2 \to \mu$ for which

$$T_\mu([1, 0, 0]) = [\langle 1, 0, 0 \rangle M_\mu] = [w_{11}, w_{12}, w_{13}, 0]$$

$$T_\mu([0, 1, 0]) = [\langle 0, 1, 0 \rangle M_\mu] = [w_{21}, w_{22}, w_{23}, 0]$$

$$T_\mu([0, 0, 1]) = [\langle 0, 0, 1\rangle M_\mu] = [x_0, y_0, z_0, 1]$$
$$T_\mu([1, 1, 1]) = [\langle h, k, 1\rangle M_\mu] = [c_1, c_2, c_3, 1]$$

By Proposition 3.2.5 there are nonzero real constants k_1, k_2, and k_3 for which

$$M_\mu = \begin{pmatrix} k_1 w_{11} & k_1 w_{12} & k_1 w_{13} & 0 \\ k_2 w_{21} & k_2 w_{22} & k_2 w_{23} & 0 \\ k_3 x_0 & k_3 y_0 & k_3 z_0 & k_3 \end{pmatrix}$$

and

$$k_1\langle w_{11}, w_{12}, w_{13}, 0\rangle + k_2\langle w_{21}, w_{22}, w_{23}, 0\rangle + k_3\langle x_0, y_0, z_0, 1\rangle = \langle c_1, c_2, c_3, 1\rangle.$$

But since the Cartesian coordinates of C_{object} are the components of the position vector

$$\mathbf{OP} + h\mathbf{w}_1 + k\mathbf{w}_2$$

$k_1 = 1, k_2 = 1,$ and $k_3 = 1.$ Thus

$$M_\mu = \begin{pmatrix} w_{11} & w_{12} & w_{13} & 0 \\ w_{21} & w_{22} & w_{23} & 0 \\ x_0 & y_0 & z_0 & 1 \end{pmatrix}$$

SPECIAL CASE 1: PERSPECTIVE PROJECTION

> The center of perspectivity is $E(0, 0, 0)$
> The image plane is $z = 1$

In this case the complete viewing operation is specified (see Figure 5.9) by

1. The center of perspectivity $E(0, 0, 0)$
2. The viewplane reference point $P(0, 0, 1)$
3. The viewplane normal vector $\mathbf{n} = \langle 0, 0, 1\rangle$
4. The viewplane up-direction vector $\mathbf{v} = \langle 0, 1, 0\rangle$
5. The window whose vertices are $A_{\text{object}}(-1, -1)$, $B_{\text{object}}(1, -1)$, $C_{\text{object}}(1, 1)$, and $D_{\text{object}}(-1, 1)$

Since the homogeneous coordinates of E are $[0, 0, 0, 1]$ and the homogeneous coordinates of μ are $[0, 0, 1, -1]$, it follows (from the case of General Perspective Projection) that the matrix representing perspective projection is

$$M_1 = \begin{pmatrix} 1 & 0 & 0 & 0 \\ 0 & 1 & 0 & 0 \\ 0 & 0 & 1 & 1 \\ 0 & 0 & 0 & 0 \end{pmatrix}$$

(*Note:* Whenever the center of perspectivity E is the origin, $E = E(0, 0, 0)$, M_1 is of the form

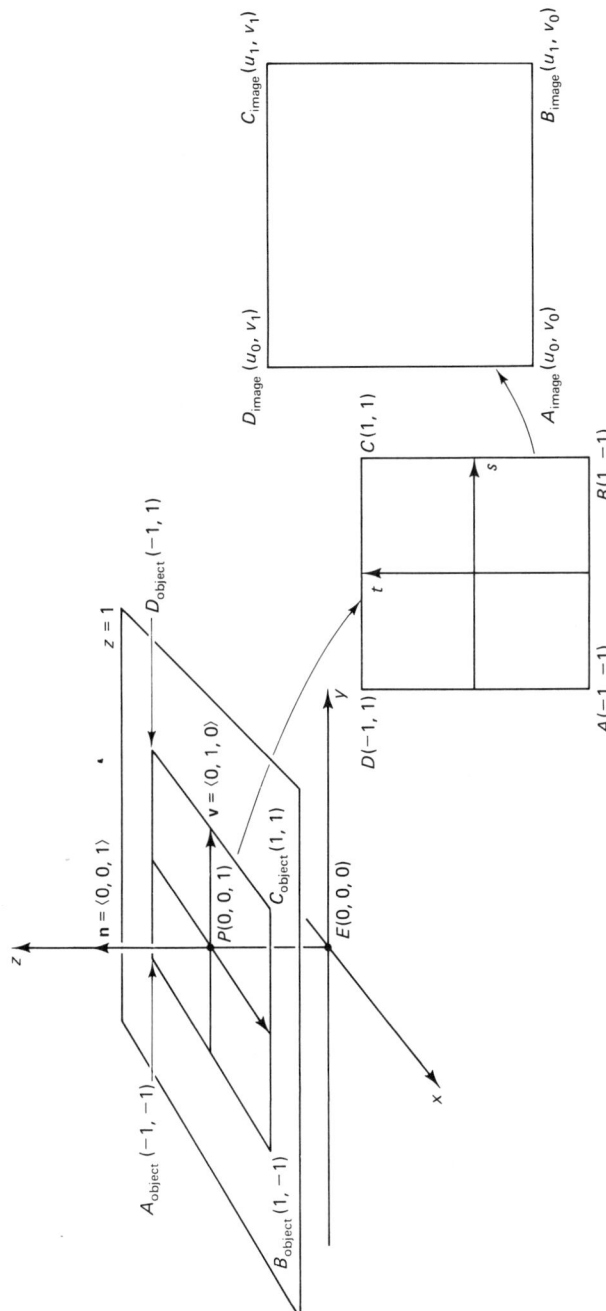

Figure 5.9 The complete viewing operation for perspective projection.

$$M_1 = \begin{pmatrix} \mu_4 & 0 & 0 & -\mu_1 \\ 0 & \mu_4 & 0 & -\mu_2 \\ 0 & 0 & \mu_4 & -\mu_3 \\ 0 & 0 & 0 & 0 \end{pmatrix}$$

where $\mu = \mu[\mu_1, \mu_2, \mu_3, \mu_4]$ is the viewplane.)

The matrix M_4 is the same as in the case of General Perspective Projection.

Using the given data and the formulas for the case of General Perspective Projection, it follows that

$$M_\mu = \begin{pmatrix} 1 & 0 & 0 & 0 \\ 0 & 1 & 0 & 0 \\ 0 & 0 & 1 & 1 \end{pmatrix}$$

so that

$$M_2 = \begin{pmatrix} 1 & 0 & 0 \\ 0 & 1 & 0 \\ 0 & 0 & \frac{1}{2} \\ 0 & 0 & \frac{1}{2} \end{pmatrix}$$

Consequently, the matrix M representing the complete viewing operation is

$$M = M_1 M_2 M_4 = \begin{pmatrix} 1 & 0 & 0 \\ 0 & 1 & 0 \\ 0 & 0 & 1 \\ 0 & 0 & 0 \end{pmatrix} \begin{pmatrix} \frac{u_1 - u_0}{2} & 0 & 0 & 0 \\ 0 & \frac{v_1 - v_0}{2} & 0 & 0 \\ \frac{u_1 + u_0}{2} & \frac{v_1 + v_0}{2} & 1 & \end{pmatrix}$$

Observe that the orientation of the coordinate system in the plane $z \doteq 1$ is left-handed when viewed from the origin. Thus, however the device coordinate system is oriented, u_0, v_0, u_1, and v_1 must be chosen so that the viewplane coordinate system is left-handed on the display device. (If a right-handed coordinate system is preferred to a left-handed one in the viewplane, we could, for example, choose the original coordinate system in three-space to be left-handed instead of right-handed.)

Note also that by performing the appropriate interchanges in M, we can find the matrix of the complete viewing operation in the case that the eye is $E(0, 0, 0)$ and the viewplane is $y = 1$, or in the case that the eye is $E(0, 0, 0)$ and the viewplane is $x = 1$.

SPECIAL CASE 2: A PERSPECTIVE PROJECTION

> The center of perspectivity is $E(0, 0, k)$
> The image plane is $z = l$

In this case the complete viewing operation is specified (see Figure 5.10) by

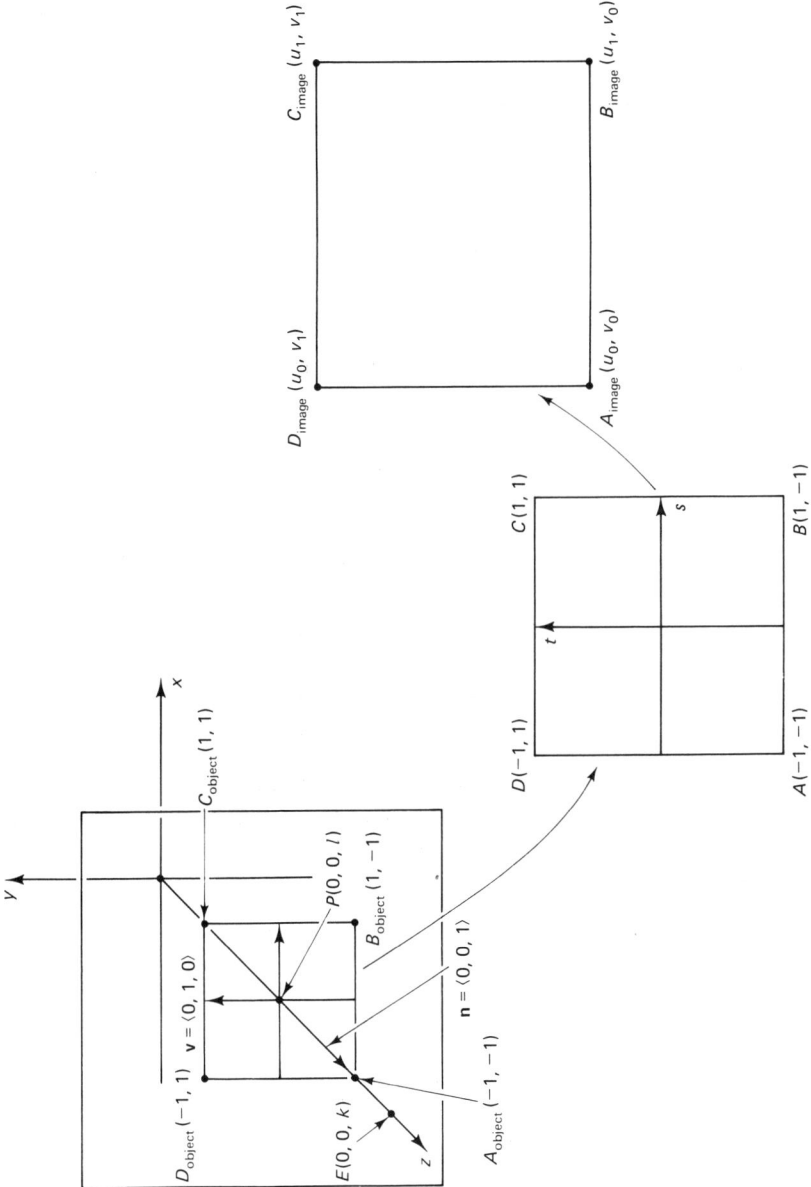

Figure 5.10 The complete viewing operation for a perspective projection.

1. The center of perspectivity $E(0, 0, k)$
2. The viewplane reference point $P(0, 0, l)$
3. The viewplane normal vector $\mathbf{n} = \langle 0, 0, 1 \rangle$
4. The viewplane up-direction vector $\mathbf{v} = \langle 0, 1, 0 \rangle$
5. The window whose vertices are $A_{\text{object}}(-1, -1)$, $B_{\text{object}}(1, -1)$, $C_{\text{object}}(1, 1)$, and $D_{\text{object}}(-1, 1)$

Since the homogeneous coordinates of E are $[0, 0, k, 1]$ and the homogeneous coordinates of μ are $[0, 0, 1, -l]$, it follows (from the case of General Perspective Projection) that the matrix representing perspective projection

$$M_1 = \begin{pmatrix} k-l & 0 & 0 & 0 \\ 0 & k-l & 0 & 0 \\ 0 & 0 & -l & -1 \\ 0 & 0 & kl & k \end{pmatrix}$$

If, for example, the viewplane is the coordinate plane $z = 0$, then

$$M_1 = \begin{pmatrix} k & 0 & 0 & 0 \\ 0 & k & 0 & 0 \\ 0 & 0 & 0 & -1 \\ 0 & 0 & 0 & k \end{pmatrix}$$

(*Note:* Whenever the viewplane is a coordinate plane such as $z = 0$, M_1 is of the form

$$M_1 = \begin{pmatrix} e_3 & 0 & 0 & 0 \\ 0 & e_3 & 0 & 0 \\ -e_1 & -e_2 & 0 & -1 \\ 0 & 0 & 0 & e_3 \end{pmatrix}$$

where $E = E[e_1, e_2, e_3, 1]$ is the center of perspectivity.)

The matrix M_4 is the same as in the case of General Perspective Projection.

Using the given data and the formulas of the case of General Perspective Projection, it follows that

$$M_\mu = \begin{pmatrix} 1 & 0 & 0 & 0 \\ 0 & 1 & 0 & 0 \\ 0 & 0 & l & 1 \end{pmatrix}$$

so that

$$M_2 = \begin{pmatrix} 1 & 0 & 0 \\ 0 & 1 & 0 \\ 0 & 0 & \dfrac{l}{l^2+1} \\ 0 & 0 & \dfrac{1}{l^2+1} \end{pmatrix}$$

Consequently, the matrix M representing the complete viewing operation is

$$M = M_1 M_2 M_4 = \begin{pmatrix} k-l & 0 & 0 & 0 \\ 0 & k-l & 0 & 0 \\ 0 & 0 & -1 & 0 \\ 0 & 0 & k & 0 \end{pmatrix} \begin{pmatrix} \dfrac{u_1 - u_0}{2} & 0 & 0 \\ 0 & \dfrac{v_1 - v_0}{2} & 0 \\ \dfrac{u_1 + u_0}{2} & \dfrac{v_1 + v_0}{2} & 1 \end{pmatrix}$$

Again, by performing the appropriate interchanges in M, we can find the matrix of the complete viewing operation in the case that the eye is $E(0, k, 0)$ and the viewplane is $y = l$ or in the case that the eye is $E(k, 0, 0)$ and the viewplane is $x = l$.

SPECIAL CASE 3: PARALLEL PROJECTION

> The direction of projection is $\mathbf{e} = \langle e_1, e_2, e_3 \rangle$
> The viewplane is $z = l$

In this case the complete viewing operation is specified (see Figure 5.11) by

1. The direction of projection $\mathbf{e} = \langle e_1, e_2, e_3 \rangle$
2. The viewplane reference point $P(0, 0, l)$

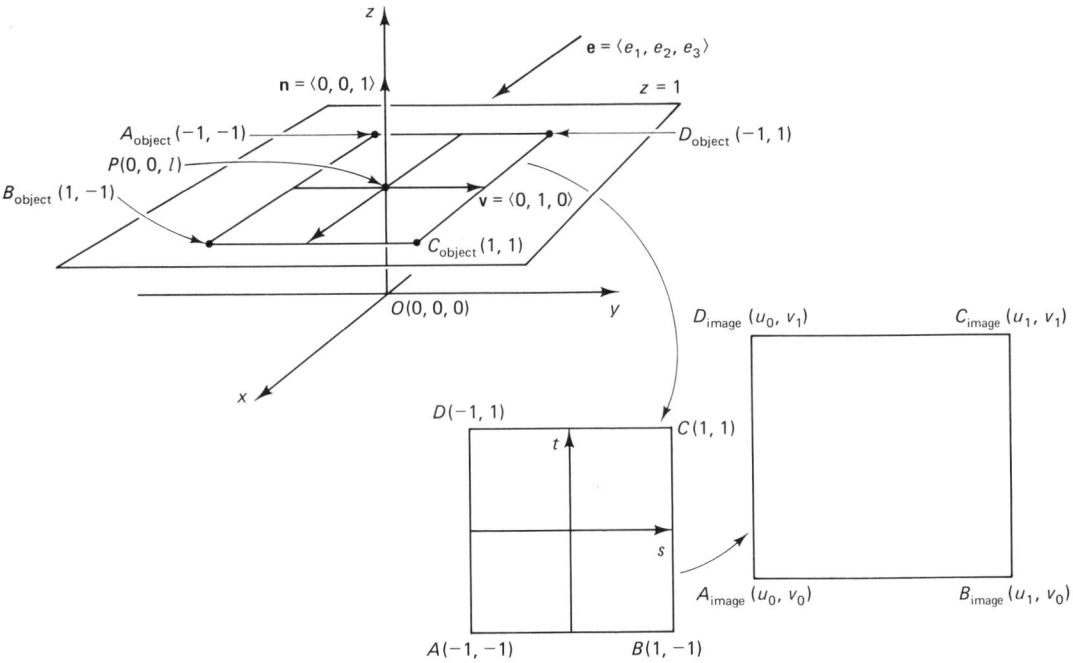

Figure 5.11 The complete viewing operation for parallel projection.

Sec. 2 The Complete Viewing Operation in Three Dimensions

3. The viewplane normal vector $\mathbf{n} = \langle 0, 0, 1\rangle$
4. The viewplane up-direction vector $\mathbf{v} = \langle 0, 1, 0\rangle$
5. The window whose vertices are $A_{\text{object}}(-1, -1)$, $B_{\text{object}}(1, -1)$, $C_{\text{object}}(1, 1)$, and $D_{\text{object}}(-1, 1)$

Since the homogeneous coordinates of $\mathbf{e}$ are $[e_1, e_2, e_3, 0]$ and the homogeneous coordinates of μ are $[0, 0, 1, -l]$, it follows (from Proposition 3.2.1) that the matrix representing parallel projection is

$$M_1 = \begin{pmatrix} e_3 & 0 & 0 & 0 \\ 0 & e_3 & 0 & 0 \\ -e_1 & -e_2 & 0 & 0 \\ e_1 l & e_2 l & e_3 l & e_3 \end{pmatrix}$$

The matrix M_4 is the same as in the case of General Perspective Projection, and the matrix M_2 is, as in the previous special case,

$$M_2 = \begin{pmatrix} 1 & 0 & 0 \\ 0 & 1 & 0 \\ 0 & 0 & \dfrac{l}{l^2 + 1} \\ 0 & 0 & \dfrac{1}{l^2 + 1} \end{pmatrix}$$

Consequently the matrix M representing the complete viewing operation

$$M = M_1 M_2 M_4 = \begin{pmatrix} e_3 & 0 & 0 & 0 \\ 0 & e_3 & 0 & 0 \\ -e_1 & -e_2 & 0 & 0 \\ e_1 l & e_2 l & e_3 l & e_3 \end{pmatrix} \begin{pmatrix} \dfrac{u_1 - u_0}{2} & 0 & 0 \\ 0 & \dfrac{v_1 - v_0}{2} & 0 \\ \dfrac{u_1 + u_0}{2} & \dfrac{v_1 + v_0}{2} & 1 \end{pmatrix}$$

Again, by performing the appropriate interchanges in M, we can find the matrix of the complete viewing operation in the case that the viewplane is $y = l$ or in the case that the viewplane is $x = l$.

What is perhaps most signficant about this special case is that analytically there is really nothing special about it—as far as projective geometry is concerned, parallel projection is treated precisely like perspective projection.

SPECIAL CASE 4: PERSPECTIVE PROJECTION

> The center of perspectivity is $E(\rho, \theta, \phi)$ in spherical coordinates
> The center of the object is the origin $O(0, 0, 0)$
> The viewplane is perpendicular to the principal line of sight and is d units from the center of the object

In this case, the complete viewing operation is specified (see Figure 5.12) by

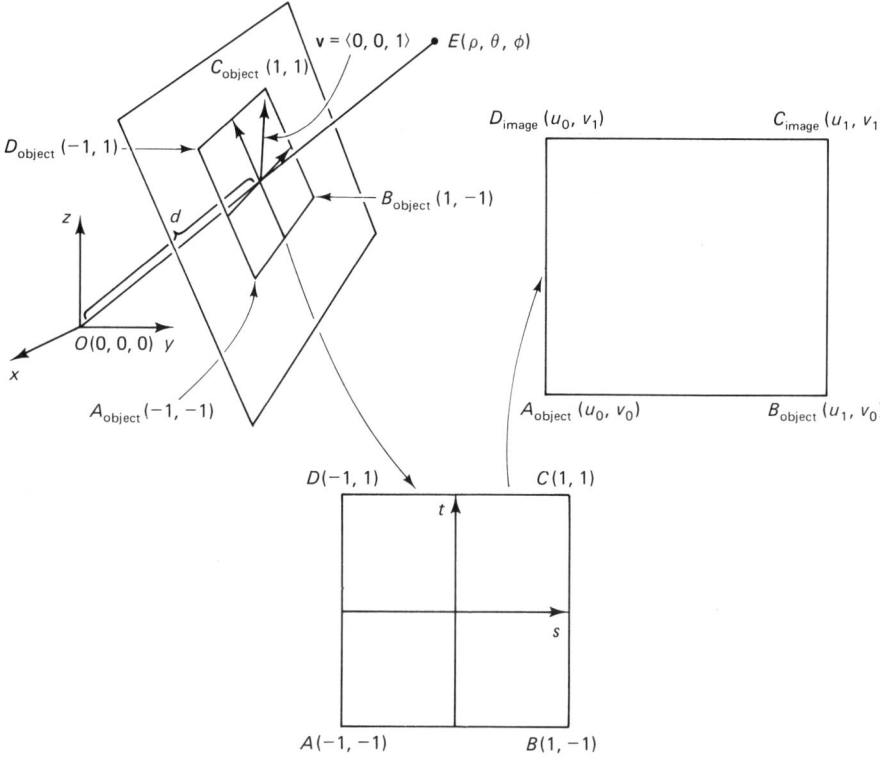

Figure 5.12 The complete viewing operation for perspective projection.

1. The center of perspectivity $E(\rho, \theta, \phi)$, given in spherical coordinates
2. The center of the object, which is the origin $O(0, 0, 0)$
3. The viewplane which is perpendicular to the principal line of sight (the line of sight that passes through the center of the object) and d units from the center of the object
4. The viewplane up-direction vector $\mathbf{v} = \langle 0, 0, 1 \rangle$
5. The window whose vertices are $A_{object}(-1, -1)$, $B_{object}(1, -1)$, $C_{object}(1, 1)$, and $D_{object}(-1, 1)$

As indicated in Section 1, spherical coordinates can be particularly useful in specifying the center of perspectivity. In particular, specification of the viewing mechanism in this manner is useful if we want to move the entire viewing mechanism about the object with as little respecification on each move as possible: Given the viewplane distance, the entire viewing mechanism is specified by the center of perspectivity and can be respecified by redefining (just) the center of perspectivity.

Since the Cartesian coordinates of E are

$$(\rho \sin \phi \cos \theta, \rho \sin \phi \sin \theta, \rho \cos \phi)$$

the homogeneous coordinates of E are

$$[\rho \sin \phi \cos \theta, \rho \sin \phi \sin \theta, \rho \cos \phi, 1]$$

To determine the equation of the viewplane μ, observe that the unit vector

$$\langle \sin \phi \cos \theta, \sin \phi \sin \theta, \cos \phi \rangle$$

is normal to μ; since μ is d units from the origin, the equation of μ is

$$\sin \phi \cos \theta \, x + \sin \phi \sin \theta \, y + \cos \phi \, z = d$$

Thus the homogeneous coordinates of μ are

$$[\sin \phi \cos \theta, \sin \phi \sin \theta, \cos \phi, -d]$$

Consequently, the matrix M_1 representing perspective projection is

$$\begin{pmatrix} \rho \sin^2\phi \sin^2\theta + \rho \cos^2\phi - d & -\rho \sin^2\phi \sin \theta \cos \theta & -\rho \sin \phi \cos \phi \cos \theta & -\sin \phi \cos \theta \\ -\rho \sin^2\phi \sin \theta \cos \theta & \rho \sin^2\phi \cos^2\theta + \rho \cos^2\phi - d & -\rho \sin \phi \cos \phi \sin \theta & -\sin \phi \sin \theta \\ -\rho \sin \phi \cos \phi \cos \theta & -\rho \sin \phi \cos \phi \sin \theta & \rho \sin^2\phi - d & -\cos \phi \\ d\rho \sin \phi \cos \theta & d\rho \sin \phi \sin \theta & d\rho \cos \phi & \rho \end{pmatrix}$$

The matrix M_4 is the same as in the case of General Perspective Projection. Using the given data and the formulas of the case of General Perspective Projection, it follows that

$$M_\mu = \begin{pmatrix} -\sin \theta & \cos \theta & 0 & 0 \\ -\cos \phi \cos \theta & -\cos \phi \sin \theta & \sin \phi & 0 \\ d \sin \phi \cos \theta & d \sin \phi \sin \theta & d \cos \phi & 1 \end{pmatrix}$$

so that

$$M_2 = \begin{pmatrix} -\sin \theta & -\cos \phi \cos \theta & \dfrac{d \sin \phi \cos \theta}{d^2 + 1} \\ \cos \theta & -\cos \phi \sin \theta & \dfrac{d \sin \phi \sin \theta}{d^2 + 1} \\ 0 & \sin \phi & \dfrac{d \cos \phi}{d^2 + 1} \\ 0 & 0 & \dfrac{1}{d^2 + 1} \end{pmatrix}$$

The matrix M representing the complete viewing operation is thus

$$M = M_1 M_2 M_4 = \begin{pmatrix} (d-\rho)\sin \theta & (d-\rho)\cos \phi \cos \theta & -\sin \phi \cos \theta \\ (\rho-d)\cos \theta & (d-\rho)\cos \phi \sin \theta & -\sin \phi \sin \theta \\ 0 & (\rho-d)\sin \phi & -\cos \phi \\ 0 & 0 & \rho \end{pmatrix} \begin{pmatrix} \dfrac{u_1 - u_0}{2} & 0 & 0 \\ 0 & \dfrac{v_1 - v_0}{2} & 0 \\ 0 & 0 & \\ \dfrac{u_1 + u_0}{2} & \dfrac{v_1 + v_0}{2} & 1 \end{pmatrix}$$

SECTION 2 EXERCISES

1. (a) Write a pseudocode program that takes as input the parameters $E(e_1, e_2, e_3)$, $P(x_0, y_0, z_0)$, $\mathbf{n} = \langle a, b, c \rangle$, $\mathbf{v} = \langle v_1, v_2, v_3 \rangle$, and h and k of the case of General

Perspective Projection and that produces as output the (entries of the) matrix M_1M_2 representing the viewing transformation.

(b) Implement the program described in part (a). (This program should be implemented on a device that supports graphics, since this program will be used in a larger program to be considered in Section 4.)

2. Use the program of Exercise 1 to compute the matrix M_1M_2 representing the viewing transformation in the case of General Perspective Projection in each case.
   (a) $E = E(8, 6, 6)$, $P = P(2, 4, 3)$, $\mathbf{n} = \langle 0, 1, 1 \rangle$, $\mathbf{v} = \langle 0, 1, 0 \rangle$, $h = 1$, and $k = 2$
   (b) $E = E(4, 0, 1)$, $P = P(3, 1, 2)$, $\mathbf{n} = \langle 2, 1, 3 \rangle$, $\mathbf{v} = \langle 1, 2, -1 \rangle$, $h = 1$, and $k = 1$
   (c) $E = E(-3, 1, 2)$, $P = P(4, 1, 3)$, $\mathbf{n} = \langle 1, 0, 1 \rangle$, $\mathbf{v} = \langle 0, 1, 5 \rangle$, $h = 3$, and $k = 1$
   (d) $E = E(1, -3, 4)$, $P = P(0, 1, 5)$, $\mathbf{n} = \langle -1, 2, 1 \rangle$, $\mathbf{v} = \langle 3, 1, 2 \rangle$, $h = 2$, and $k = 1$

3. What will be special about the form of matrices M_1, M_2, or M_1M_2 in the case of General Perspective Projection in each case?
   (a) The center of perspectivity lies along an axis
   (b) The viewplane is parallel to a coordinate plane
   (c) The viewplane is a coordinate plane

4. In the Special Cases, we assumed that the vertices of the window were $A(-1, -1)$, $B(1, -1)$, $C(1, 1)$, and $D(-1, 1)$. How should the formulas in these cases be altered if, as in the case of General Perspective Projection, the vertices of the window are to be $A(-h, -k)$, $B(h, -k)$, $C(h, k)$, and $D(-h, k)$?

5. Use the formulas of the case of General Perspective Projection to compute the matrices M_1 and M_μ in Special Case 1. Verify the computations of M_2 and M_1M_2.

6. Compute the matrices M_1 and M_2 in Special Case 1 without using projective geometry, using instead ad hoc Euclidean geometry.

7. What will happen to M_1 and M_2 in Special Case 1 if the following changes in the specifications are made?
   (a) $P = P(0, 0, k)$, $k \neq 0$,
   (b) $P = P(1, 0, 0)$, $\mathbf{n} = \langle 1, 0, 0 \rangle$
   (c) $P = P(0, 1, 0)$, $\mathbf{n} = \langle 0, 1, 0 \rangle$, $\mathbf{v} = \langle 1, 0, 0 \rangle$
   (d) $\mathbf{n} = \langle 0, 0, -1 \rangle$
   (e) $\mathbf{v} = \langle 0, -1, 0 \rangle$
   (f) $\mathbf{v} = \langle v_1, v_2, v_3 \rangle$ is arbitrary

8. How should the formulas in this section be altered if it is desired to change the orientation of each of the following?
   (a) The image   (b) The object

9. Compute the matrix M_1M_2 representing the viewing transformation in Special Case 2 in each case.
   (a) $k = 5, l = 2$ (b) $k = 10, l = 8$
   (c) $k = 4, l = 2$ (d) $k = 6, l = 1$

10. Use the formulas of the case of General Perspective Projection to compute the matrices M_1 and M_μ in Special Case 2. Verify the computations of M_2 and M_1M_2.

11. Compute the matrices M_1 and M_2 in Special Case 2 without using projective geometry, using instead ad hoc Euclidean geometry.

12. What will happen to M_1 and M_2 in Special Case 2 if the following changes in the specifications are made?
    (a) $E = E(k, 0, 0), P = P(l, 0, 0)$, $\mathbf{n} = \langle 1, 0, 0 \rangle$
    (b) $E = E(0, k, 0), P = P(0, l, 0)$, $\mathbf{n} = \langle 0, 1, 0 \rangle$, $\mathbf{v} = \langle 1, 0, 0 \rangle$

(c) $E = E(0, k, 0)$
(d) $\mathbf{n} = \langle 0, 0, -1 \rangle$
(e) $\mathbf{v} = \langle 0, -1, 0 \rangle$
(f) $\mathbf{v} = \langle v_1, v_2, v_3 \rangle$ is arbitrary

13. How do the position of the object and the relative sizes of k and l in Special Case 2 affect the image in each of the following cases?
    (a) Qualitatively (b) Quantitatively

14. Compute the matrix $M_1 M_2$ representing the viewing transformation in Special Case 3 in each case.
    (a) $\mathbf{e} = \langle 1, 2, 3 \rangle$, $l = 4$ (b) $\mathbf{e} = \langle 2, -1, 1 \rangle$, $l = 3$
    (c) $\mathbf{e} = \langle 0, 2, 1 \rangle$, $l = -2$ (d) $\mathbf{e} = \langle 1, 1, 1 \rangle$, $l = 2$

15. Use Proposition 3.2.1 and the formulas for the case of General Perspective Projection to compute the matrices M_1 and M_μ in Special Case 3. Verify the computations of M_2 and $M_1 M_2$.

16. Modify the program written in Exercise 1 so that it will work for parallel as well as perspective projection.

17. What will happen to M_1 and M_2 in Special Case 3 if the following changes in the specifications are made?
    (a) $\mathbf{e} = \langle 0, 0, 1 \rangle$ (b) $\mathbf{e} = \langle a, 0, c \rangle$
    (c) $\mathbf{n} = \langle 0, 1, 0 \rangle$ (d) $\mathbf{n} = \langle 1, 0, 0 \rangle$
    (e) The viewplane is $x = l$ (f) The viewplane is $y = l$

18. Implement a program that will take as input the parameters $E(\rho, \theta, \phi)$ and d of Special Case 4 and that will produce as output the (entries of the) matrix $M_1 M_2$ representing the viewing transformation.

19. Use the program of Exercise 18 to compute the matrix $M_1 M_2$ representing the viewing transformation in Special Case 4 in each case.
    (a) $E = E(10, 30°, 45°)$, $d = 8$ (b) $E = E(18, 45°, 45°)$, $d = 10$
    (c) $E = E(100, 90°, 30°)$, $d = 80$ (d) $E = E(20, 0°, 45°)$, $d = 10$

20. Use the formulas of the case of General Perspective Projection to compute the matrices M_1 and M_μ in Special Case 4. Verify the computations of M_2 and $M_1 M_2$.

SECTION 3: A SIMPLE THREE-DIMENSIONAL GRAPHICS LIBRARY

The goal of this section is to define a simple three-dimensional graphics library similar to the two-dimensional library presented in Section 4.2. It is difficult to make such a library complete to any degree because of the flexibility and possible variations of three-dimensional graphics. We must make some choices in presenting this library, but we hasten to point out that for a given application these choices may not be optimal. (Indeed, establishing the library itself need not be the optimal strategy: ad hoc implementation of various commands may be more appropriate.)

After the display device has been put into GraphicsMode, the command

$$\text{Initialize3DimensionalGraphics}$$

must be called to initialize the graphics environment. More specifically, this command declares the following (real) matrices: a 4×4 matrix M_0 that represents object adjustment, a 4 × 4 matrix M_1 that represents perspective or parallel projection, a 3 × 4 matrix M_μ that represents parametrization of the viewplane, a 4 × 3 matrix M_2 that represents the right inverse M_μ^+ of M_μ, a 3 × 3 matrix M_3 that represents image adjustment, a 3 × 3 matrix M_4 that represents the display transformation, and a 4 × 3 matrix M ($M = M_0 M_1 M_2 M_3 M_4$) that represents the complete viewing operation. (See Figure 5.13.) The matrices M_0 and M_3 are initialized as the 4 × 4 and 3 × 3 identity matrices, respectively. The matrices M_3 and M_4 correspond to the matrices M_0 and M_1 in the two-dimensional case.

To plot points in world coordinates we must specify the complete viewing operation. This can be done by calling one of the commands

DefineCenterOfPerspectivity(E1,E2,E3)

or

DefineDirectionOfProjection(E1,E2,E3)

which have the obvious meanings, together with

DefineViewplaneReferencePoint(X0,Y0,Z0)

DefineViewplaneNormalVector(A,B,C)

DefineViewplaneUpDirection(V1,V2,V3)

DefineWindow(MinX,MinY;MaxX,MaxY)

DefineViewport(MinU,MinV;MaxU,MaxV)

Or it can be done by calling

DefineCenterofPerspectivitySpherical(Rho,Theta,Phi)

DefineCenterOf Object(O1,O2,O3)

DefineViewplaneDistance(D)

DefineViewplaneUpDirection(V1,V2,V3)

DefineWindow(MinX,MinY;MaxX,MaxY)

DefineViewport(MinU,MinV;MaxU,MaxV)

all of which, again, have the obvious meanings.

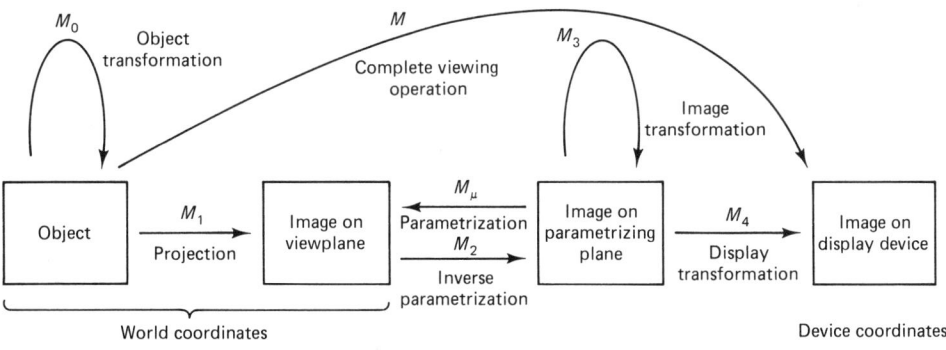

Figure 5.13 The components of the complete viewing operation.

As the appropriate calls are made, the matrices M_1, M_μ, M_2, and M_4 are automatically computed. Having made all the appropriate calls, M is automatically computed. (If any element in the viewing operation is subsequently altered, the appropriate changes are automatically made in the appropriate factor matrix or matrices, and M is automatically recomputed.)

The complete viewing operation is represented by right multiplication by M = (M[I,J]). Consequently, the complete viewing operation is implemented by

```
PROCEDURE Transform(X,Y,Z)To(U,V)
  BEGIN
    W := M[1,3]*X + M[2,3]*Y + M[3,3]*Z + M[4,3]
    U := (M[1,1]*X + M[2,1]*Y + M[3,1]*Z + M[4,1])/W
    V := (M[1,2]*X + M[2,2]*Y + M[3,2]*Z + M[4,2])/W
  END
```

and points can be plotted in world coordinates by

```
PROCEDURE Plot(X,Y,Z)
  BEGIN
    Transform(X,Y,Z)To(U,V)
    PlotInDeviceCoordinates(U,V)
  END
```

Observe that Definewindow and DefineViewport look exactly like the corresponding calls in the two-dimensional library. In fact, DefineViewport is exactly the same. However, DefineWindow is different to the extent that the vertices of the window to which it refers are points of the viewplane; nothing corresponds to this in two dimensions. (Although if the plane of the object in two dimensions is defined to be the viewplane, then the correspondence is precise. Furthermore, the transformation determined by M_{object} is a parametrization of the viewplane. In fact, various existing libraries define the elements of the viewing operation in this manner so that the two-dimensional library is a special case of the three-dimensional library.)

In the three-dimensional library we need both object and image adjustments.
The commands that control object adjustments are

$$\text{RotateObject(U1,U2,U3;Theta)}$$

which rotates world coordinates through an angle of Theta about the position vector **U** = ⟨U1, U2, U3⟩ in such a manner that if you grasp **U** with your right hand so that your fingers curl in the direction of positive Theta, your thumb will point in the direction of **U**;

$$\text{TranslateObject(H,K,L)}$$

which translates world coordinates so that the point with coordinates (H,K,L) becomes the origin;

$$\text{ScaleObject(MX,MY,MZ;MW)}$$

which scales locally by MX in the *x*-direction, MY in the *y*-direction, and MZ in the *z*-direction, and which scales globally by 1/MW;

$$\text{ShearObjectYZ(A), ShearObjectZY(B)}$$
$$\text{ShearObjectXZ(C), ShearObject ZX(D)}$$
$$\text{ShearObjectXY(E), ShearObjectYX(F)}$$

which shear by A parallel to the *yz*-plane in the *y*-direction and by B parallel to the *yz*-plane in the *z*-direction, by C parallel to the *xz*-plane in the *x*-direction and by D parallel to the *xz*-plane in the *z*-direction, and by E parallel to the *xy*-plane in the *x*-direction and by F parallel to the *xy*-plane in the *y*-direction;

$$\text{PerspectivelyProjectObject(A,B,C)}$$

which performs a perspective projection by A in the *x*-direction, B in the *y*-direction, and C in the *z*-direction, and

$$\text{GeneralObjectTransformation(A,B,C,D;E,F,G,H;I,J,K,L;M,N,O,P)}$$

which transforms by the matrix

$$\begin{pmatrix} A & B & C & D \\ E & F & G & H \\ I & J & K & L \\ M & N & O & P \end{pmatrix}$$

The commands that control image adjustment operate on the parametrizing plane, and correspond to the object adjustments in the two-dimensional library. Now, however, we insert the word Image into the procedure names so that Rotate(Theta), for example, becomes

$$\text{RotateImage(Theta)}$$

As before, all object and image adjustments can be represented by right 4×4 and 3×3 matrix multiplication and are implemented as follows: If an object adjustment, represented by the matrix M_0', is to be performed, M_0 is replaced by the product $M_0 M_0'$, and the matrix M is automatically recomputed. If an image adjustment, represented by the matrix M_3', is to be performed, M_3 is replaced by the product $M_3 M_3'$, and the matrix M is automatically recomputed.

The command for drawing a line segment in world coordinate space is implemented by the following procedure.

```
PROCEDURE DrawFrom(X0,Y0,Z0)To(X1,Y1,Z1)
  BEGIN
    Transform(X0,Y0,Z0)To(U0,V0)
    Transform(X1,Y1,Z1)To(U1,V1)
    DrawInDCFrom(U0,V0)To(U1,V1)
  END
```

Finally, we add CurrentPosition to our library to permit three-dimensional incremental techniques. The coordinates

(CurrentPositionX, CurrentPositionY, CurrentPositionZ)

are again world coordinates declared and initialized by Initialize3Dimensional-Graphics. The procedures

> MoveAbsolute(X,Y,Z) and MoveRelative(X,Y,Z)
> PlotAbsolute(X,Y,Z) and PlotRelative(X,Y,Z)
> DrawAbsolute(X,Y,Z) and DrawRelative(X,Y,Z)

are defined as before: In the case of each Absolute command, (X, Y, Z) are the world coordinates of a point with respect to which an action is performed. In the case of each Relative command, (X, Y, Z) are the coordinates of a point relative to CurrentPosition and with respect to which an action is performed.

SECTION 3 EXERCISES

1. Examine the three-dimensional graphics libraries available to you and compare them to the library we have developed so far.
2. (a) How might the viewing transformations discussed in Section 2 be implemented in this library?
   (b) What difficulties (related to specification of the viewing transformation) might one have in implementing a three-dimensional graphics library?
3. The following commands might be useful library additions. Describe the meaning and implementation of each.
   (a) PerspectivelyProject(X,Y,Z)To(U,V)
   (b) OrthogonallyProject(X,Y,Z)To(U,V)
4. What commands (in addition to those of Exercise 3) would make useful library additions? How could each be implemented? Could too many commands be added to the library?
5. In certain applications, a matrix M arises in stages as a product $M = M_1 M_2 \cdots M_n$ of invertible "simple" matrices (these simple matrixes might represent, for example, rotation, translation, scaling, shearing, and perspective projection). In such applications it is often desirable also to know M^{-1}. How can storing two matrices in memory (one representing M and the other representing M^{-1}) be used to simplify finding M^{-1}?
6. If k is the height-to-width ratio of the viewport, explain why selecting window vertices with coordinates (± 1, $\pm k$) avoids distortion caused by aspect ratio problems.

SECTION 4: A THREE-DIMENSIONAL GRAPHICS DISPLAY PROGRAM

In this section we outline a general purpose interactive three-dimensional graphics display program. This program can display, manipulate with object and image transformations and redisplay any three-dimensional object described by vertices and edges—in other words, a "wire-frame" object. This program can produce either parallel or perspective projections.

The program Graph3D outlined in this section is similar, in many respects, to Graph2D (see Section 4.3). As with Graph2D, there are different possible implementations of the procedure DisplayPicture for Graph3D. While the simplest version is adequate for some purposes, a version using DrawByParametrizing is desirable if there is a chance that any one of the points of the object is transformed to an ideal point. A hidden line algorithm is also implemented in Graph3D to provide additional depth cuing. The hidden line algorithm discussed here applies to the case in which one segment crosses over and partially hides another. In Section 5.5 we consider an algorithm for lines hidden by surfaces. For further study of hidden line and surface algorithms, see [Sutherland, Sproull, and Schumaker].

In this program we will allow the user freedom to specify all the parameters that describe the viewing mechanism. The procedure DefineViewingMechanism obtains the values of the parameters from the user and then makes the necessary calls to library procedures.

One way for DefineViewingMechanism to obtain these parameters is to ask the user a sequence of questions. A better way might be to provide a display of the current values of the parameters, initially default values, and allow the user to make changes. This method minimizes the number of parameters that need to be respecified each time and allows a naive user to use the default values. However the parameters are obtained, it must be determined whether the user wishes to use perspective or parallel projection, since the library was written to distinguish between them. If GetParameters is the name of the procedure that obtains these responses, then the procedure DefineViewingMechanism is as follows.

```
PROCEDURE DefineViewingMechanism
  BEGIN
    GetParameters
    CASE Projection OF
      Perspective: DefineCenterOfPerspectivity(E1,E2,E3)
      Parallel: DefineDirectionOfProjection(E1,E2,E3)
    ENDCASE
    DefineViewplaneReferencePoint(X0,Y0,Z0)
    DefineViewplaneNormalVector(A,B,C)
    DefineViewplaneUpDirection(V1,V2,V3)
    DefineWindow(MinX,MinY;MaxX,MaxY)
    DefineViewport(MinU,MinV;MaxU,MaxV)
  END
```

The program is

```
PROGRAM Graph3D
  BEGIN
    Initialize3DimensionalGraphics
    DefineViewingMechanism
    GetPictureFile
    REPEAT
      WRITE(menu)
      READ(Command)
      CASE Command OF
        A: DefineViewingMechanism
```

```
        B: GetPictureFile
        C: DisplayPicture
        D: Rotate3D
        E: Translate3D
        F: Scale3D
        G: Shear3D
        H: PerspectivelyProject3D
        I: Matrix3D
        J: Rotate2D
        K: Translate2D
        L: Scale2D
        M: Shear2D
        N: PerspectivelyProject2D
        O: Matrix2D
        P: Reinitialize
        Q: PrintPicture
        R: Finished
      ENDCASE
   UNTIL Finished
END
```

Procedures D–I are object transformations. Each prompts the user to supply the values of the required parameters; for example, Rotate3D prompts the user to supply the components of the direction $U = \langle U1, U2, U3 \rangle$ of the axis of rotation and the angle Theta of rotation, and then calls Rotate-Object(U1,U2,U3;Theta). Procedures J–O, similarly, are image transformations. Reinitialize resets both the object transformation matrix M_0 and the image transformation matrix M_3 to the respective identities.

Examples 4.1 and 4.2 illustrate uses of Graph3D with the simplest DisplayPicture procedure. (Other applications of Graph3D include numerous illustrations for Sections 3.3 and 5.6.)

4.1 Example. The views of the chair in Figure 5.14 were produced by Graph3D.

4.2 Example. The two pairs of views of the molecule in Figure 5.15 were

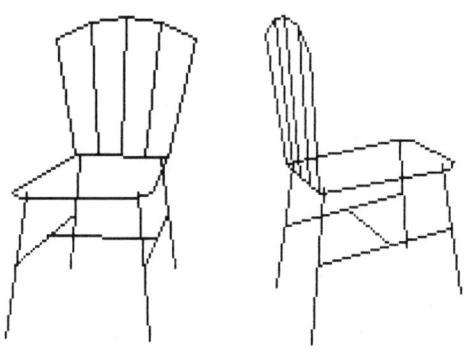

Figure 5.14 A chair.

produced by Graph3D by using two different positions for the eye in each case. The two views in each part form a *stereo pair*. If you can focus your eyes so that one eye sees each picture and then get the two images to fuse together, your brain will interpret the result as one three-dimensional molecule.

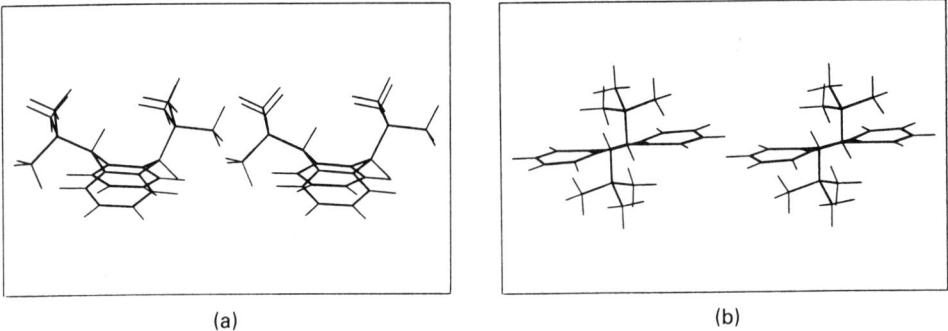

Figure 5.15 Stereo views of the molecule *trans*-9,10-di-*t*-butyl-9,10-dihydroanthracene in its (a) boat form and (b) chair form (courtesy of P. W. Rabideau and K. B. Lipkowitz, Indiana University–Purdue University at Indianapolis, and R. B. Nachbar Jr., Merck Sharp and Dohme Laboratories).

Since many people have difficulty focusing their eyes in this way, stereo pairs are generally viewed using a device that presents different pictures to each eye. Such a device may use lenses or mirrors, such as a stereopticon, or light of different colors or polarities viewed through eyeglasses with special lenses.

The two views in a stereo pair can be produced by choosing two different eye positions and a single viewplane (see Figure 5.16(a)). However (see Figure 5.16(b)), the same effects can be achieved by using a single fixed viewing mechanism and two different positions of the object (one the translation of the other).

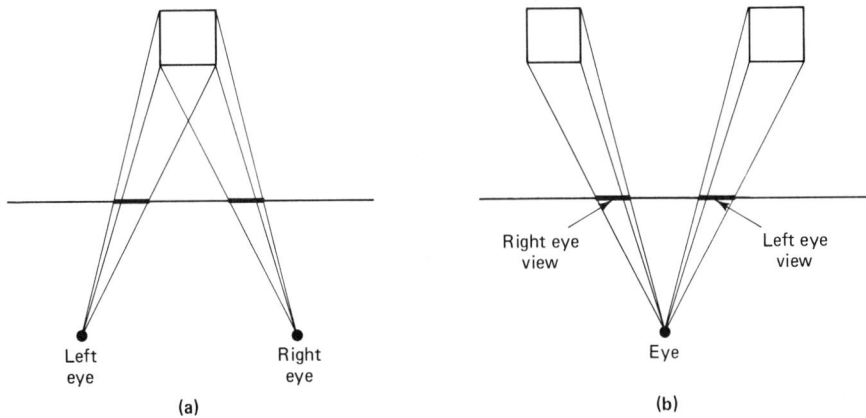

Figure 5.16 Binocular vision with a common viewplane. (a) With two eyes. (b) With translated object.

Sec. 4 A Three-Dimensional Graphics Display Program 247

In either case, we assume that there is only one viewplane. This assumption may not be reasonable if the distance separating the eyes is large compared to the distance to the object (see Figure 5.17). In this case it might be appropriate to rotate the object to prepare the view for the second eye instead of translating it. We leave it to students of vision to determine which technique more accurately represents true visual perception.

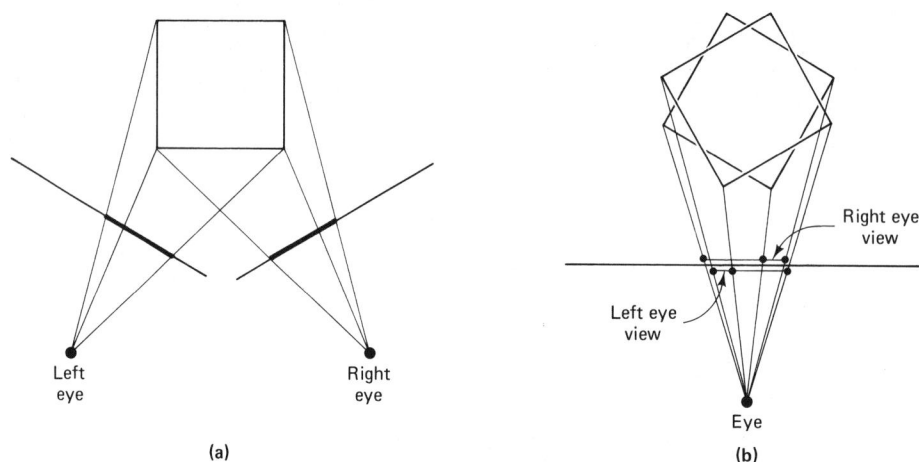

Figure 5.17 Binocular vision of nearby objects.

In Section 4.3 we observed that perspective projection could cause some points of the object to be transformed to ideal points. It was for this reason that DisplayPicture was modified to use DrawByParametrizing. You may also wish to use DrawByParametrizing with Graph3D. The reason for this is that in three dimensions, translation, scaling, and rotation can, in fact, also cause vertices to become ideal: Consider for example, the test pattern for Graph3D which is a cube whose vertices have coordinates $(\pm 1, \pm 1, \pm 1)$, viewed through the viewplane $z = 5$ with the eye at $E(0, 0, 10)$ (see Figures 5.18(a) and 5.19(a)). If the cube is translated toward the eye, it will appear to grow. Eventually it will surround the eye. Compare Figure 5.19 to Figure 5.18, which was produced by Graph3D. When the cube is moved to the other side of the eye, its image appears normal again, but the edges are drawn in the opposite directions.

We now consider a version of DisplayPicture that uses a hidden line routine. (This version was used in the production of Figure 5.23.) In order to keep it from getting overly complicated, we assume that the images it will be used to draw do not contain ideal points.

The new procedure DisplayPictureWithHiddenLines has a main subprocedure, ProcessEdges, which computes for each edge the *critical values* of the parameter for that edge—those values around which there will be a break in the image of the edge. Then another procedure DrawEdges uses these values to draw the edges. These procedures follow two initialization procedures.

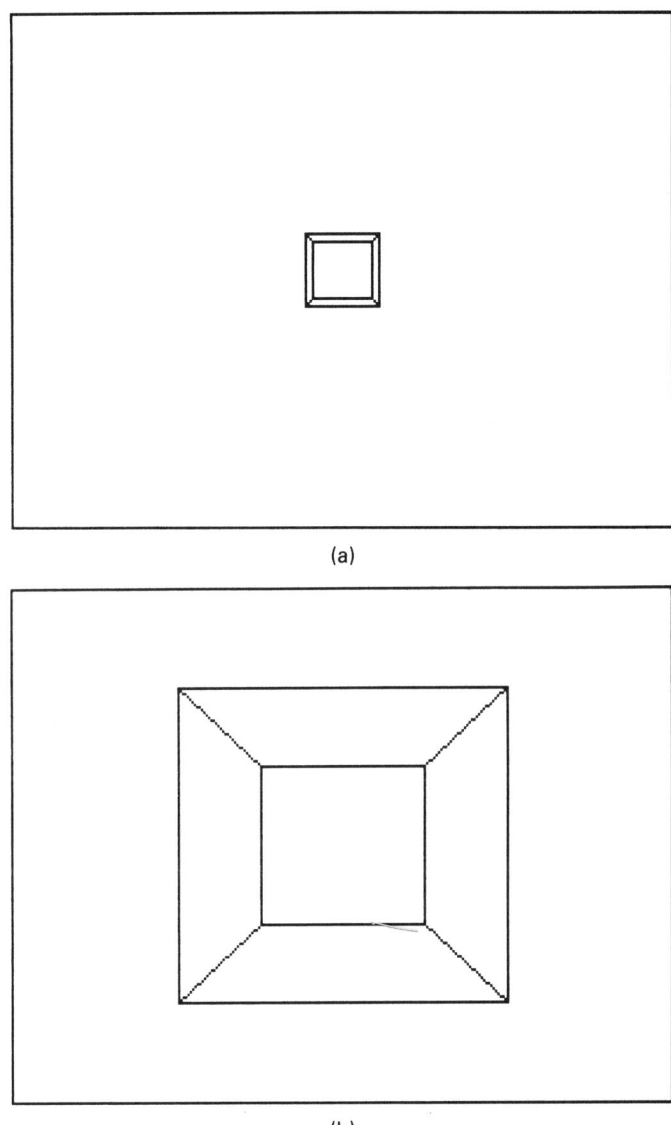

Figure 5.18 Perspective images of a cube.

```
PROCEDURE DisplayPictureWithHiddenLines
  BEGIN
    GraphicsMode
    InitializeDataStructure
    TransformVertices
    ProcessEdges
    DrawEdges
```

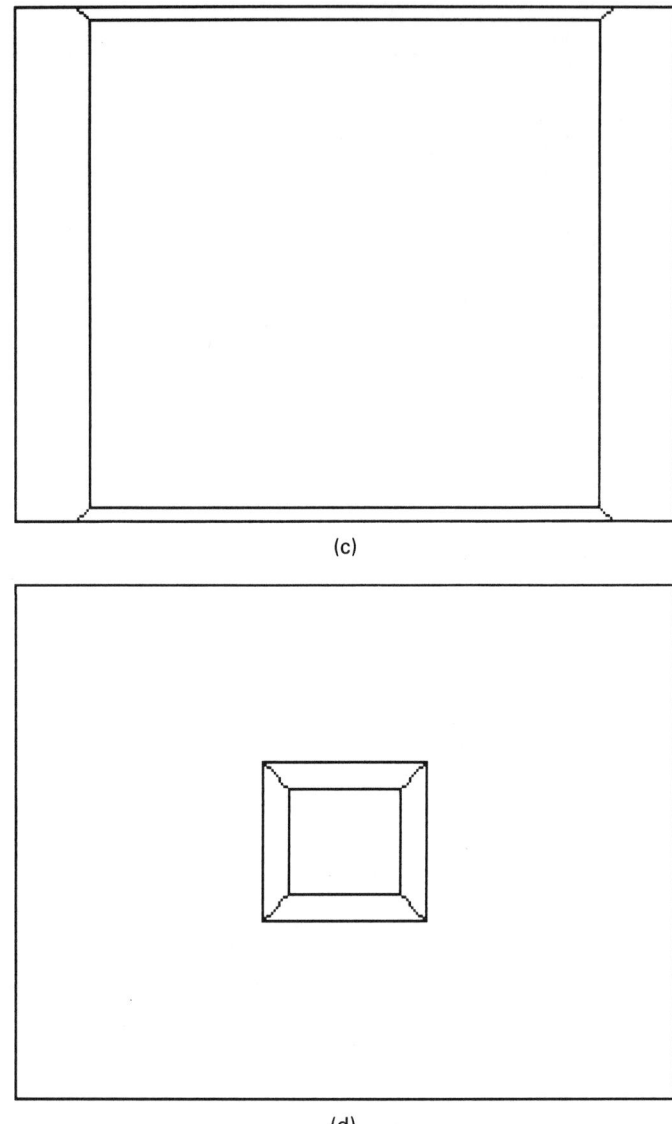

Figure 5.18 Continued.

```
    Pause for viewing picture
    TextMode
END
```

The procedure InitializeDataStructure prepares space in which to store the critical values of the parameters. The procedure TransformVertices computes the Cartesian world coordinates of the images of the vertices under the object transformation represented by the matrix M_0, and stores them in an array.

The procedure ProcessEdges considers edges two at a time. If the edges have a common vertex, nothing needs to be done. Otherwise, suppose the

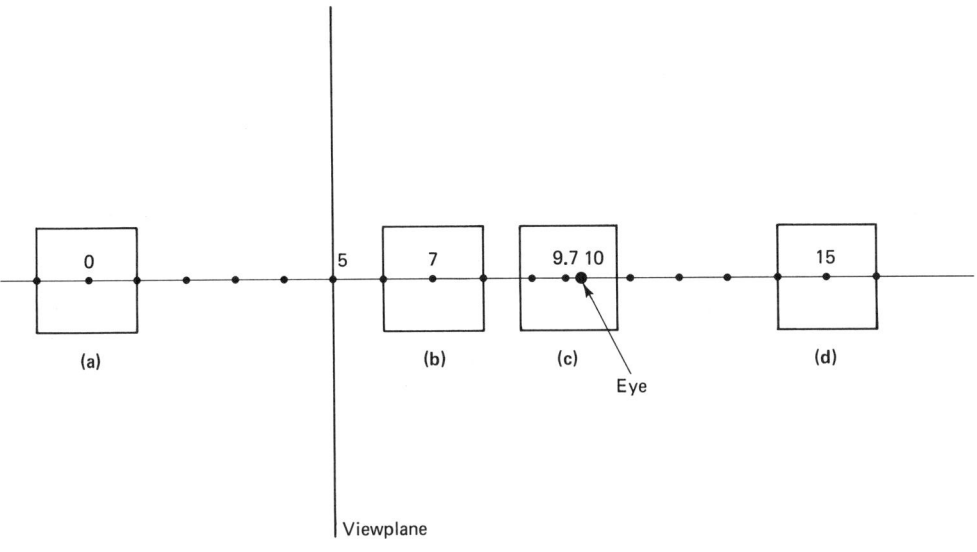

Figure 5.19 How the perspective images of a cube arise.

procedure is looking at edges I and J, which have world coordinate vertices P, Q and R, S, respectively. Let A, B, C, and D be the corresponding vertices of their images (see Figure 5.20).

The procedure ProcessEdges computes the equation $ax + by + cz + d = 0$ of the plane determined by the points P, Q, and the eye. It then evaluates the function $ax + by + cz + d$ on the coordinates of the points R and S. If the results have the same sign (see Figure 5.21(a)), the points lie on the same side of the plane, the images of the edges do not intersect, and nothing more needs to be done. If the results have different signs (see Figure 5.21(b) and (c)), the points P and Q are tested in the same way with the equation of the plane through

Figure 5.20 Two segments (a) in world coordinate space and (b) their images in device coordinate space.

R, S, and the eye. Again if these points lie on the same side of the plane (see Figure 5.21(b)), nothing more needs to be done. Otherwise the images of the two edges intersect.

Sec. 4 A Three-Dimensional Graphics Display Program

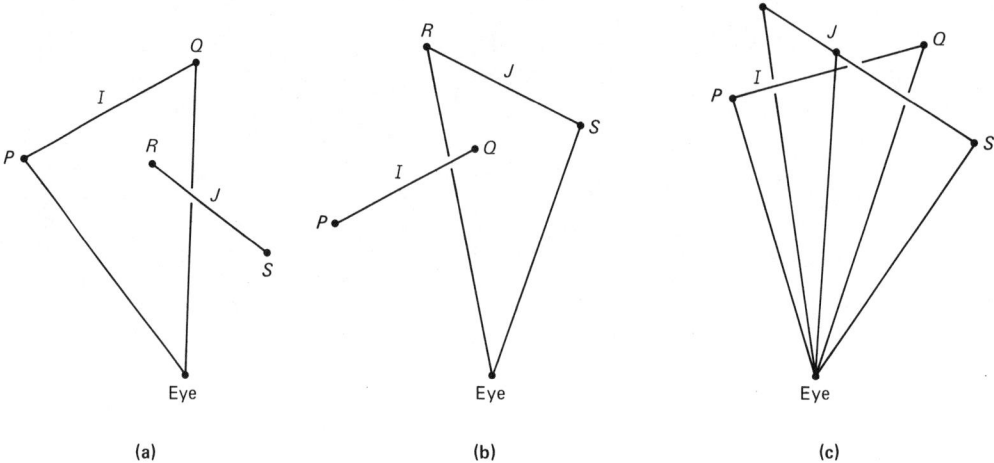

Figure 5.21 Relative positions of two line segments.

If the images of edges *I* and *J* intersect then the procedure determines which edge crosses under which. To do this it finds the equation of the plane through *P*, *Q*, and *R*. If *S* and the eye lie on the same side of this plane then edge *I* crosses under edge *J*; otherwise edge *J* crosses under edge *I* (see Figure 5.22). If edge *I* crosses under edge *J*, for example, the procedure computes and stores the critical parameter value on edge *I* for the point of intersection of edge *I* with the plane through *R*, *S*, and the eye.

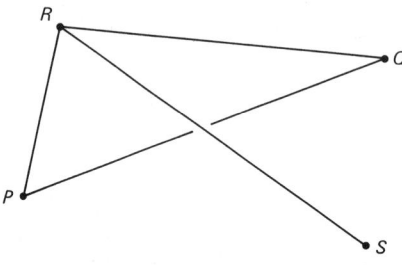

Figure 5.22 Determining which segment crosses under which.

We find the critical parameter values for the edges in world coordinate space so that successive gaps in a segment which is receding from the eye appear

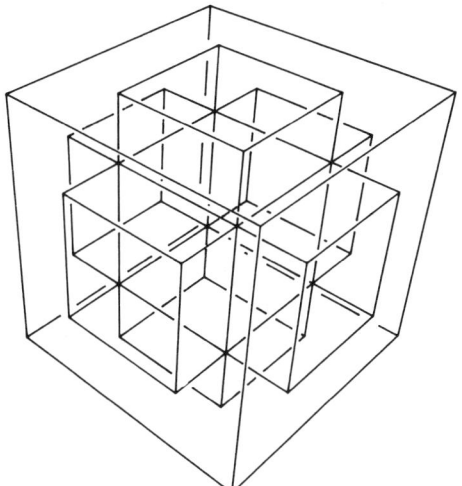

Figure 5.23 The edges of a cube with three square holes.

to shorten as the gaps recede from the eye. These are not the same critical parameter values as those of the images of the edges, since projections do not preserve the linearity of parametrization (see Section 2.4).

Graph3D can be used to draw figures that require more than one perspective projection. Such figures include Figure 5.24, a computer-generated version of Figure 1.9 (like the woodcut of Dürer—see Figure 1.63—this figure is a view in perspective of an object being viewed in perspective), as well as figures that include shadows and reflections (see Exercises 11–13). In the case of Figure 5.24, for example, the viewer in the figure can distinguish between object vertices and image vertices. All of these vertices, as well as the vertices of the viewport

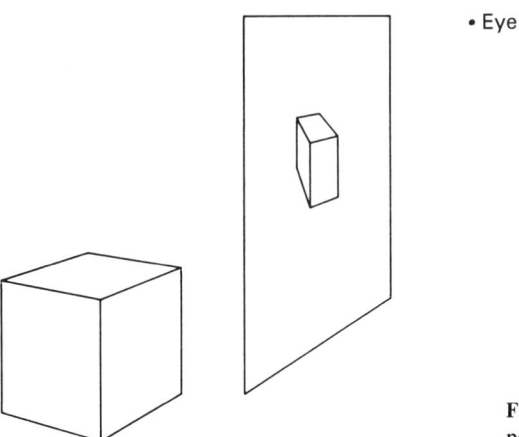

Figure 5.24 A physical model for perspective.

Sec. 4 A Three-Dimensional Graphics Display Program

and the eye, are object vertices in what we see. In particular, we use another viewplane and eye to create Figure 5.24.

SECTION 4 EXERCISES

1. Implement Graph3D on a specific graphics device.
2. Experiment with a fixed test pattern and viewplane but with the eye variable. What qualitative changes are produced in the image as the eye is moved? Where would the eye be located for the image to appear the largest?
3. Experiment with placing the eye on the same side of the viewplane as the object, both between the object and the viewplane and farther from the viewplane than the object.
4. Experimentally verify the results in Figure 5.18.
5. Explain how rotating or scaling an object in three-dimensional space can cause image points to move to infinity.
6. If Graph3D is written so that the viewplane up-direction is fixed in advance, what locations for the viewplane must be ruled out? Can arbitrary views of the object still be obtained?
7. If Graph3D is written so that the viewplane is the plane $z = 10$, the eye is restricted to lie on the z-axis, and the viewplane up-direction is the positive y-axis, can arbitrary views of the object still be obtained?
8. What are situations in which it is preferable to think of the viewing mechanism as fixed and the object as movable, and what are situations in which it is preferable to think of the object as fixed and the viewing mechanism as movable? Are there any situations in which we *must* work one way or the other? Can one way ever shorten computation time?
9. Which of the following should be considered errors in input to DefineViewingMechanism?
   (a) MinX > MaxX
   (b) The eye is on the viewplane.
   (c) The viewplane normal vector is $\langle 0, 0, 0 \rangle$.
   (d) The viewplane up-direction is $\langle 0, 0, 0 \rangle$.
   (e) The viewplane up-direction vector is perpendicular to the viewplane.
   (f) The viewplane passes through the origin of the coordinate system.
   (g) The direction of projection is $\langle 0, 0, 0 \rangle$.
   (h) The viewplane reference point is the origin.
   Which of these errors (or others) should DefineViewingMechanism try to catch?
10. Write a program that draws three-dimensional convex polyhedra. Allow the eye and viewplane to be varied. Do not draw hidden faces. (*Hint:* A face is visible if and only if an outward normal vector to the face points into the half of Euclidean three-space determined by the plane of the face, that contains the eye.)
11. Write a program that draws figures such as Figure 5.24.
12. (a) Write a program that draws a three-dimensional convex polyhedron and its shadow on a plane, assuming that the polyhedron is far enough from the plane that the polyhedron does not obscure its shadow (see Figure 5.25(a)).
    (b) Repeat part (a), but permit the polyhedron to obscure its shadow (see Figures 5.25 (b) and (c)).
13. Repeat Exercise 12 drawing reflections instead of shadows.

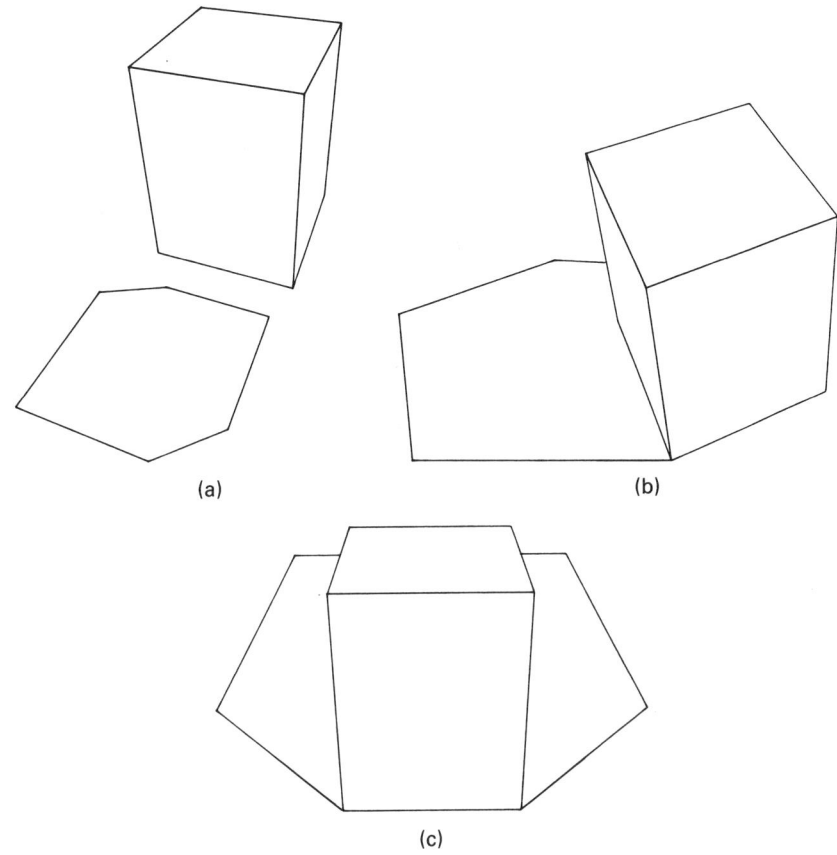

Figure 5.25 Views of a three-dimensional convex polyhedron and its shadow.

SECTION 5: GRAPHING FUNCTIONS OF TWO VARIABLES

In this section we discuss how to draw a picture of the graph of a real-valued function $z = f(x, y)$ of two real variables, x and y, in perspective. Our primary purpose is to illustrate the use of some of the features of the simple three-dimensional graphics library in the context of a useful application. At the same time, this application introduces a hidden line algorithm that illustrates how a two-dimensional graphics technique can be used to achieve a three-dimensional effect. This section is intended to be just an introduction to graphing functions of two variables and, in particular, an introduction to a particular hidden line algorithm. For further study, see [Wright], [Butland], and [Anderson]; the hidden line algorithm discussed in this section can be found in [Wright].

We will proceed in steps. First (see Figure 5.26), we consider the case in which the hidden line algorithm is not used; this step establishes the basic program structure. Next we consider the case in which the hidden line algorithm is used. The particular hidden line algorithm we use requires the use of a discrete device coordinate system. We begin by implementing the hidden line algorithm in a

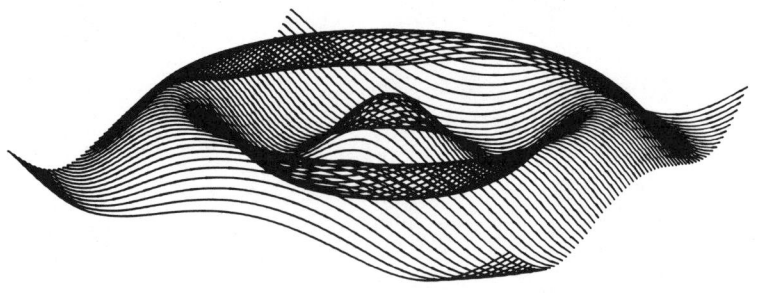

Figure 5.26 The hidden line algorithm is not used ($f(x,y) = \cos\sqrt{x^2 + y^2}$).

program that plots points in a discrete device coordinate space (see Figure 5.27). Then (see Figure 5.28) we consider the case in which the hidden line algorithm is used and curves are drawn in a continuous device coordinate space. Finally (see Figure 5.29), we consider some further modifications that can be made to improve picture quality.

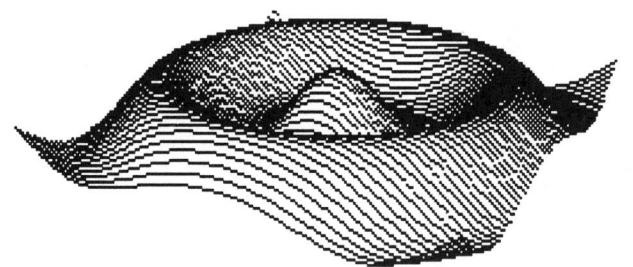

Figure 5.27 The hidden line algorithm is used and points are plotted.

To begin, we assume that we are given a (continuous) real-valued function $z = f(x, y)$ of two real variables x and y and that we wish to draw a picture of the graph of f over the rectangle

$$\{(x, y) \in R^2 \mid -L < x < L \text{ and } -W < y < W\}$$

(see Figure 5.30) for some positive real numbers L (half-length) and W (half-width).

Assuming that the elements of the complete viewing operation are as illustrated in Figure 5.31, our program begins with the following sequence of library calls:

⋮

```
GraphicsMode
Initialize3DimensionalGraphics
DefineCenterOfPerspectivitySpherical(Rho,Theta,Phi)
DefineCenterOfObject(0,0,0)
DefineViewplaneDistance(D)
DefineViewplaneUpDirection(0,0,1)
```

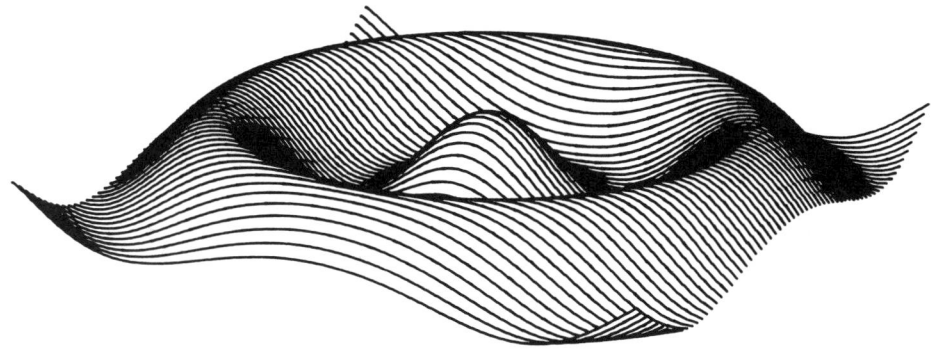

Figure 5.28 The hidden line algorithm is used and curves are drawn.

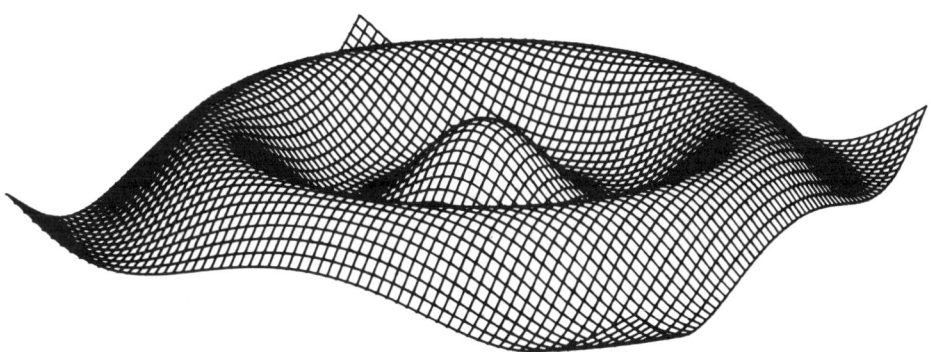

Figure 5.29 Further modifications.

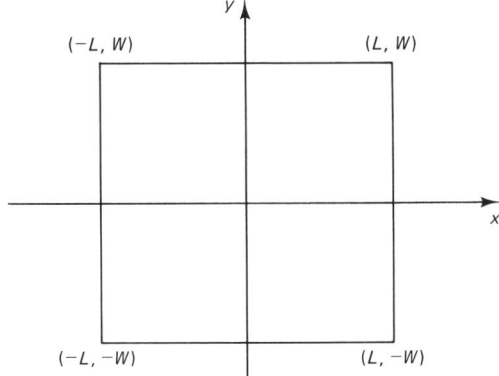

Figure 5.30 The domain of f.

Sec. 5   Graphing Functions of Two Variables   **257**

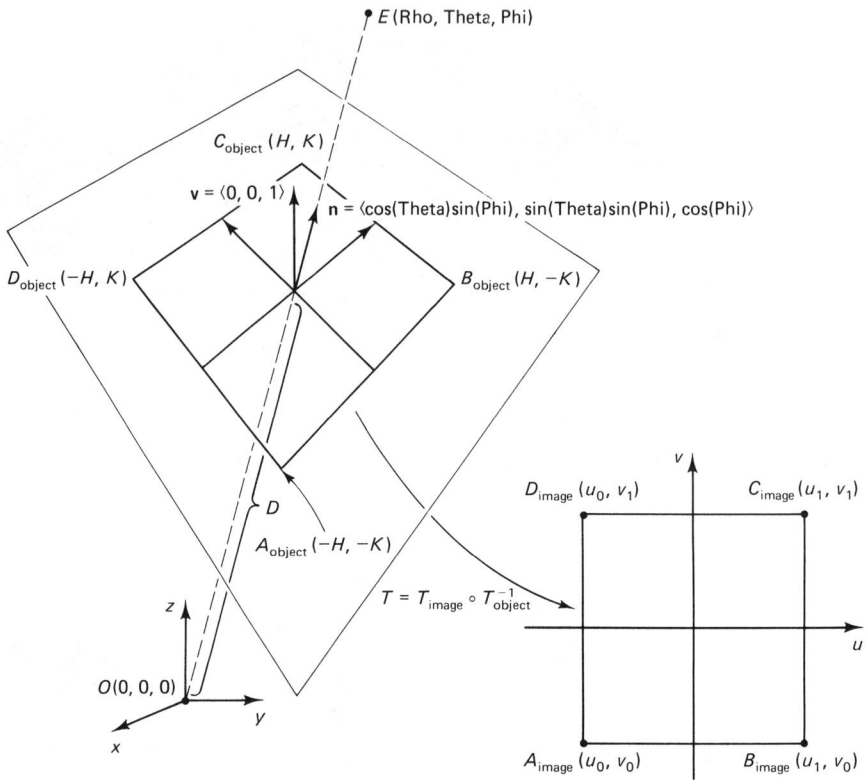

Figure 5.31 Elements of the complete viewing operation.

```
DefineWindow(-H,-K;H,K,)
DefineViewport(U0,V0;U1,V1)
    ⋮
```

These calls assume, in particular, that the center of perspectivity has spherical coordinates (Rho,Theta,Phi), that the "center" of the graph that we are to draw is the origin, and that the viewplane is perpendicular to the line of sight and D units from the origin (see Section 2, Special Case 4). The viewing mechanism is specified in this manner, since doing so simplifies respecification should it be desired to change the view of the graph: varying Rho varies the distance of the center of perspectivity from the object; varying Theta rotates the center of perspectivity about the z-axis, varying Phi changes the elevation of the center of perspectivity above or below the xy-coordinate plane. (The effects of varying these parameters are pursued in Section 6.)

Before proceeding, we introduce some terminology.

5.1 Definition. Given a continuous real-valued function $z = f(x, y)$, a curve of the form

$$\{(x, y_0, f(x, y_0)) \in R^3 \mid -L < x < L\}$$

for some fixed y_0, is an *x-curve* (see Figure 5.32) and a curve of the form
$$\{(x_0, y, f(x_0, y)) \in R^3 \mid -W < y < W\}$$
for some fixed x_0, is a *y-curve*.

We draw a picture of the graph of $z = f(x, y)$ by drawing a sequence of *x*-curves, a sequence of *y*-curves, or both. In the discussion to follow, we begin by drawing a sequence of *y*-curves.

As indicated earlier, we first consider the case in which a hidden line algorithm is not used.

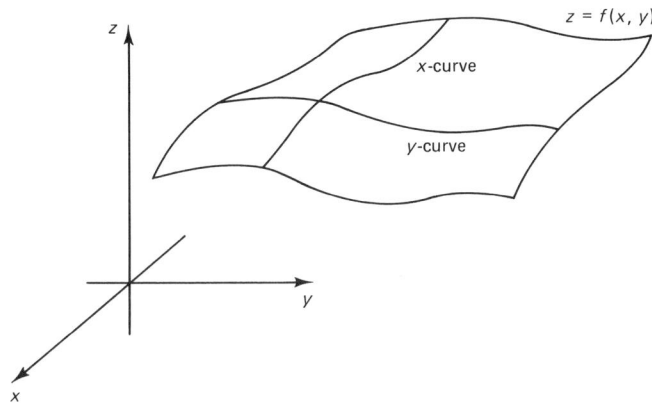

Figure 5.32 *x*-curves and *y*-curves.

Beyond the first point on each *y*-curve, plotting can be controlled by a pair of nested loops

```
FOR X := -L TO L STEP DeltaX DO
  FOR Y := -W TO W STEP DeltaY DO
    BEGIN
      Z:= f(X,Y)
      Transform(X,Y,Z)To(NewU,NewV)
      DrawFrom(OldU,OldV)To(NewU,NewV)
      (OldU,OldV) := (NewU,NewV)
    END
```

where (OldU, OldV) is somehow initially defined. The smaller the (positive) step sizes DeltaX and DeltaY, the smoother the graph but the greater the execution time.

To begin drawing each *y*-curve, (OldU, OldV) must be defined as (NewU, NewV) before drawing is done. (Not doing so would leave (OldU, OldV) undefined at the beginning of the program and, even if this were remedied, not doing so would have the unwanted effect of drawing a line segment from the end of one *y*-curve to the beginning of the next.) To do this we define the

Boolean variable FirstPointOnCurve, which is TRUE if the next point to be plotted is the first point on a *y*-curve and FALSE otherwise.

Our entire program is

```
    ⋮
    (* INITIALIZATION *)
GraphicsMode
Initialize3DimensionalGraphics
DefineCenterOfPerspectivitySpherical(Rho,Theta,Phi)
DefineCenterOfObject(0,0,0)
DefineViewplaneDistance(D)
DefineViewplaneUpDirection(0,0,1)
DefineWindow(-H,-K;H,K)
DefineViewport(U0,V0;U1,V1)

    (* PLOTTING *)
FOR X := -L TO L STEP DeltaX DO
  BEGIN
    FirstPointOnCurve := TRUE
    FOR Y := -W TO W STEP DeltaY DO
      BEGIN
        Z := f(X,Y)
        Transform(X,Y,Z)To(NewU,NewV)
        IF FirstPointOnCurve THEN (OldU,OldV) := (NewU,NewV)
        DrawFrom(OldU,OldV)To(NewU,NewV)
        (OldU,OldV) := (NewU,NewV)
        FirstPointOnCurve := FALSE
      END
  END
    ⋮
```

This program is complete except for declarations and an appropriate conclusion. An appropriate conclusion might include a closing prompt to indicate when graphing is completed (this might take the form, for example, of a message or the drawing of a border around the completed picture) and a (controlled) return to TextMode.

Points on the graph of a function $z = f(x, y)$ are often hidden from the eye by parts of the graph closer to the eye. In this event only part of the graph of the function is drawn. The decision of which points to plot and which not to plot is made on the basis of a hidden line algorithm. We use the following hidden line algorithm. It requires a discrete device coordinate space, and, in particular, two $1 \times N$ height arrays MinimumV and MaximumV, where $N = U1 - U0$ is the number of horizontal device units in the viewport (for simplicity, we will assume that $U0 < U1$ and $V0 < V1$).

The principle of the algorithm is that we draw *y*-curves in order as they recede in distance from the eye (that is, the *y*-curves closest to the eye are drawn first, and the *y*-curves farthest from the eye are drawn last) (see Figure 5.33).

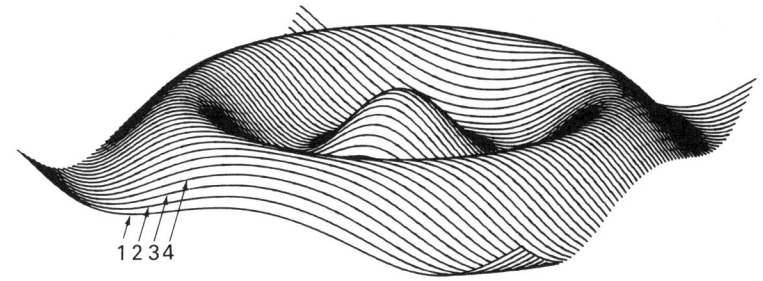

Figure 5.33 The order of curve plotting.

For each U, MinimumV[U] is the minimum of the V values for all points with device coordinates (U, V) thus far considered, and MaximumV[U] is the maximum of the V values for all points with device coordinates (U, V) thus far considered. Having determined a point with device coordinates (U, V), we perform the following visibility test: If U is not between U0 and U1, the point lies outside (to either side of) the viewport and is not plotted. If U is between U0 and U1, determine whether or not V lies between MinimumV[U] and MaximumV[U]. If it does, the point with device coordinates (U, V) corresponds to a point on the graph hidden by points of the graph considered earlier (see Figure 5.34) and hence is not plotted. If V < MinimumV[U] (the point is visible below the presently visible surface), then we update MinimumV,

$$\text{MinimumV}[U] := V$$

and plot (U, V) (if, of course, V > V0); similarly, if V > MaximumV[U] (the

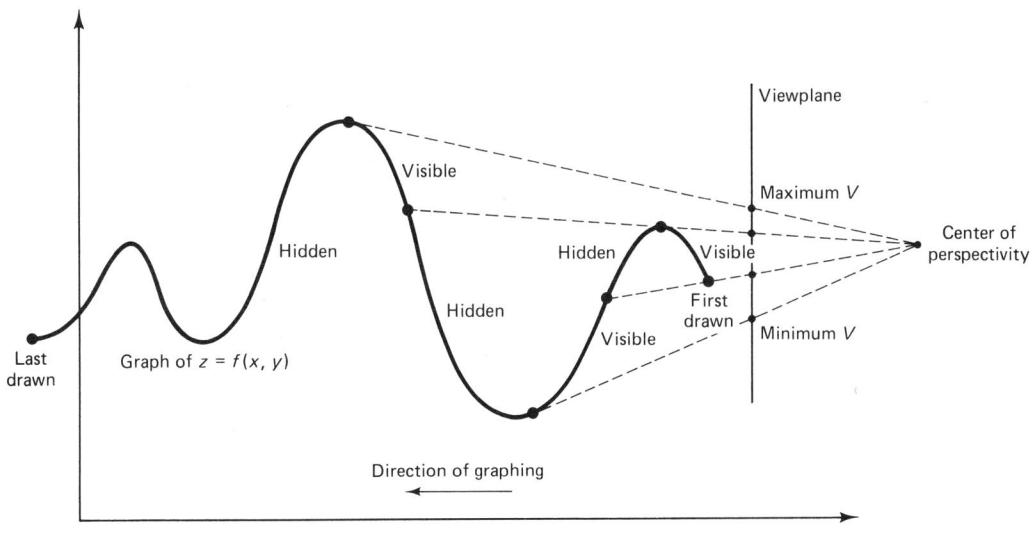

Figure 5.34 The hidden line algorithm.

Sec. 5   Graphing Functions of Two Variables

point is visible above the presently visible surface), then we update MaximumV,

$$\text{MaximumV}[U] := V$$

and plot (U, V) (if, of course, V < V1).

Implementing this algorithm requires first that the entries of MinimumV be initialized to V1, and the entries of MaximumV be initialized to V0. (For each U, MinimumV[U] is always nonincreasing, and MaximumV[U] is always nondecreasing.)

Implementing this algorithm also requires that for a given U we check (at least initially) both whether V < MinimumV[U] *and* whether V > MaximumV[U]. The reason for this is that the first time a given U occurs in a pair (U, V), both MinimumV[U] and MaximumV[U] must be reset to V.

Implementing this algorithm finally requires changing the way in which points are plotted. In particular, we cannot simply continue to use

```
    ⋮
DrawFrom(OldU,OldV)To(NewU,NewV)
    ⋮
```

The reason is that, in general, there need be no relation between the visibility of two points on a given *y*-curve and the visibility of a point between them (see Figure 5.35). Instead, we must check each (U, V) between (OldU, OldV) and (NewU, NewV) for visibility, and plot it only if it is visible.

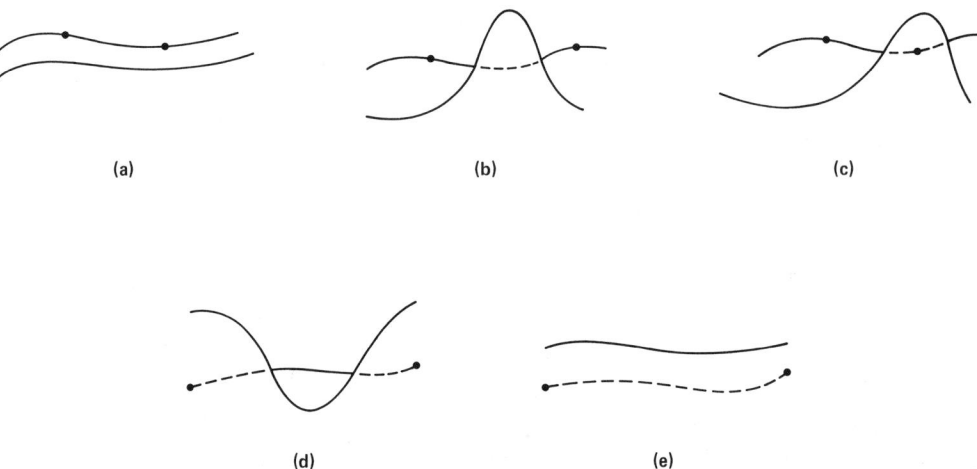

Figure 5.35 Visibility ambiguity for intermediate points.

Our earlier program modified by inclusion of the hidden line algorithm is:

```
    ⋮
(* INITIALIZATION *)
```

```
GraphicsMode
Initialize3DimensionalGraphics
DefineCenterOfPerspectivitySpherical(Rho,Theta,Phi)
DefineCenterOfObject(0,0,0)
DefineViewplane Distance(D)
DefineViewplaneUpDirection(0,0,1)
DefineWindow(-H,-K;H,K)
DefineViewport(U0,V0;U1,V1)
InitializeHeightArrays
     (* PLOTTING *)
FOR X := L TO -L STEP -DeltaX DO
  BEGIN
    FirstPointOnCurve := TRUE
    FOR Y := -W TO W STEP DeltaY DO
      BEGIN
        Z := f(X,Y)
        Transform(X,Y,Z)To(NewU,NewV)
        IF FirstPointOnCurve THEN
          BEGIN
            PlotIfVisible(NewU,NewV)
            (OldU,OldV) := (NewU,NewV)
            FirstPointOnCurve := FALSE
          END
        ELSE
          BEGIN
            DeltaU := NewU - OldU
            IF (DeltaU = 0) THEN DeltaU := 1
            Slope := (NewV - OldV)/DeltaU
            FOR PresentU := OldU TO NewU STEP 1 DO
              BEGIN
                PresentV := PresentV + Slope
                PlotIfVisible(PresentU,PresentV)
              END
            (OldU,OldV) := (NewU,NewV)
          END
      END
  END
            ⋮
     (* PLOTTING PROCEDURE *)
PROCEDURE PlotIfVisible(PresentU,PresentV)
  BEGIN
    IF (0 < PresentU) AND (PresentU < (U1-U0)) THEN
      BEGIN
        IF (PresentV > MaximumV[PresentU]) THEN
```

```
            BEGIN
              MaximumV[PresentU] := PresentV
              IF (PresentV < V1) THEN Plot(PresentU,PresentV)
            END
          IF (PresentV < MinimumV[PresentU]) THEN
            BEGIN
              MinimumV[PresentU] := PresentV
              IF (PresentV > V0) THEN Plot(PresentU,PresentV)
            END
        END
      END
        ⋮
```

This modification implements the hidden line algorithm by means of the plotting procedure PlotIfVisible. This procedure is used as follows: If (NewU, NewV) corresponds to the first point on a *y*-curve, it is plotted if it is visible. Otherwise (if (NewU, NewV) does not correspond to the first point on a *y*-curve), given (OldU, OldV) and (NewU, NewV), we first parametrize the line segment whose endpoints have device coordinates (OldU, OldV) and (NewU, NewV) by

$$\text{PresentU} = \text{PresentU}(t) = t$$
$$\text{PresentV} = \text{PresentV}(t) = t * \text{Slope} + \text{OldU}$$

where *t* is measured in discrete device units and Slope is the slope of the segment. (We are assuming that points on the display device are plotted from left to right.) Since

$$\text{PresentU}(t + 1) = t + 1 = \text{PresentU}(t) + 1$$
$$\text{PresentV}(t + 1) = (t + 1) * \text{Slope} + \text{OldV} = \text{PresentV}\, t + \text{Slope}$$

the coordinates of each point in this parametrization—beyond the first point—can be obtained from the previous coordinates by

$$\text{PresentU} := \text{PresentU} + 1$$
$$\text{PresentV} := \text{PresentV} + \text{Slope}$$

The device coordinates (PresentU, PresentV) at each stage of the parametrization from (OldU, OldV) to (NewU, NewV) are (sequentially) computed and plotted if visible.

Numerous refinements and modifications can be made to this program (their implementation is left to the exercises).

First various cumulative arithmetic errors can affect this program. For example, if our pseudocode program is to be implemented on a specific device and in a specific language, various details must be addressed. One detail of primary concern is the type of variables used in the program. In particular, variables arising from calculations (such as (NewU, NewV) and Slope) are real, while variables used for plotting might very well be restricted to integer (device units). At some point, rounding must be done to convert calculation variables into plotting variables. Even on a monitor, cumulative errors arising in this manner will contribute to (more or less) discernible breaks in the curves (see Figures 5.27).

(Another source of arithmetic error arises from the statement

$$\vdots$$
$$\text{IF (DeltaU = 0) THEN DeltaU := 1}$$
$$\vdots$$

which avoids division by 0 by cheating. We could have parametrized the segment from (OldU, OldV) to (NewU, NewV) in a manner that would have avoided this cheating. However, cheating like this makes the analysis easier; in addition, cheating like this is not seriously harmful since it does not produce errors worse than the rounding errors which will eventually arise anyway.)

To remedy this, our program can be modified to draw short segments instead of plot points (see Figure 5.36): We introduce the coordinates (PreviousU, PreviousV) and the Boolean variables PreviousVisible and PresentVisible. When plotting from (OldU, OldV) to (NewU, NewV), (PreviousU, PreviousV) represent the previous (PresentU, PresentV) values, PreviousVisible is TRUE if (PreviousU, PreviousV) was visible and FALSE otherwise, and PresentVisible is TRUE if (PresentU, PresentV) is visible and FALSE otherwise. The segment from (PreviousU, PreviousV) to (PresentU, PresentV) is visible—on a *small* scale—if and only if (PreviousU, PreviousV) and (PresentU, PresentV) are visible. Thus PlotIfVisible will

$$\vdots$$
$$\text{DrawFrom(PreviousU,PreviousV)To(PresentU,PresentV)}$$
$$\vdots$$

if PreviousVisible and PresentVisible are both TRUE.

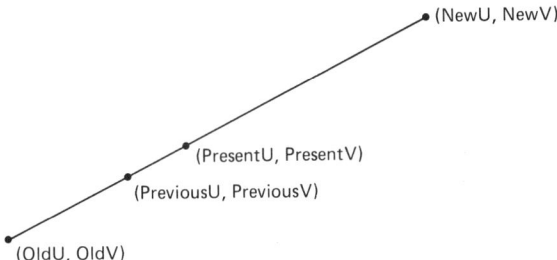

Figure 5.36 Drawing segments.

Observe that this version of the program is also necessary if output from a plotter is desired: in that case, drawing segments is preferable to plotting a *large* number of discrete points.

Second, it might also be desired to draw both *x*-curves and *y*-curves on the same graph. For simple surfaces, this can be accomplished by first drawing the *x*-curves and then drawing the *y*-curves superimposed on the *x*-curves. (*Caution:* Our program assumes that points are plotted from left to right; it must be modified if points are to be plotted from right to left. See Exercise 3.) However this technique can result in errors such as the ones illustrated in Figure 5.37.

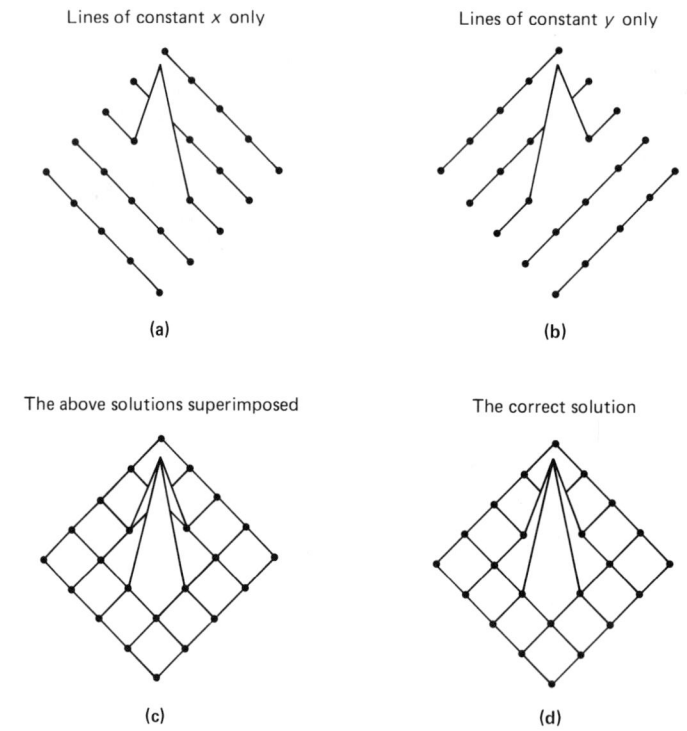

Figure 5.37 Errors arising from superposition (from [Wright]; reproduced with the permission of Thomas J. Wright).

The correct procedure to use in drawing both *x*-curves and *y*-curves on the same graph is as follows (see Figure 5.38): Assume the *y*-curve closest to the center of perspectivity is closer to the center of perspectivity than the closest *x*-curve. (A similar argument holds if the closest *x*-curve is closer than the closest *y*-curve.) First draw the closest *y*-curve. Then draw that (small) part of each *x*-curve (from nearest to the center of perspectivity to farthest from the center of perspectivity) between the first *y*-curve and the second *y*-curve. Then draw the second *y*-curve. Inductively, having drawn the *n*th *y*-curve, draw that (small) part of each *x*-curve (from nearest to the center of perspectivity to farthest from the center of perspectivity) between the *n*th *y*-curve and the $(n + 1)$st *y*-curve. Then draw the $(n + 1)$st *y*-curve.

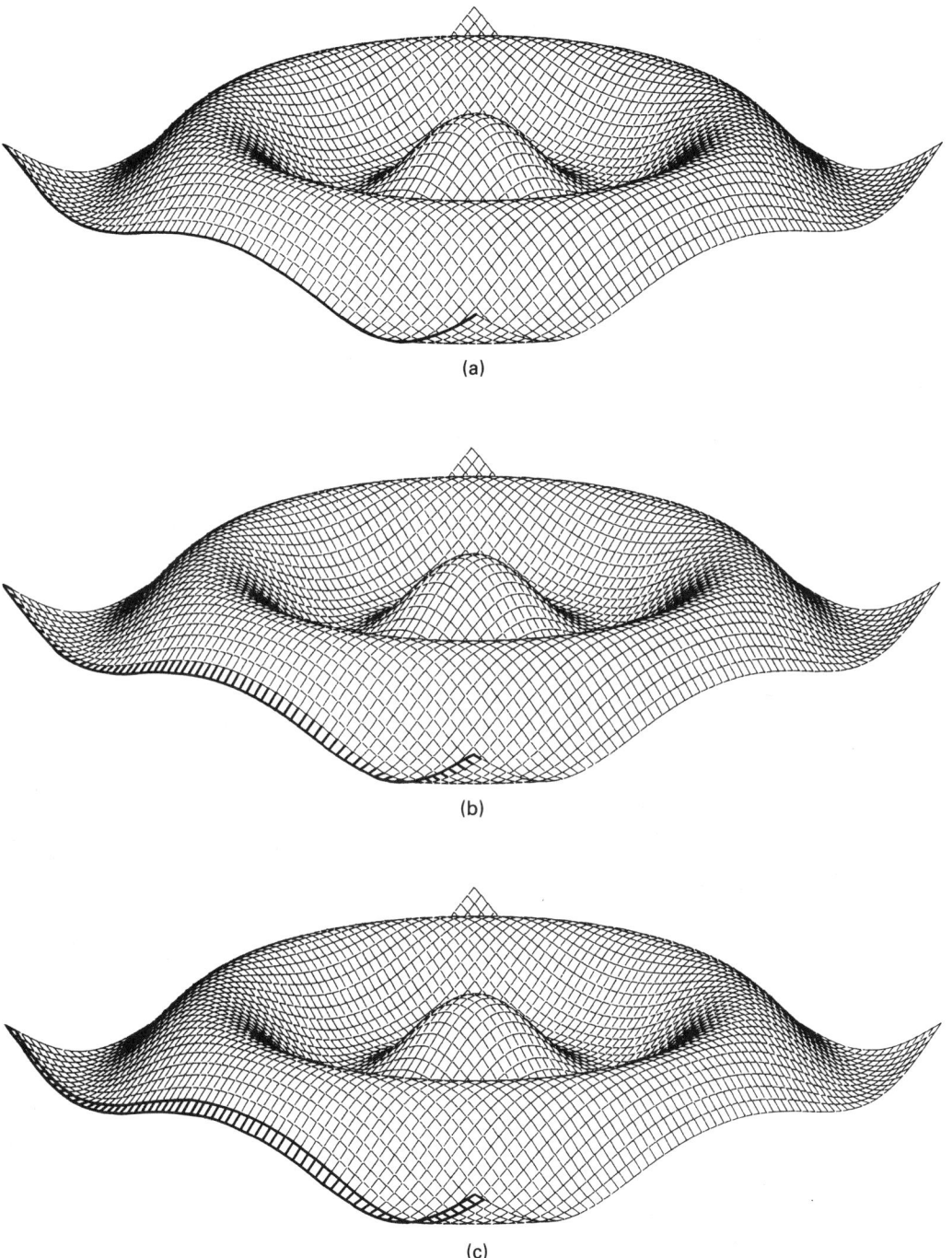

Figure 5.38 The correct procedure for drawing cross-hatched curves.

Sec. 5 Graphing Functions of Two Variables

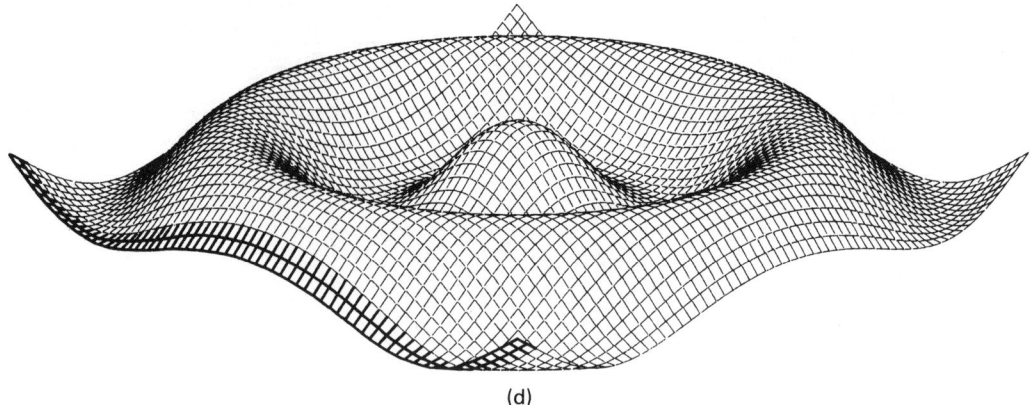

(d)

Figure 5.38 Continued.

SECTION 5 EXERCISES

1. Implement programs for drawing the graph of a function of two real variables in which each is true.
   (a) Line segments are plotted and the hidden line algorithm is not used.
   (b) Points are plotted and the hidden line algorithm is used.
   (c) Line segments are plotted and the hidden line algorithm is used.
   (d) Line segments are plotted, the hidden line algorithm is used, and both x-curves and y-curves are drawn.

2. Use the programs of Exercise 1 to draw the graphs of the following functions near the origin. (You may need to modify these functions by appropriate constants to obtain acceptable output.)
   (a) $f(x, y) = x^2 + 4y^2$ (b) $f(x, y) = x - y$
   (c) $f(x, y) = x - y^2$ (d) $f(x, y) = xy^2$
   (e) $f(x, y) = \exp(-x)\cos y$ (f) $f(x, y) = xy \cos(xy)$
   (g) $f(x, y) = \cos(x - y) \exp(-y)$ (h) $f(x, y) = \cos x + \cos y$

3. In the pseudocode program(s) presented in this section, where is the assumption being made that curves are drawn from left to right? How should the code be changed if it is desired to go from right to left instead?

4. Implement programs for drawing the graph of a function of two real variables in which each of the following is true.
   (a) Only the upper part of the surface is drawn; in other words, that part of the graph below the boundary is not drawn (this type of graph is of interest to topographers).

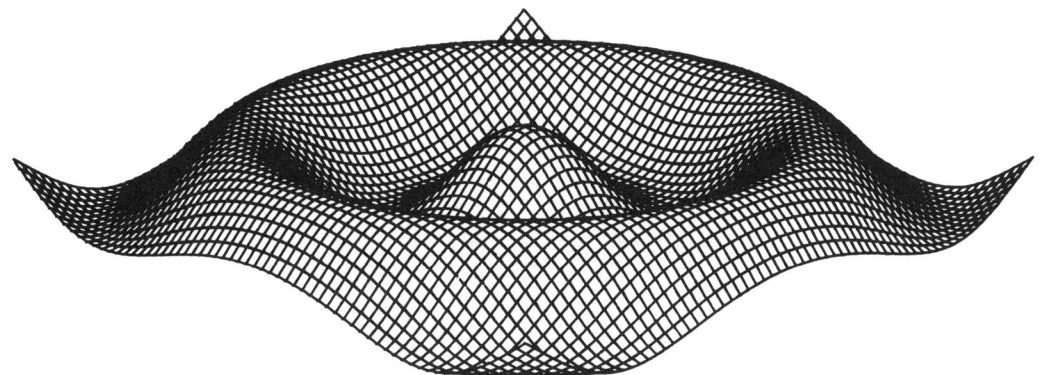

Figure 5.39 A symmetric graph.

(b) Only the boundary curves and curves of occlusion are drawn.
(c) The top of the graph is painted in one color and the bottom is painted another.
(d) Level curves are drawn (a level curve is a curve of the form
$$\{(x, y) \in R^2 \mid f(x, y) = z_0\}$$
for some fixed z_0).
(e) Contour curves are drawn (a contour curve is a curve of the form
$$\{(x, y, z_0) \in R^3 \mid f(x, y) = z_0\}$$
for some fixed z_0) and a hidden line algorithm is used.

5. Use the programs of Exercise 4 to draw the graphs of the functions of Exercise 2.
6. In this section we have assumed that functions are continuous. How would it affect the programming if functions with discontinuities were allowed? In particular, what ambiguities could result in the graph of such a function if both x-curves and y-curves are drawn? (Consider, for example, the behavior of the function $f(x, y) = (2x^2 + 3y^2)/(x^2 + y^2)$ at the origin.)
7. The graph of a function such as $f(x, y) = \cos\sqrt{x^2 + y^2}$ displays a certain amount of symmetry when viewed from the correct angle θ (see Figure 5.39). How might this symmetry be used in drawing the graph of f?
8. A function of the form
$$f(x, y) = a \exp(b(x - h)^2 + c(y - k)^2)$$
for constants a, b, c, h, and k (b and c negative) is a *bump* function: Its graph is a bump (see Figure 5.40) whose position and shape are determined by the choices of the constants.
(a) Draw the graph of a bump function.
(b) Draw the graph of a function that is the sum of several bump functions of different positions and shapes (see, for example, Figure 5.41).

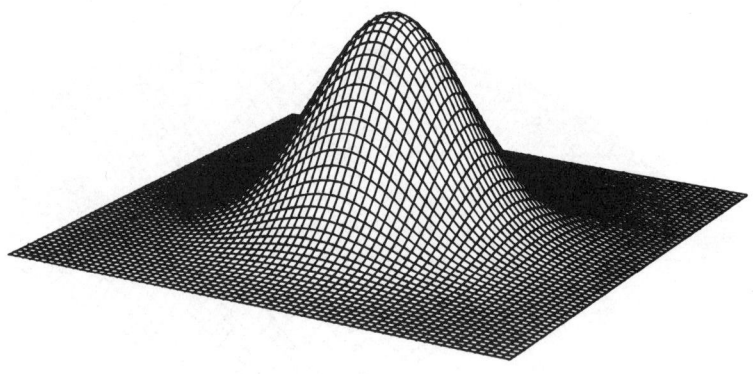

Figure 5.40 A bump function.

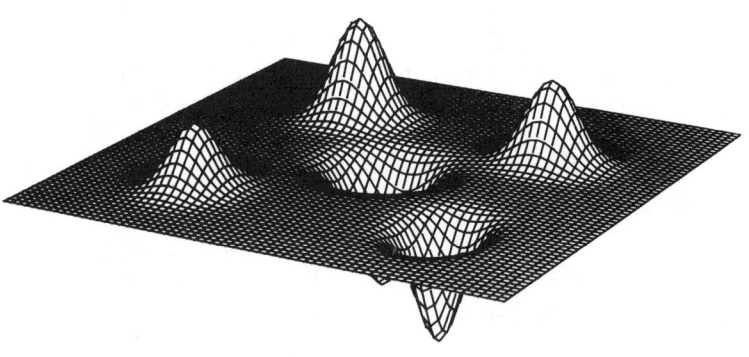

Figure 5.41 A sum of bump functions.

SECTION 6: THE AESTHETICS OF PERSPECTIVE AND PARALLEL IMAGING

In the last two sections we discussed how to obtain two-dimensional perspective and parallel images of a three-dimensional object. In this section, with this experience behind us, we discuss the aesthetics of perspective and parallel imaging. In particular, we discuss the choices we can make when locating the viewing mechanism and the object that will make the resulting image more pleasing or more useful. This section is a continuation to the study of perspective introduced in Chapter 1 but still only an introduction to a broad field. For further study see [Carlbom and Paciorek].

Frequently, in our first view of an object, we essentially guess at a reasonable viewing mechanism; then, depending on the success of the guess, we adjust the viewing mechanism, or perform an object transformation or an image transformation to alter the image. Many times more than one adjustment can achieve a given effect. For example, to decrease the size of the image, we can translate the object away from the eye and viewplane, translate the eye and viewplane away from the object, scale down the object, scale down the image, or change the size of the window. Some of these solutions are equivalent; some have side effects that should be understood. In this section we concentrate on the effects caused by varying the viewing mechanism.

Throughout this section we use the cube whose vertices have Cartesian coordinates (± 1, ± 1, ± 1) as our test pattern. A cube is sufficient to illustrate the points to be made; in fact, the discussion of three-dimensional one-, two-, or three-point perspective is meaningless unless we are talking about an object whose basic shape is that of a rectangular parallelopiped.

Since the issues to be considered with perspective projection are distinct from those related to parallel projection, the two cases are handled separately.

Perspective Projection

We begin by using the viewing mechanism of Section 5.5. The eye has spherical coordinates (ρ, θ, ϕ), the viewplane is perpendicular to the principal line of sight and d units from the origin, and the viewplane reference point is the point of intersection of the principal line of sight and the viewplane. The viewplane up-direction vector has components $\langle 0, 0, 1 \rangle$ (see Figure 5.42).

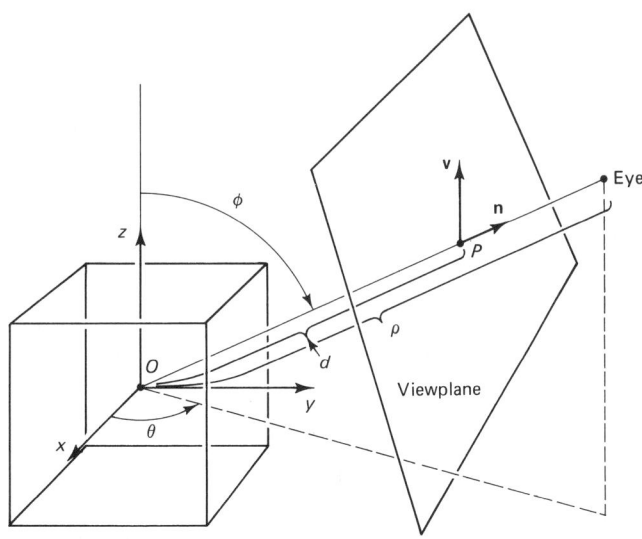

Figure 5.42 The cube and the viewing mechanism.

We first consider the effects of changing θ and ϕ. One major effect of varying these parameters is to change the number of vanishing points of the image of the cube.

For our first view of the cube, we let $\theta = 0°$, $\phi = 90°$, $\rho = 5$, and $d = 3$ (see Figure 5.43). Of the three families of parallel edges of the cube, only one is not parallel to the viewplane. Consequently, the cube appears in one-point perspective. There is one vanishing point at which the images of the lines through the edges not parallel to the viewplane meet. This point is often used by artists as the focal point of a painting (see Figure 1.62). The aerial photograph in Figure 5.44 is a particularly striking example of one-point perspective.

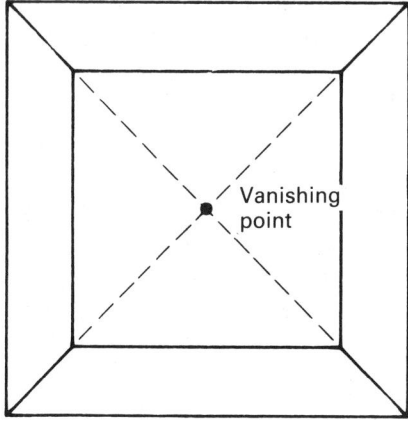

Figure 5.43 The cube viewed in one-point perspective.

If we vary θ or ϕ (but not both) from this position so that just one family of edges of the cube is parallel to the viewplane, the cube appears to us in two-point perspective (see Figure 5.45). In two-point perspective there are two vanishing points where the two families of lines through the images of the two sets of parallel edges meet. The line through these points is a vanishing line; it may serve an artist as a horizon. Figure 5.46 is an illustration from a book published in 1576 to teach artists the principles of two-point perspective.

If we vary both θ and ϕ from their initial values $\theta = 0°$, $\phi = 90°$, so that no family of edges of the cube is parallel to the viewplane, the cube appears in three-point perspective (see Figure 5.47). In three-point perspective there are three vanishing points and through them three vanishing lines. An artist may select one of these to serve as a horizon. A masterful example of three-point perspective is due to Escher (see Figure 5.48).

Finding the homogeneous coordinates of vanishing points and the equations of vanishing lines is discussed in Exercises 5 and 6.

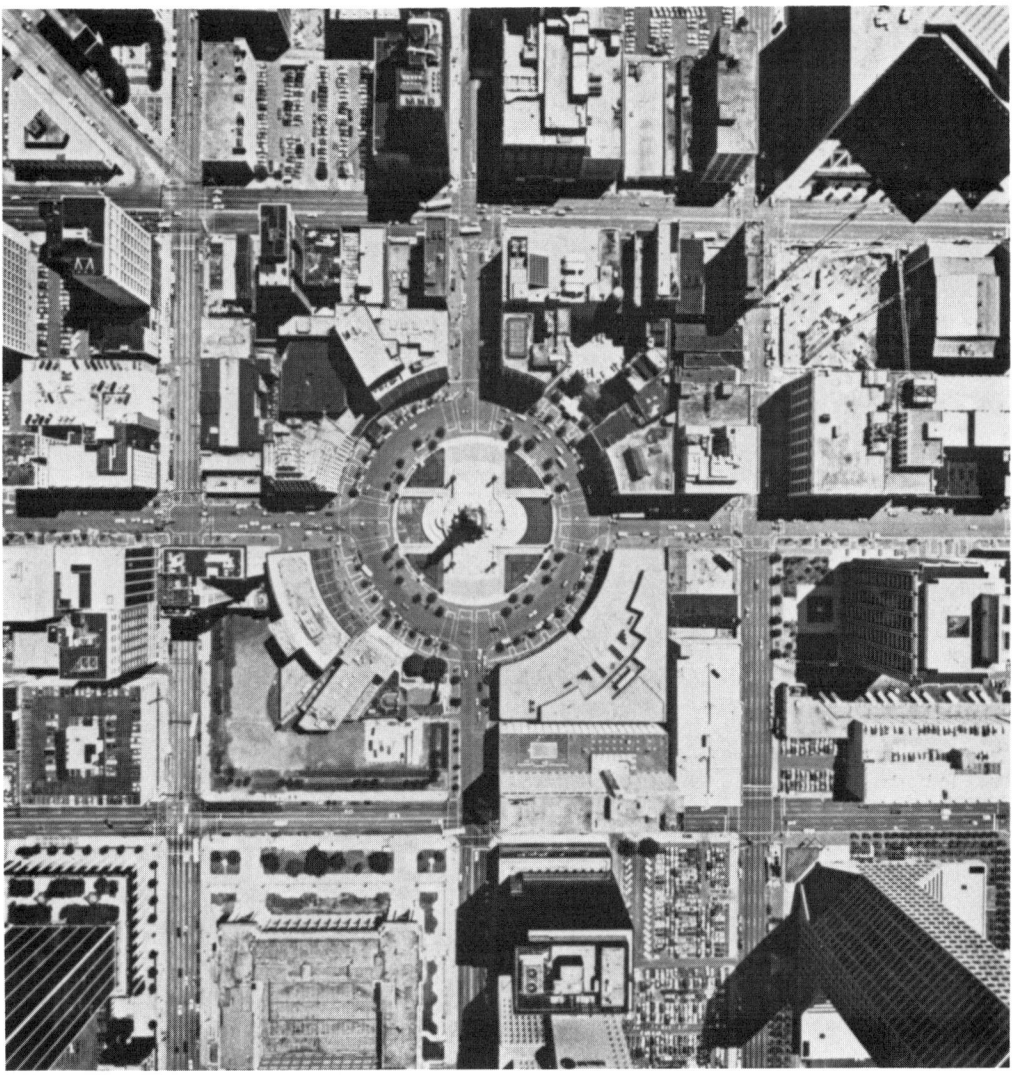

Figure 5.44 An aerial view of Monument Circle in Indianapolis (photo courtesy of Woolpert Consultants).

Sec. 6   The Aesthetics of Perspective and Parallel Imaging

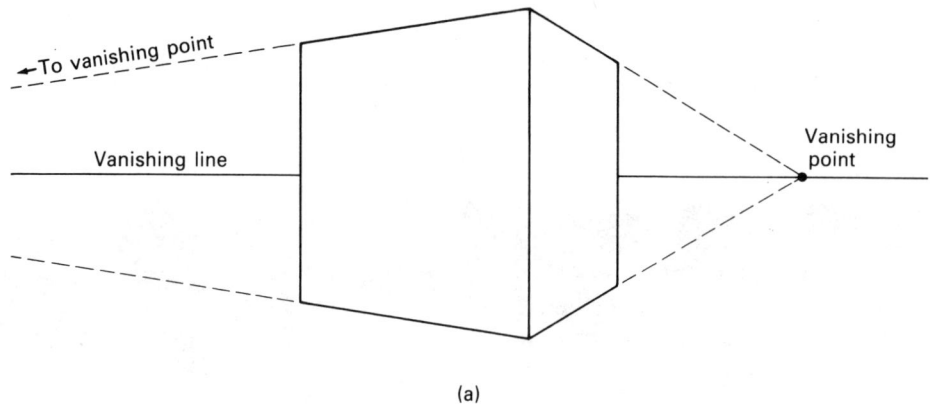

(a)

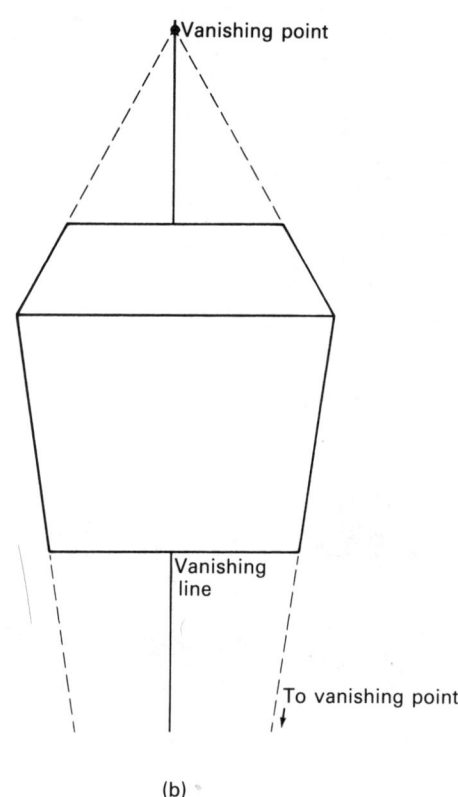

(b)

Figure 5.45 The cube viewed in two-point perspective. (a) $\theta = 30°$, $\phi = 90°$. (b) $\theta = 0°$, $\phi = 30°$.

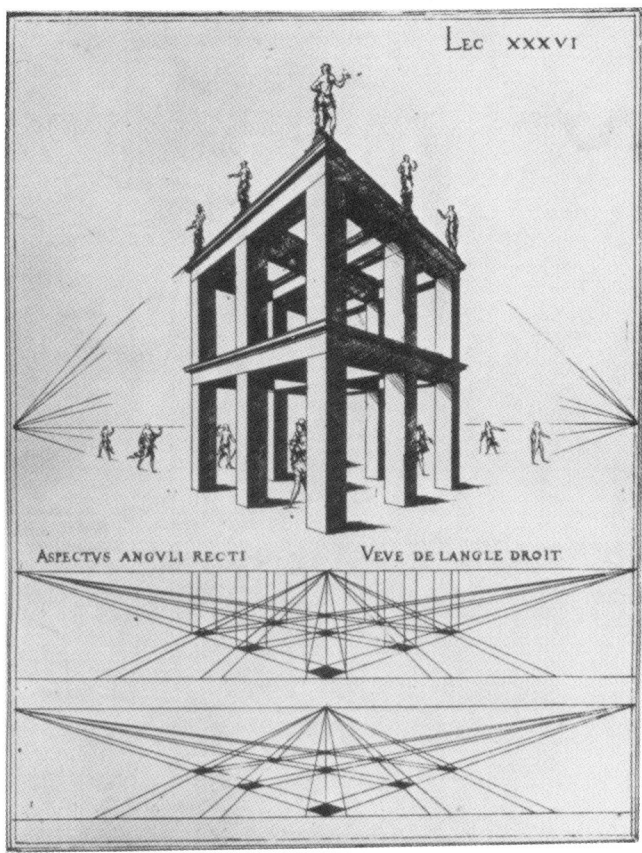

Figure 5.46 A plate from *Leçons de Perspective Positive,* by Jacques Androuet du Cerceau (Université de Paris, Bibliothèque d'Art et d'Archéologie, Fondation Jacques Doucet).

In changing θ or ϕ, we change more than just the *number* of vanishing points; we also change the *position* of the vanishing points. The amount by which θ is varied determines the amount of emphasis placed on a particular wall or side of the object (see Figure 5.49). Varying ϕ varies the amount of emphasis placed on the floor or ceiling. Varying ϕ also has the effect of raising or lowering the level of the horizon (see Figure 5.50).

Having seen the effects of varying θ and ϕ, we now fix $\theta = 30°$ and $\phi = 60°$ and discuss the effects of varying ρ and d.

If ρ is fixed and d is increased, then as long as $d < \rho$, the size of the image decreases (see Figure 5.51). If $d > \rho$ then we have image reversal (see Figure 1.10), and as d is increased the size of the image increases. Notice, however, that the three images in Figure 5.51 all have the same amount of perspective. This is because these images are all similar: The intersection of a cone and a plane is similar to the intersection of the cone and any other parallel plane (see Figure 5.52).

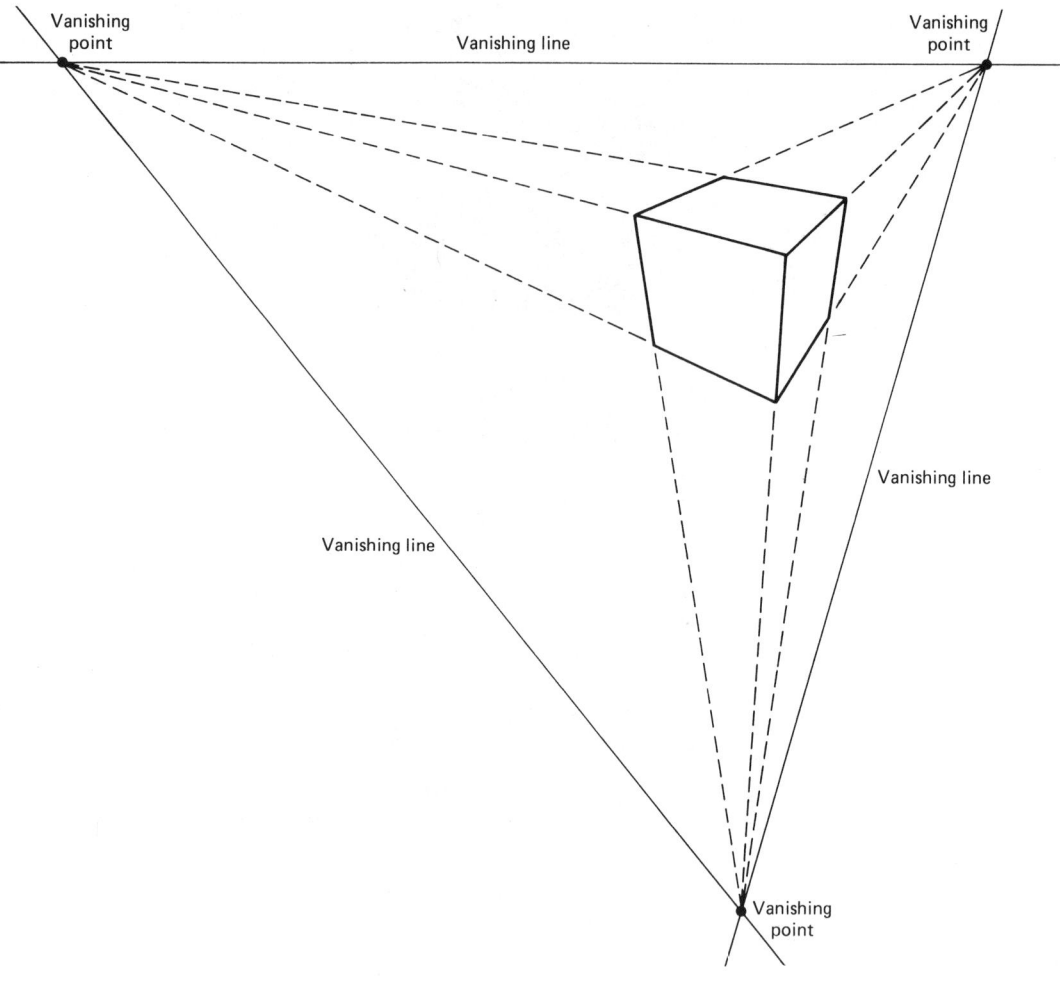

Figure 5.47 The cube viewed in three point perspective ($\theta = 30°$, $\phi = 60°$).

If d is fixed and ρ is increased from d to ∞, then the length of the image of an edge parallel to the viewplane increases from 0 to the length of the edge itself. The length of the image of an edge perpendicular to the viewplane increases from 0 to some maximum value and then decreases again (see Figures 5.53 and 5.54; note that the third dimension has been suppressed in Figure 5.53 for clarity). As we increase ρ the amount of perspective appears to decrease. In the limit, the eye is ideal, we have a parallel projection, and there is no perspective.

If both the eye and the viewplane are moved away from the object, so that ρ and d are increased but $\rho - d$ remains constant, then a combination of the

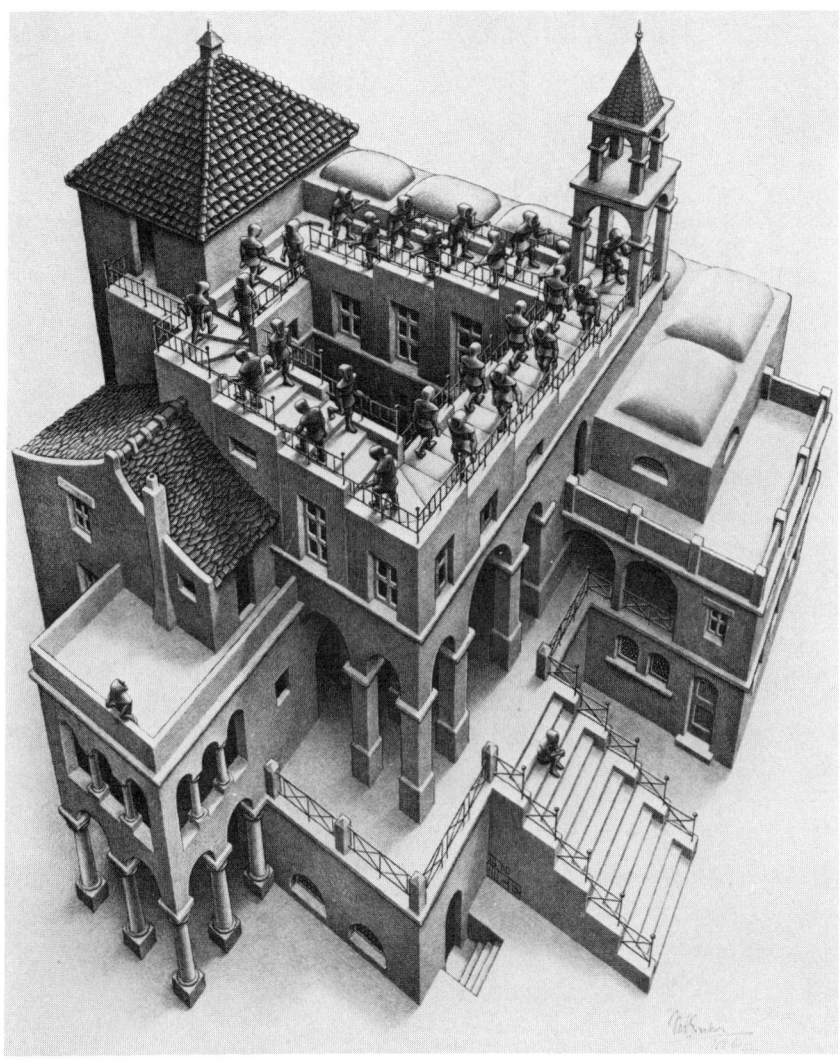

Figure 5.48 *Ascending and Descending* (© Beeldrecht, Amsterdam/VAGA, New York. Collection Haags Gemeentemuseum—The Hague).

effects described above occurs (see Figure 5.55): the size of the image and the amount of perspective both decrease. This is what you see when you walk away from an object.

We now return the viewing mechanism to the position $\theta = 0°$, $\phi = 90°$, $\rho = 5$, and $d = 3$ and investigate the effects of changing the center of the object. (Recall that the center of the object is a user defined point which is not necessarily the centroid—or any other "natural" point—associated to the object.) The center of the object can be changed by translating the cube away from the origin. We will vary the center of the object only by translating the cube parallel to the

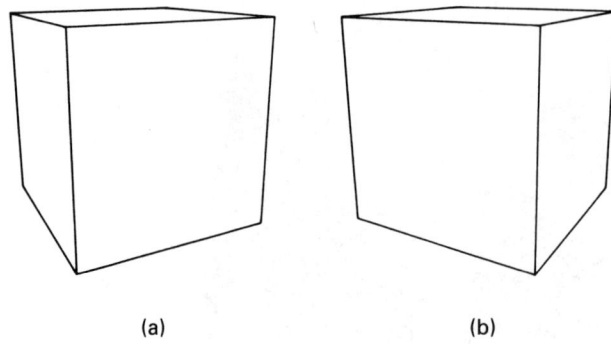

Figure 5.49 Wall emphasis. (a) Right wall ($\theta = 75°$, $\phi = 75°$). (b) Left wall ($\theta = 30°$, $\phi = 75°$).

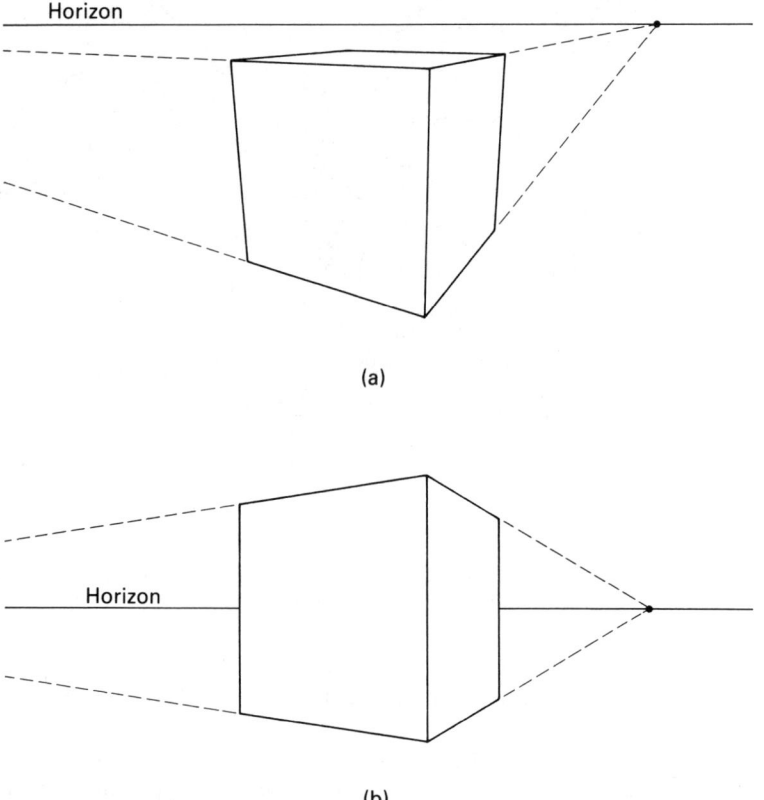

Figure 5.50 Varying the level of the horizon ($\theta = 30°$). (a) $\phi = 75°$. (b) $\phi = 90°$. (c) $\phi = 100°$.

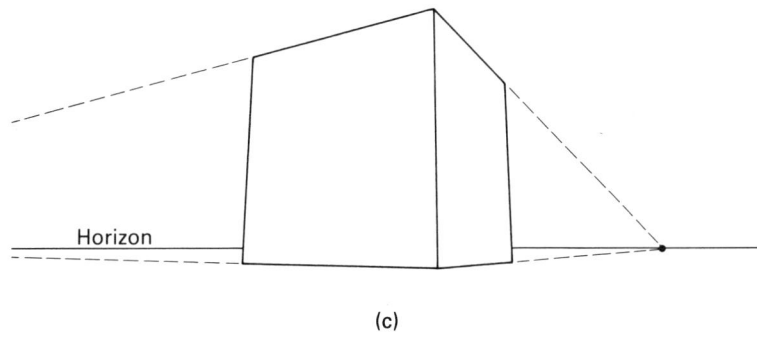

(c)

Figure 5.50 Continued.

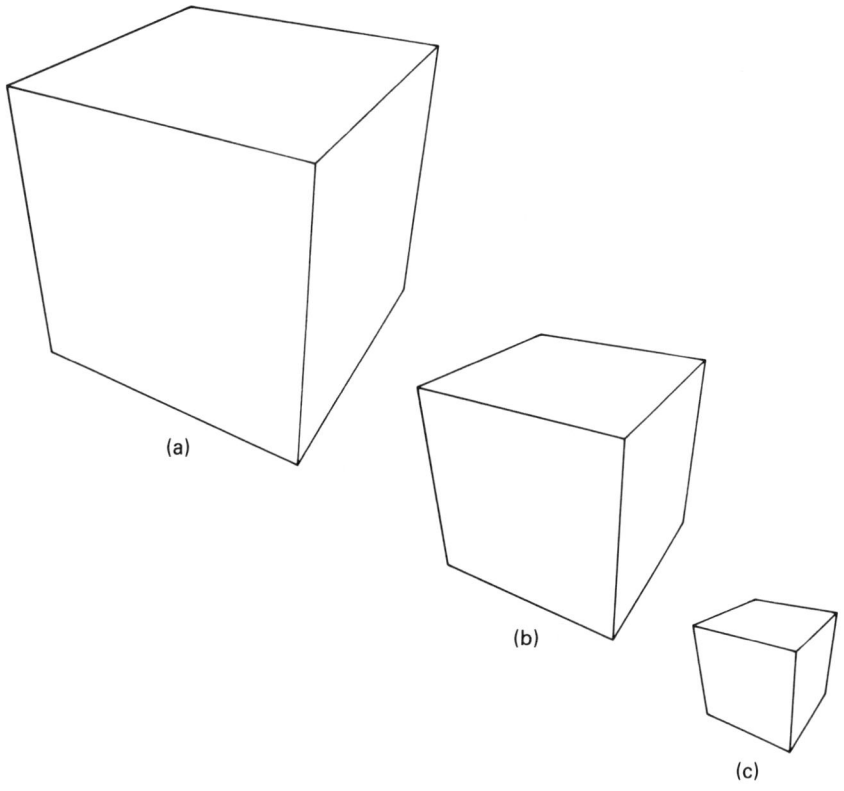

Figure 5.51 The cube viewed with the viewplane (a) near the cube ($\rho = 5$, $d = 2$), (b) at moderate distance from the cube ($\rho = 5$, $d = 3$), and (c) far from the cube ($\rho = 5$, $d = 4$).

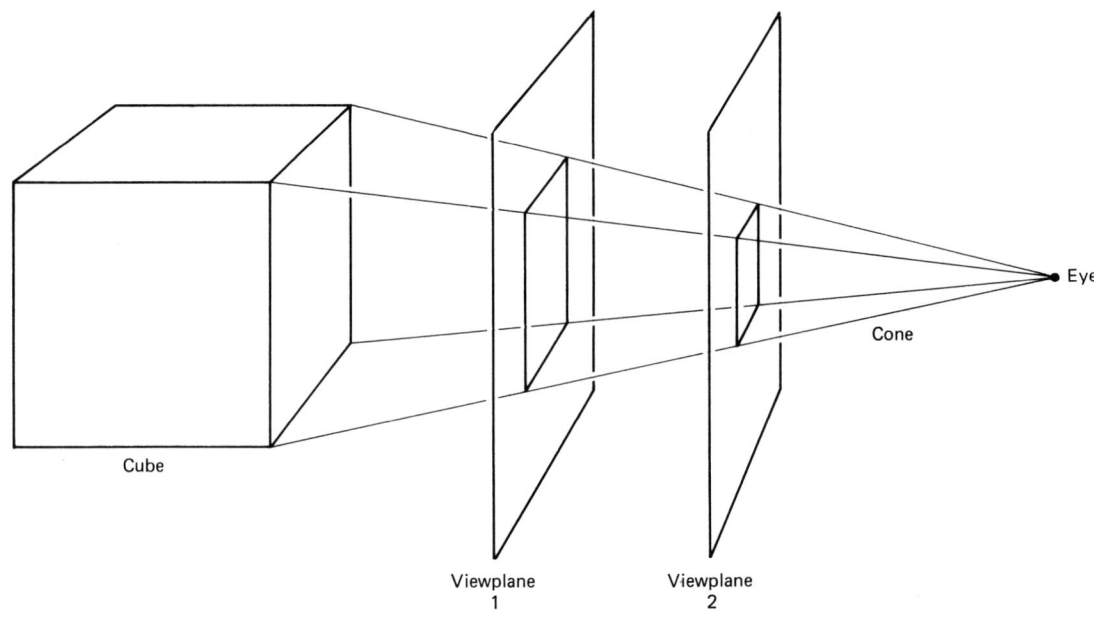

Figure 5.52 Varying the viewplane distance.

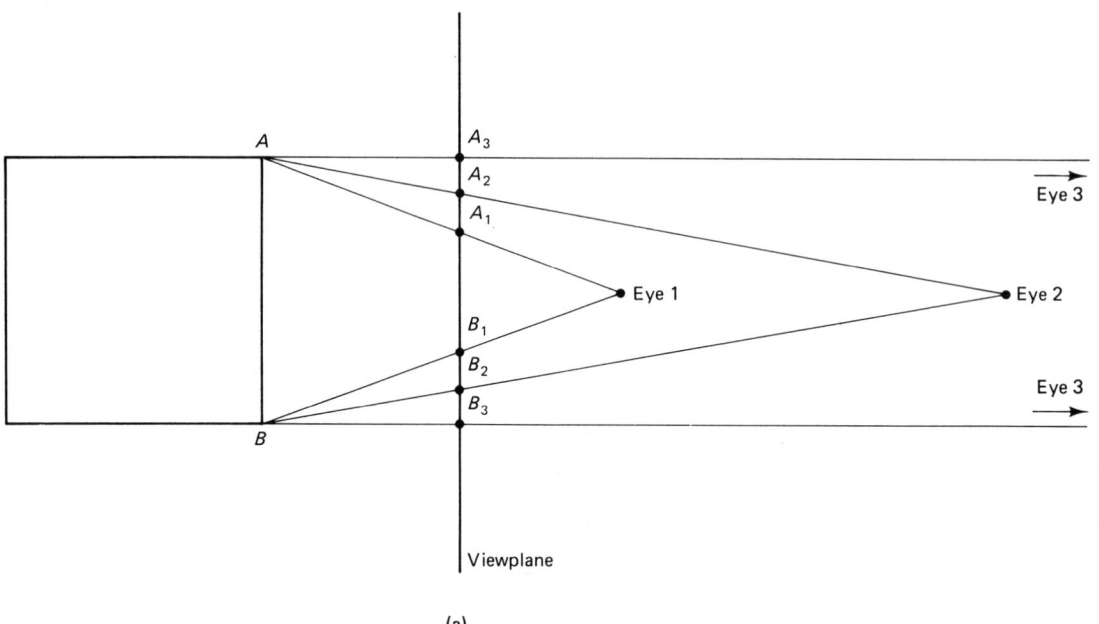

(a)

Figure 5.53 Varying the eye distance: the length of the image of an edge that is (a) parallel or (b) perpendicular to the viewplane.

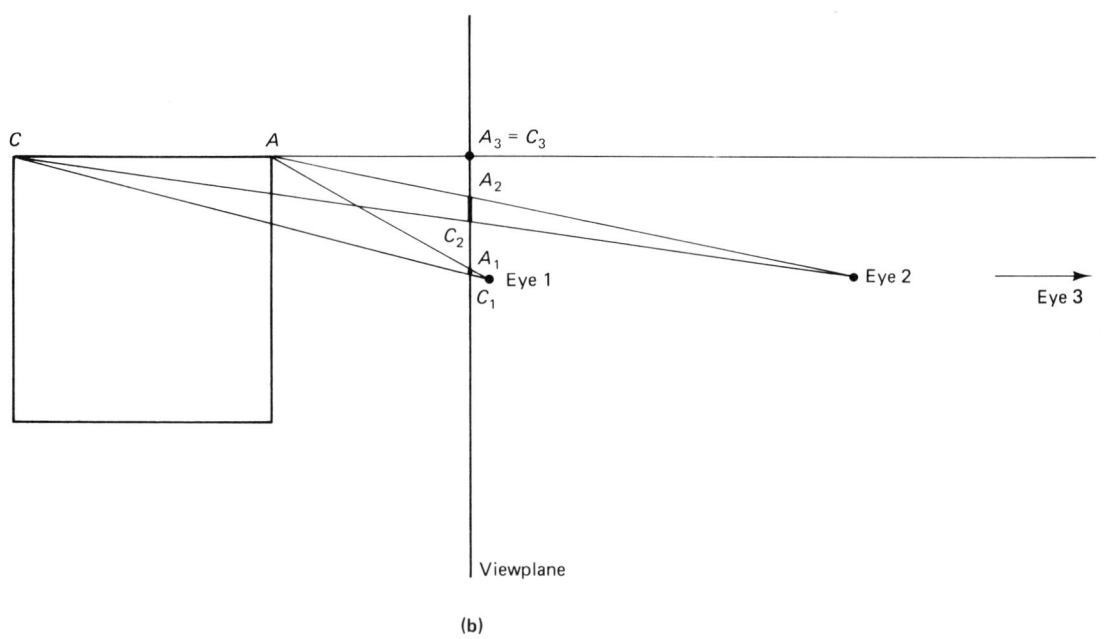

(b)

Figure 5.53 Continued.

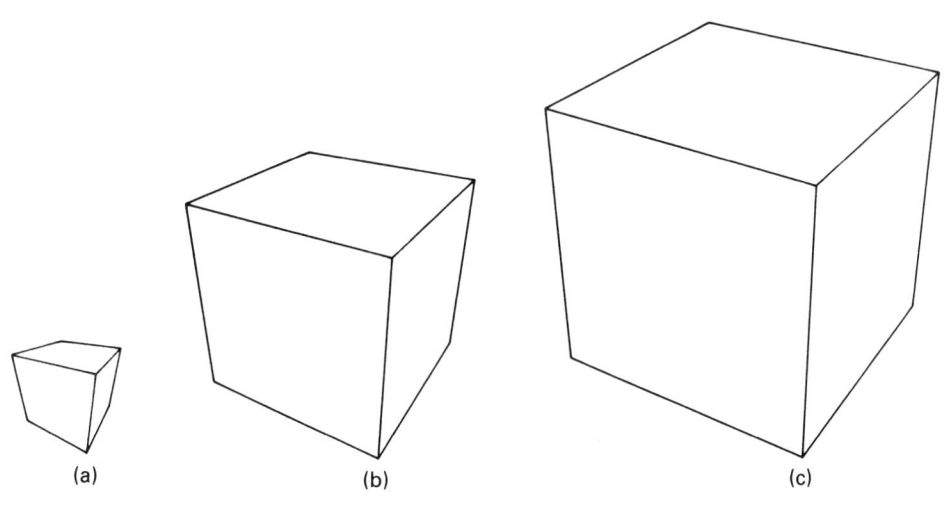

Figure 5.54 The cube viewed on a fixed viewplane with the eye (a) near the viewplane ($\rho = 3.5$, $d = 3$), (b) at moderate distance from the viewplane ($\rho = 5$, $d = 3$), and (c) far from the viewplane ($\rho = 7$, $d = 3$).

viewplane (see Figure 5.56). The view resulting from such a translation corresponds to what we would see with peripheral vision, except that now we can use an image transformation to translate the image back to the center of the display area.

Since translation does not affect the number of edges of the cube that are parallel to the viewplane, changing the center of the object does not change the number of vanishing points and therefore does not change whether the image is in one-, two-, or three-point perspective. However, translation does change the location of the vanishing points of the image and this, in turn, may change the horizon level (see Figure 5.57).

Varying the center of the object also changes the wall or ceiling and floor emphasis (see Figure 5.58).

Now let us consider how lengths change as we vary the center of the object. If an edge that is parallel to the viewplane is translated parallel to the viewplane (see Figure 5.59(a)), the length of its image remains the same. (This is proved using similar triangles; see Exercise 7.) If an edge that is perpendicular to the viewplane is translated parallel to the viewplane (see Figure 5.59(b)), its length increases as it is moved farther from the eye. The combined effects are illustrated in Figure 5.60, in which the center of the object is moved farther and farther away from the center of the cube.

Figure 5.60 illustrates several views in a continuum of pictures. Somewhere in this continuum, each person will decide that distortion has set in. The reason for this distortion is that the center of the object has been moved so far that we cannot see this image even in peripheral vision. An example of such distortion occurs in Figure 5.61, which is a "lesson" in drawing in one-point perspective. The center of the object for the viewing mechanism used to create the top part of this figure (the picture of the palace) is far above the center of the palace; the viewplane is parallel to the front of the palace rather than perpendicular to what we think should be the principal line of sight. If the viewplane were perpendicular to the desired line of sight, the palace would appear in two-point perspective (without distortion).

Being able to move an object farther from the eye and change lengths in this way may at first seem to be a paradox, but the effect does occur and it may cause distortion near the edges of a photograph or a graphics display. On the other hand, we may make use of these phenomena: notice in Figure 5.44 that the buildings appear to be leaning away from the center. (The reason for this is illustrated in Figure 5.59(b).) This *radial displacement* can, in fact, be measured and used to determine the heights of the buildings (see Section 6.4).

We now return the center of the object to the origin and consider the effects of varying several other parameters.

Changing the viewplane up-direction vector results in a rotation of the image (see Figure 5.62).

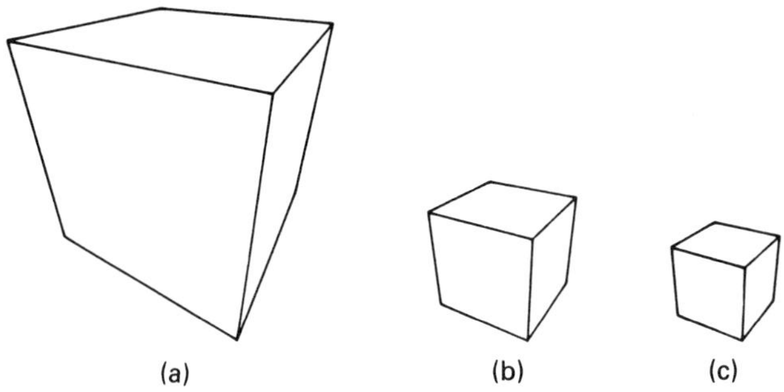

Figure 5.55 The cube viewed with eye and viewplane (a) near the cube ($\rho = 3$, $d = 2$), (b) at moderate distance from the cube ($\rho = 6$, $d = 5$), and (c) far from the cube ($\rho = 8$, $d = 7$).

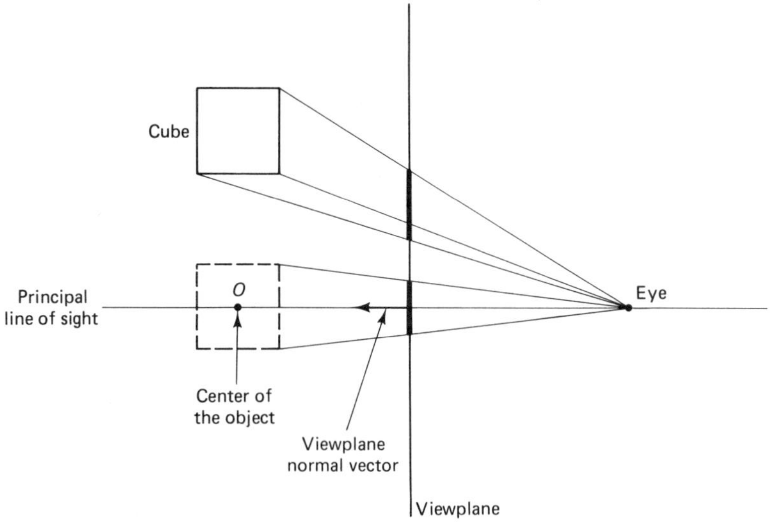

Figure 5.56 Varying the center of the object.

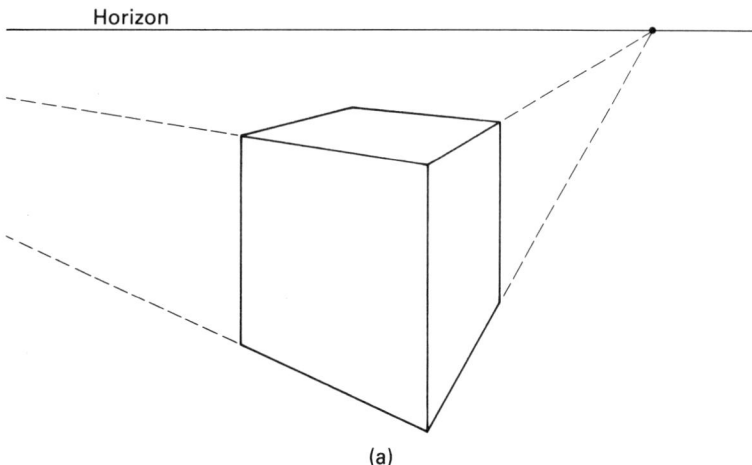

(a)

Figure 5.57 The cube viewed in two-point perspective with center (a) $C(0, 0, 2)$ above the cube, (b) $C(0, 0, 0)$ in the cube, and (c) $C(0, 0, -1.5)$ below the cube.

Sec. 6   The Aesthetics of Perspective and Parallel Imaging

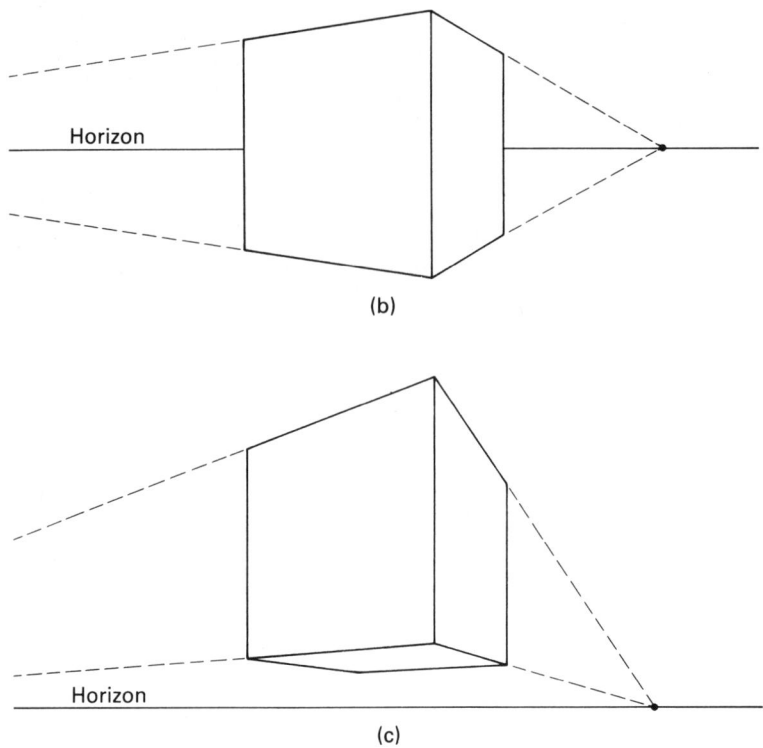

Figure 5.57 Continued.

Figure 5.58 Wall emphasis for an open box. (a) Left wall — Center $C(0, 3/4, 0)$. (b) Right wall — Center $C(0, -3/4, 0)$.

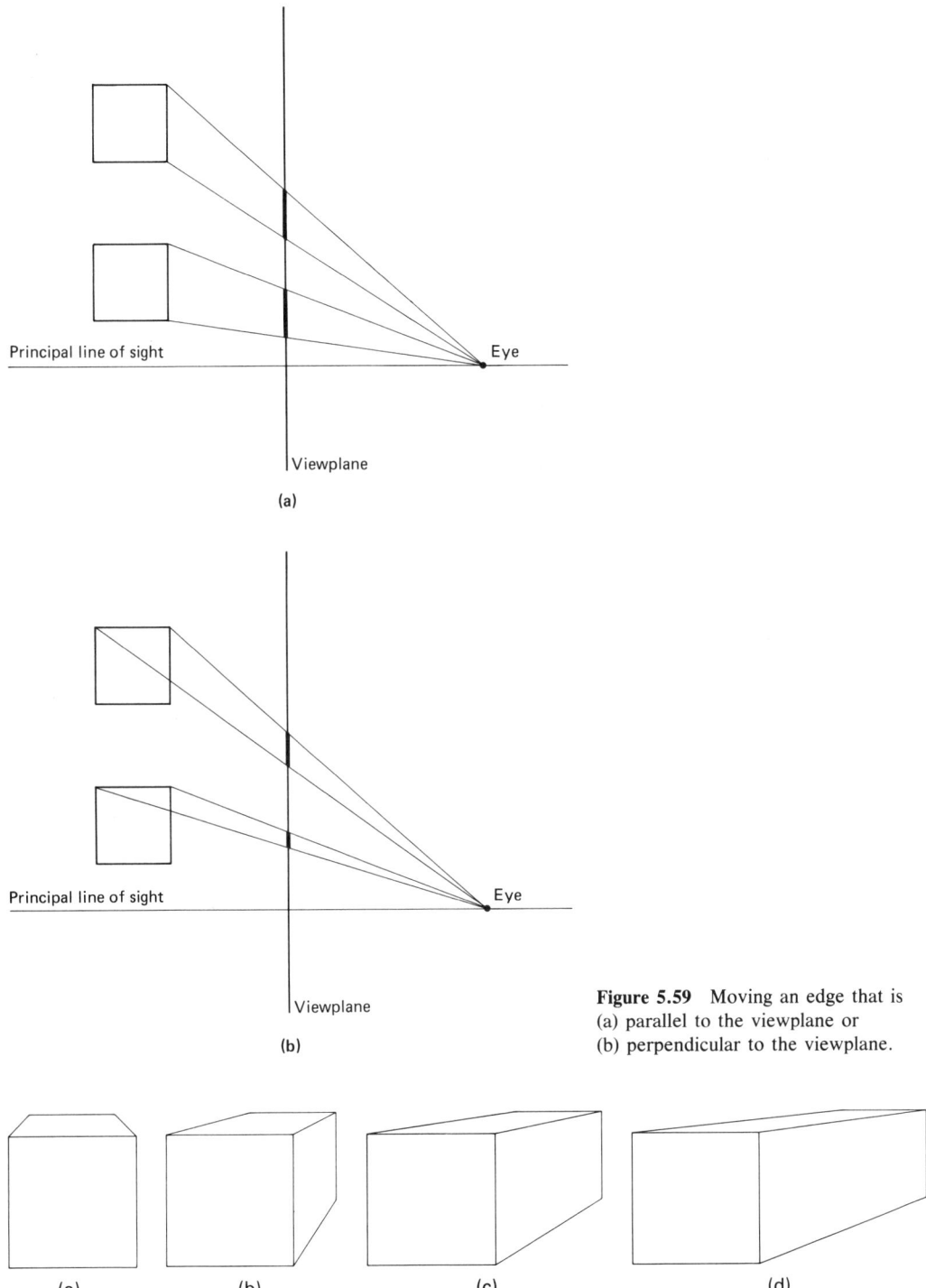

Figure 5.59 Moving an edge that is (a) parallel to the viewplane or (b) perpendicular to the viewplane.

Figure 5.60 Translating the cube parallel to the viewplane. (a) Center $C(0,0,2)$. (b) Center $C(0,3,2)$. (c) Center $C(0,6,2)$. (d) Center $C(0,9,2)$.

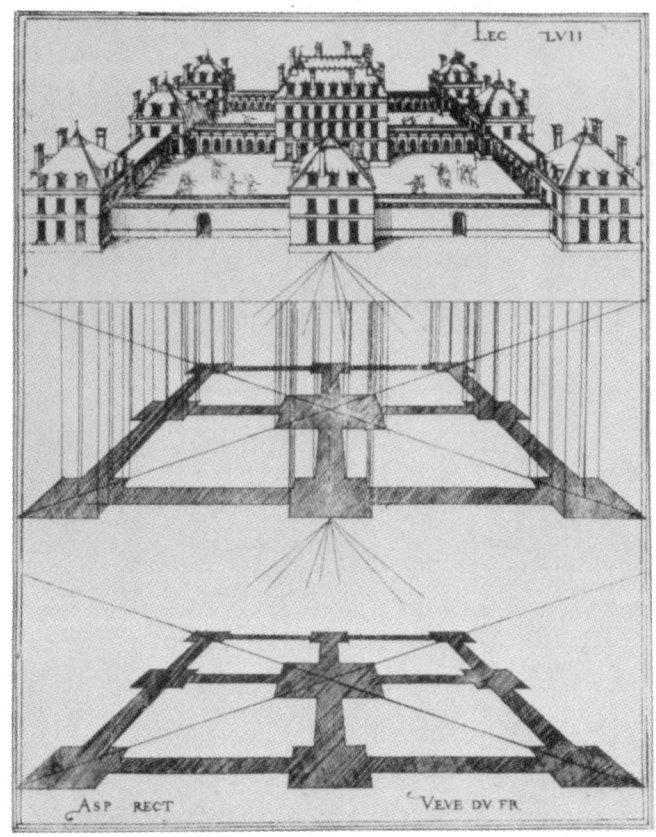

Figure 5.61 A plate from *Leçons de Perspective Position,* by Jacques Androuet du Cerceau (Université de Paris, Bibliothèque d'Art et d'Archéologie, Fondation Jacques Doucet).

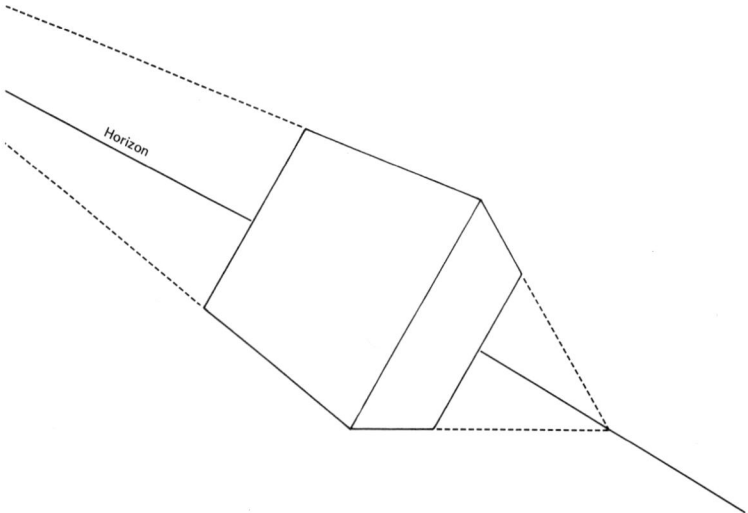

Figure 5.62 Varying the viewplane up-direction vector (from $\langle 0, 0, 1\rangle$ to $\langle 1/4, -\sqrt{3/4}, \sqrt{3/2}\rangle$—rotated 30° counterclockwise as seen from the eye).

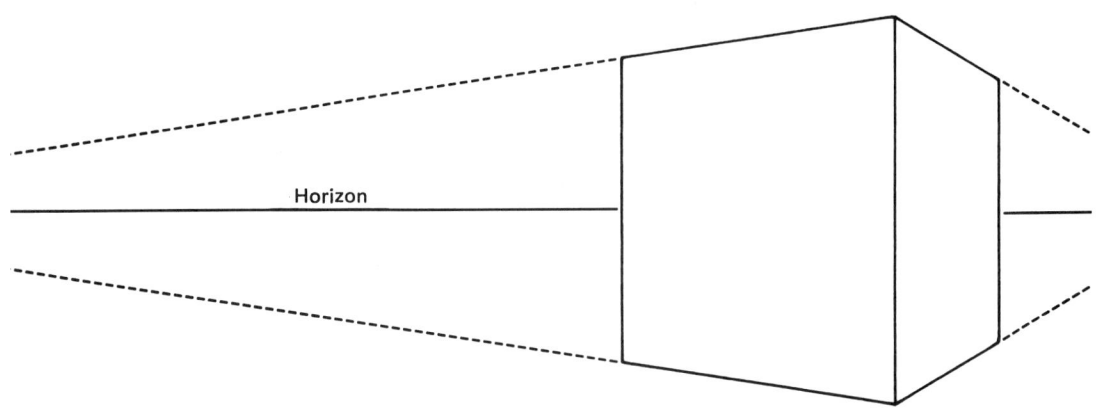

Figure 5.63 Varying the viewplane reference point (2 units to the left).

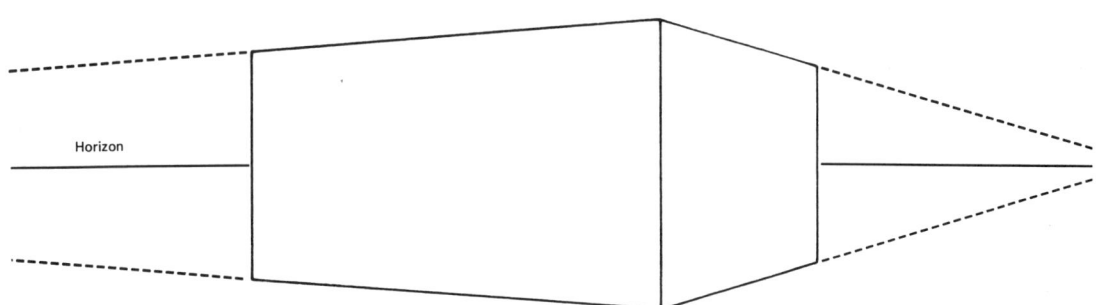

Figure 5.64 Varying the size of the window (from $(-H, -K; H, K)$ to $(-H, -2K; H, 2K)$).

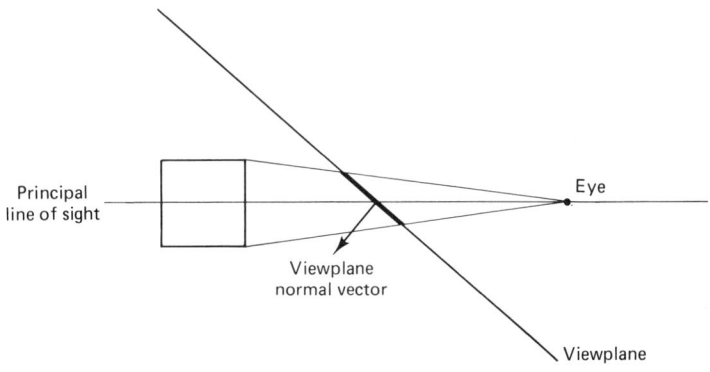

Figure 5.65 Varying the viewplane normal vector.

Changing the viewplane reference point results in a translation of the image (compare Figure 5.63 to Figure 5.57(b)).

Changing the size of the window results in a scaling of the image (see Figure 5.64).

Changing the viewplane normal vector (see Figure 5.65, which should be compared to Figure 5.56) results in a perspective elongation of the image in one direction (see Figure 5.66). A number of artists in the sixteenth and seventeenth centuries used tilted viewplanes to produce *anamorphic art*. Figure 5.67 is an example: by viewing it with your eye very close to the side, you see the mysterious lines blend together to become portraits of four men. (The portraits should be viewed alternately from the left and right sides.) The towns, hills and people visible when the picture is viewed normally recall events in the lives of the men.

More examples of anamorphic art may be found in [Baltrušaitis].

Parallel Projection

In parallel projection the eye is ideal and the lines of sight (projectors) are parallel. The parameters we may vary in this case are the direction of projection, the viewplane normal vector, the viewplane reference point, the viewplane up-direction vector, and the size of the window. (The effects of varying the last three are not discussed here, since these effects are the same as those that occur with perspective projection.)

Since the projectors in a parallel projection are parallel, the length of the image of an edge that is parallel to the viewplane will be the same as the length of the edge. In particular, the image of a face that is parallel to the viewplane has the same size and shape as the face itself (see Figure 5.68). (The image of the face on the display device will probably not have the same size but will have the same shape, assuming that any aspect ratio problems have been corrected.)

The length of the image of an edge that is not parallel to the viewplane may be different from the length of the edge itself. The ratio of the projected length of a segment to its actual length is the *foreshortening ratio* for that segment. Since two parallel edges of the object have the same foreshortening ratio (see Exercise 19), the foreshortening ratio may be used to take certain measurements directly from the image. This ease of measurement is one reason why parallel projections are often used. (Measurements *can* also be taken from a perspective image; see Chapter 6.)

In parallel projection, if the projectors are perpendicular to the viewplane, the projection is *orthographic*; otherwise it is *oblique*. We first consider orthographic projection and the effects of varying the direction of projection.

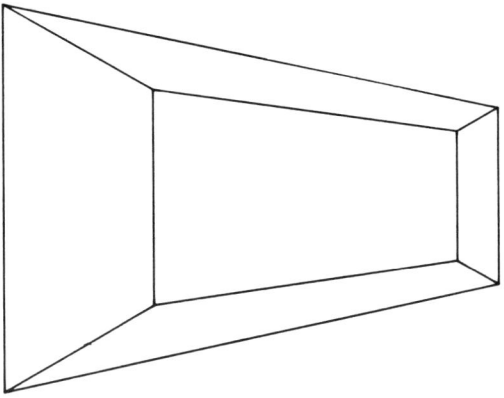

Figure 5.66 The cube viewed through a tilted viewplane ($\rho = 5$, $\theta = 0°$, $\phi = 90°$, viewplane homogeneous coordinates [2, −3, 0, 6]).

Multiview orthographic projections are sets of projections made with the directions of projection perpendicular to different faces of the object (see Figure 5.69). Multiview orthographic projections are used when we want to show the exact shape of each of several faces of an object.

Axonometric orthographic projections are made with a direction of projection from which three different faces of the object can be seen (see Figure 5.70). The choice of which axonometric projection to use in a given instance is determined by which faces of the object should be emphasized. Axonometric projections of a rectangular parallelopiped are classified in terms of the foreshortening ratios in the directions of the edges. An axonometric projection is *isometric* if the foreshortening ratios in the three principal directions are equal, *dimetric* if just two of them are equal, and *trimetric* if they are all different.

To find the foreshortening ratios for an axonometric projection, observe that if a segment makes the angle α with the viewplane its foreshortening ratio is $\cos \alpha$ (see Figure 5.71). Hence we are interested in the cosines of the angles α, β, and γ made by the directions of the edges of the cube and the viewplane.

We now compute formulas for these cosines in terms of the equation of the viewplane. We choose a coordinate system in which a vertex of the cube is the origin and the three lines determined by the adjacent edges of the cube are the coordinate axes. Suppose the equation of the viewplane is $ax + by + cz = d$, where $\mathbf{n} = \langle a, b, c \rangle$ is a unit vector (see Figure 5.72). In this case, $a = \cos \alpha'$, $b = \cos \beta'$ and $c = \cos \gamma'$, where α', β', and γ' are the direction angles of $\mathbf{n}$. Since

$$\alpha + \alpha' = \beta + \beta' = \gamma + \gamma' = 90°$$

(see Figure 5.71), it follows, for example, that $\cos \alpha = \sin \alpha' = \sqrt{1 - \cos^2 \alpha'} = \sqrt{1 - a^2}$. Thus the foreshortening ratios are $\sqrt{1 - a^2}$ in the x-direction, $\sqrt{1 - b^2}$ in the y-direction, and $\sqrt{1 - c^2}$ in the z-direction.

Sec. 6 The Aesthetics of Perspective and Parallel Imaging

Figure 5.67 *Vexierbild* (Picture Puzzle) by Erhard Schön: Emperor Charles V, Ferdinand I of Austria, Pope Paul III, and Francis I, c. 1535 (the Metropolitan Museum of Art, Harris Brisbane Dick Fund, 1929).

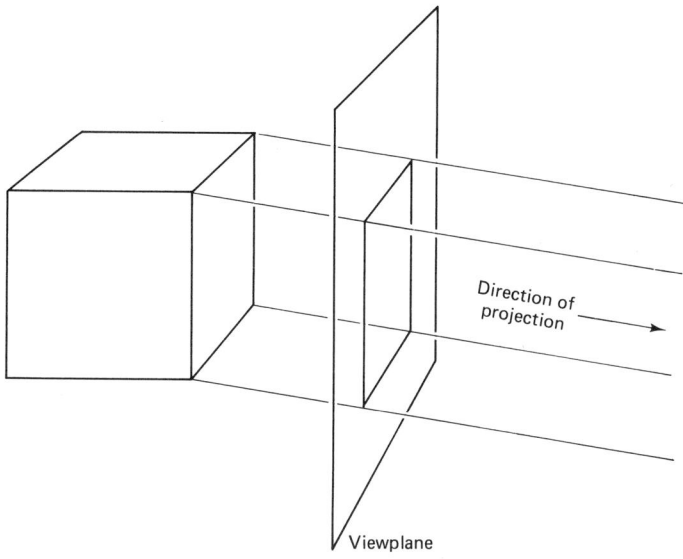

Figure 5.68 The image of a face parallel to the viewplane.

In oblique projection the projectors are not perpendicular to the viewplane. We consider here a special (but common) case of oblique projection in which the viewplane is parallel to one face of the object. Such a projection shows detail well on the one face and provides a suggestion of the three-dimensional shape of the object (see Figure 5.73).

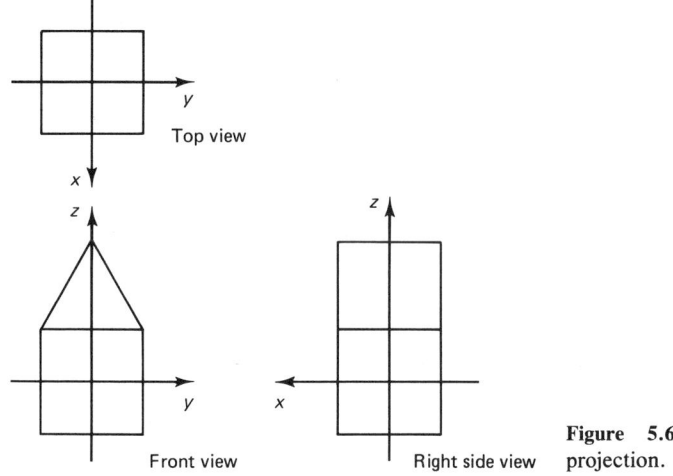

Figure 5.69 Multiview orthographic projection.

Sec. 6   The Aesthetics of Perspective and Parallel Imaging

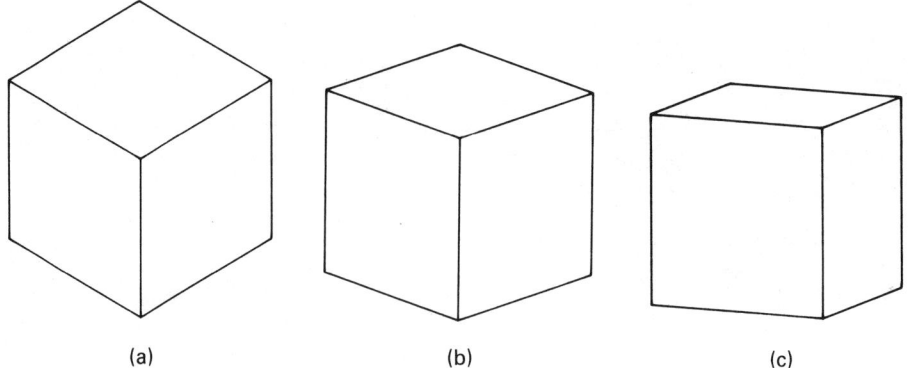

Figure 5.70 The cube viewed in (a) isometric, (b) dimetric, and (c) trimetric axonometric projection.

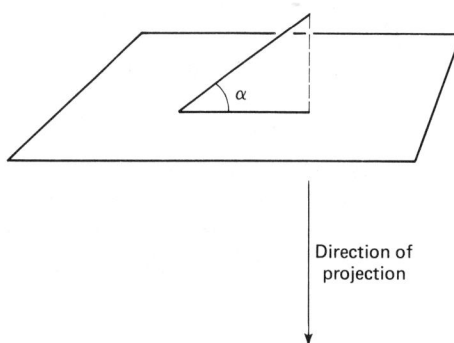

Figure 5.71 Computing the foreshortening ratio of an edge under orthographic projection.

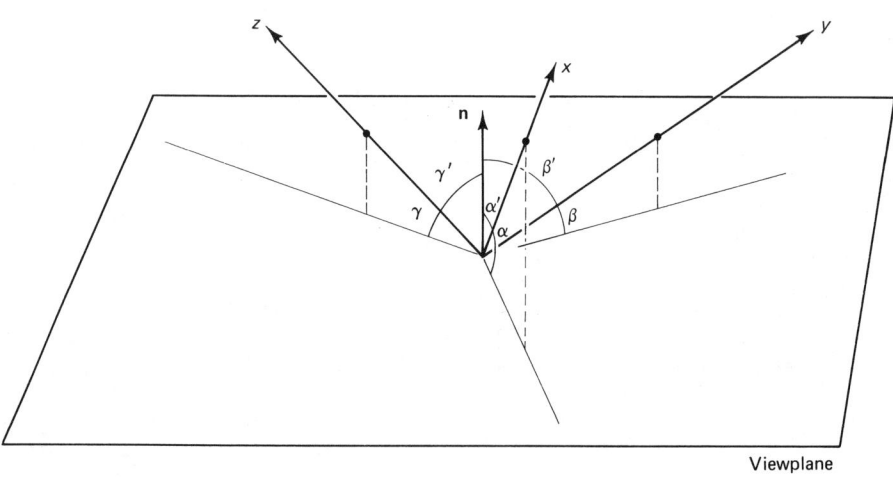

Figure 5.72 Computing the foreshortening ratios for axonometric projection.

292   Three-Dimensional Graphics   Chap. 5

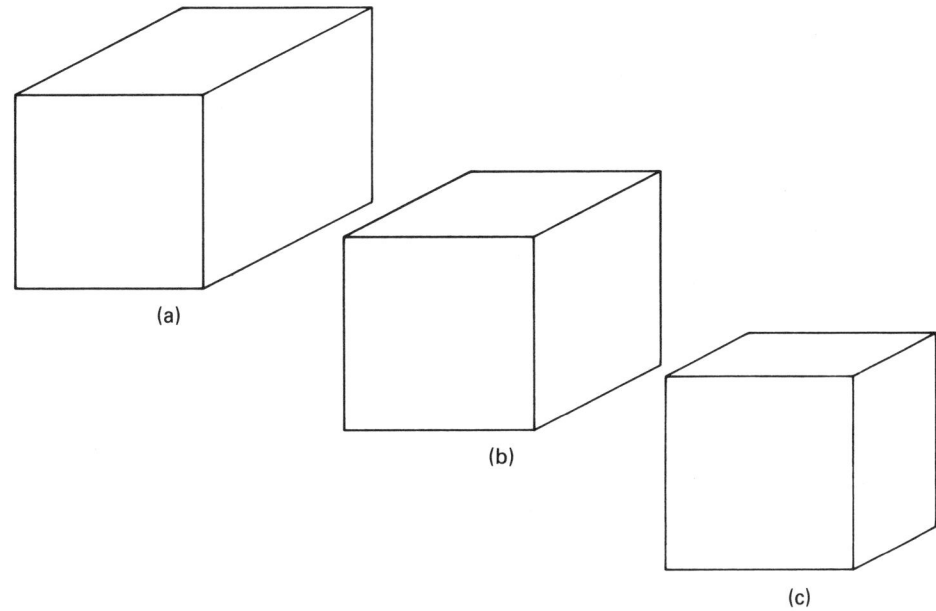

Figure 5.73 The cube viewed in oblique projection at an angle of (a) arccot(1) = 45°, (b) arccot(3/4) = 53.1°, and (c) arccot(1/2) = 63.4°.

Since one face of the object is parallel to the viewplane, we need to find the foreshortening ratio under oblique projection only for the edges that are perpendicular to the viewplane. This ratio is cot α, where α is the angle the projectors make with the viewplane (see Figure 5.74). Thus edges perpendicular to the viewplane project to segments of the same length if $\alpha = 45°$, one-half the length if $\alpha = \text{arccot}(\frac{1}{2})$, and so on.

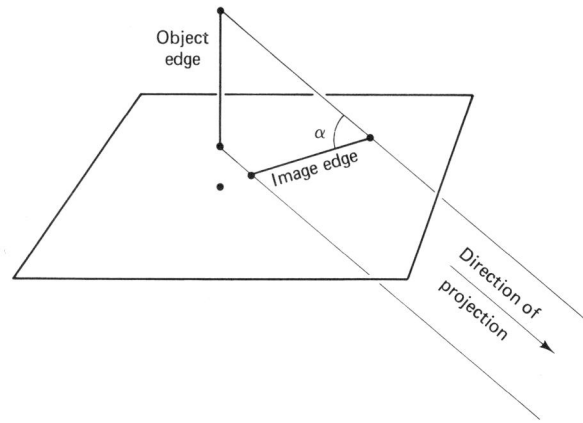

Figure 5.74 Computation of the foreshortening ratio of an edge perpendicular to the viewplane under oblique projection.

Sec. 6   The Aesthetics of Perspective and Parallel Imaging

SECTION 6 EXERCISES

1. For the works of art reproduced in Figure 5.48 and in Section 1.7, determine whether each is a perspective or parallel projection. If perspective, is it in one-, two-, or three-point perspective? Was it done properly—that is, did the artist use vanishing points and vanishing lines correctly?
2. For what values of θ and ϕ is the image of the cube whose vertices have coordinates $(\pm 1, \pm 1, \pm 1)$ in one-, two-, and three-point perspective?
3. When viewing the cube in two-point perspective, does the vanishing line through the two Euclidean vanishing points always serve as a horizon? (*Hint:* Consider Figure 5.45(b); where is the horizon in this figure?)
4. Find an object transformation that has the same effect as each of the following.
   (a) Changing θ
   (b) Changing ϕ
   (c) Moving the eye and viewplane away together, so that $\rho - d$ remains constant
   (d) Leaving d fixed but increasing ρ
   Verify your answers using Graph3D.
†5. Suppose the cube whose vertices have coordinates $(\pm 1, \pm 1, \pm 1)$ is viewed in perspective and the matrix of the complete viewing transformation is $M^{4 \times 3}$.
   (a) How can M be used to find the homogeneous coordinates of the three vanishing points?
   (b) If the direction of a line in world coordinate space has homogeneous coordinates $[a, b, c, 0]$, how can M be used to find the vanishing point on the image of this line?
   (c) Give two ways to find the homogeneous coordinates of a vanishing line in the image.
6. Suppose we use the viewing mechanism specified by ρ, θ, ϕ and d to view the cube of Exercise 5. Find the homogeneous coordinates in projective three-space of the three vanishing points and lines in the viewplane (as functions of ρ, θ, ϕ and d). If M_π^+ is a given inverse parametrization of the viewplane, how can we then find the homogeneous coordinates of the corresponding points and lines in the parametrizing plane?
7. Verify that if an edge is parallel to the viewplane, the length of its image under perspective projection is unchanged if the edge is translated parallel to the viewplane. (*Hint:* Use similar triangles in Figure 5.75 to establish $BC/EF = OB/OE = OA/OD$. Why does this suffice to prove the result?)
8. Describe how the image of an edge perpendicular to the viewplane varies in Figure 5.53(b) if the eye varies along the principal line of sight but on the *other side* of the viewplane?
9. Suppose the coordinates of the vertices of the window are initially $(\pm 3, \pm 1)$ but are changed to $(\pm 1, \pm 2)$. How does this change the image?
10. Suppose the viewplane has equation $x = 10$ and the viewplane up-direction vector is initially $\langle 0, 0, 1 \rangle$, but is changed to $\langle 1, 1, 1 \rangle$. How does this change the image?
11. Suppose a viewing mechanism is described by giving the homogeneous coordinates of the viewplane $\mu[\mu_1, \mu_2, \mu_3, \mu_4]$, the eye $E[e_1, e_2, e_3, e_4]$, and the center of the object $C[0, 0, 0, 1]$.
    (a) Describe the projection if $e_4 = 0$.

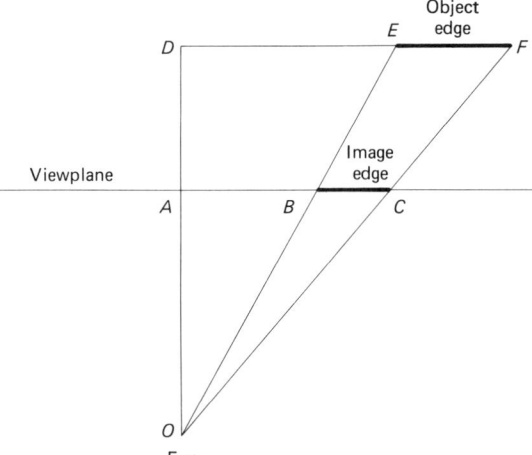

Figure 5.75 Diagram for Exercise 7.

(b) Find the viewplane normal vector.
(c) What change in this viewing mechanism is equivalent to varying ρ? Varying d?
(d) How can you tell from the homogeneous coordinates of μ and E if the principal line of sight is perpendicular to the viewplane?
(e) What change in this viewing mechanism is equivalent to varying θ? Varying ϕ?
(f) If some of these changes are difficult to make using this description of the viewing mechanism, what changes *are* easy to make and what are their effects? Verify your answers using Graph3D.

12. If the homogeneous coordinates of the viewplane, the eye, and the center of the object are $\mu[1, 1, 1, -3]$, $E[4, 4, 4, 1]$, and $C[0, 0, 0, 1]$, find the corresponding ρ, θ, ϕ and d.

13. If the viewing mechanism is described by $\rho = 5$, $\theta = 30°$, $\phi = 60°$, and $d = 3$, find the homogeneous coordinates of the viewplane and the eye.

14. Suppose we are viewing the cube in one-point perspective (see Figure 5.43) and we move the center of the object. Relate the direction in which we move the center to the direction in which the vanishing point of the image moves. If we are viewing the cube in two-point perspective, relate the direction in which we move the center to the direction in which the vanishing line moves.

15. Describe the effects of changing only the center of the object in the viewing mechanism of Exercise 11. Consider the number of vanishing points, horizon level, wall emphasis, amount of perspective, and size of the image.

16. Can the effect of viewplane tilt be achieved by image transformations? Use Graph3D to produce images on a tilted viewplane—anamorphic computer graphics.

17. The word *STOP* is to be painted on a road at an intersection so that it will read normally from a given height at a given distance. What principles must be applied in designing the letters?

18. In order to achieve the correct amount of perspective, we leave the cube centered at the origin but translate the eye and viewplane away. Unfortunately the image becomes too small. How can we correct this without changing the amount of perspective? Will it help to use an object transformation to scale up the cube? Experiment with Graph3D.

19. Show that two parallel edges have the same foreshortening ratio under parallel projection.
20. Find the foreshortening ratio of a general segment under oblique projection.
21. For an isometric projection of a cube, find the foreshortening ratios in the directions determined by the edges of the cube.
22. Can you find a quantitative measure for the amount of perspective in the image of the cube? Here are some possibilities to consider.
    (a) The ratio of the length of the image of a unit segment perpendicular to the viewplane to that of a unit segment parallel to the viewplane
    (b) d/ρ
23. Write a program to draw the cube and any visible vanishing lines. The cube may be fixed and the viewing mechanism may use fixed ρ and d, but you should allow θ and ϕ to be entered interactively. As options, consider using a hidden surface routine when drawing the cube and allow the cube to obscure portions of the vanishing lines.
24. Some artists have used a different perspective altogether, in which images of straight lines may be curves. Investigate this in M. C. Escher's *High and Low* and *House of Stairs* (see [Ernst]).

SECTION 7: DEPTH CLIPPING

In general, clipping is the screening of points from plotting (or further consideration) on the basis of a visibility criterion. In three-dimensional perspective imaging, there are two different types of clipping—lateral clipping and depth clipping. With *lateral clipping* (see Figure 5.5) a point is not plotted if it lies outside the cone of vision (the cone whose vertex is the center of perspectivity and which has the viewplane window as a cross section). With *depth clipping* (see Figure 5.6) a point is not plotted if it lies outside a specified range of distance from the center of perspectivity. The specified range of distance is specified by two planes: a *hither* (or *front*) clipping plane and a *yon* (or *back*) clipping plane. The goal of this section is to discuss the viewing operation with depth clipping.

Perhaps somewhat surprisingly, implementing the viewing operation with depth clipping is quite distinct from implementing the viewing operation without depth clipping: The geometry used to implement the viewing operation with depth clipping is illustrated in Figure 5.76. The fundamental elements of this geometry include the *view volume* (that part of the cone of vision between the hither and yon planes) and four transformations.

The first transformation T_1 is a projective transformation of projective three-space taking the view volume to the unit cube whose vertices have Cartesian coordinates (0, 0, 0), (1, 0, 0), (1, 1, 0), (0, 1, 0), (0, 0, 1), (1, 0, 1), (1, 1, 1), and (0, 1, 1); T_1 can be thought of as a change of coordinates. Since T_1 is a projective transformation, it takes the center of perspectivity to an ideal point and the viewplane to a plane perpendicular to a coordinate axis (see Exercise 2(a)). Furthermore (see Exercise 2(b)) the original perspective projection corresponds, in the new coordinate system, to projection parallel to a coordinate axis.

There are two important reasons for translating our problem into the new coordinates: First, it is easier to clip against the unit cube in the new coordinates

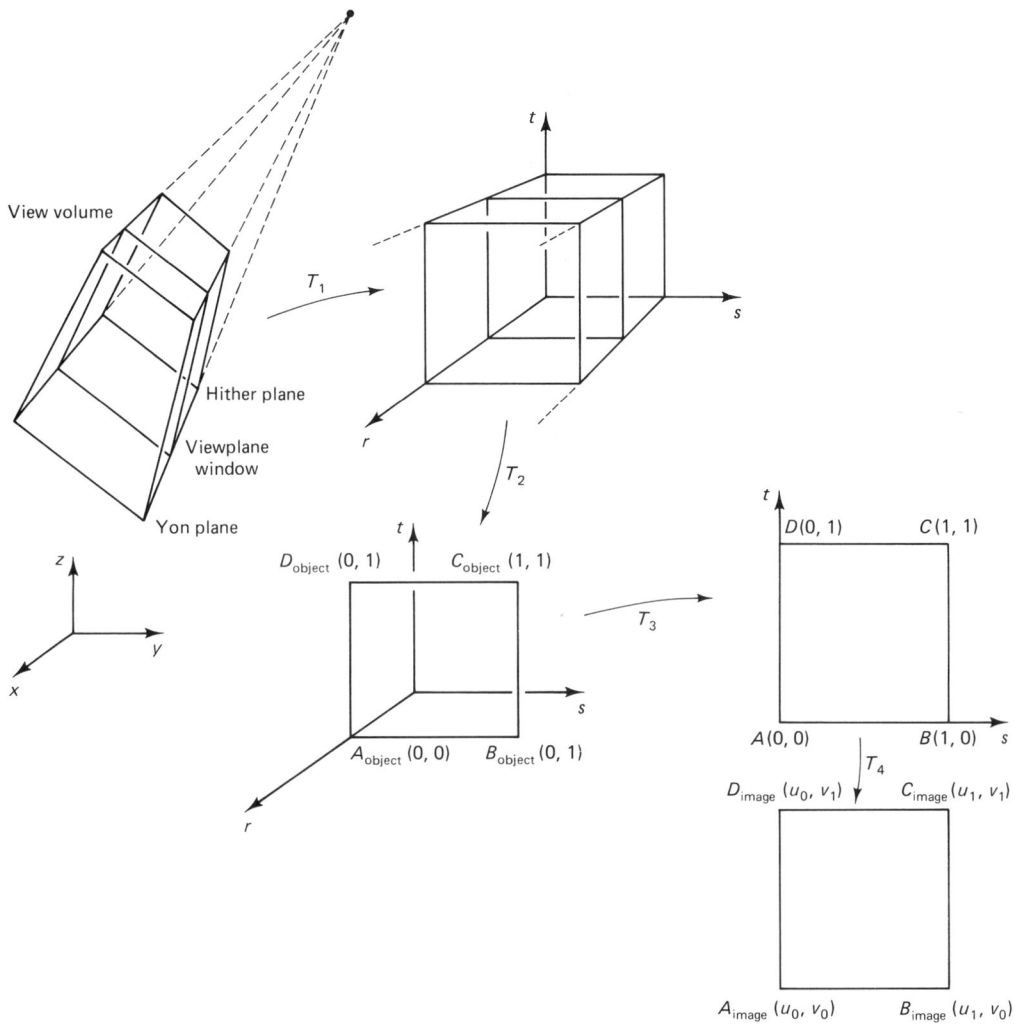

Figure 5.76 The viewing operation with depth clipping.

than it is to clip against the view volume in the original coordinates. This is because lateral *and* depth clipping in the new coordinates reduce to determining whether the coordinates of a point are between 0 and 1. Second, it is (analytically) easy to project onto the viewplane (this is what T_2 does) and find an inverse parametrization of the viewplane (this is what T_3 is). The transformation T_4 is the display transformation associated to normalized device coordinates.

While the general principles behind this construction are straightforward, specific computations are numerically tedious. (This is acceptable, however, since numerical computations can be left to the computer.) In the remainder of this section we address general principles and then provide an example to illustrate how the viewing operation with depth clipping is implemented.

First, we discuss the change of coordinates T_1 (see Figure 5.77): Given the viewplane reference point P, the viewplane normal vector $\mathbf{n}$, the viewplane up-direction vector $\mathbf{v}$, and the relative coordinates of the vertices of the viewplane window (in the coordinate system constructed in Section 1), the world coordinates of the vertices of the viewplane window can be determined (see Section 2, General Perspective Projection). Given the center of perspectivity and the equations of the hither and yon planes, the vertices and directions of the edges of the view volume can be determined. Now (see Figure 5.78) given the directions I_x, I_y, and I_z of the edges and the vertices O and U of the view volume, the Fundamental Theorem of Projective Geometry for Projective Three-Space provides an algorithm for constructing an invertible 4×4 matrix M for which

$$[\langle 1, 0, 0, 0 \rangle M] = I_x$$
$$[\langle 0, 1, 0, 0 \rangle M] = I_y$$
$$[\langle 0, 0, 1, 0 \rangle M] = I_z$$
$$[\langle 0, 0, 0, 1 \rangle M] = O$$
$$[\langle 1, 1, 1, 1 \rangle M] = U$$

Since M represents T_1^{-1}, M^{-1} represents T_1.

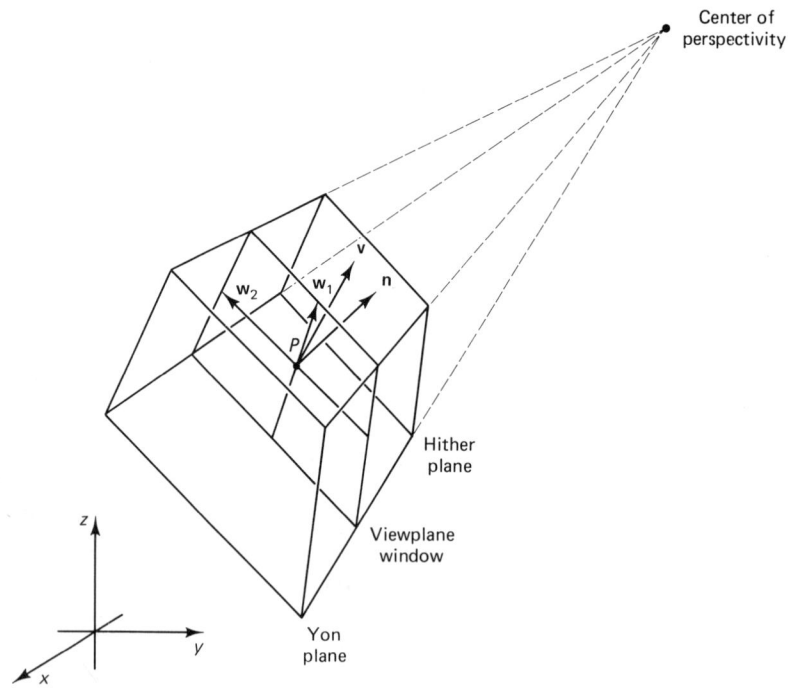

Figure 5.77 Determining the coordinates of the vertices of the viewplane window.

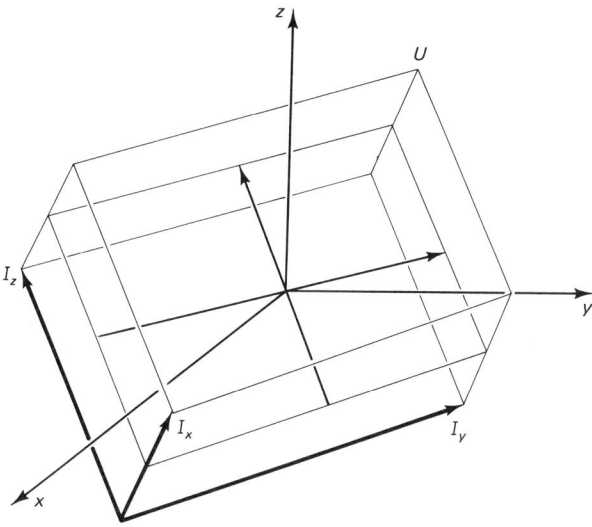

Figure 5.78 Determining the change of coordinates.

Let $[x, y, z, w]$ denote homogeneous coordinates in the original coordinate system and $[r, s, t, q]$ denote the homogeneous coordinates in the new (transformed) coordinate system.

Of the eight vertices of the view volume, we use five to compute T_1. These five vertices have the property that no four are coplanar. As long as this property is satisfied, any five of the eight vertices of the view volume can be used to construct T_1. The only fundamental difference between any of these choices is the ideal point $I_r[1, 0, 0, 0]$, $I_s[0, 1, 0, 0]$, or $I_t[0, 0, 1, 0]$ to which the center of perspectivity is mapped and the axis to which the image of the viewplane is perpendicular.

For the choice illustrated in Figure 5.78, the center of perspectivity is mapped to $I_r[1, 0, 0, 0]$, and the image of the viewplane is perpendicular to the r-axis. Thus depth clipping can be performed as follows: Given a point $P(x, y, z) = P[x, y, z, 1]$ (in the original coordinates), consider the r-coordinate of the representative $\langle r, s, t, 1 \rangle$ of $T_1(P) = [\langle x, y, z, 1 \rangle M^{-1}]$ whose last coordinate is 1. If this r-coordinate is between 0 and 1, the point should be plotted; if it is less than 0 or greater than 1, the point should not be plotted. (At no extra expense, lateral clipping can be done similarly by considering the s- and t-coordinates of $\langle r, s, t, 1 \rangle$.)

If a point is to be plotted, we could use its original coordinates and proceed as we did earlier (see Section 2); it is more efficient, however, to use the coordinates as transformed by T_1: If the Cartesian equation of the viewplane is $r = k$, then T_2 is given by

$$T_2([r, s, t, 1]) = [k, s, t, 1]$$

If T_μ is the parametrization of the viewplane given by

$$T_\mu([s, t, q]) = [k, s, t, q]$$

Sec. 7   Depth Clipping

then
$$T_3([k, s, t, 1]) = T_\mu^{-1}([k, s, t, 1]) = [s, t, 1]$$
Thus
$$T_3(T_2([r, s, t, 1])) = [s, t, 1]$$
In other words, if the x-coordinate of $T_1(P)$ is between 0 and 1, the s- and t-coordinates of $T_1(P)$ are the normalized device coordinates of the point to be plotted: If (see Section 4.1) the coordinates of the vertices of the viewport are (u_0, v_0), (u_1, v_0), (u_1, v_1) and (u_0, v_1), then the device coordinates (u, v) of the point to be plotted are
$$u = s(u_1 - u_0) + u_0$$
$$v = t(v_1 - v_0) + v_0$$

We now present a specific example illustrating how the viewing operation with depth clipping is implemented. In this example (see Figure 5.79) the viewing mechanism is specified as follows:

```
              ⋮
DefineCenterOfPerspectivitySpherical(4, π/4, π/4)
DefineCenterOfObject(0,0,0)
DefineViewplaneDistance(0)
DefineViewplaneUpDirection(0,0,1)
DefineWindow(-2,-2;2,2)
DefineViewport(U0,V0;U1,V1)
DefineHitherPlaneDistance(1)
DefineYonPlaneDistance(-1)
              ⋮
```

(the hither and yon planes are perpendicular to the principal line of sight and HitherPlaneDistance and YonPlaneDistance, respectively, from the center of the object).

The Cartesian coordinates of the center of perspectivity are $(2, 2, \sqrt{2})$. Hence the unit viewplane normal vector is $\langle \frac{1}{2}, \frac{1}{2}, \sqrt{2}/2 \rangle$. Since the viewplane up-direction is $\langle 0, 0, 1 \rangle$, the direction of the positive x-axis in the viewplane is $\langle -\sqrt{2}/2, \sqrt{2}/2, 0 \rangle$ and the direction of the positive y-axis is $\langle -\frac{1}{2}, -\frac{1}{2}, \sqrt{2}/2 \rangle$. It follows that the world coordinates of the vertices of the window are

Lower left vertex: $(1 + \sqrt{2}, 1 - \sqrt{2}, -\sqrt{2})$
Lower right vertex: $(1 - \sqrt{2}, 1 + \sqrt{2}, -\sqrt{2})$
Upper right vertex: $(-1 - \sqrt{2}, -1 + \sqrt{2}, \sqrt{2})$
Upper left vertex: $(-1 + \sqrt{2}, -1 - \sqrt{2}, \sqrt{2})$

Hence
$$I_x = I_x[1 - \sqrt{2}, 1 + \sqrt{2}, 3\sqrt{2}, 0]$$
$$I_y = I_y[-\sqrt{2}, \sqrt{2}, 0, 0]$$

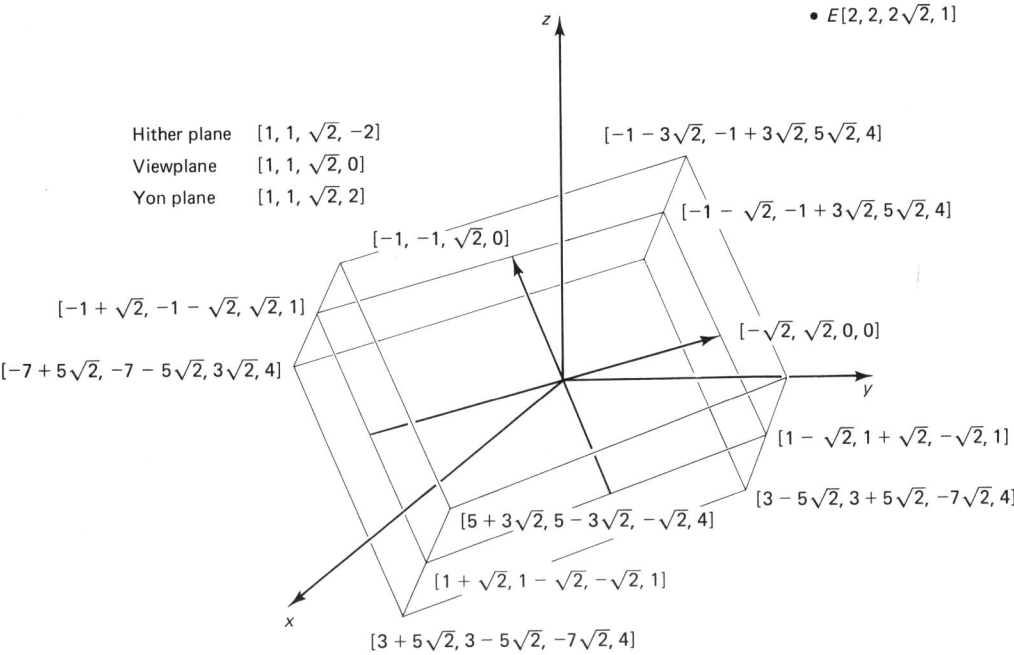

Figure 5.79 A depth clipping example.

$$I_z = I_z[-1, -1, \sqrt{2}, 0]$$

Since the equation of the hither plane is

$$\frac{x}{2} + \frac{y}{2} + \frac{\sqrt{2}z}{2} = 1$$

and the equation of the yon plane is

$$\frac{x}{2} + \frac{y}{2} + \frac{\sqrt{2}z}{2} = -1$$

it follows that

$$O = O[3 + 5\sqrt{2}, 3 - 5\sqrt{2}, -7\sqrt{2}, 4]$$
$$U = U[-1 - 3\sqrt{2}, -1 + 3\sqrt{2}, 5\sqrt{2}, 4]$$

Hence there are constants k_1, k_2, k_3, and k_4 for which, if

$$M = \begin{pmatrix} k_1(1 - \sqrt{2}) & k_1(1 + \sqrt{2}) & k_1(3\sqrt{2}) & 0 \\ k_2(-\sqrt{2}) & k_2(\sqrt{2}) & 0 & 0 \\ k_3(-1) & k_3(-1) & k_3(\sqrt{2}) & 0 \\ k_4(3 + 5\sqrt{2}) & k_4(3 - 5\sqrt{2}) & k_4(-7\sqrt{2}) & k_4(4) \end{pmatrix}$$

then $[\langle 1, 1, 1, 1\rangle M] = [-1 - 3\sqrt{2}, -1 + 3\sqrt{2}, 5\sqrt{2}, 4]$. We find that $k_1 = 2$, $k_2 = 6$, $k_3 = 6$, and $k_4 = 1$. Thus

$$M = \begin{pmatrix} 2 - 2\sqrt{2} & 2 + 2\sqrt{2} & 6\sqrt{2} & 0 \\ -6\sqrt{2} & 6\sqrt{2} & 0 & 0 \\ -6 & -6 & 6\sqrt{2} & 0 \\ 3 + 5\sqrt{2} & 3 - 5\sqrt{2} & -7\sqrt{2} & 4 \end{pmatrix}$$

and the matrix representing T_1 is

$$M^{-1} = \begin{pmatrix} 0.0625 & -0.0798 & -0.0625 & 0.0000 \\ 0.0625 & 0.0381 & -0.0625 & 0.0000 \\ 0.0884 & -0.0295 & 0.0295 & 0.0000 \\ 0.1250 & 0.1667 & 0.1667 & 0.2500 \end{pmatrix}$$

In other words, $T_1([x, y, z, 1]) = [\langle x, y, z, 1\rangle M^{-1}] = [r, s, t, 1]$, where

$$r = 0.2500x + 0.2500y + 0.3536z + 0.5000$$
$$s = -0.3190x + 0.1524y - 0.1179z + 0.6667$$
$$t = -0.2500x - 0.2500y + 0.1179z + 0.6667$$

Given $P(x, y, z)$, if one of r, s, or t is less than 0 or greater than 1, no point is to be plotted. (Checking r is depth clipping; checking s and t is lateral clipping.) If r, s, and t are all between 0 and 1, plot the point with device coordinates

$$u = s(u_1 - u_0) + u_0$$
$$v = t(v_1 - v_0) + v_0$$

SECTION 7 EXERCISES

1. In each of the following cases, derive the depth clipping formulas described in this section.
   (a) CenterOfPerspectivitySpherical$(10, 0, \pi/2)$,
       CenterOfObject$(0,0,0)$,
       ViewplaneDistance(0),
       ViewplaneUpDirection$(0,0,1)$,
       Window$(-2,-2;2,2)$,
       HitherPlaneDistance(4),
       YonPlaneDistance(-4)
   (b) CenterOfPerspectivitySpherical$(0,0,0)$,
       CenterOfObject$(0,0,10)$,
       ViewplaneDistance(0),
       ViewplaneUpDirection$(1,0,0)$,
       Window$(-2,-2;2,2)$,
       HitherPlaneDistance(4),
       YonPlaneDistance(-4)
   (c) CenterOfPerspectivitySpherical$(10, \pi/4, \pi/4)$,

```
            CenterOfObject(0,0,0),
            ViewplaneDistance(1),
            ViewplaneUpDirection(0,0,1),
            Window(-2,-2;2,2),
            HitherPlaneDistance(4),
            YonPlaneDistance(-4)
        (d) CenterOfPerspectivitySpherical(10,π/4,π/4),
            CenterofObject(0,0,0)
            ViewplaneDistance(0),
            ViewplaneUpDirection(1,0,1),
            Window(-2,-2;2,2),
            HitherPlaneDistance(4),
            YonPlaneDistance(-4)
```

2. (a) Provide a convincing argument that explains why T_1 takes the center of perspectivity to an ideal point and the viewplane to a plane perpendicular to a coordinate axis.
   (b) Provide a convincing argument that explains why the original perspective projection corresponds, in the new coordinate system, to projection parallel to a coordinate axis.
3. (a) Write a pseudocode program whose input is the information necessary to perform depth clipping and whose output is the depth clipping transformation.
   (b) Implement the pseudocode program written in part (a).
4. How could the method presented in this section be modified to allow for a different rectangle in the parametrizing plane? In particular, how could it be modified to allow for a rectangle in the parametrizing plane whose vertices have coordinates $(-1, -1)$, $(1, -1)$, $(1, 1)$, and $(-1, 1)$?

SECTION 8: AN ATTITUDE INDICATOR SIMULATION

The *attitude indicator* of an aircraft (also known as the *artificial horizon indicator*) is an instrument that displays an artificial horizon and some "grid lines on the ground." An attitude indicator is actually a float that is maintained level with respect to the ground by a gyroscope. The artificial horizon is painted on this float, and by referring to it the pilot can judge the inclination and angle of bank of the aircraft. The grid lines, which are also painted on this float, indicate which side of the horizon is ground and which is sky. The standard attitude indicator does not give any information about altitude or the direction of flight.

An enhanced version of the attitude indicator is also present on some military aircraft. It does give information about altitude and the direction of flight. It displays the horizon on a small CRT on the instrument panel and also lines on the ground side of the horizon that approximate what the pilot would see if he or she were flying over flat country marked with grid lines. These lines become closer together as the pilot climbs, and they turn as the pilot turns. This enhanced attitude indicator gets its information from inertial guidance equipment (gyroscopes) and radar. (It can also display additional information about the local terrain, a landing strip, or a target, but we ignore this here.)

In this section we develop a program to simulate the portion of an enhanced attitude indicator that draws the horizon and the images of the grid lines. This program differs from a standard flight simulator in that it does not draw the image of any feature on the ground other than the grid lines. The program accepts as input the parameters describing the current position and attitude of the aircraft and produces a display that would be seen by the pilot (see Figure 5.80).

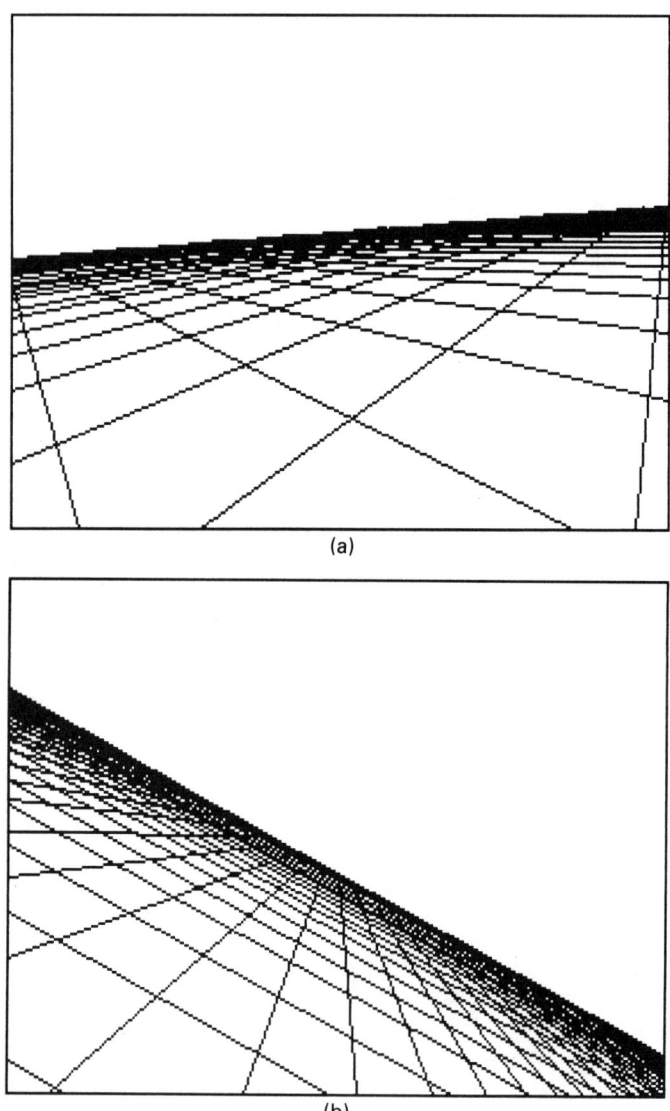

Figure 5.80 Several views produced by our simulated enhanced attitude indicator. (a) $X_0 = 0$, $Y_0 = 0$, $A = 5,000$, $\psi = 45°$, $\theta = -10°$, $\phi = 5°$. (b) $X_0 = 1,000$, $Y_0 = 3,000$, $A = 10,000$, $\psi = 0°$, $\theta = 0°$, $\phi = -30°$. (c) $X_0 = 0$, $Y_0 = 0$, $A = 20,000$, $\psi = 30°$, $\theta = 0°$, $\phi = 100°$. (d) $X_0 = 3,000$, $Y_0 = 2,000$, $A = 15,000$, $\psi = -20°$, $\theta = -110°$, $\phi = 0°$.

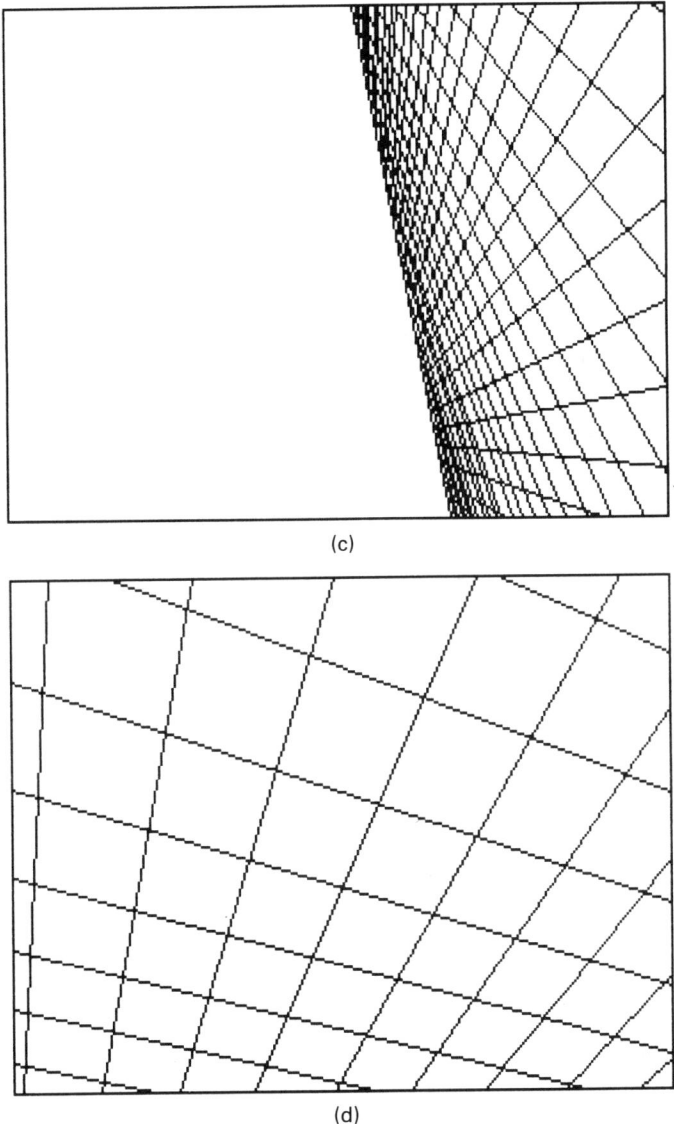

Figure 5.80 Continued.

This program might be used as a subroutine by another program that simulates the position and attitude of the aircraft. This latter program might, in turn, be "flown" by joysticks. With this consideration in mind, the issue of response time becomes important. Thus many of our design decisions are based on the desire to display a picture as fast as possible. In particular, we perform most of the computations in closed form and leave only drawing to the computer. (As a consequence, the program uses only the two-dimensional graphics library.)

The approach taken in this section differs from that of other sections in that we first find the homogeneous coordinates of the lines to be drawn rather than the points to be plotted. Only in the very last step do we return to the previous approach to draw the visible portion of a line by connecting two points on it.

We assume the earth is flat and that the only visible features on it are grid lines every 5000 ft running north-south and east-west. This *ground plane* is seen through a window in a viewplane π, which is an imaginary plane one foot in front of the pilot's eye. The window is 2 ft square, with the origin in the center. Figure 5.81 shows the tranformations whose matrices we must find. The matrix M_G represents a parametrization of the ground plane, M_1 represents perspective projection to the viewplane, M_π^+ represents inverse parametrization of the viewplane, and $M = M_G M_1 M_\pi^+$.

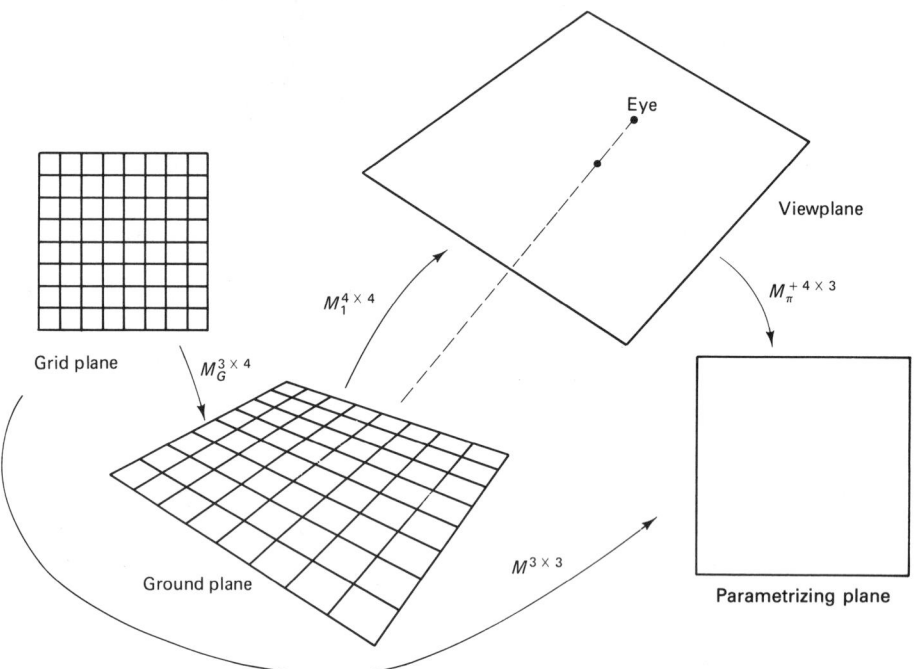

Figure 5.81 The composite transformation M from the grid plane to the parametrizing plane.

We begin by finding the matrices M_1 and M_π^+ with respect to a coordinate system for Euclidean three-space called the *reference frame* (see Figure 5.82(a)). This coordinate system has its origin at the pilot's eye, its positive x-axis north, its positive y-axis east, and its positive z-axis down. A second coordinate system, which is also useful, is the *aircraft frame* (see Figure 5.82(b)). The origin of the aircraft frame is the center of gravity of the aircraft. For simplicity we assume that the pilot's eye is the center of gravity, so that the origins of the aircraft and reference frames are the same point. The aircraft frame has its positive x-

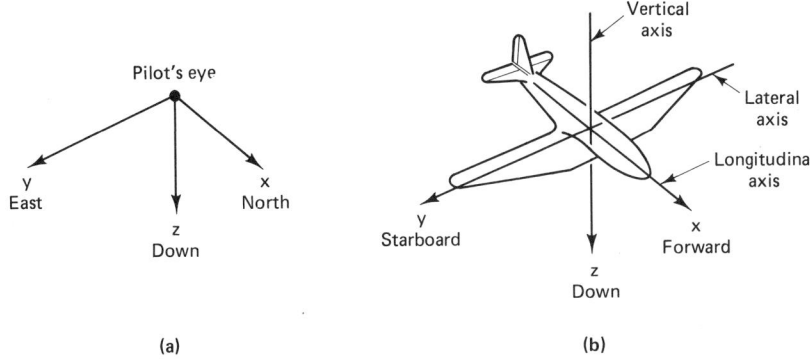

Figure 5.82 The coordinate frames.

axis (the *longitudinal axis*) forward, its positive *y*-axis (the *lateral axis*) starboard, and its positive *z*-axis (the *vertical axis*) down, relative to the aircraft. The units for both frames are feet.

The *attitude* of the aircraft is the state of the aircraft frame relative to the reference frame. We describe a specific attitude by three *Euler angles* (see Figure 5.83). The first Euler angle ψ is measured from the north toward the east in the horizontal plane. The second angle θ is measured up from the horizontal to the longitudinal axis. The third angle ϕ is measured about the longitudinal axis.

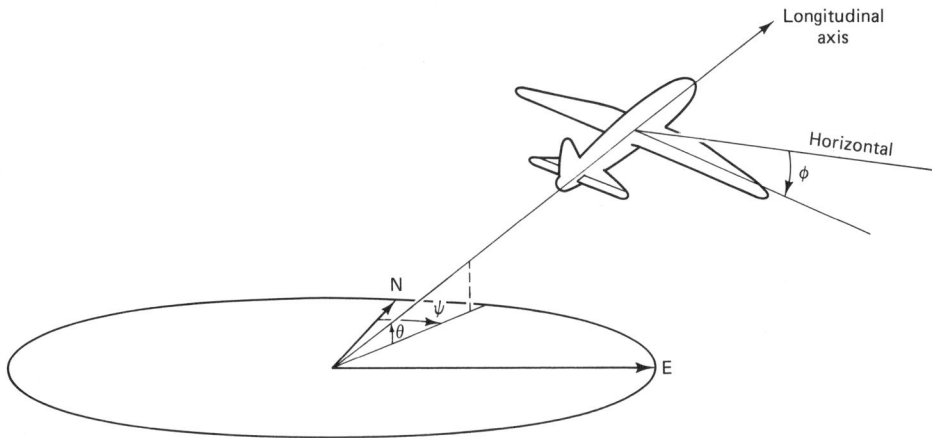

Figure 5.83 The Euler angles.

The Euler angles uniquely determine the attitude of the aircraft, but the attitude does not uniquely determine the Euler angles. In particular, if the aircraft is traveling straight up or straight down, its attitude is determined by *any* angle ψ, $\theta = \pm 90°$, and the appropriate angle ϕ. The program we are to write must accept any three angles ψ, θ, and ϕ that determine a given attitude.

Sec. 8   An Attitude Indicator Simulation

Changes in attitude are achieved by rotating the aircraft relative to an axis in the aircraft frame: Positive *yaw* turns right, positive *pitch* moves the nose up, and positive *roll* moves the right wing down (see Figure 5.84).

The Euler angles describe a sequence of rotations that transform the aircraft from the attitude of level flight headed straight north to an arbitrary attitude. Beginning with the aircraft aligned with the reference frame, we make an initial yaw through an angle ψ, then an initial pitch through an angle θ, and then an initial roll through an angle ϕ. These transformations must be made in this order, since they do not commute. (Of course it is not actually possible to fly by making such simple attitude changes; in actual flight, attitude changes must be coordinated with each other. If the aircraft is in an arbitrary attitude rather than aligned with the reference frame, the effect of a yaw, pitch, or roll on the Euler angles can be quite complicated.)

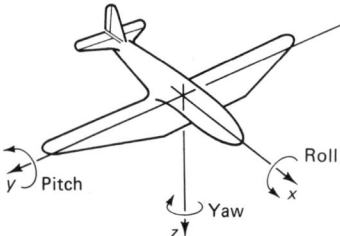

Figure 5.84 Changing the attitude.

The matrix that represents an initial yaw through the angle ψ from the reference frame is

$$M_\psi = \begin{pmatrix} \cos\psi & \sin\psi & 0 & 0 \\ -\sin\psi & \cos\psi & 0 & 0 \\ 0 & 0 & 1 & 0 \\ 0 & 0 & 0 & 1 \end{pmatrix}$$

As a result of such a yaw, the aircraft has new lateral and longitudinal axes.

After making the initial yaw, we wish to pitch through the angle θ about the new lateral axis. Since the axis of rotation is now not a coordinate axis, however, we must use the technique of Exercise 3.3.22 to find the matrix that represents this transformation. The transformation can be expressed as the composition of three transformations whose matrices we know: yawing back to the original attitude, pitching through the angle θ about the original y-axis, and then yawing forward again. Thus the matrix that represents pitching through the angle θ is $M_\psi^{-1} M_\theta M_\psi$, where

$$M_\theta = \begin{pmatrix} \cos\theta & 0 & -\sin\theta & 0 \\ 0 & 1 & 0 & 0 \\ \sin\theta & 0 & \cos\theta & 0 \\ 0 & 0 & 0 & 1 \end{pmatrix}$$

The composition of yawing and pitching is therefore represented by the matrix $M_\psi(M_\psi^{-1}M_\theta M_\psi) = M_\theta M_\psi$. As a result of yawing and pitching, the aircraft has new lateral, longitudinal, and vertical axes.

The matrix that represents a roll through the angle ϕ can be found the same way we found the matrix that represents the pitch. This tranformation is also the composition of restoring the original attitude, rolling through the angle ϕ about the x-axis, and then moving forward to the desired attitude. Thus the initial roll through the angle ϕ is represented by $(M_\theta M_\psi)^{-1}M_\phi(M_\theta M_\psi)$, where

$$M_\phi = \begin{pmatrix} 1 & 0 & 0 & 0 \\ 0 & \cos\phi & \sin\phi & 0 \\ 0 & -\sin\phi & \cos\phi & 0 \\ 0 & 0 & 0 & 1 \end{pmatrix}$$

The matrix M_R that represents the composition of yaw, pitch, and roll through the Euler angles is therefore

$$M_R = M_\phi M_\theta M_\psi$$

the product of the matrices that represent yaw, pitch, and roll about the reference frame axes but in the *reverse* order.

We now compute

$$M_R = \begin{pmatrix} \cos\psi\cos\theta & \sin\psi\cos\theta & -\sin\theta & 0 \\ \cos\psi\sin\theta\sin\phi - \sin\psi\cos\phi & \sin\psi\sin\theta\sin\phi + \cos\psi\cos\phi & \cos\theta\sin\phi & 0 \\ \cos\psi\sin\theta\cos\phi + \sin\psi\sin\phi & \sin\psi\sin\theta\cos\phi - \cos\psi\sin\phi & \cos\theta\cos\phi & 0 \\ 0 & 0 & 0 & 1 \end{pmatrix}$$

Since M_R is orthogonal, $M_R^{-1} = M_R^T$.

The viewplane π would have the equation $x = 1$ and homogeneous coordinates $[1, 0, 0, -1]$ if the aircraft were in level flight headed straight north. In general, the viewplane is the image of the plane whose homogeneous coordinates are $[1, 0, 0, -1]$, under rotation by the transformation represented by M_R. So the homogeneous coordinates (with respect to the reference frame) of the viewplane are

$$\left[M_R^{-1}\begin{pmatrix} 1 \\ 0 \\ 0 \\ -1 \end{pmatrix}\right] = \begin{bmatrix} \cos\psi\cos\theta \\ \sin\psi\cos\theta \\ -\sin\theta \\ -1 \end{bmatrix}$$

To find the projection matrix M_1, we use Proposition 3.2.1, the homogeneous coordinates of the eye, $[0, 0, 0, 1]$, and the homogeneous coordinates of the viewplane to obtain

$$M_1 = \begin{pmatrix} 1 & 0 & 0 & \cos\psi\cos\theta \\ 0 & 1 & 0 & \sin\psi\cos\theta \\ 0 & 0 & 1 & -\sin\theta \\ 0 & 0 & 0 & 0 \end{pmatrix}$$

The plane $x = 1$ is parametrized (see Figure 5.85) by

Sec. 8 An Attitude Indicator Simulation

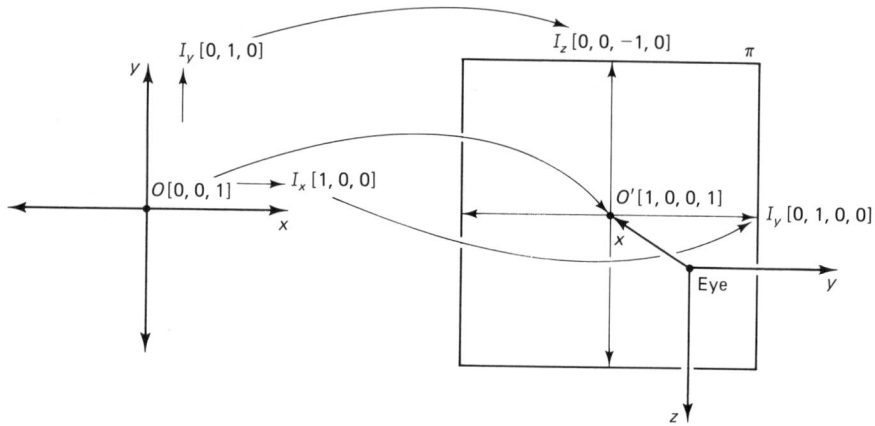

Figure 5.85 Parametrizing the plane whose equation is $x = 1$.

$$M_x = \begin{pmatrix} 0 & 1 & 0 & 0 \\ 0 & 0 & -1 & 0 \\ 1 & 0 & 0 & 1 \end{pmatrix}$$

The viewplane π is therefore parametrized by $M_\pi = M_x M_R$:

$$M_\pi = \begin{pmatrix} \cos\psi \sin\theta \sin\phi - \sin\psi \cos\phi & \sin\psi \sin\theta \sin\phi + \cos\psi \cos\phi & \cos\theta \sin\phi & 0 \\ -\cos\psi \sin\theta \cos\phi - \sin\psi \sin\phi & -\sin\psi \sin\theta \cos\phi + \cos\psi \sin\phi & -\cos\theta \cos\phi & 0 \\ \cos\psi \cos\theta & \sin\psi \cos\theta & -\sin\theta & 1 \end{pmatrix}$$

The inverse parametrization M_π^+ of the viewplane is thus given by $M_\pi^+ = M_R^{-1} M_x^+$. Since $M_x^+ = M_x^T (M_x M_x^T)^{-1}$,

$$M_x^+ = \begin{pmatrix} 0 & 0 & \frac{1}{2} \\ 1 & 0 & 0 \\ 0 & -1 & 0 \\ 0 & 0 & \frac{1}{2} \end{pmatrix}$$

and

$$M_\pi^+ = \begin{pmatrix} \cos\psi \sin\theta \sin\phi - \sin\psi \cos\phi & -\cos\psi \sin\theta \cos\phi - \sin\psi \sin\phi & \frac{1}{2}\cos\psi \cos\theta \\ \sin\psi \sin\theta \sin\phi + \cos\psi \cos\phi & -\sin\psi \sin\theta \cos\phi + \cos\psi \sin\phi & \frac{1}{2}\sin\psi \cos\theta \\ \cos\theta \sin\phi & -\cos\theta \cos\phi & -\frac{1}{2}\sin\theta \\ 0 & 0 & \frac{1}{2} \end{pmatrix}$$

We now find the matrix M_G of the parametrization of the ground plane by the grid plane (see Figure 5.86). We wish to map the origin of the grid plan to the point on the ground plane directly beneath the pilot's eye, the positive x-axis of the grid plane to north, and the positive y-axis to east. If the altitude of the aircraft is A, the ground plane has the equation $z = A$, so

$$M_G = \begin{pmatrix} 1 & 0 & 0 & 0 \\ 0 & 1 & 0 & 0 \\ 0 & 0 & A & 1 \end{pmatrix}$$

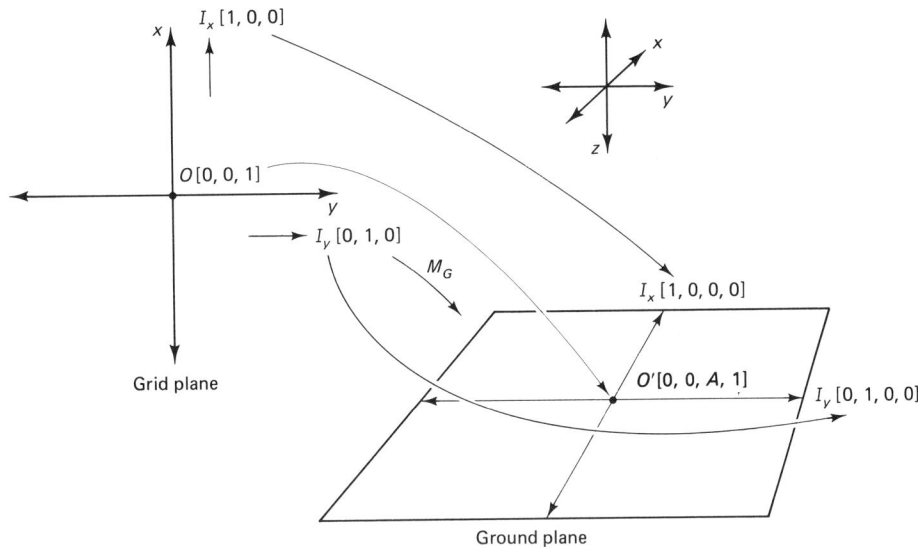

Figure 5.86 Parametrizing the ground plane.

Finally, we compute the matrix M that represents the composite transformation from the grid plane to the parametrizing plane.

$$M = \begin{pmatrix} \cos\psi\sin\theta\sin\phi - \sin\psi\cos\phi & -\cos\psi\sin\theta\cos\phi - \sin\psi\sin\phi & \cos\psi\cos\theta \\ \sin\psi\sin\theta\sin\phi + \cos\psi\cos\phi & -\sin\psi\sin\theta\cos\phi + \cos\psi\sin\phi & \sin\psi\cos\theta \\ A\cos\theta\sin\phi & -A\cos\theta\cos\phi & -A\sin\theta \end{pmatrix}$$

We also need the inverse matrix of M:

$$M^{-1} = \begin{pmatrix} \cos\psi\sin\theta\sin\phi - \sin\psi\cos\phi & \sin\psi\sin\theta\sin\phi + \cos\psi\cos\phi & \dfrac{1}{A}\cos\theta\sin\phi \\ -\cos\psi\sin\theta\cos\phi - \sin\psi\sin\phi & -\sin\psi\sin\theta\cos\phi + \cos\psi\sin\phi & -\dfrac{1}{A}\cos\theta\cos\phi \\ \cos\psi\cos\theta & \sin\psi\cos\theta & -\dfrac{1}{A}\sin\theta \end{pmatrix}$$

Using M and M^{-1} we can compute the homogeneous coordinates of the vanishing points of the grid lines, the homogeneous coordinates of the horizon, and the homogeneous coordinates of the grid lines themselves, all with respect to the reference frame.

The vanishing points in the parametrizing plane of the grid lines are the images under M of the ideal points in the grid plane whose homogeneous coordinates are [1, 0, 0] and [0, 1, 0]. Hence the north-south and east-west vanishing points are

$VP_{NS}[\cos\psi\sin\theta\sin\phi - \sin\psi\cos\phi, -\cos\psi\sin\theta\cos\phi - \sin\psi\sin\phi, \cos\psi\cos\theta]$

$VP_{EW}[\sin\psi\sin\theta\sin\phi + \cos\psi\cos\phi, -\sin\psi\sin\theta\cos\phi + \cos\psi\sin\phi, \sin\psi\cos\theta]$

The horizon is the image under M of the ideal line in the grid plane. Thus the homogeneous coordinates of the horizon in the parametrizing plane are

$$\left[M^{-1}\begin{pmatrix}0\\0\\1\end{pmatrix}\right] = \begin{bmatrix}\cos\theta\sin\phi\\-\cos\theta\cos\phi\\-\sin\theta\end{bmatrix}$$

A north-south grid line in the grid plane has an equation of the form $y = m$ and homogeneous coordinates $[0, 1, -m]$. The homogeneous coordinates of the image of this grid line in the parametrizing plane are

$$\left[M^{-1}\begin{pmatrix}0\\1\\-m\end{pmatrix}\right] = \begin{bmatrix}\sin\psi\sin\theta\sin\phi + \cos\psi\cos\phi - \dfrac{m}{A}\cos\theta\sin\phi\\-\sin\psi\sin\theta\cos\phi + \cos\psi\sin\phi + \dfrac{m}{A}\cos\theta\cos\phi\\\sin\psi\cos\theta + \dfrac{m}{A}\sin\theta\end{bmatrix}$$

The homogeneous coordinates of the images of the east-west grid lines can be found similarly. These coordinates are not actually used in the program, however, since the routine that draws the first set of grid lines can be reused to draw the second set merely by increasing the angle ψ by 90°.

The grid lines in the grid plane have the equations $x = -X_0 + 5000K$ and $y = -Y_0 + 5000K$, where $0 \leq -X_0, Y_0 < 5000$ are the Euclidean coordinates of a point of intersection of two grid lines and K is an integer (see Figure 5.87). Figure 5.88 illustrates a typical picture we now see if we graph the horizon and the images of the grid lines for $K = -25, \ldots, 25$.

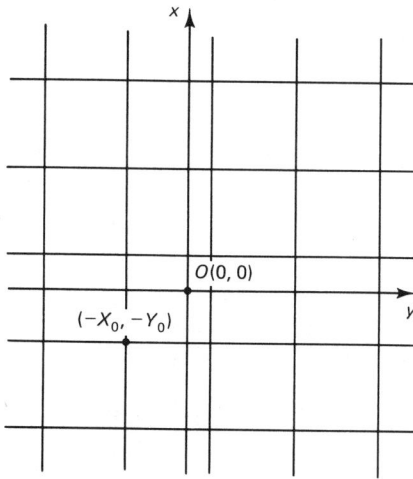

Figure 5.87 The grid lines.

Figure 5.88 suggests two problems we must solve.

1. We need to distinguish between sky and ground and not draw those portions of the images of the grid lines that lie in the sky.
2. Even with the images of 51 grid lines in each direction, there are unsightly

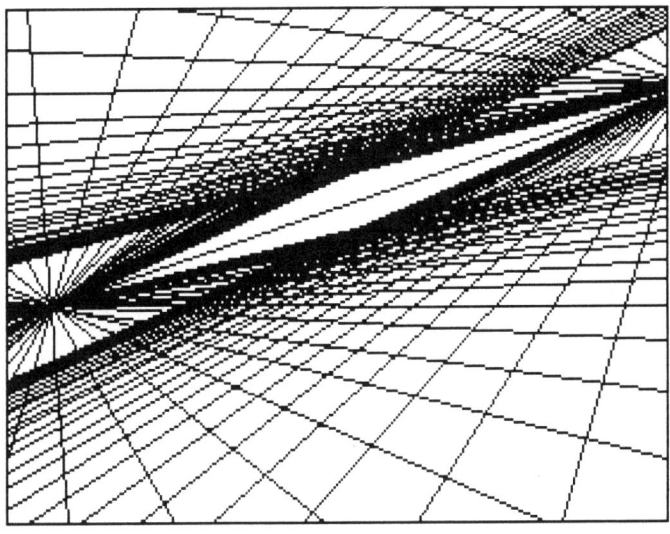

Figure 5.88 Preliminary graph of the horizon and the images of the grid lines ($X_0 = 0$, $Y_0 = 0$, $A = 15000$, $\psi = 40°$, $\theta = -10°$, $\phi = 20°$).

blank spaces formed at the sides of the display under the horizon, which would be filled in only if we were to graph many more grid lines.

Speed constraints make it impossible to solve the second problem by drawing enough grid lines to fill in the blank spaces, so we instead replace the actual horizon by an artificial horizon. The artificial horizon is sufficiently displaced on the ground side of the true horizon that the only blank spaces are in the sky and hence are not visible (after we stop drawing lines in the sky). The artificial horizon is the image of an artificial ideal line in the grid plane, which—by experimentation—we place at a distance of 100,000 ft, or approximately 20 mi, from the origin. (This solution to the second problem somewhat counteracts the assumption of a flat earth made earlier, although the resulting picture becomes less realistic above approximately 40,000 ft.)

The equation of the artificial ideal line in the grid plane (see Figure 5.89) is

$$x \cos \psi + y \sin \psi = 100{,}000$$

The homogeneous coordinates of the artificial horizon in the parametrizing plane are thus

$$\left[M^{-1} \begin{pmatrix} \cos \psi \\ \sin \psi \\ -100{,}000 \end{pmatrix} \right] = \begin{bmatrix} -\sin \phi \\ \cos \phi \\ -\mathrm{DAH} \end{bmatrix}$$

where DAH, the distance from the origin to the artificial horizon in the parametrizing plane, is

$$\mathrm{DAH} = \frac{A \cos \theta + 100{,}000 \sin \theta}{A \sin \theta - 100{,}000 \cos \theta}$$

Sec. 8 An Attitude Indicator Simulation

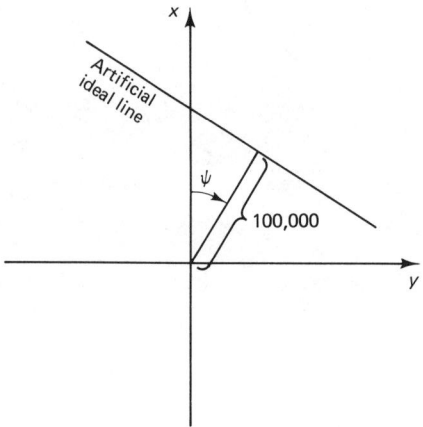

Figure 5.89 The artificial ideal line in the grid plane.

Thus the equation of the artificial horizon in the parametrizing plane is

$$-x \sin \phi + y \cos \phi = \text{DAH}$$

Having obtained Figure 5.88 and the equation of the artificial horizon, we have gone about as far as we can using projective geometry.

The first problem described above, that of differentiating between sky and ground, is a Euclidean problem, which we can solve by using the equation of the artificial horizon. The vector $\mathbf{u}' = \langle -\sin \phi, \cos \phi \rangle$ is perpendicular to the artificial horizon, the vector $\mathbf{u} = \langle \cos \phi, \sin \phi \rangle$ is parallel to the artificial horizon, and—because of the way the Euler angles were defined—$\mathbf{u}'$ points into the sky (see Figure 5.90). The point on the artificial horizon nearest the origin has the position vector $\text{DAH}\mathbf{u}'$. For any other point $P(x, y)$, its position vector $\mathbf{OP}$ can be expressed

$$\mathbf{OP} = \text{DAH}\mathbf{u}' + a\mathbf{u} + b\mathbf{u}'$$

for some scalars a and b (see Figure 5.91). The point P is in the sky if $b > 0$ and on the ground if $b < 0$. Since

$$\mathbf{OP} \cdot \mathbf{u}' = \text{DAH} + b$$

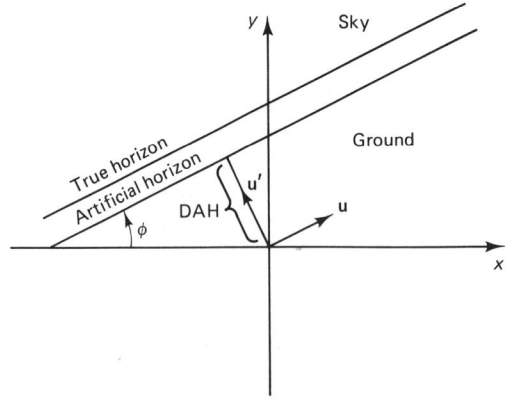

Figure 5.90 The artificial horizon in the parametrizing plane.

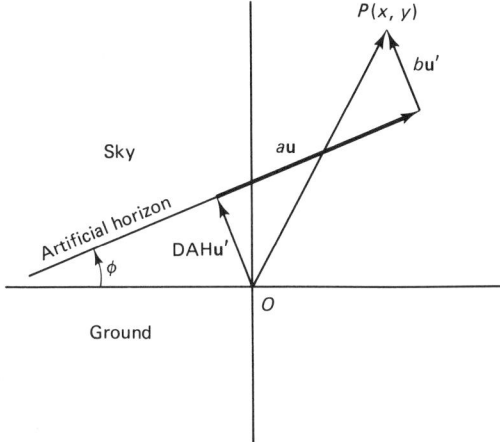

Figure 5.91 Determining sky and ground.

we obtain

$$b = -x \sin \phi + y \cos \phi - \text{DAH}$$

In summary, $P(x, y)$ is in the sky if $-x \sin \phi + y \cos \phi > \text{DAH}$ and on the ground if $-x \sin \phi + y \cos \phi < \text{DAH}$.

There is one more subject to discuss before beginning the program. As indicated earlier, different Euler angles can describe the same attitude for the aircraft. For example, the position of the aircraft in Figure 5.92 is described either by $\psi = 0°$, $\theta = -135°$, $\phi = 0°$ or by $\psi = 180°$, $\theta = -45°$, $\phi = 180°$. The program is much easier to write if we have control over the angle θ. We can insure that $-90° \leq \theta \leq 90°$ if we first reduce θ to an angle between $0°$ and $360°$ and then, depending on this value of θ, make substitutions for ψ, θ and ϕ

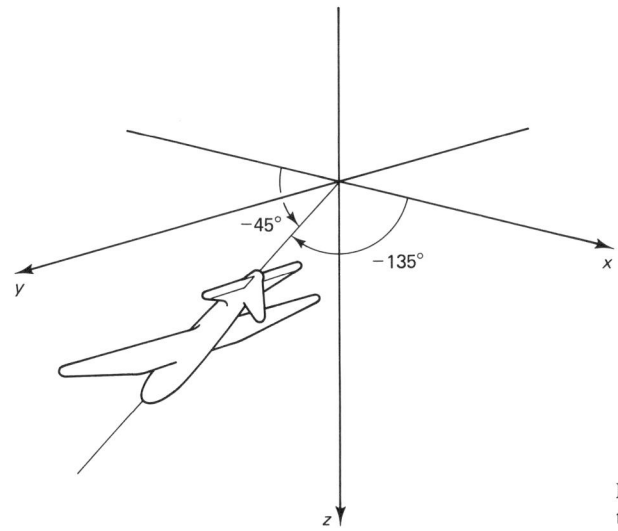

Figure 5.92 A possible attitude for the aircraft.

Sec. 8   An Attitude Indicator Simulation

from the table in Figure 5.93. (If θ is one of the boundary values, it does not matter which substitution is used.)

Our main program, for which we still need to write several subroutines, is the following. It begins with the user inputting values for the six parameters. The parameters X_0, Y_0, and A describe the location of the aircraft relative to an intersection point of two grid lines. We allow the user to input arbitrary values for X_0 and Y_0 and reduce them in the program to equivalent values between 0 and 5000.

```
PROGRAM AttitudeIndicator
  BEGIN
    Read(X0,Y0,A,Psi,Theta,Phi)
    Normalize
    ComputeTrig
    Initialize2DimensionalGraphics
    DefineWindow(-1,-K;1,K)
    DefineViewport(U0,V0;U1,V1)
    GraphicsMode
    IF Theta < 60° THEN (* the view is not necessarily all sky *)
      BEGIN
        DAH := (A*CosTheta + 100,000*SinTheta)/
               (A*SinTheta - 100,000*CosTheta)
        IF DAH > -√2 THEN (* the view is not necessarily all sky *)
          BEGIN
            IF DAH > √2 THEN (* the view is all ground *)
              BEGIN
                DrawGridLines(-Y0,CosPsi,SinPsi)
                DrawGridLines(-X0,-SinPsi,CosPsi)
              END
            ELSE (* the artificial horizon is visible *)
              BEGIN
                DrawArtificialHorizon
                DrawPartialGridLines(-Y0,CosPsi,SinPsi)
                DrawPartialGridLines(-X0,-SinPsi,CosPsi)
              END
          END
      END
    Border
    Pause for viewing picture
    TextMode
  END
```

The procedure Normalize modifies the Euler angles according to Figure 5.93, reduces X0 and Y0 to equivalent values between 0 and 5000, and checks that A is at least 1 ft and no more than 55,000 ft (for reasons explained below).

	New θ	New ψ	New ϕ
360° – 270°	$\theta - 360°$	ψ	ϕ
270° – 180°	$180° - \theta$	$\psi + 180°$	$\phi + 180°$
180° – 90°	$180° - \theta$	$\psi + 180°$	$\phi + 180°$
90° – 0°	θ	ψ	ϕ
0° – –90°	θ	ψ	ϕ
–90° – –180°	$-180° - \theta$	$\psi + 180°$	$\phi + 180°$
–180° – –270°	$-180° - \theta$	$\psi + 180°$	$\phi + 180°$
–270° – –360°	$\theta + 360°$	ψ	ϕ

Figure 5.93 Table for normalizing the angle θ.

ComputeTrig evaluates the sine and cosine of each Euler angle.

Testing whether or not Theta < 60° is a first quick check to decide if the picture is all sky. This, together with the upper bound of 55,000 ft on the altitude, guarantees that the denominator of DAH is not zero.

The flow of the program depends upon the size of DAH. The window in the parametrizing plane is contained in a circle of radius $\sqrt{2}$ (see Figure 5.94). If $-\sqrt{2} < \text{DAH} < \sqrt{2}$, then the artificial horizon crosses this circle and also possibly the display area. If $\text{DAH} < -\sqrt{2}$, the view is all sky, while if $\text{DAH} > \sqrt{2}$, the view is all ground.

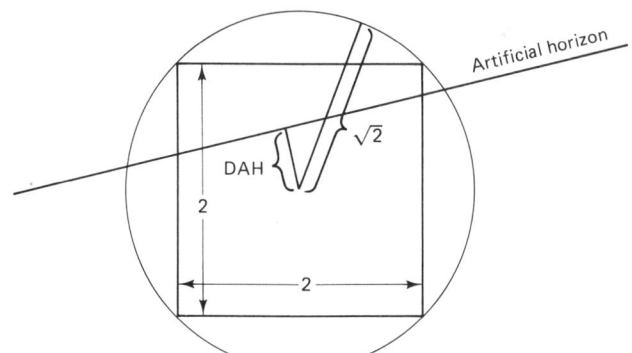

Figure 5.94 The window contained in a circle.

The remainder of the program consists of drawing the two sets of images of the grid lines, with the artificial horizon either off screen or visible. In each case, one subroutine suffices to draw grid lines, with the angle Psi increased by 90° for the second set; since $\cos(\psi + 90°) = -\sin\psi$ and $\sin(\psi + 90°) = \cos\psi$, no new trigonometric computations need to be made. Each of the procedures DrawGridLines and DrawPartialGridLines uses the common values of SinTheta, CosTheta, SinPhi and CosPhi but uses the values SinPsi and CosPsi passed to it as parameters.

The procedure DrawGridLines, used when the view is all ground, begins by computing constants used by the equations of the images of all 51 grid lines. Then the program converts the equation of the image of each grid line to polar normal form

$$c_1 x + c_2 y = c_3$$

and draws the line if it is visible. (This equation is in polar normal form if $c_1 = -\sin\theta$ and $c_2 = \cos\theta$ where θ is the inclination of the line, $0 \leq \theta < \pi$. In this case, c_3 is the directed distance from the line to the origin: if c_3 is positive then the origin and the normal vector $\langle -\sin\theta, \cos\theta \rangle$ are on opposite sides of the line, while if c_3 is negative they are on the same side. The equation of any line can be put into polar normal form by multiplying by $\pm 1/\sqrt{c_1^2 + c_2^2}$.) The procedure ConvertToPolarNormalForm also needs to trap for the ideal line and insure that the procedure DrawGridLines does not try to draw it. This can be done, for example, by returning some large number for c_3.)

```
PROCEDURE DrawGridLines(Z,CosPsi,SinPsi)
  BEGIN
    A1 := SinPsi*SinTheta*SinPhi + CosPsi*CosPhi
    B1 := -CosTheta*SinPhi
    A2 := -SinPsi*SinTheta*CosPhi + CosPsi*SinPhi
    B2 := CosTheta*CosPhi
    A3 := SinPsi*CosTheta
    B3 := SinTheta
    FOR K := -25 TO 25 DO
      BEGIN
        R := (Z + 5000*K)/A
        Convert(A1 + R*B1,A2 + R*B2,A3 + R*B3)
          ToPolarNormalForm(C1,C2,C3)
        IF |C3| < √2 THEN
          DrawFrom(C3*C1 + 2*C2,C3*C2 - 2*C1)
            To(C3*C1 - 2*C2,C3*C2 + 2*C1)
      END
  END
```

The idea used for drawing the image of a grid line is illustrated in Figure 5.95. The vector $\langle c_1, c_2 \rangle$ is perpendicular to the image of the grid line and the vector $\langle c_2, -c_1 \rangle$ is parallel to the image of the grid line. The vector $c_3 \langle c_1, c_2 \rangle$ is the position vector of the nearest point P_0 on the image of the grid line to the origin. The point P_1 with position vector $c_3 \langle c_1, c_2 \rangle + 2 \langle c_2, -c_1 \rangle$ is on the line but offscreen on one side of P_0, while the point P_2 with position vector $c_3 \langle c_1, c_2 \rangle - 2 \langle c_2, -c_1 \rangle$ is offscreen on the other side of P_0. Drawing the segment between P_1 and P_2 draws the portion of the image of the grid line which is on the screen. (Note we depend on automatic clipping here.)

The procedure DrawPartialGridLines is used when the artificial horizon is visible. It begins with the same computations as DrawGridLines but computes

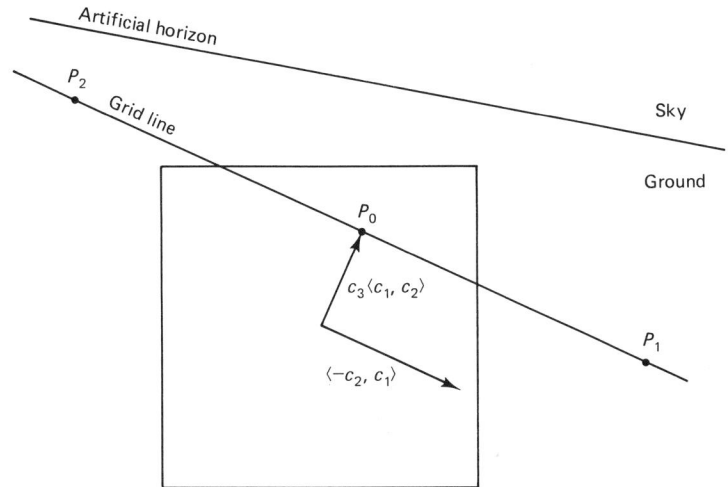

Figure 5.95 Drawing the image of a grid line when the artificial horizon is offscreen.

homogeneous coordinates of the point of intersection P of the grid line and the artificial horizon (see Proposition 2.2.7). If P is onscreen (see Figure 5.96), DrawPartialGridLines calls DrawWithPVisible; otherwise (see Figure 5.97), it calls DrawWithPNotVisible.

```
PROCEDURE DrawPartialGridLines(Z,CosPsi,SinPsi)
  BEGIN
    A1 := SinPsi*SinTheta*SinPhi + CosPsi*CosPhi
```

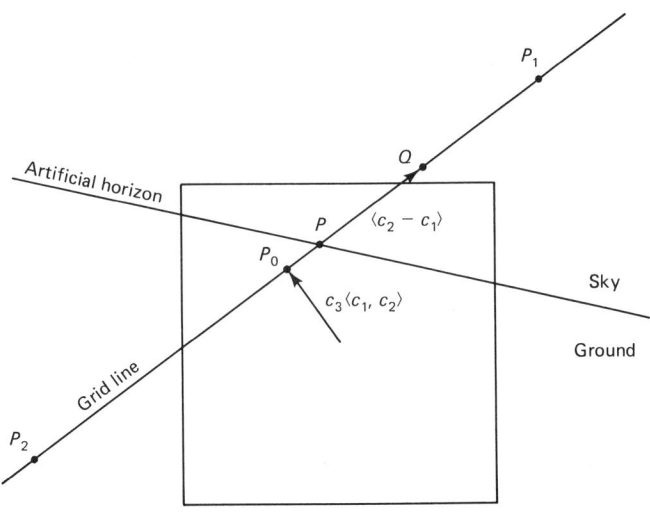

Figure 5.96 Drawing the image of a grid line that crosses the visible artificial horizon onscreen.

Sec. 8   An Attitude Indicator Simulation

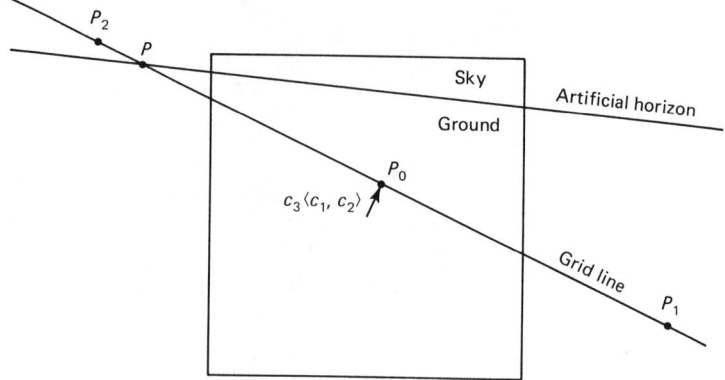

Figure 5.97 Drawing the image of a grid line that crosses the visible artificial horizon offscreen.

```
B1 := -CosTheta*SinPhi
A2 := -SinPsi*SinTheta*CosPsi + CosPsi*SinPhi
B2 := CosTheta*CosPhi
A3 := SinPsi*CosTheta
B3 := SinTheta
FOR K := -25 TO 25 DO
  BEGIN
    R := (Z + 5000*K)/A
    Convert(A1 + R*B1,A2 + R*B2, A3 + R*B3)
      ToPolarNormalForm(C1,C2,C3)
    IF |C3| < √2 THEN
      BEGIN
        P1 := -C2*DAH + C3*CosPhi
        P2 := C1*DAH + C3*SinPhi
        P3 := C1*CosPhi + C2*SinPhi
        IF P3 ≠ 0 THEN
          BEGIN
            P1 := P1/P3
            P2 := P2/P3
          END
        IF P3 ≠ 0 AND (-1 ≤ P1 ≤ 1) AND (-1 ≤ P2 ≤ 1)
          THEN DrawWithPVisible
          ELSE DrawWithPNotVisible
      END
  END
END
```

The procedure DrawWithPVisible examines the terminal point Q of the vector with components $\langle c_2, -c_1 \rangle$ and initial point P (see Figure 5.96). If Q is on the ground, then $\langle c_2, -c_1 \rangle$ points in the ground direction and the procedure draws from P to the point P_1 with position vector $c_3\langle c_1, c_2\rangle + 2\langle c_2, -c_1\rangle$; otherwise it draws from P to the point P_2 with position vector $c_3\langle c_1, c_2\rangle - 2\langle c_2, -c_1\rangle$.

The procedure DrawWithPNotVisible checks the point P_0 with position vector $c_3\langle c_1, c_2\rangle$ to determine whether the image of the grid line crosses the screen in the ground or in the sky (see Figure 5.97). If the line is on the ground, it is then drawn from P_1 to P_2 as in procedure DrawGridLines.

```
PROCEDURE DrawWithPVisible
  BEGIN
    IF -(P1 + C2)*SinPhi + (P2 - C1)*CosPhi < DAH THEN
      (* Q is on the ground *)
      DrawFrom(P1,P2)To(C3*C1 + 2*C2,C3*C2 - 2*C1)
    ELSE(* Q is in the sky *)
      DrawFrom(P1,P2)To(C3*C1 - 2*C2,C3*C2 + 2*C1)
  END

PROCEDURE DrawWithPNotVisible
  BEGIN
    IF -(C3*C1)*SinPhi + (C3*C2)*CosPhi < DAH THEN
      DrawFrom(C3*C1 + 2*C2,C3*C2 - 2*C1)
        To(C3*C1 - 2*C2,C3*C2 + 2*C1)
  END
```

SECTION 8 EXERCISES

1. Write a program to simulate the standard attitude indicator.
2. Write a program similar to the program Attitude Indicator, which draws only two parallel lines on the ground (cut off by the horizon) that simulate a runway.
3. Implement the program Attitude Indicator on a specific graphics device.
4. How could you use a "double-buffering" technique to improve the response time of the program Attitude Indicator? (In double buffering, you display one picture while preparing the next. This cannot be done with some hardware.)
5. Suppose the aircraft is flying straight up with the right wing pointed northeast. If the first Euler angle $\psi = 150°$, find values of θ and ϕ that will describe the given attitude.
6. Flying straight north, you begin changing θ and ϕ together: $\theta = 5°$, $\phi = 5°$, $\theta = 10°$, $\phi = 10°, \ldots, \theta = 180°$, $\phi = 180°$. As you do this, the horizon disappears off one corner of the screen and then reappears on another. Which corners are these? What trajectory would the aircraft be following?

Reconstruction

INTRODUCTION

In this chapter our primary interest is in object reconstruction and camera calibration. In object reconstruction we compute the coordinates of a point in projective or Euclidean three-space using information obtained from several views of the point. In camera calibration (which we view as camera reconstruction), we compute a matrix that represents a complete viewing operation using knowledge of how a small set of known points is transformed. Interestingly, both object reconstruction and camera calibration may be used in a single application: In some applications, there is a feedback loop between the two.

Reconstruction techniques are used, for example, in medical tomography to obtain three-dimensional data about an organ of the body given X-ray or ultrasound projections of the organ. They are also used in robotics, in missile guidance, and in photogrammetry.

In order to relate object reconstruction and camera calibration to computer graphics, consider the equation

$$[\mathbf{a}^{1\times 4} M^{4\times 3}] = [\mathbf{b}^{1\times 3}]$$

where $A[\mathbf{a}]$ is an object point in projective three-space and $B[\mathbf{b}]$ is its image in the projective plane under the complete viewing operation represented by the matrix M. In computer graphics, $\mathbf{b}$ is computed given $\mathbf{a}$ and M. In object reconstruction, $\mathbf{a}$ is computed given $\mathbf{b}$ and M. (Actually this is not possible with just one equation, but it is possible if two or more sets of values of $\mathbf{b}$ and M are given.) In camera calibration, M is computed given $\mathbf{a}$ and $\mathbf{b}$. (This again is not actually possible with just one equation, but it is possible if at least six sets of values of **a** and corresponding values of **b** are given.)

In Section 1 we illustrate how to obtain the least squares solution to an overdetermined and inconsistent system $\mathbf{a}M = \mathbf{b}$ of linear equations using the right inverse M^+ of the coefficient matrix M. In Section 2 we use least squares to solve the object reconstruction problem. In Section 3 we use least squares to solve the camera calibration problem. In Section 4 we bring these reconstruction methods together with other topics of projective geometry and apply them to the study of photogrammetric rectification: In photogrammetric rectification, we begin with a photograph taken with a camera that was tilted and remove the effects of the tilt.

The complete theory of reconstruction requires statistical and mathematical analysis beyond the scope of this book. This chapter is intended just as an introduction to this subject. For further reference consult [Ballard and Brown] and [Barnard and Fischler].

SECTION 1: LEAST SQUARES AND THE RIGHT INVERSE

Reconstruction techniques ultimately require the least squares solution of overdetermined and inconsistent systems of linear equations. Such a system is of the form

$$m_{11}a_1 + \cdots + m_{1k}a_k = b_1$$
$$\vdots$$
$$m_{N1}a_1 + \cdots + m_{Nk}a_k = b_N$$

where $N > k$. In matrix form, this equation becomes

$$(a_1 \cdots a_k)\begin{pmatrix} m_{11} & \cdots & m_{N1} \\ \vdots & & \vdots \\ m_{1k} & \cdots & m_{Nk} \end{pmatrix} = (b_1 \cdots b_N)$$

or $\mathbf{a}M = \mathbf{b}$, where

$$\mathbf{a} = (a_1 \cdots a_k)$$

$$M = \begin{pmatrix} m_{11} & \cdots & m_{N1} \\ & & \vdots \\ m_{1k} & \cdots & m_{Nk} \end{pmatrix}$$

and

$$\mathbf{b} = (b_1 \cdots b_N)$$

Since this system is inconsistent, it has no solution in the standard sense of the term solution. The object of least squares is to find the best possible solution $\mathbf{a} = \mathbf{a}_0$ of such a system.

There are several different ways to interpret what best possible solution means.

One way is to think of right multiplication by M as a transformation from Euclidean k-space to Euclidean N-space. The image of (all of) Euclidean k-

space under this transformation is a linear subspace π of Euclidean N-space. Since $\mathbf{a}M = \mathbf{b}$ is inconsistent, $\mathbf{b}$ does not lie in π (see Figure 6.1). The best possible solution is that $\mathbf{a}_0$ for which $\mathbf{a}_0 M$ is the orthogonal projection of $\mathbf{b}$ onto π: This is the solution for which the difference (or error) $|\mathbf{a}M - \mathbf{b}|$ over all $\mathbf{a}$ in Euclidean k-space is minimal. (Recall that the minimal distance between a point and a plane is the perpendicular distance from the point to the plane.)

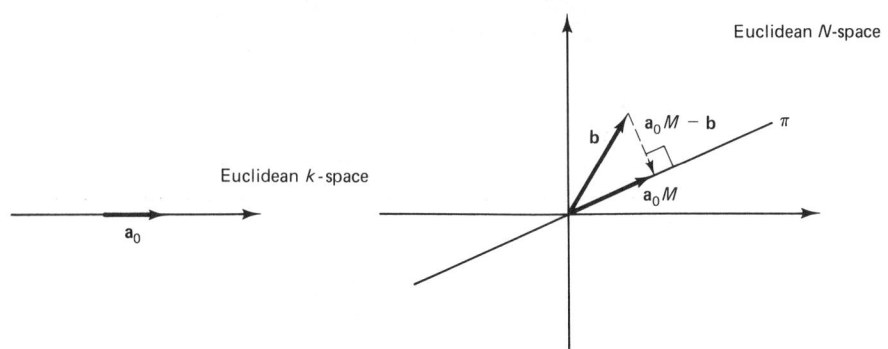

Figure 6.1 Least squares — smallest error vector.

Another way to interpret best possible solution is to observe that the equation of a k-plane in Euclidean $(k + 1)$-space that passes through the origin has an equation of the form

$$z = a_1 y_1 + \cdots + a_k y_k$$

for constants $a_1, \ldots, a_k$. If we think of

$$\{(m_{i1}, \ldots, m_{ik}, b_i) \mid i = 1, \ldots, N\}$$

as the coordinates of N data points in Euclidean $(k + 1)$-space, then the best possible solution corresponds to the k-plane (that is, the a_i's) that passes through the origin and best approximates the data in the sense that the sum of the squares of the vertical distances from each point to the k-plane is minimal (see Figure 6.2).

These interpretations are the same since in both cases we are trying to minimize

$$|\mathbf{a}M - \mathbf{b}| = \sqrt{\sum_{i=1}^{N} [b_i - (m_{i1} a_1 + \cdots + m_{ik} a_k)]^2}$$

The following result is the primary result of this section.

1.1 Proposition. The least squares solution $\mathbf{a}_0$ to the overdetermined and inconsistent inhomogeneous system $\mathbf{a}M = \mathbf{b}$ satisfies the equation

$$\mathbf{a}_0 M M^T = \mathbf{b} M^T$$

If M has rank k, then MM^T is invertible and

$$\mathbf{a}_0 = \mathbf{b} M^T (M M^T)^{-1} = \mathbf{b} M^+$$

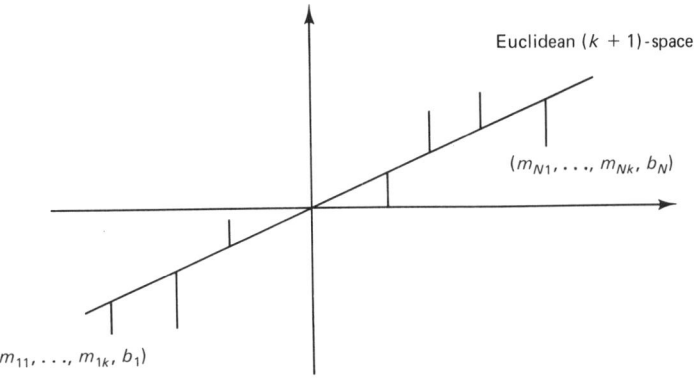

Figure 6.2 Least squares — best fitting k-plane.

Before proving this result we consider an example.

1.2 Example. The system
$$(a_1 \ a_2)\begin{pmatrix} 1 & 0 & -1 & 1 & 1 \\ 0 & 1 & 1 & 1 & 1 \end{pmatrix} = (2 \ 3 \ 1 \ 4 \ 6)$$
is overdetermined and inconsistent. It is inconsistent since the fourth and fifth equations are
$$a_1 + a_2 = 4$$
$$a_1 + a_2 = 6$$
Geometrically solving this system can be interpreted as finding the plane
$$z = a_1 y_1 + a_2 y_2$$
In Euclidean three-space that passes through the origin and best matches the data
$$\{(1, 0, 2), (0, 1, 3), (-1, 1, 1), (1, 1, 4), (1, 1, 6)\}$$
Since M has rank 2, Proposition 1.1 applies:
$$M^+ = \begin{pmatrix} 0.2667 & -0.0667 \\ -0.0667 & 0.2667 \\ -0.3333 & 0.3333 \\ 0.2000 & 0.2000 \\ 0.2000 & 0.2000 \end{pmatrix}$$
and
$$\mathbf{a}_0 = \mathbf{b}M^+ = (2 \ \ 3)$$

Proof (of Proposition 1.1). Let π denote the image of Euclidean k-space under multiplication by M. The least squares solution $\mathbf{a}_0$ of $\mathbf{a}M = \mathbf{b}$ is the vector in π for which the difference $|\mathbf{a}M - \mathbf{b}|$ over all $\mathbf{a}$ in Euclidean k-space is minimal. Since $|\mathbf{a}M - \mathbf{b}|$ is minimal if and only if $\mathbf{a}M - \mathbf{b}$ is perpendicular to π, it follows that
$$(\mathbf{a}_0 M - \mathbf{b})(\mathbf{a}M)^T = 0$$

Sec. 1   Least Squares and the Right Inverse

for every **a**, or
$$(\mathbf{a}_0 MM^T - \mathbf{b}M^T)\mathbf{a}^T = 0$$
This can be true for every **a** if and only if
$$\mathbf{a}_0 MM^T - \mathbf{b}M^T = \mathbf{0}$$
or
$$\mathbf{a}_0 MM^T = \mathbf{b}M^T$$
The remainder of the result follows immediately. ∎

As the next example illustrates, least squares can produce undesirable results.

1.3 Example. Finding the line $y = mx + b$ that best matches the data
$$\{(0, 2), (0, 1), (0, 0), (0, -1), (0, -2), (1, 2), (1, 1), (1, 0), (1, -1), (1, -2)\}$$
(see Figure 6.3) reduces to solving the overdetermined and inconsistent system
$$(2 \ 1 \ 0 \ -1 \ -2 \ 2 \ 1 \ 0 \ -1 \ -2)$$
$$= (m \ b)\begin{pmatrix} 0 & 0 & 0 & 0 & 0 & 1 & 1 & 1 & 1 & 1 \\ 1 & 1 & 1 & 1 & 1 & 1 & 1 & 1 & 1 & 1 \end{pmatrix}$$
The least squares solution to this system is
$$(m \ b) = (0 \ 0)$$
so the line that this solution indicates best fits this data is $y = 0$. However, the line $x = \frac{1}{2}$ might intuitively seem to better fit the data. The problem here is created by the fact that it was assumed in the beginning that the line that best fits this data has an equation of the form $y = mx + b$ (that is, the line is nonvertical); this assumption biases the solution of the problem.

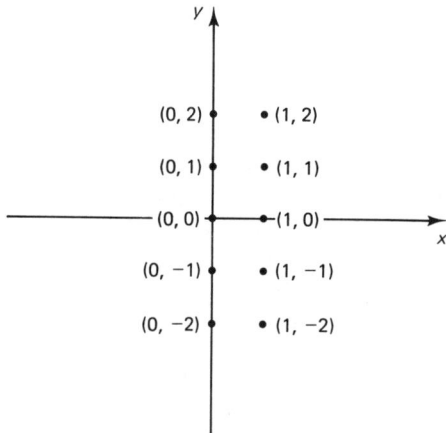

Figure 6.3 A "failing" of least squares.

Until now, we have considered only overdetermined and inconsistent *inhomogeneous* systems $\mathbf{a}M = \mathbf{b}$. We now consider overdetermined *homogeneous* systems $\mathbf{a}M = \mathbf{0}$.

A homogeneous system such as $\mathbf{a}M = \mathbf{0}$ can be overdetermined, but it cannot be inconsistent: it always has the solution $\mathbf{a} = \mathbf{0}$. What normally stands in the way of finding a solution to such a system are restrictions that might be placed on $\mathbf{a}$. (For example, it might be required that $\mathbf{a}$ cannot be $\mathbf{0}$.) To find the least squares solution to an overdetermined homogeneous system $\mathbf{a}M = \mathbf{0}$, an assumption must thus be made. (For example, it might be assumed that at least one component of $\mathbf{a}$ is not 0.) Unfortunately, it is only with respect to this assumption that the least squares solution is well defined; a different assumption might well lead to a different solution.

1.4 Example. The system

$$(a_1 \quad a_2 \quad a_3) \begin{pmatrix} 1 & 0 & 2 & 1 \\ 2 & -1 & 1 & 1 \\ 1 & 3 & 1 & 4 \end{pmatrix} = (0 \quad 0 \quad 0 \quad 0)$$

is overdetermined and homogeneous. To solve this system, which in expanded form is

$$a_1 + 2a_2 + a_3 = 0$$
$$0a_1 - a_2 + 3a_3 = 0$$
$$2a_1 + a_2 + a_3 = 0$$
$$a_1 + a_2 + 4a_3 = 0$$

subject to the constraint that $\mathbf{a} = \langle a_1, a_2, a_3 \rangle$ cannot be $\mathbf{0} = \langle 0, 0, 0 \rangle$, observe that *if* a_3 is nonzero, then this system can be rewritten

$$a_1' + 2a_2' = -1$$
$$0a_1' - a_2' = -3$$
$$2a_1' + a_2' = -1$$
$$a_1' + a_2' = -4$$

where $a_1' = a_1/a_3$ and $a_2' = a_2/a_3$. In matrix form this system is

$$(a_1' \quad a_2') \begin{pmatrix} 1 & 0 & 2 & 1 \\ 2 & -1 & 1 & 1 \end{pmatrix} = (-1 \quad -3 \quad -1 \quad -4)$$

It follows from application of the right inverse technique to this overdetermined and inconsistent *inhomogeneous* system that $a_1' = -1.7059$ and $a_2' = 0.6471$. Since the original system was homogeneous, a_1, a_2, and a_3 can be determined only up to a scalar multiple. Thus if we assume a_3 is nonzero, the solution is, up to a scalar multiple,

$$(a_1 \quad a_2 \quad a_3) = (-1.7059 \quad 0.6471 \quad 1.0000)$$

If, on the other hand, we assume a_2 is nonzero, the solution is, up to a scalar multiple,

$$(a_1 \quad a_2 \quad a_3) = (-0.9469 \quad 1.0000 \quad 0.0973)$$

Rewriting this solution (so that we can compare it with our first solution) by dividing each term by 0.0973, we obtain

$$(a_1 \quad a_2 \quad a_3) = (-9.7273 \quad 10.2727 \quad 1.0000)$$

This solution is *not* the same as our first solution! To repeat what we said earlier, it is only with respect to a given assumption that the least squares solution to an overdetermined homogeneous system $\mathbf{a}M = \mathbf{0}$ is well defined: A different assumption might well lead to a different solution. In this example, in particular, the assumptions a_2 *is nonzero* and a_3 *is nonzero* lead to different solutions.

Care must be taken in interpreting a least squares solution to an overdetermined and inconsistent system $\mathbf{a}M = \mathbf{b}$: In general, (see Exercise 7) the least squares solution to $\mathbf{a}M = \mathbf{b}$ depends on the precise form in which the system is written. In other words, equivalent systems can have different least squares solutions! (Two systems are *equivalent* if and only if one can be obtained from the other through a combination of multiplying an equation by a nonzero constant, adding a multiple of one equation to another, and interchanging two equations.) The reason for this is that the transformation of Euclidean space that takes the points determined by one system to the points determined by the other system does not necessarily preserve vertical distances. (Many things we know about consistent systems of linear equations—such as the fact that we can replace one such system with another such system without changing the solution set—are different when we consider inconsistent systems.) Not only, in fact, can equivalent systems have different least squares solutions, but they can also have different errors $|\mathbf{a}M - \mathbf{b}|$.

SECTION 1 EXERCISES

1. Implement a program that will find the least squares solution to an overdetermined and inconsistent inhomogeneous system.

2. Use the program of Exercise 1 to find the least squares solution to the following overdetermined and inconsistent inhomogeneous systems:

    (a) $a_1 + 2a_2 = -1$
    $a_1 + 3a_2 = 0$
    $0a_1 + a_2 = 2$
    $4a_1 + 2a_2 = -2$

    (b) $a_1 + 2a_2 + a_3 = 2$
    $0a_1 + a_2 + 3a_3 = 0$
    $2a_1 - 2a_2 + a_3 = 1$
    $0a_1 + a_2 + 0a_3 = 1$

    (c) $0a_1 + 0a_2 = 1$
    $0a_1 + a_2 = 2$
    $a_1 + 2a_2 = -3$
    $-a_1 + 0a_2 = 0$
    $a_1 - a_2 = 1$

    (d) $a_1 + 3a_2 + 2a_3 = 1$
    $4a_1 + 5a_2 - 6a_3 = 5$
    $-a_1 + 3a_2 + 2a_3 = 3$
    $0a_1 + a_2 + 3a_3 = 0$
    $2a_1 + a_2 + 6a_3 = 2$

3. Find the equation of the plane $z = a_1 y_1 + a_2 y_2$ in Euclidean three-space determined by the data $\{(y_1, y_2, z)\}$:
    (a) $\{(2, 1, 3), (1, 2, 1)\}$
    (b) $\{(3, 1, 2), (1, 3, 3)\}$

4. Find the equation of the plane $z = a_1 y_1 + a_2 y_2$ in Euclidean three-space that best matches the data $\{(y_1, y_2, z)\}$
    (a) $\{(-1, 0, 3), (0, 2, 1), (1, 1, 1)\}$
    (b) $\{(1, 2, 4), (3, 1, 1), (-1, 0, 2), (1, 3, 2)\}$
    (c) $\{(-1, 2, 1), (2, 3, 1), (1, 3, 4), (2, 1, 1)\}$
    (d) $\{(1, 0, 1), (1, 3, 1), (1, 2, 0), (1, 0, 2), (2, 3, -1)\}$

5. For each part of Exercise 4, find the equation of the plane $z = a_1y_1 + a_2y_2 + a_3$ in Euclidean three-space that best matches the data $\{(y_1, y_2, z)\}$.

6. For each $i = 1, 2, 3$, find the least squares solution to the overdetermined homogeneous system

$$a_1 + 2a_2 - a_3 = 0$$
$$2a_1 + a_2 + a_3 = 0$$
$$a_1 + 2a_2 + 2a_3 = 0$$
$$3a_1 - 2a_2 - 2a_3 = 0$$

obtained by assuming that a_i is nonzero.

7. Find the least squares solution to each of the following systems of equations:

$$\begin{array}{ll} x + y = 3 & 2x + 2y = 6 \\ -x + y = 1 \quad \text{and} & -x + y = 1 \\ 2x + y = 3 & 2x + y = 3 \end{array}$$

Compute the length of the error vector $\mathbf{a}M - \mathbf{b}$ in each case. *Note:* This exercise illustrates that the least squares solution to the system $\mathbf{a}M = \mathbf{b}$ and the error $|\mathbf{a}M - \mathbf{b}|$ depend on the precise form in which the system is written. The reason for this is that in this case we are trying to find the planes $z = a_1y_1 + a_2y_2$ in Euclidean three-space that pass through the origin and best approximate the data

$$\{(1, -1, 2), (1, 1, 1), (3, 1, 3)\} \quad \text{and} \quad \{(2, -1, 2), (2, 1, 1), (6, 1, 3)\}$$

respectively. The transformation of Euclidean three-space that takes the first set of points to the second is given by $T(x, y, z) = (2x, y, z)$, but this transformation does not preserve vertical distances. For example, the vertical distance from the point $P(1, 1, 2)$ to the plane π whose equation is $z = x$ is 1; but the vertical distance from $T(P)$, whose coordinates are $(2, 1, 2)$ to the plane $T(\pi)$, whose equation is $u = 2w$, is $2\sqrt{5}/5 = 0.8944$.

SECTION 2: OBJECT RECONSTRUCTION

In this section we consider reconstruction of the coordinates of object points, given the coordinates of the corresponding image points under at least two (known) complete viewing operations. For simplicity we restrict ourselves to object and image points that are Euclidean.

The reason that at least two views are needed is that one view determines only a line on which the object point must lie (see Figure 6.4(a)). A second view determines a second line, and if we have made all measurements perfectly, these two lines intersect in three-space at the object point (see Figure 6.4(b)). Since we cannot make all measurements perfectly, the two lines probably do not, in fact, intersect (see Figure 6.4(c)), and we must use least squares to solve the problem.

We begin by recalling some facts about homogeneous coordinates and projective transformations.

Let A be an object point and let B be its image under a complete viewing operation T. We can choose a representative $\mathbf{a}$ of the homogeneous coordinates

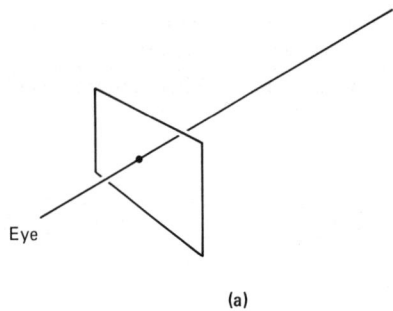

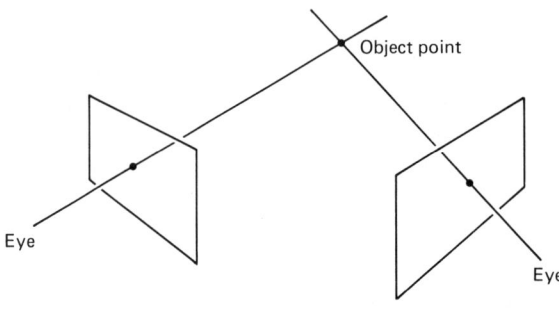

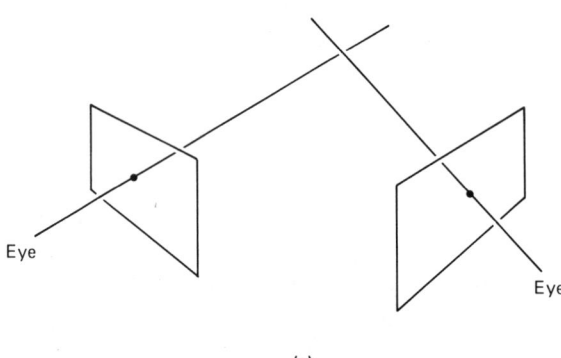

Figure 6.4 Geometric interpretation of object reconstruction. (a) One view. (b) Two views, exact measurements. (c) Two views, inexact measurements.

of A, $\mathbf{b}$ of B, and a matrix M that represents T for which $[\mathbf{a}M] = [\mathbf{b}]$. With an arbitrary choice of M, we may not have $\mathbf{a}M = \mathbf{b}$. For example, if $\mathbf{a} = \langle 4, -2, 3, 1 \rangle$, $\mathbf{b} = \langle -1, 2, 1 \rangle$, and

$$M = \begin{pmatrix} 2 & 0 & 1 \\ 1 & 0 & 2 \\ -2 & 1 & 1 \\ -2 & 1 & -1 \end{pmatrix}$$

then
$$\langle 4, -2, 3, 1\rangle \begin{pmatrix} 2 & 0 & 1 \\ 1 & 0 & 2 \\ -2 & 1 & 1 \\ -2 & 1 & -1 \end{pmatrix} = \langle -2, 4, 2\rangle$$

If we wish to have $\mathbf{a}M = \mathbf{b}$ (rather than just $[\mathbf{a}M] = [\mathbf{b}]$), we are not free to choose the representatives $\mathbf{a}$, $\mathbf{b}$, and M arbitrarily: One choice is limited by the other two. In this case, if

$$M = \begin{pmatrix} 1 & 0 & \frac{1}{2} \\ \frac{1}{2} & 0 & 1 \\ -1 & \frac{1}{2} & \frac{1}{2} \\ -1 & \frac{1}{2} & -\frac{1}{2} \end{pmatrix}$$

then
$$\langle 4, -2, 3, 1\rangle \begin{pmatrix} 1 & 0 & \frac{1}{2} \\ \frac{1}{2} & 0 & 1 \\ -1 & \frac{1}{2} & \frac{1}{2} \\ -1 & \frac{1}{2} & -\frac{1}{2} \end{pmatrix} = \langle -1, 2, 1\rangle$$

If we are given the homogeneous coordinates of an object point $A[a_1, a_2, a_3, 1]$ and the images $B'[b'_1, b'_2, 1]$ and $B''[b''_1, b''_2, 1]$ of A under two different projective transformations (compete viewing operations) T' and T'', we can choose matrices M' and M'' that represent T' and T'', respectively, and for which

$$\langle a_1, a_2, a_3, 1\rangle M' = \langle b'_1, b'_2, 1\rangle \quad \text{and} \quad \langle a_1, a_2, a_3, 1\rangle M'' = \langle b''_1, b''_2, 1\rangle$$

By amalgamating these equations, we obtain a system of linear equations from which we can retrieve $\langle a_1, a_2, a_3, 1\rangle$.

2.1 Example. If $\mathbf{a} = \langle 0, -2, 3, 1\rangle$, $\mathbf{b'} = \langle 2, 0, 1\rangle$, $\mathbf{b''} = \langle -2, -1, 1\rangle$,

$$M' = \begin{pmatrix} 0 & -1 & 3 \\ 1 & 2 & -1 \\ 1 & 1 & 0 \\ 1 & 1 & -1 \end{pmatrix} \quad \text{and} \quad M'' = \begin{pmatrix} 1 & 0 & 1 \\ 1 & 1 & 1 \\ 0 & 0 & 1 \\ 0 & 1 & 0 \end{pmatrix}$$

then $\mathbf{a}M' = \mathbf{b'}$ and $\mathbf{a}M'' = \mathbf{b''}$. To retrieve the coordinates of A we amalgamate the two equations into one equation

$$\mathbf{a}\begin{pmatrix} 0 & -1 & 3 & 1 & 0 & 1 \\ 1 & 2 & -1 & 1 & 1 & 1 \\ 1 & 1 & 0 & 0 & 0 & 1 \\ 1 & 1 & -1 & 0 & 1 & 0 \end{pmatrix} = \langle 2, 0, 1, -2, -1, 1\rangle$$

which we write

$$\mathbf{a}N = \mathbf{c}$$

Sec. 2   Object Reconstruction

Since T' and T'' were chosen so that N has rank 4, the best possible solution to this overdetermined inhomogeneous system is
$$\mathbf{c}N^+ = \langle 0, -2, 3, 1 \rangle$$

In object reconstruction we are given the representatives $\mathbf{b}' = \langle b'_1, b'_2, 1 \rangle$ and $\mathbf{b}'' = \langle b''_1, b''_2, 1 \rangle$ of the homogeneous coordinates of the images of some unknown object point $A[\mathbf{a}] = A[a_1, a_2, a_3, 1]$ under two known complete viewing operations T' and T'', which are represented by the matrices M' and M'',
$$[\mathbf{a}M'] = [\mathbf{b}'] \quad \text{and} \quad [\mathbf{a}M''] = [\mathbf{b}'']$$
and we are to reconstruct $\mathbf{a}$. Now, however, $\mathbf{a}$ is unknown, so it cannot be used to choose the correct matrices M' and M'' that represent T' and T''. Thus we must use arbitrary matrices that represent T' and T''. Since the matrices representing T' and T'' are unique up to scalar multiples (see Exercise 8), replacing the known matrices M' and M'' by arbitrary representative matrices in the equations
$$\mathbf{a}M' = \langle b'_1, b'_2, 1 \rangle \quad \text{and} \quad \mathbf{a}M'' = \langle b''_1, b''_2, 1 \rangle$$
is equivalent to rewriting these equations as
$$\mathbf{a}M' = \langle h'b'_1, h'b'_2, h' \rangle \quad \text{and} \quad \mathbf{a}M'' = \langle h''b''_1, h''b''_2, h'' \rangle$$
where h' and h'' are unknown. We use a solution similar to the one presented in Example 2.1 to find $\mathbf{a}$ after we eliminate h' and h'' from these nonlinear equations.

We eliminate h' and h'' and form a linear system as follows: Let M be either M' or M'' and let $\mathbf{m}_1$, $\mathbf{m}_2$ and $\mathbf{m}_3$ be the columns of M. The system $\mathbf{a}M = \langle hb_1, hb_2, h \rangle$ can be written
$$\langle \mathbf{a} \cdot \mathbf{m}_1, \mathbf{a} \cdot \mathbf{m}_2, \mathbf{a} \cdot \mathbf{m}_3 \rangle = \langle hb_1, hb_2, h \rangle$$
Thus $h = \mathbf{a} \cdot \mathbf{m}_3$ and
$$\langle \mathbf{a} \cdot \mathbf{m}_1, \mathbf{a} \cdot \mathbf{m}_2, \mathbf{a} \cdot \mathbf{m}_3 \rangle = \langle \mathbf{a} \cdot \mathbf{m}_3\, b_1, \mathbf{a} \cdot \mathbf{m}_3\, b_2, \mathbf{a} \cdot \mathbf{m}_3 \rangle$$
$$\langle \mathbf{a} \cdot \mathbf{m}_1, \mathbf{a} \cdot \mathbf{m}_2 \rangle = \langle \mathbf{a} \cdot \mathbf{m}_3\, b_1, \mathbf{a} \cdot \mathbf{m}_3\, b_2 \rangle$$
or
$$\langle \mathbf{a} \cdot (\mathbf{m}_1 - b_1 \mathbf{m}_3), \mathbf{a} \cdot (\mathbf{m}_2 - b_2 \mathbf{m}_3) \rangle = \mathbf{0}$$
In expanded form, this is the linear system
$$\langle a_1, a_2, a_3, 1 \rangle \begin{pmatrix} m_{11} - b_1 m_{31} & m_{21} - b_2 m_{31} \\ m_{12} - b_1 m_{32} & m_{22} - b_2 m_{32} \\ m_{13} - b_1 m_{33} & m_{23} - b_2 m_{33} \\ m_{14} - b_1 m_{34} & m_{24} - b_2 m_{34} \end{pmatrix} = \langle 0, 0 \rangle$$
Thus
$$\langle a_1, a_2, a_3 \rangle \begin{pmatrix} m_{11} - b_1 m_{31} & m_{21} - b_2 m_{31} \\ m_{12} - b_1 m_{32} & m_{22} - b_2 m_{32} \\ m_{13} - b_1 m_{33} & m_{23} - b_2 m_{33} \end{pmatrix} = \langle b_1 m_{34} - m_{14}, b_2 m_{34} - m_{24} \rangle$$

Amalgamating the two matrix equations, we have

$$\langle a_1, a_2, a_3 \rangle \begin{pmatrix} m'_{11} - b'_1 m'_{31} & m'_{21} - b'_2 m'_{31} & m''_{11} - b''_1 m''_{31} & m''_{21} - b''_2 m''_{31} \\ m'_{12} - b'_1 m'_{32} & m'_{22} - b'_2 m'_{32} & m''_{12} - b''_1 m''_{32} & m''_{22} - b''_2 m''_{32} \\ m'_{13} - b'_1 m'_{33} & m'_{23} - b'_2 m'_{33} & m''_{13} - b''_1 m''_{33} & m''_{23} - b''_2 m''_{33} \end{pmatrix}$$

$$= \langle b'_1 m'_{34} - m'_{14}, b'_2 m'_{34} - m'_{24}, b''_1 m''_{34} - m''_{14}, b''_2 m''_{34} - m''_{24} \rangle$$

which we write

$$\langle a_1, a_2, a_3 \rangle N = \mathbf{c}$$

This is an overdetermined—and almost certainly inconsistent—system of four linear equations in three unknowns. Again, if T' and T'' are chosen so that N has rank 4, then the best possible solution to this system is $\mathbf{c}N^+$.

2.2 Example. Graph3D was used to produce two views of the cube whose vertices are ($\pm 1, \pm 1, \pm 1$) (see Figure 6.5). Matrices that represent the complete viewing operations T' and T'' (as reported by a modified Graph3D) are, respectively,

$$M' = \begin{pmatrix} 781.20 & 0.00 & 0.00 \\ 0.00 & 657.00 & 0.00 \\ -362.70 & -377.07 & -1.00 \\ 1953.00 & 1337.00 & 14.00 \end{pmatrix} \text{ and } M'' = \begin{pmatrix} 350.51 & 111.28 & 0.43 \\ -69.56 & 196.46 & -0.50 \\ 62.12 & -193.46 & -0.75 \\ 837.00 & 573.00 & 6.00 \end{pmatrix}$$

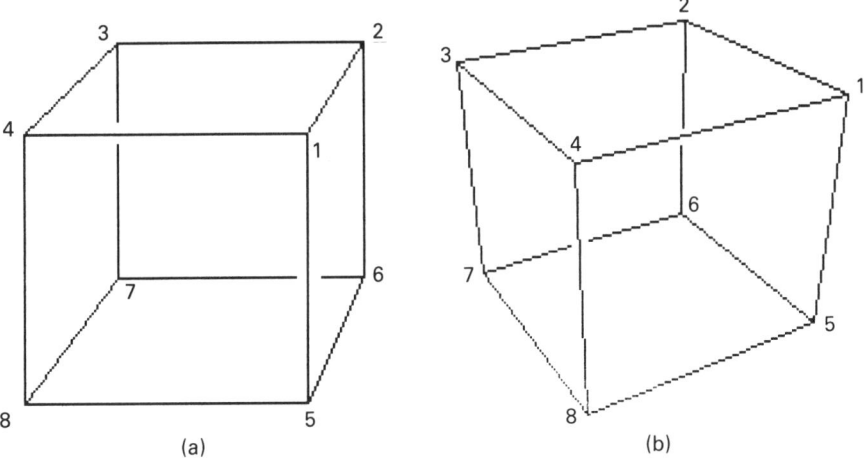

Figure 6.5 Two perspective views of a cube.

We reconstruct the coordinates of the vertices of the cube as follows: First we label corresponding vertices of each image of the cube. Then we measure the device coordinates of each image vertex. (This is easy to do if a short program is first written to create a pair of *pixel rulers*—a horizontal line and a vertical line with a mark every 10 pixels.) The values obtained are given in Figure 6.6. Next we compute, for each pair of image vertices, the matrices N

$$\begin{array}{c}\begin{array}{c}1\\2\\3\\4\\5\\6\\7\\8\end{array}\begin{pmatrix}182 & 124 & 1\\206 & 158 & 1\\102 & 158 & 1\\62 & 124 & 1\\182 & 23 & 1\\206 & 70 & 1\\102 & 70 & 1\\62 & 23 & 1\end{pmatrix}\begin{pmatrix}228 & 133 & 1\\158 & 161 & 1\\61 & 146 & 1\\111 & 108 & 1\\213 & 48 & 1\\156 & 89 & 1\\72 & 67 & 1\\116 & 14 & 1\end{pmatrix}\\\text{(a)}\qquad\qquad\text{(b)}\end{array}$$

Figure 6.6 Image point matrices for the views of the cube in Figure 6.5.

and N^+, the vectors **c** and $\mathbf{c}N^+$, and the coordinates **a** of the object point A. For example, for the first vertex,

$$N = \begin{pmatrix} 781.20 & 0.00 & 252.47 & 54.09 \\ 0.00 & 657.00 & 44.44 & 262.96 \\ -180.70 & -253.07 & 233.12 & -93.71 \end{pmatrix}$$

$$N^+ = \begin{pmatrix} 0.001035 & -0.000429 & -0.001050 \\ -0.000077 & 0.001223 & -0.000244 \\ 0.000745 & 0.001213 & 0.003236 \\ 0.000068 & 0.000542 & 0.000063 \end{pmatrix}$$

$$\mathbf{c} = \langle 595.00, 399.00, 531.00, 225.00 \rangle$$

$$\mathbf{c}N^+ = \langle 0.996, 0.998, 1.010 \rangle$$

and

$$\mathbf{a} = \langle 0.996, 0.998, 1.010, 1 \rangle$$

The results of all these computations are given in Figure 6.7.

$$\begin{array}{c}\begin{array}{c}1\\2\\3\\4\\5\\6\\7\\8\end{array}\begin{pmatrix}1 & 1 & 1 & 1\\1 & 1 & -1 & 1\\-1 & 1 & -1 & 1\\-1 & 1 & 1 & 1\\1 & -1 & 1 & 1\\1 & -1 & -1 & 1\\-1 & -1 & -1 & 1\\-1 & -1 & 1 & 1\end{pmatrix}\begin{pmatrix}0.996 & 0.998 & 1.010 & 1\\0.993 & 1.003 & -0.990 & 1\\-1.003 & 1.001 & -0.991 & 1\\-1.001 & 0.997 & 1.006 & 1\\0.993 & -1.006 & 0.996 & 1\\0.996 & -0.997 & -0.977 & 1\\-1.008 & -1.014 & -1.007 & 1\\-1.004 & -1.006 & 0.997 & 1\end{pmatrix}\\\text{(a)}\qquad\qquad\text{(b)}\end{array}$$

Figure 6.7 (a) Original object point matrix. (b) Reconstructed object point matrix.

Object reconstruction admits some error, as is evident in Figure 6.7. The error in reconstruction was zero in Example 2.1, since the coordinates of **b**′ and **b**″ were computed rather than measured. The error in reconstruction in Example 2.2 was caused by errors in measurement: We were limited in measuring accuracy to the size of a pixel. Error in reconstruction resulting from error in measurement can be reduced by providing more than two views (see Exercises 5 and 6).

SECTION 2 EXERCISES

1. Let
$$M' = \begin{pmatrix} 0 & 0 & 2 \\ 1 & 0 & 0 \\ 0 & 1 & 0 \\ 0 & 0 & -1 \end{pmatrix} \quad \text{and} \quad M'' = \begin{pmatrix} 1 & 0 & 1 \\ 1 & 1 & 0 \\ 0 & 0 & 1 \\ 0 & 1 & 0 \end{pmatrix}$$

   For each object point A, find its images B' and B'' under the complete viewing operations represented by M' and M''. Then use the method of Example 2.2 to reconstruct the coordinates of A.
   (a) $A[1, 0, 0, 1]$ (b) $A[1, 1, 1, 1]$
   (c) $A[1, 0, 1, -2]$ (d) $A[2, 1, 0, 1]$

2. Verify the computations of N, N^+, $\mathbf{c}$, and $\mathbf{a}$ for the first vertex in Example 2.2. Then do the same computations for the second vertex.

3. Write a computer program to compute the coordinates of an object point, given the matrices of two viewing transformations and the corresponding coordinates of the two image points.

4. Graph3D was used to produce the two views of the cube in Figure 6.8. The matrices of the complete viewing operations are, respectively,
$$M_1 = \begin{pmatrix} 253.77 & 87.95 & 0.43 \\ -69.56 & 115.10 & -0.50 \\ 6.47 & -152.89 & -0.75 \\ 837.00 & 573.00 & 6.00 \end{pmatrix} \quad \text{and} \quad M_2 = \begin{pmatrix} 585.90 & 0.00 & 0.00 \\ 0.00 & 492.75 & 0.00 \\ -4.29 & -360.83 & -1.00 \\ 1813.50 & 1241.50 & 13.00 \end{pmatrix}$$

   and the image point matrices of the vertices are, respectively,
$$\begin{pmatrix} 198 & 120 & 1 \\ 152 & 139 & 1 \\ 87 & 129 & 1 \\ 120 & 104 & 1 \\ 199 & 64 & 1 \\ 150 & 91 & 1 \\ 95 & 77 & 1 \\ 124 & 41 & 1 \end{pmatrix} \quad \text{and} \quad \begin{pmatrix} 200 & 114 & 1 \\ 172 & 150 & 1 \\ 88 & 150 & 1 \\ 102 & 114 & 1 \\ 200 & 32 & 1 \\ 172 & 79 & 1 \\ 88 & 79 & 1 \\ 102 & 32 & 1 \end{pmatrix}$$

   Use your program of Exercise 3 to reconstruct the coordinates of the vertices of the

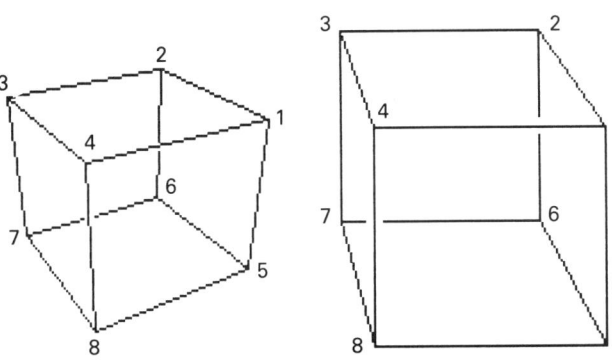

Figure 6.8 Two perspective views of a cube.

cube. Does this pair of views allow for better or worse reconstruction than the views used in Example 2.2?

5. Modify the construction given in this section to use the information provided by a third viewing transformation and image.

6. Write a computer program that uses information from three views of an object point to reconstruct its coordinates. Use your program with M' and M'' of Example 2.2 and M_1 of Exercise 4 to reconstruct the coordinates of the vertices of the cube. Are the results more satisfactory than when only two views are used?

7. What choices of a second view of a scene would provide no help in object reconstruction?

†8. Show that the matrix M representing a complete viewing operation T is unique up to a scalar multiple. (*Hint:* Begin by showing that each row of M is unique up to a scalar multiple, and then show that there is a unique dependence relation between the rows of M.)

9. In this section we assumed that the object point A and the image points B' and B'' are Euclidean.
   (a) How should the technique presented in this section be modified if B', B'', or both B' and B'' are ideal?
   (b) Do Exercise 1 using $A[0, 1, 0, 1]$.
   (c) How should the technique presented in this section be modified if A is not known to be Euclidean?
   (d) Do Exercise 1 using $A[0, 1, 1, 0]$.

SECTION 3: CAMERA CALIBRATION

Suppose T is a complete viewing operation that transforms points from three-dimensional world coordinate space to two-dimensional device coordinate space. The goal of camera calibration is to compute the components of some matrix $M^{4\times 3}$ that represents T from the knowledge of how T transforms the coordinates of a small number of known points. After discussing a method for solving this problem, we will see that the same method lets us reconstruct a representative matrix $M^{3\times 3}$ of a complete viewing operation from two-dimensional world coordinate space to two-dimensional device coordinate space.

If $n \geq 4$ and we are given the homogeneous coordinates of n object points $A_i[a_{i1}, a_{i2}, a_{i3}, a_{i4}]$ and the images $B_i[b_{i1}, b_{i2}, 1]$ of each A_i under a projective transformation T, which is represented by a matrix M, we can choose representatives $\langle a_{i1}, a_{i2}, a_{i3}, a_{i4} \rangle$ of each $[a_{i1}, a_{i2}, a_{i3}, a_{i4}]$ for which

$$\langle a_{i1}, a_{i2}, a_{i3}, a_{i4} \rangle M = \langle b_{i1}, b_{i2}, 1 \rangle$$

By amalgamating these equations, we obtain a system of linear equations from which we can retrieve M.

3.1 Example. If

$$\mathbf{a}_1 = \langle 0, 0, -1, 1 \rangle \quad \text{and} \quad \mathbf{b}_1 = \langle -1, -3, 1 \rangle$$
$$\mathbf{a}_2 = \langle -1, 1, 0, 1 \rangle \quad \text{and} \quad \mathbf{b}_2 = \langle -1, -1, 1 \rangle$$
$$\mathbf{a}_3 = \langle 0, 1, 1, 1 \rangle \quad \text{and} \quad \mathbf{b}_3 = \langle 2, 4, 1 \rangle$$

$$\mathbf{a}_4 = \langle 1, 2, 2, 2 \rangle \quad \text{and} \quad \mathbf{b}_4 = \langle 5, 10, 1 \rangle$$
$$\mathbf{a}_5 = \langle -1, 1, 2, 0 \rangle \quad \text{and} \quad \mathbf{b}_5 = \langle 2, 5, 1 \rangle$$

and

$$M = \begin{pmatrix} 1 & 2 & -1 \\ -1 & 1 & -2 \\ 2 & 3 & 1 \\ 1 & 0 & 2 \end{pmatrix}$$

then $\mathbf{a}_i M = \mathbf{b}_i$ for each i. To retrieve M, we amalgamate the six equations into the one equation

$$\begin{pmatrix} 0 & 0 & -1 & 1 \\ -1 & 1 & 0 & 1 \\ 0 & 1 & 1 & 1 \\ 1 & 2 & 2 & 2 \\ -1 & 1 & 2 & 0 \end{pmatrix} M = \begin{pmatrix} -1 & -3 & 1 \\ -1 & -1 & 1 \\ 2 & 4 & 1 \\ 5 & 10 & 1 \\ 2 & 5 & 1 \end{pmatrix}$$

which we write

$$AM = B$$

Since the points A_i were chosen so that A has rank 4, we can retrieve M by multiplying both sides of this equation by the *left* inverse $(A^T A)^{-1} A^T$ of A. (The matrix $(A^T A)^{-1} A^T$ is a left inverse of A since $((A^T A)^{-1} A^T) A = (A^T A)^{-1} A^T A = I$.) In this case

$$(A^T A)^{-1} A^T = \begin{pmatrix} -0.5833 & 0.0000 & -0.2500 & 0.4167 & -0.5833 \\ -2.3333 & 2.0000 & -1.0000 & 0.6667 & -1.3333 \\ 0.8333 & -1.0000 & 0.5000 & -0.1667 & 0.8333 \\ 1.7500 & -1.0000 & 0.7500 & -0.2500 & 0.7500 \end{pmatrix}$$

so that

$$M = (A^T A)^{-1} A^T B = \begin{pmatrix} 1 & 2 & -1 \\ -1 & 1 & -2 \\ 2 & 3 & 1 \\ 1 & 0 & 2 \end{pmatrix}$$

In camera calibration we are given the representatives $\mathbf{b}_i = \langle b_{i1}, b_{i2}, 1 \rangle$ of the images of n known object points $A_i[\mathbf{a}_i] = A_i[a_{i1}, a_{i2}, a_{i3}, a_{i4}]$ under a complete viewing operation T represented by the unknown matrix M:

$$[\mathbf{a}_i M] = [\mathbf{b}_i]$$

and we are to reconstruct M (up to a scalar multiple). Now, however, M is unknown, so it cannot be used to choose the "correct" representatives $\langle a_{i1}, a_{i2}, a_{i3}, a_{i4} \rangle$ of the homogeneous coordinates of A_i. Thus we must use arbitrary representatives of the homogeneous coordinates of the A_i. Replacing the known representatives $\langle a_{i1}, a_{i2}, a_{i3}, a_{i4} \rangle$ by arbitrary representatives in the equations

$$\langle a_{i1}, a_{i2}, a_{i3}, a_{i4} \rangle M = \langle b_{i1}, b_{i2}, 1 \rangle$$

is equivalent to rewriting these equations as
$$\langle a_{i1}, a_{i2}, a_{i3}, a_{i4} \rangle M = \langle h_i b_{i1}, h_i b_{i2}, h_i \rangle$$
where the h_i are unknown. We use least squares to find M after we eliminate the h_i's from these nonlinear equations.

Suppose
$$M = \begin{pmatrix} m_{11} & m_{21} & m_{31} \\ m_{12} & m_{22} & m_{32} \\ m_{13} & m_{23} & m_{33} \\ m_{14} & m_{24} & m_{34} \end{pmatrix}$$

and $\mathbf{m}_1$, $\mathbf{m}_2$ and $\mathbf{m}_3$ are the column vectors of M. Let
$$A = \begin{pmatrix} a_{11} & a_{12} & a_{13} & a_{14} \\ a_{21} & a_{22} & a_{23} & a_{24} \\ \vdots & \vdots & \vdots & \vdots \\ a_{n1} & a_{n2} & a_{n3} & a_{n4} \end{pmatrix}$$

be the object point matrix, so that the ith row $\mathbf{a}_i = \langle a_{i1}, a_{i2}, a_{i3}, a_{i4} \rangle$ is the representative of the homogeneous coordinates of the ith of n known object points. Let $\langle b_{i1}, b_{i2}, 1 \rangle$ be the representative of the homogeneous coordinates of the corresponding ith known image point. As we saw in Section 2, for each object point and its image there is some h_i such that $\mathbf{a}_i M = \langle h_i b_{i1}, h_i b_{i2}, h_i \rangle$. After h_i is eliminated from this equation, we have—as in Section 2—
$$\langle \mathbf{a}_i \cdot (\mathbf{m}_1 - b_{i1}\mathbf{m}_3), \mathbf{a}_i \cdot (\mathbf{m}_2 - b_{i2}\mathbf{m}_3) \rangle = \mathbf{0}$$
which may be rewritten
$$\mathbf{m}_1 \cdot \mathbf{a}_i - \mathbf{m}_3 \cdot (b_{i1} \mathbf{a}_i) = 0$$
$$\mathbf{m}_2 \cdot \mathbf{a}_i - \mathbf{m}_3 \cdot (b_{i2} \mathbf{a}_i) = 0$$
or, in matrix form,

$$\langle \mathbf{m}_1, \mathbf{m}_2, \mathbf{m}_3 \rangle \begin{pmatrix} \mathbf{a}_i & \mathbf{0} \\ \mathbf{0} & \mathbf{a}_i \\ -b_{i1}\mathbf{a}_i & -b_{i2}\mathbf{a}_i \end{pmatrix} = \langle \mathbf{m}_1, \mathbf{m}_2, \mathbf{m}_3 \rangle \begin{pmatrix} a_{i1} & 0 \\ a_{i2} & 0 \\ a_{i3} & 0 \\ a_{i4} & 0 \\ 0 & a_{i1} \\ 0 & a_{i2} \\ 0 & a_{i3} \\ 0 & a_{i4} \\ -b_{i1}a_{i1} & -b_{i2}a_{i1} \\ -b_{i1}a_{i2} & -b_{i2}a_{i2} \\ -b_{i1}a_{i3} & -b_{i2}a_{i3} \\ -b_{i1}a_{i4} & -b_{i2}a_{i4} \end{pmatrix} = \langle 0, 0 \rangle$$

where
$$\mathbf{m} = \langle \mathbf{m}_1, \mathbf{m}_2, \mathbf{m}_3 \rangle$$
$$= \langle m_{11}, m_{12}, m_{13}, m_{14}, m_{21}, m_{22}, m_{23}, m_{24}, m_{31}, m_{32}, m_{33}, m_{34} \rangle$$

is a vector with 12 components. We amalgamate these n matrix equations into the single matrix equation

$$\mathbf{m}\begin{pmatrix} \mathbf{a}_1 & \mathbf{a}_2 & \cdots & \mathbf{a}_n & \mathbf{0} & \mathbf{0} & \cdots & \mathbf{0} \\ \mathbf{0} & \mathbf{0} & \cdots & \mathbf{0} & \mathbf{a}_1 & \mathbf{a}_2 & \cdots & \mathbf{a}_n \\ -b_{11}\mathbf{a}_1 & -b_{21}\mathbf{a}_2 & \cdots & -b_{n1}\mathbf{a}_n & -b_{12}\mathbf{a}_1 & -b_{22}\mathbf{a}_2 & \cdots & -b_{n2}\mathbf{a}_n \end{pmatrix} = \mathbf{0}$$

If $C^{12 \times 2n}$ is the matrix above, the system we are studying is $\mathbf{m}C = \mathbf{0}$ subject to the constraint that $\mathbf{m} \neq \mathbf{0}$. If $n \geq 6$, this system is overdetermined and almost certainly inconsistent. It is homogeneous, and we seek the best possible nonzero solution $\mathbf{m}$ in the sense of least squares.

3.2 Example. Graph3D was used to produce the view of the cube in Figure 6.5(a). A matrix M' that represents the complete viewing operation is given in Example 2.2, the object point matrix A is given in Figure 6.7(a), and the image point matrix B is given in Figure 6.6(a). In Example 2.2 this data was used to reconstruct A; here we use it to reconstruct M.

The matrix $C^{12 \times 16}$ is

$$\begin{pmatrix}
1 & 1 & -1 & -1 & 1 & 1 & -1 & -1 & 0 & 0 & 0 & 0 & 0 & 0 & 0 & 0 \\
1 & 1 & 1 & 1 & -1 & -1 & -1 & -1 & 0 & 0 & 0 & 0 & 0 & 0 & 0 & 0 \\
1 & -1 & -1 & 1 & 1 & -1 & -1 & 1 & 0 & 0 & 0 & 0 & 0 & 0 & 0 & 0 \\
1 & 1 & 1 & 1 & 1 & 1 & 1 & 1 & 0 & 0 & 0 & 0 & 0 & 0 & 0 & 0 \\
0 & 0 & 0 & 0 & 0 & 0 & 0 & 0 & 1 & 1 & -1 & -1 & 1 & 1 & -1 & -1 \\
0 & 0 & 0 & 0 & 0 & 0 & 0 & 0 & 1 & 1 & 1 & 1 & -1 & -1 & -1 & -1 \\
0 & 0 & 0 & 0 & 0 & 0 & 0 & 0 & 1 & -1 & -1 & 1 & 1 & -1 & -1 & 1 \\
0 & 0 & 0 & 0 & 0 & 0 & 0 & 0 & 1 & 1 & 1 & 1 & 1 & 1 & 1 & 1 \\
-182 & -206 & 102 & 62 & -182 & -206 & 102 & 62 & -124 & -158 & 158 & 124 & -23 & -70 & 70 & 23 \\
-182 & -206 & -102 & -62 & 182 & 206 & 102 & 62 & -124 & -158 & -158 & -124 & 23 & 70 & 70 & 23 \\
-182 & 206 & 102 & -62 & -182 & 206 & 102 & -62 & -124 & 158 & 158 & -124 & -23 & 70 & 70 & -23 \\
-182 & -206 & -102 & -62 & -182 & -206 & -102 & -62 & -124 & -158 & -158 & -124 & -23 & -70 & -70 & -23
\end{pmatrix}$$

The least squares solution to the system $\mathbf{m}C = \mathbf{0}$ depends, of course, on which of the 12 components of M is assumed to be nonzero. In the analogous situation in Section 2 there was a natural choice: we chose $a_4 = 1$ (that is, we chose A to be Euclidean). Now, however, there is no natural choice. If M_i is the matrix obtained by assuming that the ith component of $\mathbf{m}$ is nonzero, then we obtain (after normalizing), for example

$$M_{12} = \begin{pmatrix} 780.7072 & 0.4396 & 0.0042 \\ -0.7546 & 657.7872 & -0.0056 \\ -359.8826 & -375.8426 & -0.9842 \\ 1948.0188 & 1332.1840 & 14.0000 \end{pmatrix}$$

The 12 solutions $M_1, \ldots, M_{12}$ are all different! Of course, they cannot all be equally satisfactory solutions; in fact, in this case we might expect a problem with M_9 since the corresponding entry of M' is zero.

To compare these solutions, we compute, for each M_i, the square root of the sum of the squares of the distances from each image point B_j to the image of A_j under the transformation represented by M_i. The values of these errors are tabulated in Figure 6.9 for the 12 matrices M_i. The largest errors correspond

i	error
1	0.2585
2	1.1394
3	0.2586
4	0.2586
5	1.5944
6	0.2586
7	0.2586
8	0.2586
9	1.3695
10	1.0367
11	0.2586
12	0.2586

Figure 6.9 The errors for the matrices M_i.

to those matrices M_i for which the corresponding entry of M' is zero. In general, of course, the matrix M' will not be known in advance. In this case it would be reasonable to choose as an approximation to M' one of the M_i for which the error is small.

Now let us consider the problem of reconstructing a matrix $M^{3\times 3}$ that represents a transformation of the coordinates $\mathbf{a}_i = \langle a_{i1}, a_{i2}, a_{i3} \rangle$ of n object points in the projective plane to the coordinates $\langle b_{i1}, b_{i2}, 1 \rangle$ of n image points. Using the methods above, we reduce the problem to finding the invariant least squares solution to $\mathbf{m}C = \mathbf{0}$, where $C^{9\times 2n}$ is as before, but $\mathbf{m}$ now has only 9 components.

If $n > 4$, then $\mathbf{m}C = \mathbf{0}$ is overdetermined, and we must resort to least squares as in the case when M is 4×3. An application is presented in Section 4.

If $n = 4$ and no three points in either of the two given sets of four points are collinear, there is a unique matrix (see Theorem 3.2.9) which represents the transformation from one set of four points to the other. One method for finding this matrix was given in Example 3.2.10 (and again in Section 3.7). We now have a second method for finding this matrix.

3.3 Example. To find the 3×3 nonsingular matrix M taking $A_1[1, 2, 1]$ to $B_1[0, 1, 1]$, $A_2[0, 1, 2]$ to $B_2[1, -1, 1]$, $A_3[-1, 4, 2]$ to $B_3[1, 1, 1]$ and $A_4[1, 1, 2]$ to $B_4[0, 2, 1]$, let

$$C^{9\times 8} = \begin{pmatrix} 1 & 0 & -1 & 1 & 0 & 0 & 0 & 0 \\ 2 & 1 & 4 & 1 & 0 & 0 & 0 & 0 \\ 1 & 2 & 2 & 2 & 0 & 0 & 0 & 0 \\ 0 & 0 & 0 & 0 & 1 & 0 & -1 & 1 \\ 0 & 0 & 0 & 0 & 2 & 1 & 4 & 1 \\ 0 & 0 & 0 & 0 & 1 & 2 & 2 & 2 \\ 0 & 0 & 1 & 0 & -1 & 0 & 1 & -2 \\ 0 & -1 & -4 & 0 & -2 & 1 & -4 & -2 \\ 0 & -2 & -2 & 0 & -1 & 2 & -2 & -4 \end{pmatrix}$$

Since $mC = 0$ is a system of eight linear equations in nine unknowns, it has a nonzero solution. This solution $\langle 3, -1, -1, 9, -1, 2, 9, 1, -2 \rangle$, reassembled as a matrix, yields

$$M = \begin{pmatrix} 3 & 9 & 9 \\ -1 & -1 & 1 \\ -1 & 2 & -2 \end{pmatrix}$$

SECTION 3 EXERCISES

1. Write a computer program to implement the algorithm for camera calibration illustrated in Example 3.2.
2. Use your program of Exercise 1, the object point matrix A of Figure 6.7(a), and the image point matrix of Figure 6.6(b) to reconstruct the matrix M'' of Example 2.2.
3. Use your program of Exercise 1, the object point matrix A of Figure 6.7(a), and the image point matrices of Exercise 2.4 to reconstruct the matrices M_1 and M_2 of Exercise 2.4.
4. Write a computer program to implement the algorithm illustrated in Example 3.3.
5. Use the algorithm of Example 3.3 or your program of Exercise 4 to find each nonsingular 3×3 matrix M described.
   (a) Taking $A_1[1, 0, 1]$ to $B_1[-4, 0, 1]$, $A_2[0, 1, 1]$ to $B_2[2, 1, 1]$, $A_3[1, -1, 1]$ to $B_3[-5, 1, 1]$, and $A_4[2, 0, 1]$ to $B_4[-2, -2, 1]$
   (b) Taking $A_1[1, 0, 0]$ to $B_1[2, 1, 1]$, $A_2[0, 1, 0]$ to $B_2[0, -3, 1]$, $A_3[0, 0, 1]$ to $B_3[4, -1, 1]$, and $A_4[1, 1, 1]$ to $B_4[2, -1, 1]$
   (c) Taking $A_1[1, 2, 1]$ to $B_1[0, 1, 1]$, $A_2[0, 1, 2]$ to $B_2[-1, 1, 1]$, $A_3[-1, 4, 2]$ to $B_3[1, 1, 1]$, and $A_4[1, 1, 2]$ to $B_4[0, 2, 1]$
   (d) Taking $A_1[2, 0, 1]$ to $B_1[-2, 1, 2]$, $A_2[1, 0, 0]$ to $B_2[3, 0, 1]$, $A_3[0, 2, 1]$ to $B_3[1, 1, 1]$, and $A_4[3, 2, 2]$ to $B_4[1, 1, 2]$
6. Develop an algorithm similar to that of Example 3.3 that will work even if some of the image points are ideal.
7. To what extent can a viewing mechanism be completely determined by a 4×3 matrix M that represents a complete viewing operation?

SECTION 4: PHOTOGRAMMETRIC RECONSTRUCTION

In this section we study various methods of projective geometry that are used in photogrammetry to study aerial photographs such as Figure 6.10. This section is only an introduction to some of the mathematical aspects of photogrammetry; for further study see [American Society of Photogrammetry].

Figure 6.11 illustrates the geometry of aerial photography. The point O (the center of perspectivity) is the lens of the camera, f is the focal length of the camera, and h is the flying height. The original impression is on the negative plane, but the print appears on the *photo plane*. The angle between the photo plane and the horizontal plane is the *tilt*. An aerial photograph in which the tilt is zero is *vertical* (the camera axis is vertical).

Figure 6.10 An aerial photograph of Indianapolis (photo courtesy of Woolpert Consultants).

In an aerial photograph, there are normally two different causes for displacement of an image point from where it would appear on a map; *displacement due to relief* and *displacement due to tilt*. Displacements due to relief occur because of the varying heights of the ground and of objects on the ground, and because a photograph is a perspective projection rather than an orthogonal projection (see Figure 6.12(a).) Displacement due to tilt is illustrated in Figure 6.12(b). The advantage to working with a vertical photograph (as opposed to one in which there is tilt) is that the only displacements in a vertical photograph are those due to relief, and over level ground these are uniform radial displacements along

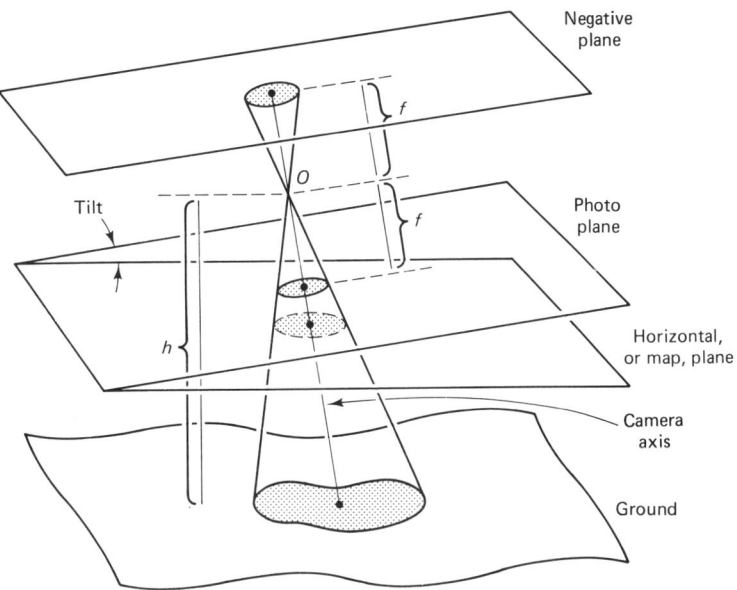

Figure 6.11 The geometry of aerial photography.

lines that pass through the image of the point directly below the camera (see Figure 5.44).

The process of projecting a tilted photograph from the photo plane to a horizontal plane in order to remove displacements due to tilt is *rectification*. The horizontal plane onto which we project the photo plane during rectification is the *map plane*. It is important to observe, however, that after rectification we do not really have a map on the map plane, since a map is an orthogonal projection of the ground, while the image on the map plane is a perspective projection, which retains displacements due to relief.

Rectified photographs are used in the production of topographical maps, orthographic maps, and mosaics. To see how displacements due to relief in a rectified photograph can be used, we consider the determination of heights: Suppose we are given a rectified (or vertical) photograph (see Figure 6.13) of a scene in which B is directly over A (perhaps they are points at the top and bottom of a building) and O is not collinear with A and B. The images b and a of B and A are distinct, and the distance d_2 is the displacement due to relief. If the height h of O above A is known, then the height g of B above A can be computed: Since $d_2/r = d_1/R = g/h$, $g = hd_2/r$. (In practice, two different rectified photographs of the same area are viewed through a stereoplotter to determine elevations.)

In this section we study methods of rectification and map production. These methods of photographic reconstruction can be classified as graphical or analytical. In either graphical or analytical photographic reconstruction, we must be given a certain number of *control points*—points on the ground whose coordinates

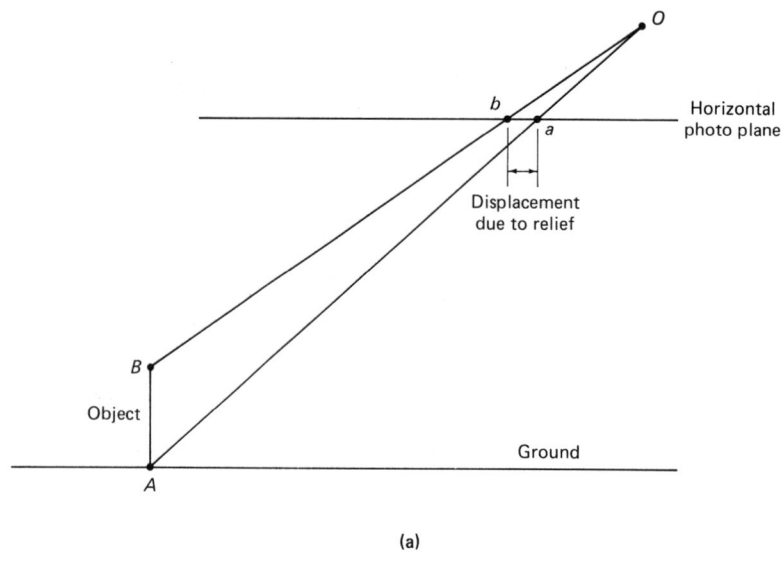

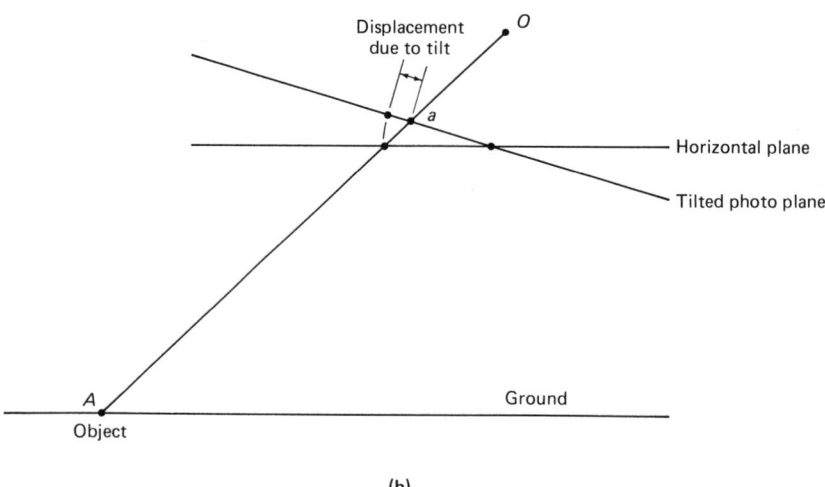

Figure 6.12 Image displacement due to (a) relief and (b) tilt.

are known and whose images appear in the photograph. Having selected a coordinate system in the map plane, we can immediately transform the control points to points in the map plane.

We begin our discussion by showing how to use the cross ratio (see Sections 1.6 and 3.6) in linear photogrammetric reconstruction. If we have located three collinear points A, B, and C in the map plane that correspond to three collinear

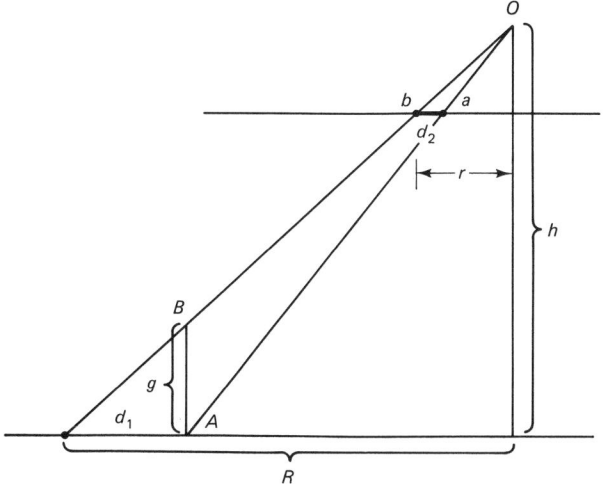

Figure 6.13 Computing heights from a single photograph.

points a, b, and c in the photo plane, then we can find the image D of any other point d on the line in the photo plane through a, b, and c.

4.1 Example. First recall that

$$R(A, B, C, D) = \frac{\text{dist}(A, C) \times \text{dist}(B, D)}{\text{dist}(B, C) \times \text{dist}(A, D)}$$

where $\text{dist}(A, B)$ denotes the directed Euclidean distance from the point A to the point B. Given the distances in Figure 6.14 and using the fact that $R(A, B,$

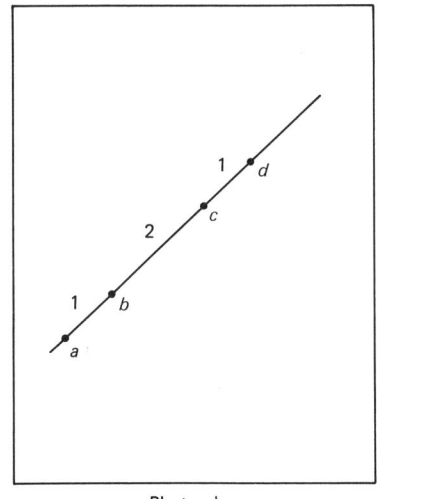
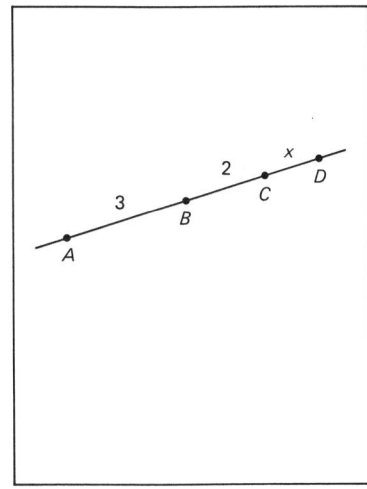

Photo plane   Map plane

Figure 6.14 The use of the cross ratio in linear photogrammetric reconstruction.

Sec. 4 Photogrammetric Reconstruction

$C, D) = R(a, b, c, d)$ (since the transformations describing rectification and map production are projective transformations), it follows that

$$\frac{5(2 + x)}{2(5 + x)} = \frac{3 \times 3}{2 \times 4}$$

and hence $x = \frac{5}{11}$.

We now show how to use the cross ratio and generalized homogeneous coordinates in planar photogrammetric reconstruction. If we have located four points A, B, C, and D, no three of which are collinear, in the map plane that correspond to four points a, b, c, and d, no three of which are collinear, in the photo plane, then we can find the image E of any other point e in the photo plane as follows (see Figure 6.15): Think of a, b, c, and d as the ideal points i_x and i_y, the origin o, and the unit point u of a generalized homogeneous coordinate system in the photo plane. The unit points u_x and u_y are the intersections of the lines ac and bd and the lines bc and ad, respectively. Given any point e in the photo plane, we let e_x and e_y be the intersections of the lines ac and be and the lines bc and ae, respectively, and we compute the cross ratios $R(i_x, o, u_x, e_x)$ and $R(i_y, o, u_y, e_y)$. Similarly, think of A, B, C, and D as the ideal points I_x and I_y, the origin O, and the unit point U of a generalized homogeneous coordinate system in the map plane. We locate the unit points U_x and U_y as the intersections of the lines AC and BD and the lines BC and AD, respectively, and then the points E_x and E_y by requiring that

$$R(I_x, O, U_x, E_x) = R(i_x, o, u_x, e_x) \quad \text{and} \quad R(I_y, O, U_y, E_y) = R(i_y, o, u_y, e_y)$$

The desired point E is the intersection of the lines BE_x and AE_y.

Alternately, we can use the following technique, which works well if one or more of the construction points u_x, u_y, e_x, or e_y is in a position that makes measuring distances difficult: The four lines ab, ac, ae, and ad intersect the line bd in the four points b, u, v, and d (see Figure 6.16(b)). The corresponding line BD intersects AB, AC, and AD at B, U, and D. The cross ratio can be used to find V (see also Exercise 2). The line AV must pass through E. We now repeat this process using the lines ba, be, bd, and bc and the line ac to locate the point W and the line BW (see Figure 6.16(c)). The desired point E is the intersection of the lines AV and BW.

To do photogrammetric reconstruction analytically we must first find the matrix $M^{3 \times 3}$ representing the projective transformation from the photo plane to the map plane. We do this by choosing coordinate systems in the photo and map planes (see Figure 6.17). We can either find the coordinates of the four control points and their images and use the method of Example 3.2.10, or, if we wish more accuracy, we can use more than four control points and the method of Example 3.3.

4.2 Example. Figure 6.18 was made by the attitude indicator simulation program of Section 5.8 with $X_0 = Y_0 = 0$, altitude $h = 10,000$, $\psi = 30°$, $\theta = -75°$ and $\phi = 10°$. To compute M we choose a coordinate system in each plane (see

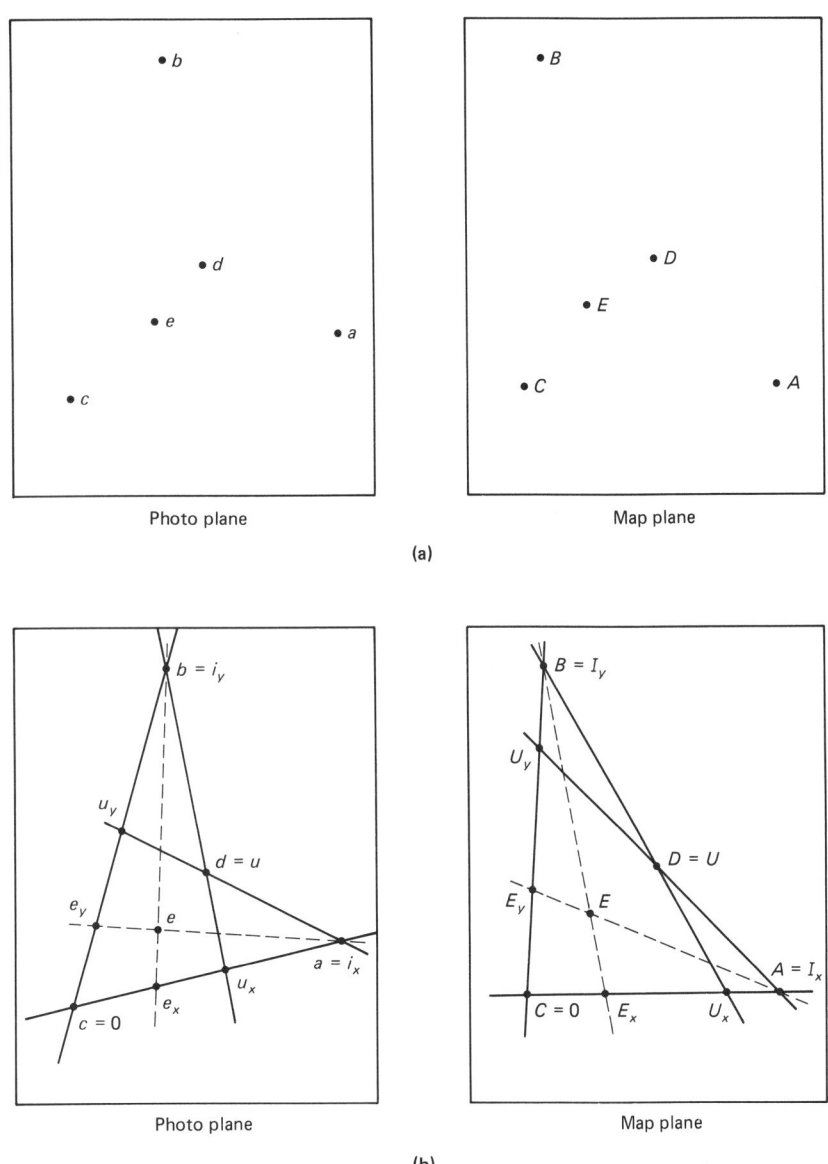

Figure 6.15 The use of the cross ratio and generalized homogeneous coordinates in planar photogrammetric reconstruction.

Figure 6.19) and select eight control points. The coordinates of the points plotted in Figure 6.19(a) and (b) are listed in Figure 6.20. The measurements made in the photo plane were converted to feet, using the fact that the pilot's window in the attitude indicator simulation program is 2 ft square. The matrix M obtained by using the technique of Example 3.3 is

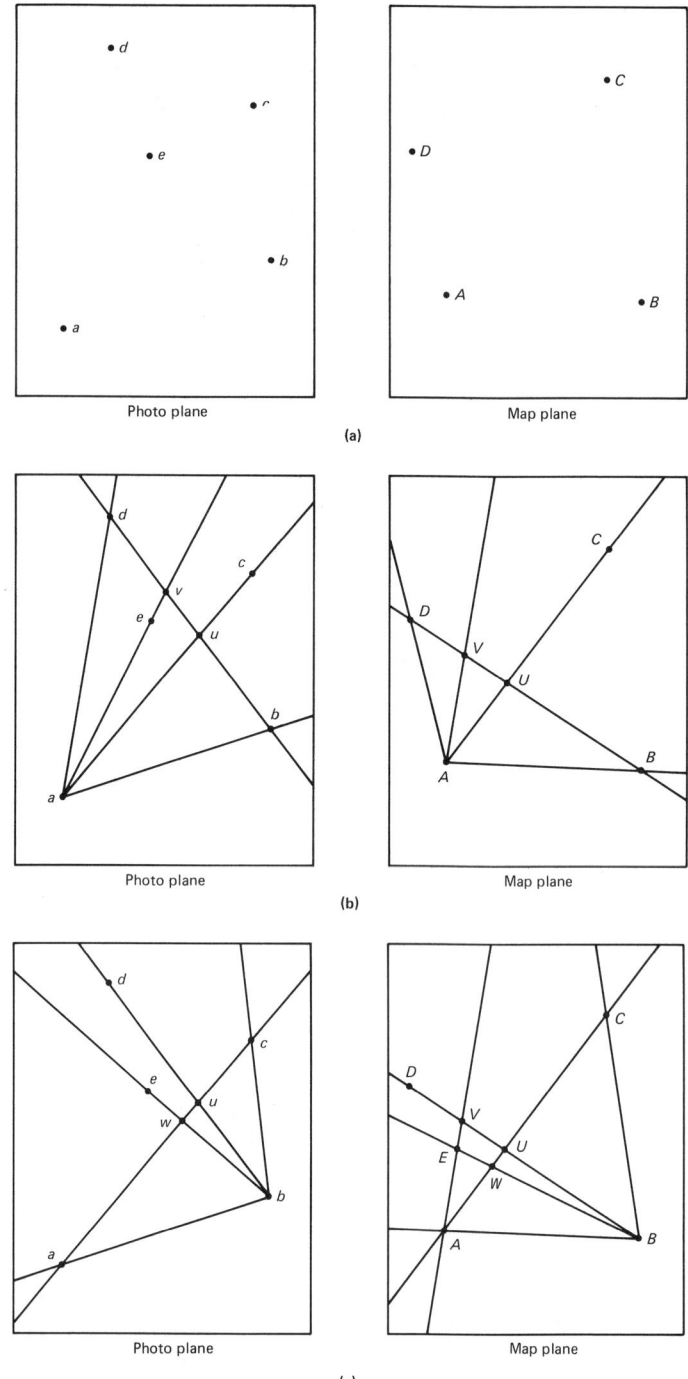

Figure 6.16 Another approach to photogrammetric reconstruction.

348　　　　　　　　　　　　　　　　　　　　　　Reconstruction　　Chap. 6

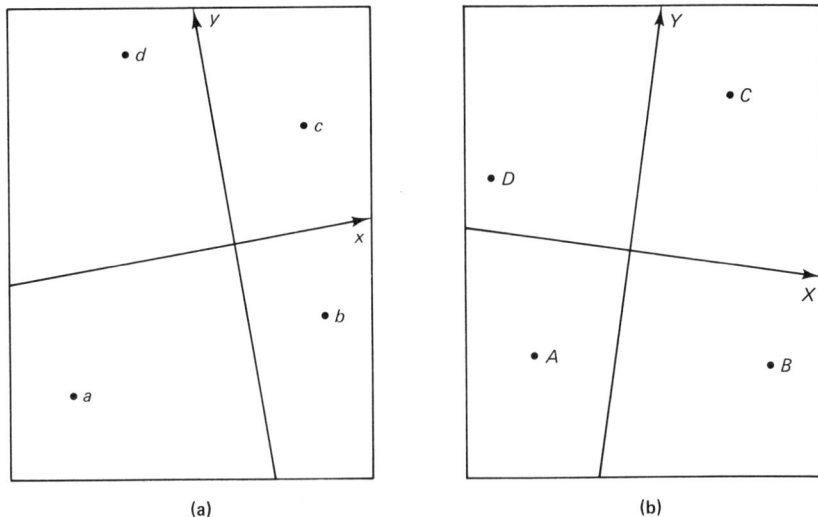

Figure 6.17 Coordinate systems in (a) the photo plane and (b) the map plane.

$$M = \begin{pmatrix} 4084. & -3392. & 0.0248 \\ 3331. & 3911. & -0.1358 \\ -6707. & 666.7 & 0.6241 \end{pmatrix}$$

Using either graphic or analytic techniques, we can rectify as many points as we wish. In practice, however, this is inefficient. Instead, after the tilt and certain other parameters describing orientation elements of the photograph have been found, the tilted photograph can be rephotographed in such a way as to remove displacement due to tilt. To be more specific we need some additional terminology.

Figure 6.18 A simulated aerial photograph.

Sec. 4   Photogrammetric Reconstruction

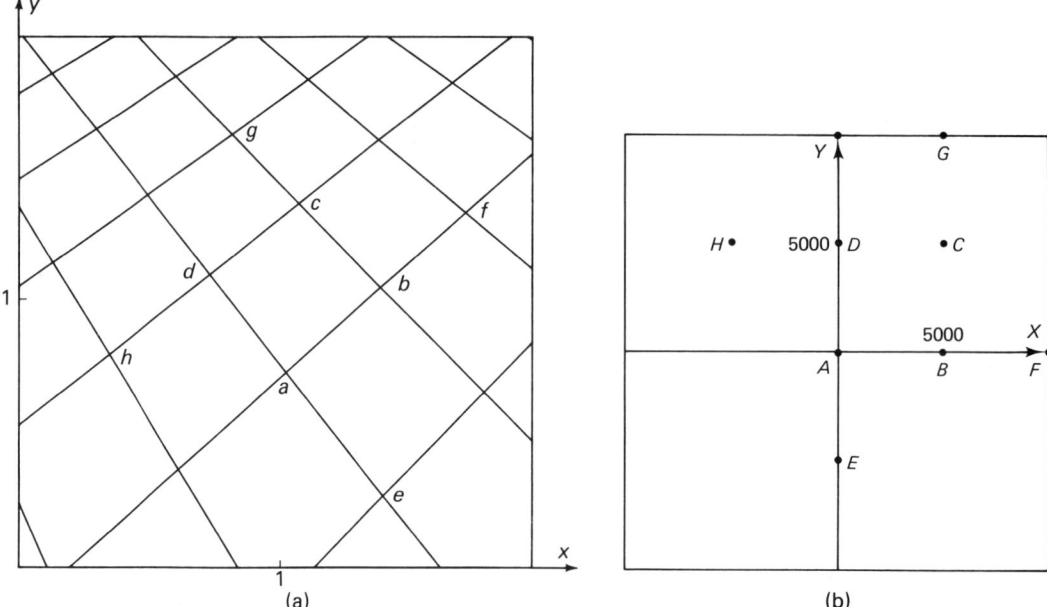

Figure 6.19 Coordinate systems for (a) the photo plane and (b) the map plane.

A number of *orientation elements*—distinguished points and lines—can be identified in a tilted photo plane and in the map plane (see Figure 6.21). The line through O perpendicular to the photo plane meets the photo plane and the map plane at the *principal points* p and P, respectively. The line through O perpendicular to the map plane meets the photo plane and the map plane at the *nadir points* n and N, respectively. The lines pn and PN are the *principal lines* in the photo and map planes, respectively. The plane OPN is the *principal plane* (see Figure 6.22). The tilt t, again, is $\angle PON$, which is the same as $\angle OWN$. The line in the principal plane that bisects $\angle PON$ intersects the principal lines at the *isocenters* i and I, respectively. The line in the photo plane that passes through i and is perpendicular to the principal line is the *isometric parallel*. The *horizon* in the photo plane is the intersection of the photo plane with the plane that passes through O and is parallel to the map plane. The *horizon point* v is the point at which the principal line in the photo plane intersects the horizon.

a	1.0415	0.7337	A	0	0
b	1.4137	1.0558	B	5000	0
c	1.0916	1.3708	C	5000	5000
d	0.7445	1.1024	D	0	5000
e	1.4245	0.2684	E	0	−5000
f	1.7394	1.3386	F	10000	0
g	0.8304	1.6321	G	5000	10000
h	0.3472	0.8017	H	−5000	5000
	(a)			(b)	

Figure 6.20 Coordinates of the points in Figure 6.19.

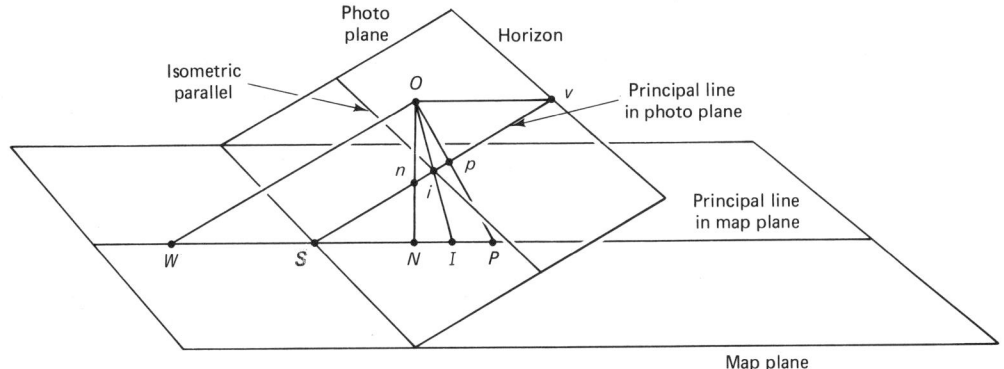

Figure 6.21 Orientation elements in the photo plane and the map plane.

The points W and S and distances f' and h' are also labeled in Figures 6.21 and 6.22.

Here is how the orientation elements in the photo plane are found graphically.

4.3 Example. To locate the orientation elements in Figure 6.18, we first find the principal point p, which is, in this case, the center of the picture. (In general it is the intersection of lines joining the *fiducial marks,* which are made by the aerial camera on the edges of the photograph; see Figure 6.10.) We select a quadrilateral *abcd* in the photograph that is the image of a parallelogram (in fact a square) on the ground. (The fact that the points A, B, C, and D in the map plane form a parallelogram is a simplifying assumption. Without this assumption, the constructions below become much more difficult.)

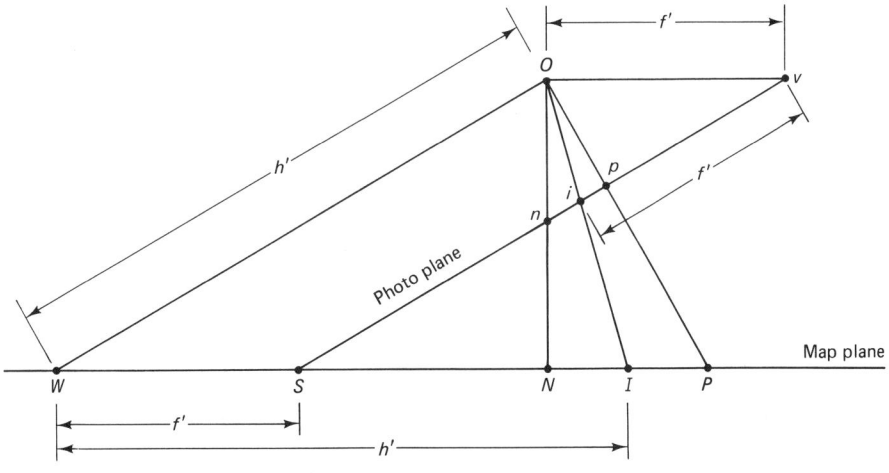

Figure 6.22 The principal plane.

Sec. 4   Photogrammetric Reconstruction

We now construct Figure 6.23. Since the vanishing points of the parallelogram *ABCD* on the ground are ideal, they determine the ideal line. Therefore, the vanishing points of the quadrilateral on the photo plane determine the horizon. The principal line is the line that passes through *p* and is perpendicular to the horizon. The horizon point *v* is the intersection of the horizon and the principal line.

Next we find the line that passes through *p* and is perpendicular to the principal line. On it, at a distance *f* from *p*, we find the point O'; O' is the image of the point *O* in Figure 6.21 under a 90° rotation of the principal plane about the principal line in the photo plane. Since $\angle vOn$ (see Figure 6.21) is a right angle, we can find the nadir point *n* by constructing the right triangle $vO'n$. The isocenter *i* is found by bisecting $\angle pO'n$, and the isometric parallel is the line that passes through *i* and is perpendicular to the principal line.

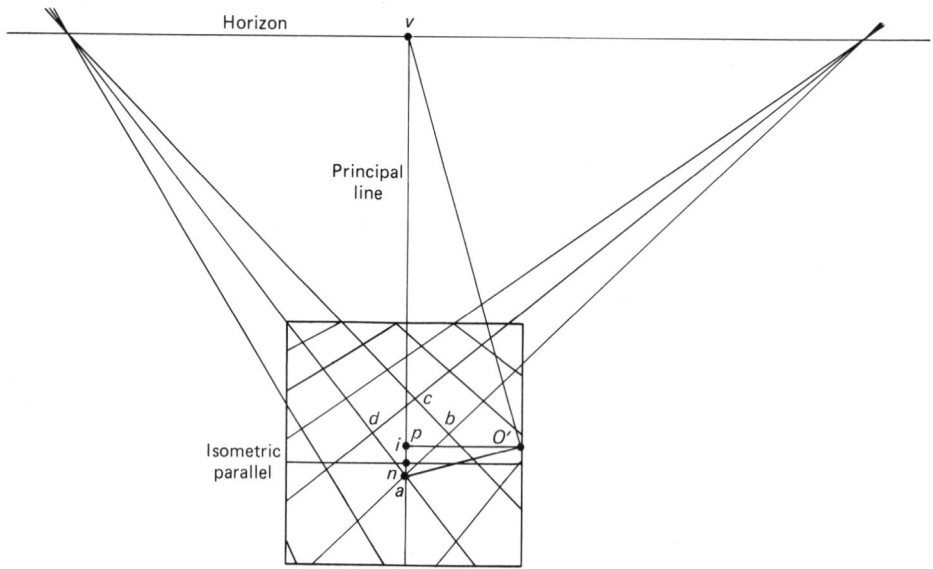

Figure 6.23 Locating the orientation elements in the photo plane.

The orientation elements in the map plane can be found using graphical rectification, as discussed earlier (see Figure 6.24).

To describe the orientation elements analytically, measurements can be made in the photo and map planes. However, it is possible to obtain greater accuracy by computing these numerical values from the components of the rectification matrix. To do this, we use a second method for computing a matrix of the rectification transformation.

The rectification transformation can be written as a composition of five elementary transformations (see [American Society of Photogrammetry]): a rotation

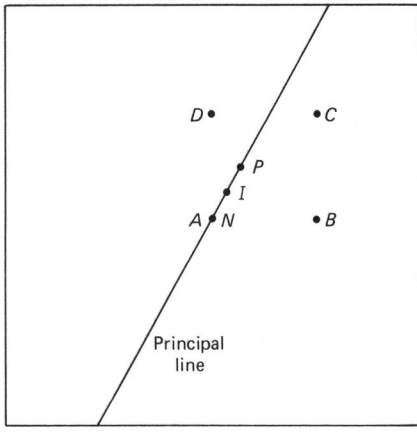

Figure 6.24 The orientation elements in the map plane.

and a translation of the photo plane, a projection to the map plane, and a rotation and translation of the map plane. In each plane, a rectangular coordinate system has the origin at the isocenter with the y'- and Y'-axes along the principal lines and with the positive directions toward the principal points (see Figure 6.25). The parameters that describe the rotations and translations are labeled in Figure 6.25: In the photo plane, (x'_t, y'_t) are the coordinates of the x, y-origin with respect to the x', y'-axes, and the *swing angle s* is the directed angle from the negative y'-axis to the positive y-axis. In the map plane, (X_T, Y_T) are the coordinates of I with respect to the X, Y-axes and the *azimuth angle* α is the directed angle from the positive Y'-axis to the positive Y-axis.

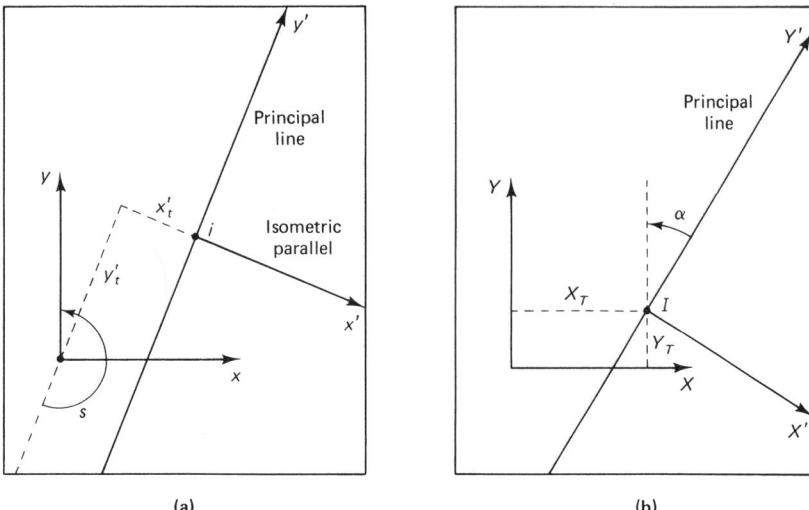

Figure 6.25 Principal coordinate systems in (a) the photo plane and (b) the map plane.

Sec. 4 Photogrammetric Reconstruction

The components of the matrix $M = (m_{ij})$ obtained by the second method, expressed in terms of these parameters, are:

$$m_{11} = -h'\cos(s - \alpha) + X_T \sin s$$
$$m_{12} = -h'\sin(s - \alpha) + Y_T \sin s$$
$$m_{13} = \sin s$$
$$m_{21} = h'\sin(s - \alpha) + X_T \cos s$$
$$m_{22} = -h'\cos(s - \alpha) + Y_T \cos s$$
$$m_{23} = \cos s$$
$$m_{31} = h'(x'_t \cos \alpha + y'_t \sin \alpha) + X_T(f' - y'_t)$$
$$m_{32} = -h'(x'_t \sin \alpha - y'_t \cos \alpha) + Y_T(f' - y'_t)$$
$$m_{33} = f' - y'_t$$

Observe that M is completely determined, not just up to scalar multiplication, if we use these specific five matrices.

The matrix obtained by the first method can be normalized by multiplying by

$$k = \frac{1}{\sqrt{m_{13}^2 + m_{23}^2}}$$

so that the resulting m_{13} and m_{23} are the sine and cosine of an angle. We then have, at least up to sign, the matrix constructed by the second method. We can then use the formulas above to solve for the parameters that describe the orientation elements. In particular, we find:

$$\sin s = m_{13}$$
$$\cos s = m_{23}$$
$$f' - y'_t = m_{33}$$
$$X_T = m_{13}(m_{11} - m_{22}) + m_{23}(m_{12} + m_{21})$$
$$Y_T = m_{13}(m_{12} + m_{21}) - m_{23}(m_{11} - m_{22})$$
$$\tan(s - \alpha) = \frac{m_{13}(m_{13}m_{21} - m_{23}m_{11}) + m_{23}(m_{13}m_{22} - m_{23}m_{12})}{-m_{13}(m_{13}m_{22} - m_{23}m_{12}) + m_{23}(m_{13}m_{21} - m_{23}m_{11})}$$
$$h' = \frac{m_{13}(m_{13}m_{21} - m_{23}m_{11}) + m_{23}(m_{13}m_{22} - m_{23}m_{12})}{\sin(s - \alpha)}$$
$$x'_t = \frac{m_{33}}{h'}\left[\left(\frac{m_{31}}{m_{33}} - X_T\right)\cos \alpha - \left(\frac{m_{32}}{m_{33}} - Y_T\right)\sin \alpha\right]$$
$$y'_t = \frac{m_{33}}{h'}\left[\left(\frac{m_{31}}{m_{33}} - X_T\right)\sin \alpha + \left(\frac{m_{32}}{m_{33}} - Y_T\right)\cos \alpha\right]$$

The tilt can be computed from the formulas

$$\cos t = \frac{f' - y'_p}{f'} \quad \text{and} \quad \sin t = \frac{f}{f'}$$

where y'_p is the y'-coordinate of the principal point, and the height can be computed from the formula

$$h = h'\sin t = \frac{h'f}{f'}$$

4.4 Example. For the matrix found in Example 4.2, $k = 7.244$, so

$$M = \begin{pmatrix} 29584. & -24572. & 0.1800 \\ 24130. & 28331. & -0.9837 \\ -48586. & 4830. & 4.521 \end{pmatrix}$$

Hence $s = 169.6°$, $f' - y'_t = 4.521$, $X_T = 660$, $Y_T = 1153$, $\tan(s - \alpha) = -0.8410$, $\alpha = 29.7°$, $h' = 38{,}470$, $x'_t = -1.16$, $y'_t = -0.67$, $f' = 3.85$, $t = 15.05°$, and $h = 9992$. It is instructive to compare these values with those values used to produce Figure 6.18.

Having determined the parameters which describe the orientation elements of a photograph with tilt, the photograph can be rephotographed in such a way as to remove displacement due to tilt.

SECTION 4 EXERCISES

1. Verify (see Figure 6.13) the equations $d_2/r = d_1/R = g/h$.
2. Draw two lines and pick three points a, b, and c on one and three points A, B, and C on the other. Now choose any fourth point d on the first line and use the cross ratio to produce the image D of d under the projective transformation that takes a to A, b to B, and c to C.
3. Pick four points a, b, c, and d, no three of which are collinear, in one copy of the plane and four points A, B, C, and D, no three of which are collinear, in another copy of the plane. Now choose any fifth point e in the first copy of the plane and use the cross ratio and generalized homogeneous coordinates to produce the image E of e under the projective transformation that takes a to A, b to B, c to C, and d to D.
4. Repeat Exercise 3 using the alternate technique for planar photogrammetric reconstruction.
5. In the alternate technique for planar photogrammetric reconstruction, the cross ratio was used to locate the point V and the ray AE (and later W and the ray BE). Explain why the following *paper strip technique* also works: Lay a paper strip across rays ab, ac, ae, and ad and mark the points b_1, c_1, e_1, and d_1 of intersection (see Figure 6.26). Then lay the strip across the rays AB, AC, and AD so that b_1 lies on AB, c_1 lies on AC, and d_1 lies on AD. Then e_1 determines the ray AE.
6. Why is the nadir point in Figure 6.23 the intersection of the images of two grid lines?
7. What special difficulties would arise in trying to locate the orientation elements in Figure 6.23 in each case?
   (a) The tilt had been a much smaller angle
   (b) The four control points on the ground had not been the vertices of a parallelogram
8. Draw an arbitrary quadrilateral framed in a square, and an arbitrary square framed in a square. Assuming that the framed quadrilateral represents a photograph of the

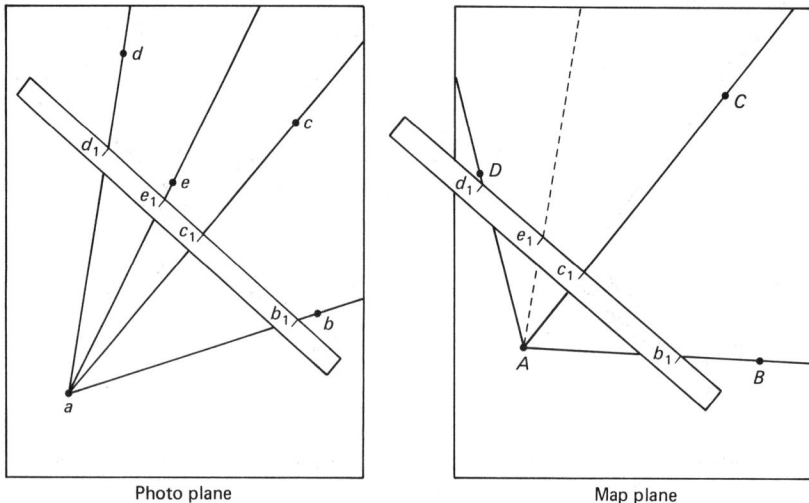

Figure 6.26 The paper strip technique.

framed square, construct the orientation elements in the photo plane and in the map plane.

9. Find the orientation element parameters associated to the matrix
$$M = \begin{pmatrix} 1 & 0 & 1 \\ -1 & -1 & 1 \\ 1 & 2 & -2 \end{pmatrix}$$

10. Find the matrix in Example 4.2 using only the first four control points, and use it to compute the orientation element parameters as in Example 4.4. Compare your results to those obtained using all eight control points.

11. Use the attitude indicator simulation program of Section 5.8 to create a simulated aerial photograph. (Be extremely careful to correct any aspect ratio problems.)
    (a) Locate the orientation elements as in Example 4.3.
    (b) Choose coordinates and determine a representative matrix M of the rectification transformation as in Example 4.2.
    (c) Normalize your matrix M and compute numerical values of the orientation elements as in Example 4.4.
    (d) Verify your computations by making measurements in your drawing of part (a).

7

Conics in Design

INTRODUCTION

This chapter is devoted to the study of the projective theory of conics and applications of this theory to problems of curve fitting. The reason for using projective geometry to study conics is that all nondegenerate conics (circles, ellipses, parabolas, and hyperbolas) and degenerate conics (one line, two lines, or a point) can be considered in a consistent fashion. In Section 1 we see how conics are described projectively by quadratic forms. Using this description we show that the nondegenerate conics are all projectively equivalent. In Section 2 we solve the problem of finding the equation of the conic that passes through five given points. This problem is solved in two different ways, the second way (due to Liming) being adapted in Section 3 to solve five additional conic-fitting problems.

SECTION 1: THE PROJECTIVE THEORY OF CONICS

In this section we consider the projective theory of conics. Applications of this theory of conics to design are presented in Sections 2 and 3.

Discussion of the projective theory of conics requires the use of quadratic forms.

1.1 Definition. A *quadratic form* in three variables x, y, and z is a homogeneous polynomial of degree 2:

$$Ax^2 + By^2 + Cz^2 + Dxy + Exz + Fyz$$

Every quadratic form can be represented as a product $\mathbf{p}Q\mathbf{p}^T$ where Q is a symmetric 3×3 matrix and $\mathbf{p} = \langle x, y, z \rangle$:

$$Ax^2 + By^2 + Cz^2 + Dxy + Exz + Fyz = (x\ y\ z)\begin{pmatrix} A & \frac{1}{2}D & \frac{1}{2}E \\ \frac{1}{2}D & B & \frac{1}{2}F \\ \frac{1}{2}E & \frac{1}{2}F & C \end{pmatrix}\begin{pmatrix} x \\ y \\ z \end{pmatrix} = \mathbf{p}Q\mathbf{p}^T$$

Conversely, every product of the form $\mathbf{p}Q\mathbf{p}^T$, Q symmetric, represents a quadratic form.

1.2 Example. The quadratic form
$$3x^2 - y^2 + 2z^2 + xy + 2xz - 3yz$$
is the product
$$(x\ y\ z)\begin{pmatrix} 3 & \frac{1}{2} & 1 \\ \frac{1}{2} & -1 & -\frac{3}{2} \\ 1 & -\frac{3}{2} & 2 \end{pmatrix}\begin{pmatrix} x \\ y \\ z \end{pmatrix}$$

1.3 Definition. A *conic* is the set of all points $P = P[x, y, z]$ in the projective plane whose homogeneous coordinates $[x, y, z]$ satisfy a homogeneous equation of degree 2:
$$Ax^2 + By^2 + Cz^2 + Dxy + Exz + Fyz = 0$$

Because of the homogeneity of the equation of a conic, the set of points described in Definition 1.3 is well defined (see Exercise 2.2.21).

The equation of a conic can be written in matrix form
$$\mathbf{p}Q\mathbf{p}^T = 0$$
where Q is symmetric and $\mathbf{p}$ is a representative of $[\mathbf{p}]$. Observe that the same conic is defined by the matrix Q and by the matrix tQ for any $t \neq 0$. In particular, Q and $-Q$ define the same conic.

To relate the conics of projective geometry to the conics of Euclidean geometry, recall that if $z \neq 0$, then $[x, y, z] = [x/z, y/z, 1]$. Given the equation of a projective conic
$$Ax^2 + By^2 + Cz^2 + Dxy + Exz + Fyz = 0$$
divide by z^2 to obtain
$$A\left(\frac{x}{z}\right)^2 + B\left(\frac{y}{z}\right)^2 + C + D\left(\frac{x}{z}\right)\left(\frac{y}{z}\right) + E\left(\frac{x}{z}\right) + F\left(\frac{y}{z}\right) = 0$$
If
$$x' = \frac{x}{z} \quad \text{and} \quad y' = \frac{y}{z}$$
this equation becomes
$$A(x')^2 + B(y')^2 + C + Dx'y' + Ex' + Fy' = 0$$
This is the equation of a Euclidean conic. Reversing this process, we see that the equation of every Euclidean conic corresponds to the equation of a projective conic. Passing back and forth between the homogeneous and Cartesian equations of a conic is simply a matter of substitution and multiplying or dividing by z^2. (There is one exception: the projective conic whose equation is $z^2 = 0$ has no

corresponding Euclidean conic. See Exercise 6.) Alternately, if the homogeneous equation of a conic has the form
$$\langle x, y, z \rangle Q \langle x, y, z \rangle^T = 0$$
then the Cartesian equation of the conic has the form
$$\langle x', y', 1 \rangle Q \langle x', y', 1 \rangle^T = 0$$

As we shall see after Proposition 1.16, a nondegenerate projective conic may contain one or two ideal points, which, of course, do not lie on its Euclidean counterpart.

1.4 Example. The circle with the Cartesian equation $x^2 + y^2 = 1$, the *unit circle*, has the homogeneous equation $x^2 + y^2 - z^2 = 0$ and the quadratic form
$$Q = \begin{pmatrix} 1 & 0 & 0 \\ 0 & 1 & 0 \\ 0 & 0 & -1 \end{pmatrix}$$
The ellipse with the Cartesian equation $x^2/a^2 + y^2/b^2 = 1$ has the homogeneous equation $x^2/a^2 + y^2/b^2 - z^2 = 0$ and the quadratic form
$$Q = \begin{pmatrix} \frac{1}{a^2} & 0 & 0 \\ 0 & \frac{1}{b^2} & 0 \\ 0 & 0 & -1 \end{pmatrix}$$
The parabola with the Cartesian equation $y = x^2$ has the homogeneous equation $x^2 - yz = 0$ and the quadratic form
$$Q = \begin{pmatrix} 1 & 0 & 0 \\ 0 & 0 & -\frac{1}{2} \\ 0 & -\frac{1}{2} & 0 \end{pmatrix}$$

We now digress briefly to discuss projective equivalence. Recall that a projective transformation is a perspective transformation or a composite of more than one perspective transformation.

1.5 Definition. Two planar figures in projective three-space are *projectively equivalent* if and only if there is a projective transformation taking one figure to the other.

For example, in Section 1.3 we observed that any pair of lines is projectively equivalent to any other pair of lines. Also, the Fundamental Theorem of Projective Geometry allows us to define a transformation that takes the vertices of a square to the vertices of any other quadrilateral; thus every quadrilateral is projectively equivalent to a square (see also Exercise 1.3.7). Furthermore, any two conics on the same cone in Euclidean three-space are projectively equivalent. In this latter case, the projective equivalence is established by a perspective transformation whose center is the vertex of the cone.

We now consider (general) figures in the projective plane.

1.6 Definition. Two figures C_1 and C_2 in the projective plane are *projectively equivalent* if and only if there are planes π and μ in projective three-space, parametrizations T_π of π and T_μ of μ, and a projective transformation T of projective three-space taking $T_\pi(C_1)$ to $T_\mu(C_2)$ (see Figure 7.1).

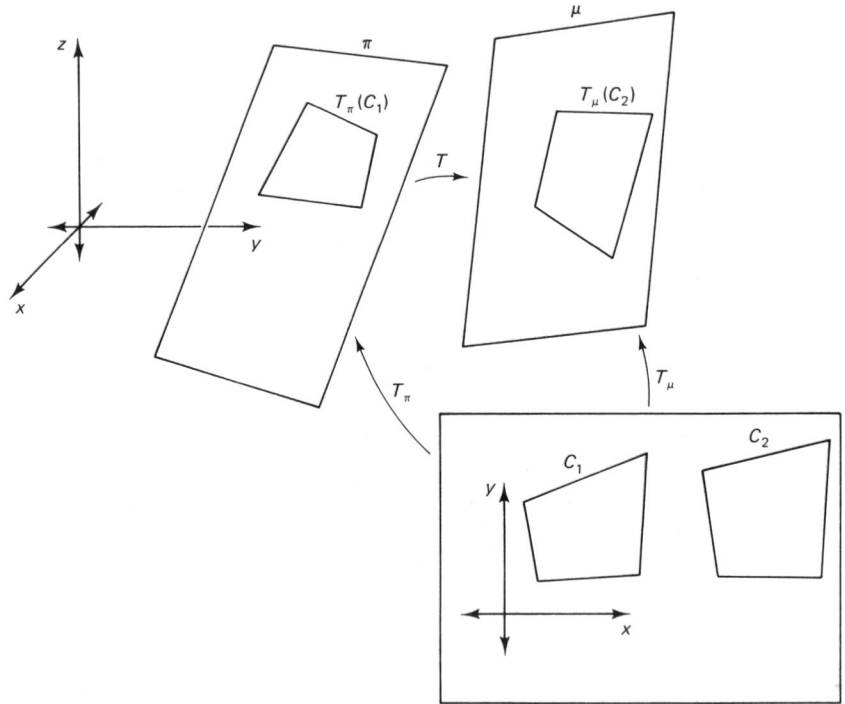

Figure 7.1 Projective equivalence of two figures in the projective plane.

As a consequence of Proposition 3.2.7, two figures in the projective plane are projectively equivalent if and only if there is a projective transformation of the projective plane that takes one figure to the other figure. Furthermore, since a projective transformation of the projective plane is represented by right multiplication by an invertible 3×3 matrix, we have the following result.

1.7 Proposition. Two figures in the projective plane are projectively equivalent if and only if there is an invertible matrix $M^{3\times3}$ such that the points of one figure correspond to the points of the other under right multiplication by M.

We now return to conics defined by quadratic forms.

1.8 Proposition. If Q represents a quadratic form and $M^{3\times3}$ is an invertible matrix, MQM^T also represents a quadratic form. The conics determined by Q and by $\pm MQM^T$ are projectively equivalent, and the matrix M represents the transformation that defines the equivalence.

Proof. The matrix MQM^T represents a quadratic form because it is symmetric. If $\mathbf{p}$ satisfies $\mathbf{p}Q\mathbf{p}^T = 0$ and $\mathbf{p} = \mathbf{p}'M$, then $\mathbf{p}'$ satisfies the equation
$$(\mathbf{p}'M)Q(\mathbf{p}'M)^T = 0$$
or
$$\mathbf{p}'(MQM^T)\mathbf{p}'^T = 0$$
Thus right multiplication by M represents a projective transformation that maps the points $P'[\mathbf{p}']$ of the conic determined by MQM^T to the points $P[\mathbf{p}]$ of the conic determined by Q. Hence these conics are projectively equivalent. ∎

1.9 Example. Let M be the matrix of translation
$$M = \begin{pmatrix} 1 & 0 & 0 \\ 0 & 1 & 0 \\ h & k & 1 \end{pmatrix}$$
and Q the quadratic form that defines the unit circle:
$$Q = \begin{pmatrix} 1 & 0 & 0 \\ 0 & 1 & 0 \\ 0 & 0 & -1 \end{pmatrix}$$
Then
$$MQM^T = \begin{pmatrix} 1 & 0 & h \\ 0 & 1 & k \\ h & k & h^2 + k^2 - 1 \end{pmatrix}$$
The homogeneous equation of the corresponding conic $\mathbf{p}'(MQM^T)\mathbf{p}'^T = 0$ is
$$(x')^2 + 2hx'z' + h^2(z')^2 + (y')^2 + 2ky'z' + k^2(z')^2 - (z')^2 = 0$$
or
$$(x' + hz')^2 + (y' + kz')^2 = (z')^2$$
This is the homogeneous equation of a translated circle (see Figure 7.2).

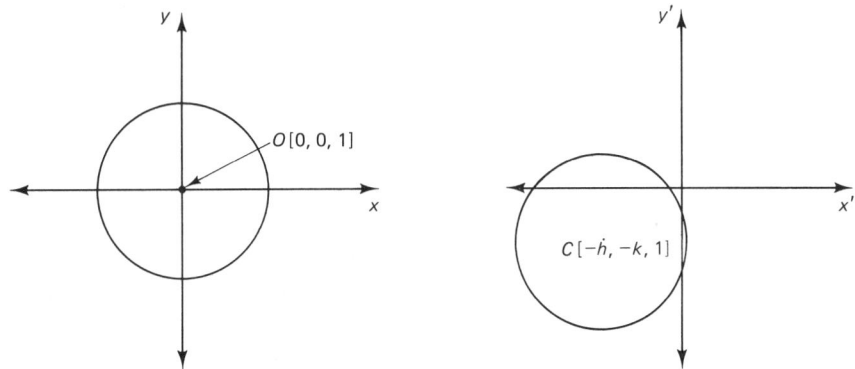

Figure 7.2 A circle and a projectively equivalent translated circle.

1.10 Example. Let M be the scaling transformation

$$M = \begin{pmatrix} \dfrac{1}{a} & 0 & 0 \\ 0 & \dfrac{1}{b} & 0 \\ 0 & 0 & 1 \end{pmatrix}$$

and Q again be the quadratic form for the unit circle. Then

$$MQM^T = \begin{pmatrix} \dfrac{1}{a^2} & 0 & 0 \\ 0 & \dfrac{1}{b^2} & 0 \\ 0 & 0 & -1 \end{pmatrix}$$

The corresponding conic is an ellipse (see Figure 7.3). By Proposition 1.8, this ellipse and the unit circle are projectively equivalent.

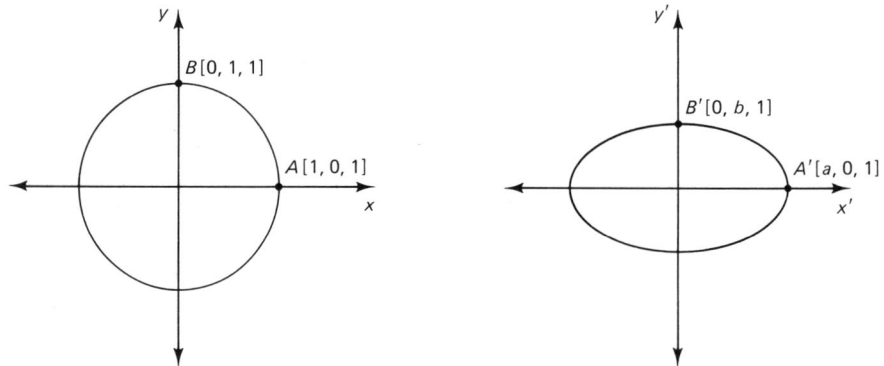

Figure 7.3 A circle and a projectively equivalent ellipse.

We now develop an algorithm for simplifying quadratic forms. We use this algorithm to demonstrate that all nondegenerate conics are projectively equivalent.

Suppose we are given a symmetric matrix Q. When an invertible matrix M is chosen correctly, the matrix MQM^T is simpler than Q. In fact, we can choose M so that the matrix MQM^T has nonzero entries only on the diagonal; furthermore, the diagonal entries are either 1, -1 or 0. The quadratic form represented by MQM^T is simpler than the quadratic form represented by Q.

We build up M as a product of elementary matrices. (Recall that an elementary matrix is formed from the identity matrix by one of the following elementary row operations: multiplying a row by a nonzero constant, interchanging two rows, or adding a multiple of one row to another.) If E is an elementary matrix, the matrix EQ is the result of performing an elementary row operation on Q. The matrix EQE^T is the result of performing the same *column* operation on EQ. The following example illustrates one such pair of operations.

1.11 Example. If
$$Q = \begin{pmatrix} 0 & 1 & 2 \\ 1 & 0 & 4 \\ 2 & 4 & 0 \end{pmatrix}$$
we can add the second row to the first row and the second column to the first column:
$$\begin{pmatrix} 1 & 1 & 0 \\ 0 & 1 & 0 \\ 0 & 0 & 1 \end{pmatrix} \begin{pmatrix} 0 & 1 & 2 \\ 1 & 0 & 4 \\ 2 & 4 & 0 \end{pmatrix} \begin{pmatrix} 1 & 0 & 0 \\ 1 & 1 & 0 \\ 0 & 0 & 1 \end{pmatrix} = \begin{pmatrix} 2 & 1 & 6 \\ 1 & 0 & 4 \\ 6 & 4 & 0 \end{pmatrix}$$

1.12 Proposition. By a sequence of identical row and column operations a symmetric matrix,
$$Q = \begin{pmatrix} a & d & e \\ d & b & f \\ e & f & c \end{pmatrix}$$
can be reduced to a matrix of the form
$$\begin{pmatrix} p & 0 & 0 \\ 0 & q & 0 \\ 0 & 0 & r \end{pmatrix}$$
where each of p, q, and r is either $+1$, -1 or 0.

Proof. If $Q = 0$ we are through. Hence assume $Q \neq 0$. If the diagonal elements of Q are all zero, we perform the operation illustrated in Example 1.11 to create a matrix with a nonzero diagonal entry. If this nonzero entry is not in the (1, 1) position of Q, we can move it there by interchanging rows and columns. Hence we may assume $a \neq 0$. By adding a multiple of row 1 to row 2 and of column 1 to column 2, we obtain
$$\begin{pmatrix} a & 0 & e' \\ 0 & b' & f' \\ e' & f' & c' \end{pmatrix}$$
Using rows 1 and 3 and columns 1 and 3, we obtain
$$\begin{pmatrix} a & 0 & 0 \\ 0 & b'' & f'' \\ 0 & f'' & c'' \end{pmatrix}$$
We continue in this way, now using operations that involve neither row 1 nor column 1, to obtain a matrix of the form
$$\begin{pmatrix} a & 0 & 0 \\ 0 & b & 0 \\ 0 & 0 & c \end{pmatrix}$$
If a is not $+1$, -1, or 0, we multiply on both sides by
$$\begin{pmatrix} \frac{1}{\sqrt{|a|}} & 0 & 0 \\ 0 & 1 & 0 \\ 0 & 0 & 1 \end{pmatrix}$$

We do the same for b and c. ∎

1.13 Example. To simplify the quadratic form $3x^2 - y^2 + 10z^2 + 12xz - 4yz$, we first find its matrix

$$Q = \begin{pmatrix} 3 & 0 & 6 \\ 0 & -1 & -2 \\ 6 & -2 & 10 \end{pmatrix}$$

We subtract twice the first row from the third and then twice the first column from the third to obtain

$$\begin{pmatrix} 3 & 0 & 0 \\ 0 & -1 & -2 \\ 0 & -2 & -2 \end{pmatrix}$$

Now we subtract twice the second row from the third and twice the second column from the third to obtain

$$\begin{pmatrix} 3 & 0 & 0 \\ 0 & -1 & 0 \\ 0 & 0 & 2 \end{pmatrix}$$

We multiply on each side by

$$\begin{pmatrix} \frac{1}{\sqrt{3}} & 0 & 0 \\ 0 & 1 & 0 \\ 0 & 0 & \frac{1}{\sqrt{2}} \end{pmatrix}$$

to obtain

$$\begin{pmatrix} 1 & 0 & 0 \\ 0 & -1 & 0 \\ 0 & 0 & 1 \end{pmatrix}$$

We may also interchange the second and third rows and columns to obtain

$$\begin{pmatrix} 1 & 0 & 0 \\ 0 & 1 & 0 \\ 0 & 0 & -1 \end{pmatrix}$$

Thus the simplified quadratic form is $x^2 + y^2 - z^2$: In this case

$$M = \begin{pmatrix} 1 & 0 & 0 \\ 0 & 0 & 1 \\ 0 & 1 & 0 \end{pmatrix} \begin{pmatrix} \frac{\sqrt{3}}{3} & 0 & 0 \\ 0 & 1 & 0 \\ 0 & 0 & \frac{\sqrt{2}}{2} \end{pmatrix} \begin{pmatrix} 1 & 0 & 0 \\ 0 & 1 & 0 \\ 0 & -2 & 1 \end{pmatrix} \begin{pmatrix} 1 & 0 & 0 \\ 0 & 1 & 0 \\ -2 & 0 & 1 \end{pmatrix}$$

$$= \begin{pmatrix} \frac{\sqrt{3}}{3} & 0 & 0 \\ -\sqrt{2} & -\sqrt{2} & \frac{\sqrt{2}}{2} \\ 0 & 1 & 0 \end{pmatrix}$$

and

$$\begin{pmatrix} 1 & 0 & 0 \\ 0 & 1 & 0 \\ 0 & 0 & -1 \end{pmatrix} = MQM^T$$

Given a quadratic form, we have just seen that a sequence of identical row and column operations can reduce its matrix Q to a diagonal matrix MQM^T whose diagonal entries are $+1$, -1, or 0. Since the corresponding conics are projectively equivalent and since Q and $-Q$ determine the same conic, it follows that every conic is projectively equivalent to a conic whose matrix has one of the following forms:

$$\begin{pmatrix} 1 & 0 & 0 \\ 0 & 1 & 0 \\ 0 & 0 & 1 \end{pmatrix} \quad \begin{pmatrix} 1 & 0 & 0 \\ 0 & 1 & 0 \\ 0 & 0 & -1 \end{pmatrix} \quad \begin{pmatrix} 1 & 0 & 0 \\ 0 & 1 & 0 \\ 0 & 0 & 0 \end{pmatrix} \quad \begin{pmatrix} 1 & 0 & 0 \\ 0 & -1 & 0 \\ 0 & 0 & 0 \end{pmatrix} \quad \begin{pmatrix} 1 & 0 & 0 \\ 0 & 0 & 0 \\ 0 & 0 & 0 \end{pmatrix} \quad \begin{pmatrix} 0 & 0 & 0 \\ 0 & 0 & 0 \\ 0 & 0 & 0 \end{pmatrix}$$
(a) (b) (c) (d) (e) (f)

There are no points in the conic determined by (a). Form (b) is the form for the unit circle. The conic determined by (f) consists of all points in the projective plane. The conics determined by (c), (d), and (e) are illustrated in Figure 7.4.

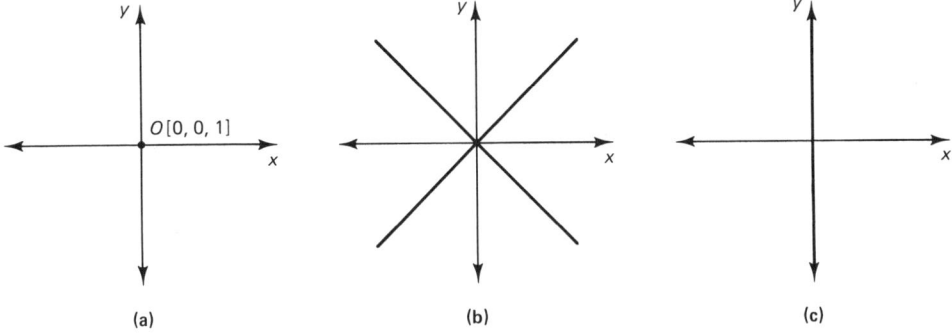

Figure 7.4 Degenerate conics. (a) A point $(x^2 + y^2 = 0)$. (b) Two lines $(x^2 - y^2 = 0)$. (c) One line $(x^2 = 0)$.

Those conics whose quadratic form is of type (b) include all the familiar examples: circles, ellipses, parabolas, and hyperbolas. They are the *nondegenerate* conics. Their graphs include no straight lines or isolated points. Their quadratic forms have nonzero determinants. They are all projectively equivalent to the unit circle. We have thus established the following result.

1.14 Proposition. All nondegenerate conics are projectively equivalent.

One important application of this result is that given the matrix M of a nondegenerate conic whose quadratic form is represented by the matrix Q, there is a natural parametrization of the conic. Since

$$\{P[\cos \theta, \sin \theta, 1] \mid \theta \in [0, 2\pi]\}$$

is a parametrization of the unit circle and since the conic is the image of the unit circle under the projective transformation represented by M (see Propositions 1.7 and 1.8),

$$\{P[\langle \cos \theta, \sin \theta, 1 \rangle M] \mid \theta \in [0, 2\pi]\}$$

is a parametrization of the conic.

1.15 Example. The conic $3x^2 - y^2 + 10z^2 + 12xz - 4yz = 0$ of Example 1.13 has the parametrization

$$\{P[\langle \cos \theta, \sin \theta, 1 \rangle M] \mid \theta \in [0, 2\pi]\}$$
$$= \{P[2\sqrt{3} \cos \theta - 6\sqrt{2} \sin \theta, -6\sqrt{2} \sin \theta + 6, 3\sqrt{2} \sin \theta] \mid \theta \in [0, 2\pi]\}$$

We end this section by discussing the points of intersection of a line and a nondegenerate conic. First observe that the three possible relationships of a line and the unit circle are illustrated in Figure 7.5 (see Exercise 20). The line that meets the unit circle in one point is tangent to the circle at that point.

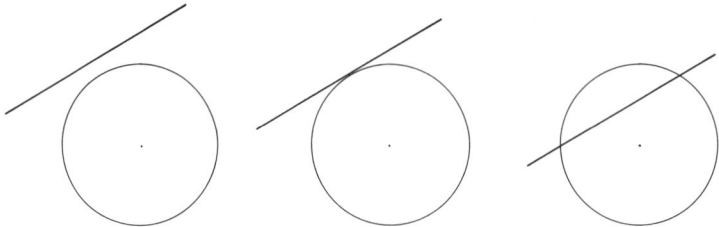

Figure 7.5 A line and a circle intersect in zero, one, or two points.

1.16 Proposition. A nondegenerate conic and a line in the projective plane either

(a) Have no points in common
(b) Have one point in common (in which case the line is tangent to the conic)
(c) Have two points in common

Proof. Let C_0 be the unit circle with quadratic form Q_0 and let C be a nondegenerate conic with quadratic form Q. By Proposition 1.14, $Q_0 = MQM^T$, where M is the matrix of a projective transformation. Since M is invertible, C and any line $l[\mathbf{l}]$ have the same number of points of intersection as the transformed conic C_0 (the unit circle) and the transformed line whose homogeneous coordinates are $[M^{-1}l]$. This number, as we noted above, is zero, one or two. ∎

This result explains the number of ideal points on certain conics: A parabola is tangent to the ideal line, and it contains one ideal point. A hyperbola crosses the ideal line in two points.

We discuss tangents to conics further in Section 3.

SECTION 1 EXERCISES

1. Find the matrix of each of the following quadratic forms.
   (a) $x^2 - 3y^2 - 2z^2 + 4xy - 2xz - 10yz$
   (b) $x^2 + y^2 + z^2$
   (c) $xy - xz$

2. What homogeneous equation corresponds to each of the following Cartesian equations?
   (a) $3x^2 + 7xy + y^2 + 2x - y = 8$
   (b) $x^2 + 4x - 8 = 0$
   (c) $x^2 + 4x - 8 = y$
   (d) $x + 2y = 3$

3. Verify that if one choice of representative homogeneous coordinates for P satisfies
   $$Ax^2 + By^2 + Cz^2 + Dxy + Exz + Fyz = 0$$
   then so does any other.

4. Verify that if the symmetric matrices Q and Q' determine the same conic, then $Q = tQ'$ for some scalar t.

5. Verify that MQM^T is symmetric if Q is symmetric.

6. What Euclidean equation corresponds to the projective equation $z^2 = 0$? What is the graph of $z^2 = 0$ in the projective plane? What is the graph of $1 = 0$ in the Euclidean plane?

7. Reduce each of the following symmetric matrices to the simplified form of Proposition 1.12.

   (a) $\begin{pmatrix} 1 & 0 & 1 \\ 0 & -1 & -2 \\ 1 & -2 & -3 \end{pmatrix}$
   (b) $\begin{pmatrix} 1 & 2 & 0 \\ 2 & 14 & 3 \\ 0 & 3 & 1 \end{pmatrix}$
   (c) $\begin{pmatrix} 0 & 0 & 1 \\ 0 & 1 & 0 \\ 1 & 0 & 0 \end{pmatrix}$
   (d) $\begin{pmatrix} 1 & 1 & 1 \\ 1 & 1 & 1 \\ 1 & 1 & 1 \end{pmatrix}$

8. For each part of Exercise 7, find a parametrization of the conic whose quadratic form is represented by the given matrix.

9. Implement a program that will use the parametrization of a conic (similar to the one discussed, for example, in Example 1.15) to draw the conic. For each part of Exercise 7, use this program to draw the conic whose quadratic form is represented by the given matrix.

10. Find the quadratic form for the circle whose center is the origin and radius is r. Find the matrix of a projective equivalence between this circle and the unit circle.

11. In each case, use the method illustrated in Example 1.15 to show that the parametrization
$$\{P[\cos \theta, \sin \theta, 1] \mid \theta \in [0, 2\pi]\}$$
of the unit circle gives rise to a parametrization similar to the given parametrization. (Such parametrizations are not unique. Why?)
    (a) $\{P[b \cos \theta, a \sin \theta, ab] \mid \theta \in [0, 2\pi]\}$ of the ellipse whose Cartesian equation is $x^2/a^2 + y^2/b^2 = 1$
    (b) $\{P[\cos \theta, 1 - \sin \theta, 1 + \sin \theta] \mid \theta \in [0, 2\pi]\}$ of the parabola whose Cartesian equation is $y = x^2$
    (c) $\{P[\sec \theta, \tan \theta, 1] \mid \theta \in [0, 2\pi]\}$ of the hyperbola whose Cartesian equation is $x^2 - y^2 = 1$

12. Find the matrix of a projective equivalence between each of the following.
    (a) The parabola $y = 4 - x^2$ and the circle $x^2 + y^2 = 1$
    (b) The hyperbola $xy = 1$ and the circle $x^2 + y^2 = 1$
    (c) The above parabola and the above hyperbola

13. Find the matrix of a projective equivalence between the ellipses $x^2/9 + y^2/16 = 1$ and $x^2/16 + y^2/25 = 1$.

14. Show that the directions of the asymptotes of the hyperbola
$$\frac{x^2}{a^2} - \frac{y^2}{b^2} = 1$$
are the ideal points of the hyperbola (see Exercise 2.1.10).

15. Find the ideal point of the parabola $y = x^2$.

16. The union of the points on two lines l_1 and l_2 is a degenerate conic. If the lines have equations $ax + by + c = 0$ and $dx + ey + f = 0$, show that the equation of this conic is
$$C(x, y) = (ax + by + c)(dx + ey + f) = 0$$
If
$$\mathbf{l}_1 = \begin{pmatrix} a \\ b \\ c \end{pmatrix} \quad \text{and} \quad \mathbf{l}_2 = \begin{pmatrix} d \\ e \\ f \end{pmatrix}$$
are representative homogeneous coordinates of l_1 and l_2 (written as column matrices), show that the matrix of the degenerate quadratic form corresponding to C is
$$Q = \tfrac{1}{2}(\mathbf{l}_1 \mathbf{l}_2^T + \mathbf{l}_2 \mathbf{l}_1^T)$$

17. (Continuation of Exercise 16). For each of the following pairs of lines, find the matrix of the degenerate conic they determine. Show that each matrix simplifies to the matrix
$$\begin{pmatrix} 1 & 0 & 0 \\ 0 & -1 & 0 \\ 0 & 0 & 0 \end{pmatrix}$$
    (a) $x = 4$, $y = -5$ (b) $x = 4$, $x = 6$
    (c) $x + 3y = 6$, $x - y = 0$ (d) $x = 0$ and the ideal line

18. The points on the line $l[\mathbf{l}]$ determine a degenerate conic. If the equation of the line is $ax + by + c = 0$, the equation of the conic is $(ax + by + c)^2 = 0$. Show that the matrix of the quadratic form for this conic is $Q = \mathbf{l}\mathbf{l}^T$.

19. (Continuation of Exercise 18). For each of the following lines, find the matrix of the degenerate conic it determines. Show that each matrix simplifies to

$$\begin{pmatrix} 1 & 0 & 0 \\ 0 & 0 & 0 \\ 0 & 0 & 0 \end{pmatrix}$$

(a) $x = 0$ (b) $x = 6$
(c) $y = 0$ (d) $3x - 5y + 1 = 0$
(e) The ideal line

20. Verify that a line intersects a circle in zero, one, or two points and that if it intersects in one point the line is tangent to the circle. (*Hint:* Show that the points of intersection are determined as the roots of a quadratic equation.)

21. Find the points where the following lines and conics intersect. Each equation is given in its homogeneous form.
    (a) $x = 0$, $x^2 - y^2 - z^2 = 0$ (b) $y = 0$, $x^2 - y^2 - z^2 = 0$
    (c) $z = 0$, $x^2 - y^2 - z^2 = 0$ (d) $y - 2z = 0$, $xz - y^2 = 0$
    (e) $x - z = 0$, $x^2 + y^2 - z^2 = 0$ (f) $z = 0$, $x^2 + y^2 - z^2 = 0$
    (g) $z = 0$, $x^2 - yz = 0$

SECTION 2: THE CONIC THROUGH FIVE POINTS

Designers must often find the equation of a conic that satisfies certain requirements. The requirements, in the form of points that must lie on the conic or lines that must be tangent to the conic, are used to determine (up to a common factor) the coefficients of the left-hand side of the equation

$$Ax^2 + By^2 + Cz^2 + Dxy + Exz + Fyz = 0$$

The problem we consider in this section is how to find the equation of the conic that passes through the five points $P_1[x_1, y_1, z_1]$, $P_2[x_2, y_2, z_2]$, $P_3[x_3, y_3, z_3]$, $P_4[x_4, y_4, z_4]$ and $P_5[x_5, y_5, z_5]$. The existence and uniqueness of a conic that solves this problem, as well as the first method of solving the problem, follow from the next proposition.

2.1 Proposition. There exists a unique conic passing through any given five points, no four of which are collinear. The conic is nondegenerate if no three of the points are collinear; otherwise it is degenerate.

Proof. A homogeneous quadratic equation is given by

$$\begin{vmatrix} x^2 & y^2 & z^2 & xy & xz & yz \\ x_1^2 & y_1^2 & z_1^2 & x_1y_1 & x_1z_1 & y_1z_1 \\ x_2^2 & y_2^2 & z_2^2 & x_2y_2 & x_2z_2 & y_2z_2 \\ x_3^2 & y_3^2 & z_3^2 & x_3y_3 & x_3z_3 & y_3z_3 \\ x_4^2 & y_4^2 & z_4^2 & x_4y_4 & x_4z_4 & y_4z_4 \\ x_5^2 & y_5^2 & z_5^2 & x_5y_5 & x_5z_5 & y_5z_5 \end{vmatrix} = 0$$

This equation is satisfied by the coordinates of each of the five points, since the determinant has two identical rows when it is evaluated at any of the points. It is necessary, however, for us to verify that not all of the coefficients of the quadratic equation are zero. From the five points we can select four, no three of which are collinear, and choose general homogeneous coordinates so that these four points are $I_x[1, 0, 0]$, $I_y[0, 1, 0]$, $O[0, 0, 1]$ and $U[1, 1, 1]$. Let $[p, q, r]$ be homogeneous coordinates for the fifth point. Substituting the coordinates into the equation, we obtain

$$\begin{vmatrix} x^2 & y^2 & z^2 & xy & xz & yz \\ p^2 & q^2 & r^2 & pq & pr & qr \\ 1 & 0 & 0 & 0 & 0 & 0 \\ 0 & 1 & 0 & 0 & 0 & 0 \\ 0 & 0 & 1 & 0 & 0 & 0 \\ 1 & 1 & 1 & 1 & 1 & 1 \end{vmatrix} = 0$$

or

$$r(p - q)xy + q(r - p)xz + p(q - r)yz = 0$$

The three coefficients of the left-hand side are all zero only if one of eight conditions holds; but, in fact, the conditions all fail:

$r = 0$ and $q = 0$ and $p = 0$ (Impossible)
$r = 0$ and $q = 0$ and $q = r$ (Ruled out because $[p, q, r] \neq [1, 0, 0]$)
⋮
$p = q$ and $r = p$ and $q = r$ (Ruled out because $[p, q, r] \neq [1, 1, 1]$)

Hence there is a conic, given by the determinant equation, that satisfies the given conditions.

To see that the conic through the five points is unique, suppose

$$Ax^2 + By^2 + Cz^2 + Dxy + Exz + Fyz = 0$$

and

$$A'x^2 + B'y^2 + C'z^2 + D'xy + E'xz + F'yz = 0$$

are the equations of two conics, both of which pass through the same five points, no four of which are collinear. We wish to show that $A = kA'$, $B = kB'$, ..., $F = kF'$ for some $k \neq 0$. Again, we may assume the points are the five points used above. We find immediately that $A = A' = 0$, $B = B' = 0$, and $C = C' = 0$. Using the unit point we find

$$D + E + F = 0 \quad \text{and} \quad D' + E' + F' = 0$$

or

$$F = -D - E \quad \text{and} \quad F' = -D' - E'$$

Using the fifth point we find

$$Dpq + Epr + Fqr = 0 \quad \text{and} \quad D'pq + E'pr + F'qr = 0$$

Substituting for F and F' in these equations,

$$Dq(p - r) + Er(p - q) = 0 \quad \text{and} \quad D'q(p - r) + E'r(p - q) = 0$$

Hence both of the vectors $\langle D, E \rangle$ and $\langle D', E' \rangle$ are perpendicular to $\mathbf{v} = \langle q(p - r), r(p - q) \rangle$. We must show $\mathbf{v}$ is not the zero vector. The two components of $\mathbf{v}$ are both zero only if one of the following four conditions holds, but again the conditions all fail:

$q = 0$ and $r = 0$ (Ruled out because $[p, q, r] \neq [1, 0, 0]$)

$q = 0$ and $p = q$ (Ruled out because $[p, q, r] \neq [0, 0, 1]$)

$p = r$ and $r = 0$ (Ruled out because $[p, q, r] \neq [0, 1, 0]$)

$p = r$ and $p = q$ (Ruled out because $[p, q, r] \neq [1, 1, 1]$)

So $\langle D, E \rangle$ and $\langle D', E' \rangle$ are both perpendicular to the nonzero vector **v**, and, because $F = -D - E$ and $F' = -D' - E'$, neither $\langle D, E \rangle$ nor $\langle D', E' \rangle$ is the zero vector. It follows that $\langle D, E \rangle = k \langle D', E' \rangle$ for some $k \neq 0$. From this,

$$D = kD', \quad E = kE', \quad \text{and} \quad F = kF'$$

The statements in the proposition about degeneracy and nondegeneracy follow from the classification of conics in Section 1. ∎

2.2 Example. The conic that passes through the points $P_1[1, 1, 1]$, $P_2[-1, 1, 1]$, $P_3[-1, -1, 1]$, $P_4[1, -1, 1]$ and $P_5[0, 2, 1]$ (see Figure 7.6) has the equation

$$\begin{vmatrix} x^2 & y^2 & z^2 & xy & xz & yz \\ 1 & 1 & 1 & 1 & 1 & 1 \\ 1 & 1 & 1 & -1 & -1 & 1 \\ 1 & 1 & 1 & 1 & -1 & -1 \\ 1 & 1 & 1 & -1 & 1 & -1 \\ 0 & 4 & 1 & 0 & 0 & 2 \end{vmatrix} = 0$$

which reduces to $3x^2 + y^2 - 4z^2 = 0$. This equation has the inhomogeneous form $3x^2 + y^2 = 4$. This conic is nondegenerate.

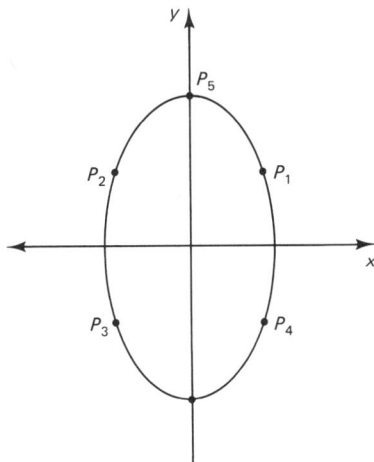

Figure 7.6 The conic through $P_1[1, 1, 1]$, $P_2[-1, 1, 1]$, $P_3[-1, -1, 1]$, $P_4[1, -1, 1]$, and $P_5[0, 2, 1]$.

If all the given five points are Euclidean, an inhomogeneous form of the determinant yields the inhomogeneous equation directly.

2.3 Example. The conic that passes through $P_1(1, 0)$, $P_2(2, 0)$, $P_3(3, 0)$, $P_4(1, 1)$ and $P_5(2, 2)$ (see Figure 7.7) has the equation

$$\begin{vmatrix} x^2 & xy & y^2 & x & y & 1 \\ 1 & 0 & 0 & 1 & 0 & 1 \\ 4 & 0 & 0 & 2 & 0 & 1 \\ 9 & 0 & 0 & 3 & 0 & 1 \\ 1 & 1 & 1 & 1 & 1 & 1 \\ 4 & 4 & 4 & 2 & 2 & 1 \end{vmatrix} = 0$$

which reduces to $y(y - x) = 0$. This conic is degenerate.

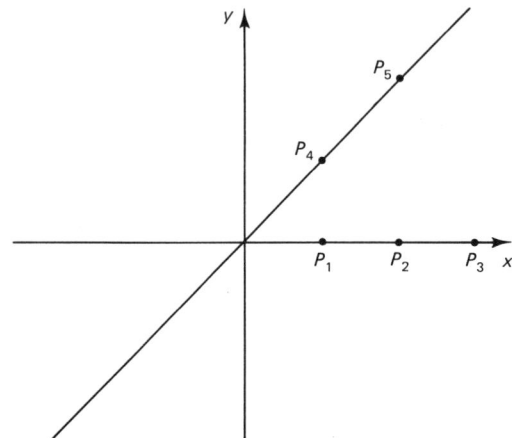

Figure 7.7 The conic through $P_1(1, 0)$, $P_2(2, 0)$, $P_3(3, 0)$, $P_4(1, 1)$, and $P_5(2, 2)$.

The second method for finding the equation of the conic through five points was suggested by Liming (see [Faux and Pratt] or [Liming]). This method is also applied to other problems in Section 3.

2.4 Proposition. Let $C_1(x, y) = 0$ and $C_2(x, y) = 0$ be the (Cartesian) equations of two conics. Then for any extended real number λ,

$$C_\lambda(x, y) = \lambda C_1(x, y) + (1 - \lambda)C_2(x, y) = 0$$

is the equation of a conic that passes through the points of intersection of C_1 and C_2. If C_1 and C_2 have exactly four points of intersection, then every conic through these four points is in this family and is determined by specifying any fifth point on the conic.

Proof. For each λ, the equation $C_\lambda(x, y) = 0$ is quadratic and, therefore, is the equation of some conic. If the coordinates (x, y) of a point satisfy $C_1(x, y) = 0$ and $C_2(x, y) = 0$, then they satisfy $C_\lambda(x, y) = 0$, so the conic passes through the points of intersection of C_1 and C_2.

If the conics C_1 and C_2 intersect in exactly four points, we observe first that no three of these points can be collinear. (If three were collinear, each of the conics would be degenerate, the line through the three points would be part of each conic, and the conics would therefore have more than four points of intersection.) Thus when we include the fifth point, no four of the points can be collinear, so by Proposition 2.1 there is a unique conic through the five points. The value of λ that determines the equation of the conic is that value of λ for which $C_\lambda(x_5, y_5) = \lambda C_1(x_5, y_5) + (1 - \lambda)C_2(x_5, y_5) = 0$; that is,

$$\lambda = \begin{cases} \dfrac{C_2(x_5, y_5)}{C_2(x_5, y_5) - C_1(x_5, y_5)} & \text{if } C_1(x_5, y_5) \neq C_2(x_5, y_5) \\ \pm \infty, & \text{if } C_1(x_5, y_5) = C_2(x_5, y_5) \end{cases}$$ ■

2.5 Example. The two degenerate conics $y^2 - 1 = 0$ and $x^2 - 1 = 0$ have four points of intersection with coordinates $(1, 1)$, $(1, -1)$, $(-1, 1)$ and $(-1, -1)$. Members of the family $C_\lambda(x, y) = \lambda(y^2 - 1) + (1 - \lambda)(x^2 - 1) = 0$ are illustrated in Figure 7.8.

2.6 Example. We use the method of Proposition 2.4 to repeat Example 2.2. Of the first four points given in Example 2.2, no three are collinear. The easiest way to find two conics for which these four points are the points of intersection is to choose degenerate conics—in this case they are the conics $y^2 - 1 = 0$ and $x^2 - 1 = 0$ of Example 2.5. Substituting $(x_5, y_5) = (0, 2)$ into

$$\lambda(y^2 - 1) + (1 - \lambda)(x^2 - 1) = 0$$

we find $\lambda = \frac{1}{4}$. The equation is thus $3x^2 + y^2 = 4$.

2.7 Example. We repeat Example 2.3. Of the five points given in Example 2.3, we select four, no three of which are collinear: $P_1(1, 0)$, $P_2(2, 0)$, $P_4(1, 1)$ and $P_5(2, 2)$. If we choose $C_1(x, y) = (x - 1)(x - 2) = 0$ and $C_2(x, y) = (x + y - 2)(2x - y - 2) = 0$ (see Figure 7.9), then

$$C_\lambda(x, y) = \lambda(x - 1)(x - 2) + (1 - \lambda)(x + y - 2)(2x - y - 2) = 0$$

Substituting the coordinates $(x_3, y_3) = (3, 0)$, we find $\lambda = 2$. The equation is $y(y - x) = 0$.

In Section 1 conics were described by quadratic forms. We can fit Proposition 2.4 into that context. (This is not just an academic exercise: On one hand, introducing homogeneous coordinates allows some of the points P_i to be ideal. On the other hand, this also makes it easier to adapt the method to the other problems we address in Section 3.)

Suppose the degenerate conic C_1 is the union of lines l_1 and l_2, while the degenerate conic C_2 is the union of lines l_3 and l_4. Then the quadratic forms for C_1 and C_2 are (see Exercise 1.16)

$$Q_1 = \tfrac{1}{2}(l_1 l_2^T + l_2 l_1^T) \quad \text{and} \quad Q_2 = \tfrac{1}{2}(l_3 l_4^T + l_4 l_3^T)$$

The quadratic form corresponding to C_λ is $Q_\lambda = \lambda Q_1 + (1 - \lambda)Q_2$ if $\lambda \neq \pm\infty$ or $Q_1 - Q_2$ if $\lambda = \pm\infty$. The value of λ that corresponds to the conic passing through the fifth point P_5 can be determined by the requirement that the *homogeneous* coordinates of P_5 satisfy

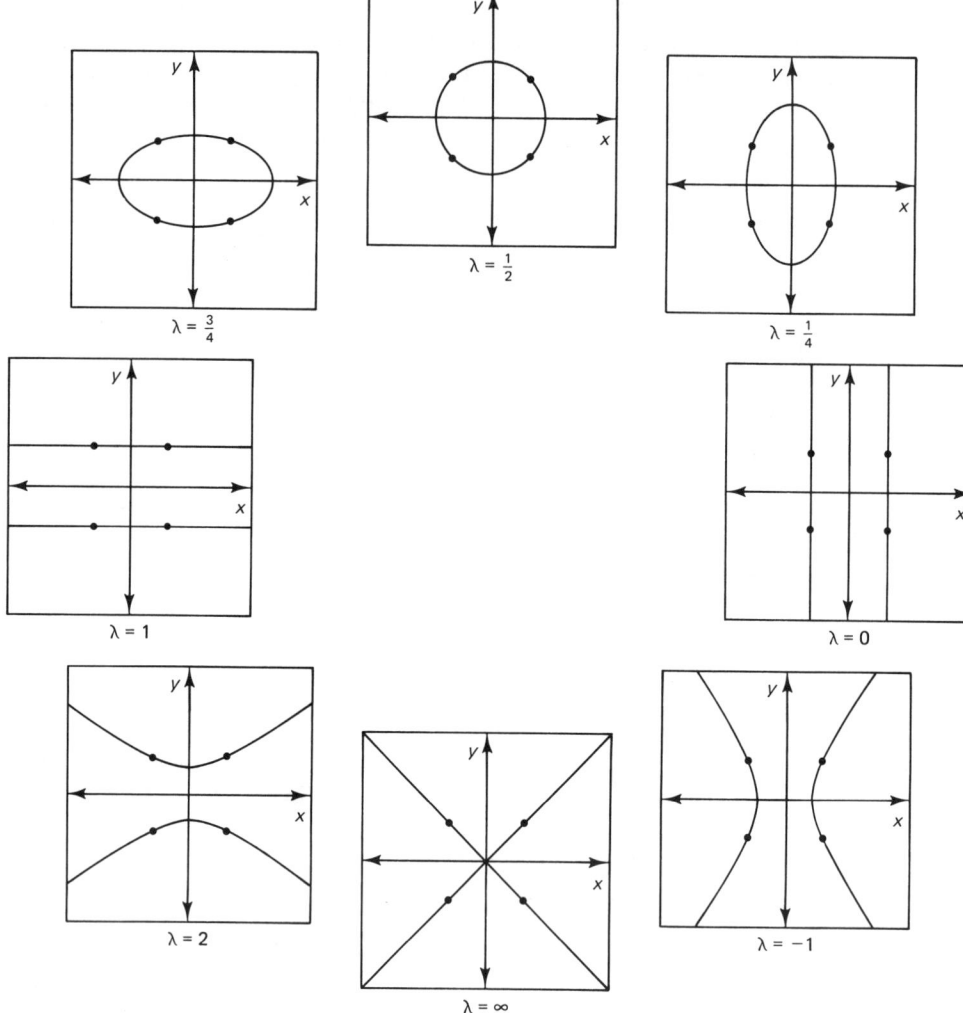

Figure 7.8 Conics in the family $\lambda(y^2 - 1) + (1 - \lambda)(x^2 - 1) = 0$.

$$\mathbf{p}_5 Q_\lambda \mathbf{p}_5^T = 0$$

2.8 Example. We repeat Example 2.2 using homogeneous coordinates. The points $P_1[1, 1, 1]$, $P_2[-1, 1, 1]$, $P_3[-1, -1, 1]$ and $P_4[1, -1, 1]$ determine the lines

$$l_1 \begin{bmatrix} 1 \\ 0 \\ -1 \end{bmatrix}, \quad l_2 \begin{bmatrix} 1 \\ 0 \\ 1 \end{bmatrix}, \quad l_3 \begin{bmatrix} 0 \\ 1 \\ -1 \end{bmatrix}, \quad l_4 \begin{bmatrix} 0 \\ 1 \\ 1 \end{bmatrix}$$

Thus

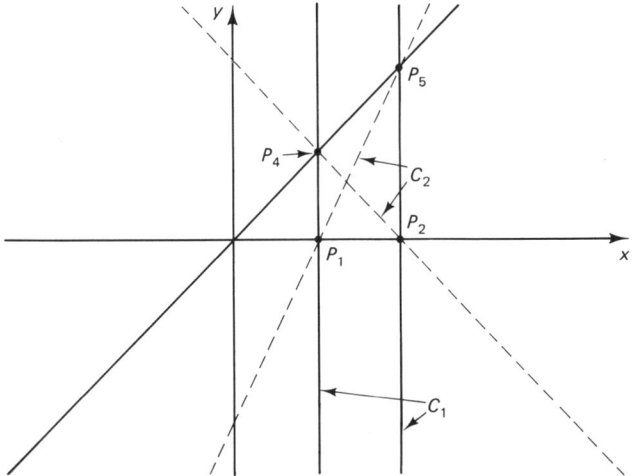

Figure 7.9 Two conics that intersect at $P_1(1, 0)$, $P_2(2, 0)$, $P_4(1, 1)$, and $P_5(2, 2)$.

$$Q_1 = \begin{pmatrix} 1 & 0 & 0 \\ 0 & 0 & 0 \\ 0 & 0 & -1 \end{pmatrix}, \quad Q_2 = \begin{pmatrix} 0 & 0 & 0 \\ 0 & 1 & 0 \\ 0 & 0 & -1 \end{pmatrix}$$

and

$$Q_\lambda = \begin{pmatrix} \lambda & 0 & 0 \\ 0 & 1-\lambda & 0 \\ 0 & 0 & -1 \end{pmatrix}$$

Since we are looking for the conic that passes through $P_5[0, 2, 1]$, we must have

$$(0 \ \ 2 \ \ 1) \begin{pmatrix} \lambda & 0 & 0 \\ 0 & 1-\lambda & 0 \\ 0 & 0 & -1 \end{pmatrix} \begin{pmatrix} 0 \\ 2 \\ 1 \end{pmatrix} = 0$$

It follows that $\lambda = \frac{3}{4}$. Thus

$$Q_\lambda = \begin{pmatrix} \frac{3}{4} & 0 & 0 \\ 0 & \frac{1}{4} & 0 \\ 0 & 0 & -1 \end{pmatrix}$$

The homogeneous equation of the conic is

$$\tfrac{3}{4}x^2 - \tfrac{1}{4}y^2 - z^2 = 0$$

and the Cartesian equation of the conic is

$$3x^2 - y^2 = 4$$

SECTION 2 EXERCISES

1. Find the equation of the conic through the given five points.
   (a) $P_1(1, 1)$, $P_2(1, -1)$, $P_3(-1, 1)$, $P_4(-1, -1)$, $P_5(\sqrt{7}, \tfrac{1}{2})$
   (b) $P_1[2, 0, 1]$, $P_2[-2, 0, 1]$, $P_3[2, 3, 0]$, $P_4[2, -3, 0]$, $P_5[4, 3\sqrt{3}, 1]$

(c) $P_1[0, 0, 1]$, $P_2[4, 0, 1]$, $P_3[2, 4, 1]$, $P_4[0, 1, 0]$, $P_5[3, 3, 1]$
(d) $P_1(2, 2)$, $P_2(-2, -2)$, $P_3(1, 4)$, $P_4(-4, -1)$, $P_5(8, \frac{1}{2})$

2. Show that five points, four of which are collinear, do not determine a unique conic.

3. If conics C_1 and C_2 have only two points of intersection, is every conic that passes through the two points in the family $\lambda C_1 + (1 - \lambda)C_2$?

4. Sketch the conics in the family determined by $C_1(x, y) = (x + 2)^2 + y^2 - 1 = 0$ and $C_2(x, y) = (x - 2)^2 + y^2 - 1 = 0$ that correspond to the values $\lambda = 0, 1, 2, \pm \infty$, and -1. For which values of λ is there no conic?

5. Sketch conics in the family $\lambda(x^2 - y^2) + (1 - \lambda)((x - 2)^2 + y^2 - 1) = 0$ determined by a pair of intersecting lines and a circle. Geometrically determine which conics correspond to values of λ in each of the following intervals: $(0, \frac{1}{4})$, $(-\infty, 0) \cup (1, \infty]$, $(\frac{1}{2}, 1)$ and $(\frac{1}{4}, \frac{1}{2})$.

6. Write a program to graph a sequence of conics in a family C_λ.

7. Write a program that will accept the coordinates of five points as input and will provide as output the equation of the conic through the points and the graph of the conic.

8. It is not usually possible to find a conic that passes through six or more given points. However, using the method of least squares (see Section 6.1) a conic can be found that best matches the data. Find the equation $Ax^2 + Bxy + Cy^2 + Dx + Ey + F = 0$ of the conic that best matches the following data.
(a) $P_1(-1, -1)$, $P_2(1, -1)$, $P_3(1, 1)$, $P_4(-1, 1)$, $P_5(\frac{1}{2}, 2)$, $P_6(-\frac{1}{2}, \frac{3}{2})$
(b) $P_1(-4, -4)$, $P_2(-2, 0)$, $P_3(-4, 4)$, $P_4(2, 0)$, $P_5(2, 3)$, $P_6(4, 4)$
(c) $P_1(-1, -5)$, $P_2(0, 0)$, $P_3(2, 4)$, $P_4(3, 3)$, $P_5(4, 0)$, $P_6(5, -5)$
(d) $P_1(0, 1)$, $P_2(0, -1)$, $P_3(-0.4, 0)$, $P_4(-2.6, 0)$, $P_5(1, 2.8)$, $P_6(1, -2.3)$

9. Find the equation $x^2 + y^2 + Dx + Ey + F = 0$ of the circle determined by the data $\{(x, y)\}$.
(a) $\{(2, 1), (5, -2), (1, 4)\}$
(b) $\{(-4, 0), (4, 1), (1, 5)\}$

10. Find the equation $x^2 + y^2 + Dx + Ey + F = 0$ of the circle that best matches the data $\{(x, y)\}$.
(a) $\{(-1, 0), (1, 1), (1, 2), (2, -2), (3, 0)\}$
(b) $\{(-7, 0), (-5, 8), (-3, -6), (0, -6), (0, -7), (4, 6), (6, 0)\}$
(c) $\{(-5, 0), (-4, -3), (0, -4), (0, -6), (0, 6), (0, 7), (2, 6), (6, 0)\}$
(d) $\{(-7, 4), (-6, -2), (-2, -4), (0, 8), (0, -3), (4, 4), (5, 7), (5, -2), (7, 3)\}$

11. Find the polynomial $z = p(x, y) = a_0 x^2 + a_1 xy + a_2 y^2$ determined by the data $\{(x, y, z)\}$.
(a) $\{(2, 1, 0), (1, 3, 2), (0, 2, 2)\}$
(b) $\{(1, 0, 1), (2, 1, 2), (3, 1, 1)\}$

12. Find the polynomial $z = p(x, y) = a_0 x^2 + a_1 xy + a_2 y^2$ that best matches the data $\{(x, y, z)\}$.
(a) $\{(1, -2, 0), (-1, 2, 0), (0, 2, 3), (1, 0, 1), (1, 2, 2)\}$
(b) $\{(1, -2, 0), (-1, 0, 2), (0, 0, 3), (1, 3, 1), (2, 2, -1), (3, 0, 2)\}$
(c) $\{(2, -3, 0), (1, -2, 1), (-1, 0, 1), (0, 0, 0), (3, 1, -1), (-3, 2, -1), (2, 3, 0)\}$
(d) $\{(-3, -3, 0), (-2, -2, 1), (1, -1, 2), (0, -2, 3), (2, 1, 2), (2, -1, 1), (3, 1, 0)\}$

13. Find the polynomial $y = p(x) = a_0 + a_1 x + a_2 x^2 + a_3 x^3$ determined by the data $\{(x, y)\}$.

(a) $\{(-1, 2), (1, 2), (2, 0), (3, 4)\}$
(b) $\{(1, 2), (2, 2), (3, 0), (4, 4)\}$

14. Find the polynomial $y = p(x) = a_0 + a_1 x + a_2 x^2 + a_3 x^3$ that best matches the data $\{(x, y)\}$.
    (a) $\{(-2, 0), (-1, 0), (0, 3), (1, 1), (2, 2)\}$
    (b) $\{(-2, 0), (-1, 0), (0, 3), (1, 1), (2, 2), (3, 0)\}$
    (c) $\{(-3, 0), (-2, 1), (-1, 1), (0, 0), (1, -1), (2, -1), (3, 0)\}$
    (d) $\{(-3, 0), (-2, 1), (-1, 2), (0, 3), (1, 2), (2, 1), (3, 0)\}$

SECTION 3: ADDITIONAL CONIC-FITTING PROBLEMS

In this section we address five additional conic-fitting problems. One such problem is to find the equation of the conic that is tangent to two given lines at two given points and also passes through a third given point (see Figure 7.10). Such a curve might form the outline for part of a fender of a car or part of the fuselage of an airplane. This is Problem 2, given below. The problems vary depending upon how many points and how many tangent lines are given. The technical conditions in the statements of the problems are explained below.

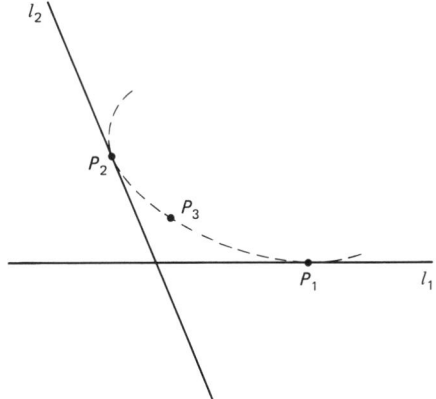

Figure 7.10 Data for conic-fitting problem 2.

Five Conic-Fitting Problems

1. Find the equation of the conic that passes through four given points and is tangent to a given line at one of these points (see Figure 7.11(a)). (At least two of the points must not lie on the line (see Figure 7.11(b)), and if three of the points lie off the line, then no two of them may be collinear with the point of tangency (see Figure 7.11(c)).)

2. Find the equation of the conic that passes through three given points and is tangent to two given lines at two of these points (see Figure 7.12(a)). (The three points may not be collinear (see Figure 7.12(b)), and neither

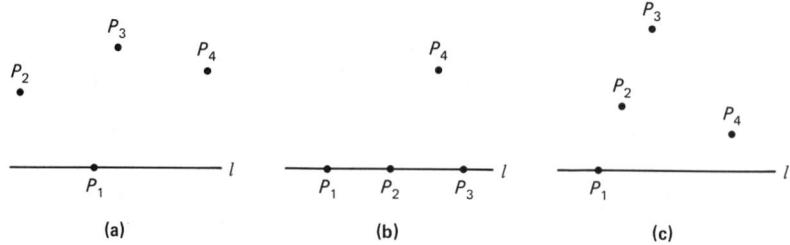

Figure 7.11 Data for conic-fitting problem 1: (a) allowed, (b) and (c) not allowed.

point of tangency may be the point of intersection of the lines (see Figure 7.12(c)).)

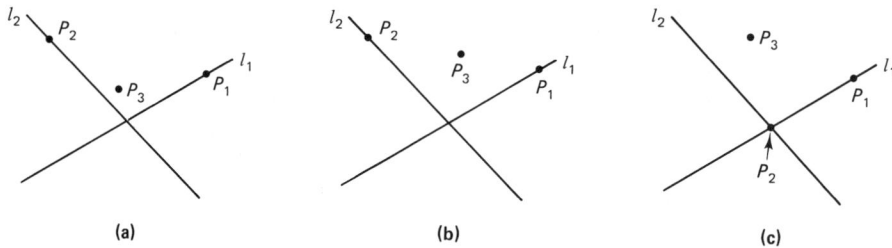

Figure 7.12 Data for conic-fitting problem 2: (a) allowed, (b) and (c) not allowed.

3. Find the equation of the conic that passes through two given points, is tangent to two given lines at these points and is tangent to one other given line (see Figure 7.13(a)). (The three lines may not be concurrent (see Figure 7.13(b)), and neither point may be the point of intersection of the first two lines (see Figure 7.13(c)).)

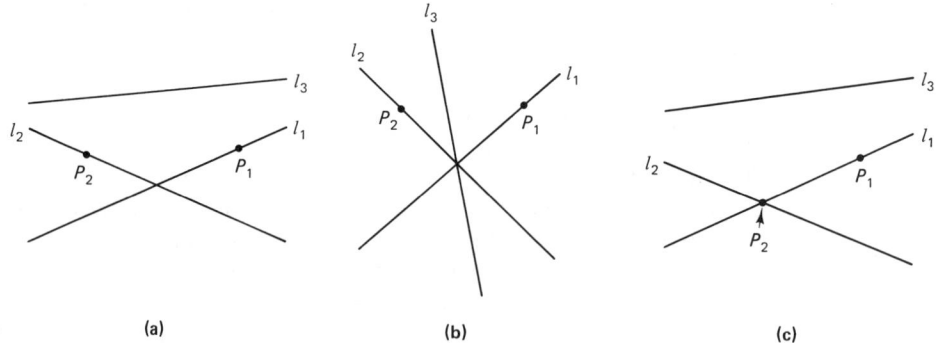

Figure 7.13 Data for conic-fitting problem 3: (a) allowed, (b) and (c) not allowed.

4. Find the equation of the conic that passes through one given point, has a given tangent line at this point, and is tangent to three other given lines (see Figure 7.14(a)). (No three of the lines may be concurrent (see Figure 7.14(b)), and only the tangent line may pass through the given point (see Figure 7.14(c)).)

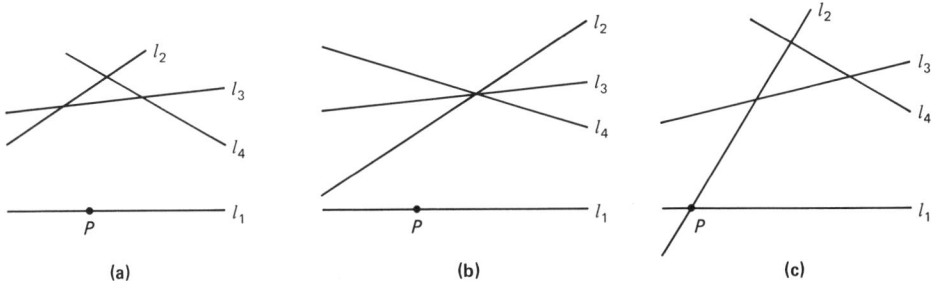

Figure 7.14 Data for conic-fitting problem 4: (a) allowed, (b) and (c) not allowed.

5. Find the equation of the conic that is tangent to five given lines (see Figure 7.15(a)). (No three of the lines may be concurrent (see Figure 7.15(b)).)

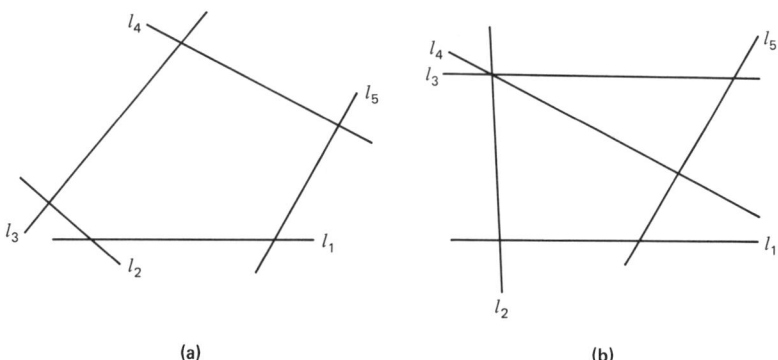

Figure 7.15 Data for conic-fitting problem 5: (a) allowed, (b) not allowed.

To solve these problems we need to know how to compute the equation of a tangent line to a conic. If P_0 is a point on the conic whose matrix is Q, then
$$\mathbf{p}_0 Q \mathbf{p}_0^T = 0$$
Thus the coordinates of P_0 satisfy the equation
$$\mathbf{p} Q \mathbf{p}_0^T = 0$$
This is a linear equation whose graph is a line through P_0. This line is, in fact, the tangent line to the conic at the point P_0.

Sec. 3   Additional Conic-Fitting Problems

3.1 Proposition. The tangent line l_0 to the nondegenerate conic with the equation
$$\mathbf{p}Q\mathbf{p}^T = 0$$
at a point P_0 that lies on the conic has the equation
$$\mathbf{p}Q\mathbf{p}_0^T = 0$$
Thus representative homogeneous coordinates for l_0 (written as a column matrix) are $\mathbf{l}_0 = Q\mathbf{p}_0^T$.

Proof. Suppose that the line whose equation is $\mathbf{p}Q\mathbf{p}_0^T = 0$ also passes through another point P_1 on the conic. Then $\mathbf{p}_0 Q \mathbf{p}_0^T = 0$, $\mathbf{p}_1 Q \mathbf{p}_1^T = 0$ and $\mathbf{p}_1 Q \mathbf{p}_0^T = 0$. Consequently, every point on the line through P_0 and P_1 will also lie on the conic:
$$(t\mathbf{p}_0 + (1-t)\mathbf{p}_1)Q(t\mathbf{p}_0 + (1-t)\mathbf{p}_1)^T = 0$$
This is a contradiction, since we assumed the conic to be nondegenerate. Hence the line whose equation is $\mathbf{p}Q\mathbf{p}_0^T = 0$ meets the conic only at the point P_0 and thus (by Proposition 1.16) must be the tangent line at P_0. ■

3.2 Example. The point $P_0(4, 3)$ lies on the hyperbola
$$\frac{x^2}{8} - \frac{y^2}{9} = 1$$
The homogeneous equation of the tangent line l_0 to the hyperbola at P_0 is, therefore,
$$(x, y, z) \begin{pmatrix} \frac{1}{8} & 0 & 0 \\ 0 & -\frac{1}{9} & 0 \\ 0 & 0 & -1 \end{pmatrix} \begin{pmatrix} 4 \\ 3 \\ 1 \end{pmatrix} = 0$$
or $3x - 2y - 6z = 0$. The Cartesian equation of l_0 is $3x - 2y = 6$. The homogeneous coordinates of l_0 are
$$\begin{pmatrix} \frac{1}{8} & 0 & 0 \\ 0 & -\frac{1}{9} & 0 \\ 0 & 0 & -1 \end{pmatrix} \begin{pmatrix} 4 \\ 3 \\ 1 \end{pmatrix} = \begin{pmatrix} \frac{1}{2} \\ -\frac{1}{3} \\ -1 \end{pmatrix}$$
or $[3, -2, -6]$.

We now discuss Problem 2.

Problem 2: Find the equation of the conic that passes through three given points and is tangent to two given lines at two of these points. (The three points may not be collinear, and neither point of tangency may be the point of intersection of the lines.)

We find the equation of the conic tangent to l_1 at P_1, tangent to l_2 at P_2, and passing through P_3, first using the method of Proposition 2.4. Let l_3 be the line through P_1 and P_2 (see Figure 7.16). One quadratic equation $C_1(x, y) = 0$ is determined by the equations of l_1 and l_2; it has the matrix
$$Q_1 = \tfrac{1}{2}(\mathbf{l}_1 \mathbf{l}_2^T + \mathbf{l}_2 \mathbf{l}_1^T)$$

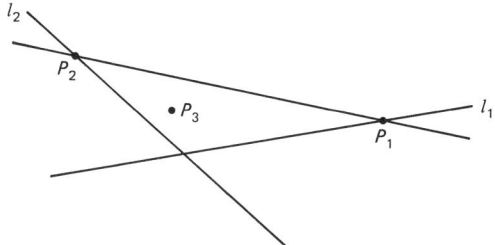

Figure 7.16 Data for problem 2.

Another quadratic equation is determined by the equation of l_3 alone. It has the matrix

$$Q_2 = \mathbf{l}_3 \mathbf{l}_3^T$$

(see Exercise 1.18). The Cartesian coordinates of P_3 can be used to solve for λ in

$$C_\lambda = \lambda C_1 + (1 - \lambda)C_2 = 0$$

or the homogeneous coordinates of P_3 can be used to solve for λ in

$$\lambda Q_1 + (1 - \lambda)Q_2 = 0$$

In both cases, we obtain the same value for λ.

To verify that l_1, for example, is tangent to the graph of $C_\lambda(x, y) = 0$ at P_1, we must show that l_1 has representative homogeneous coordinates given by $Q_\lambda \mathbf{p}_1^T$. Evaluating,

$$\begin{aligned} Q_\lambda \mathbf{p}_1^T &= \lambda(\tfrac{1}{2})(\mathbf{l}_1 \mathbf{l}_2^T + \mathbf{l}_2 \mathbf{l}_1^T)\,\mathbf{p}_1^T + (1-\lambda)\,\mathbf{l}_3 \mathbf{l}_3^T \mathbf{p}_1^T \\ &= \lambda(\tfrac{1}{2})\mathbf{l}_1 \mathbf{l}_2^T \mathbf{p}_1^T \qquad \text{(since } P_1 \text{ lies on } l_1 \text{ and } l_3) \\ &= \lambda(\tfrac{1}{2})(\mathbf{p}_1 \mathbf{l}_2)^T \mathbf{l}_1 \end{aligned}$$

Since $\mathbf{p}_1 \mathbf{l}_2$ is a nonzero scalar (P_1 does not lie on l_2) and since $\lambda \neq 0$ (P_1, P_2, and P_3 are not collinear), we do, indeed, have representative homogeneous coordinates for l_1.

3.3 Example. We find the equations of the conics tangent to the x-axis at $P_1(8, 0)$, tangent to the y-axis at $P_2(0, 4)$ and passing through (a) $P_3(3, 1)$, (b) $P_4(2, 1)$, (c) $P_5(1, 1)$, (d) $P_6(2, 2)$, and (e) $P_7(2, -2)$ (see Figure 7.17).

The line through P_1 and P_2 has the equation $x + 2y - 8 = 0$, so in each case we are to solve for λ in

$$\lambda xy + (1 - \lambda)(x + 2y - 8)^2 = 0$$

In (a), $\lambda = \tfrac{3}{2}$. The equation is thus

$$x^2 + xy + 4y^2 - 16x - 32y + 64 = 0$$

The discriminant of this equation is -15, which is negative, so the conic is an ellipse (see Figure 7.18(a)). In (b), $\lambda = \tfrac{8}{7}$ and the conic is a parabola (see Figure 7.18(b)). In (c), $\lambda = \tfrac{25}{24}$ and the conic is a hyperbola (see Figure 7.18(c)). In (d), the equation reduces to $4 = 0$. We have stumbled upon the conic in the family that corresponds to $\lambda = \pm \infty$, and we therefore use

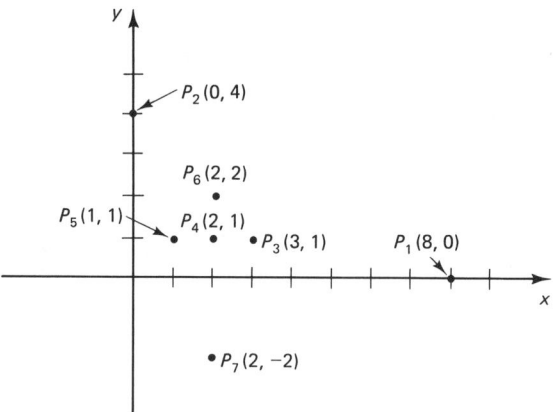

Figure 7.17 Data for five examples of problem 2.

$$xy - (x + 2y - 8)^2 = 0$$

whose graph is an ellipse (see Figure 7.18(d)). Case (e) demonstrates that the third point does not need to lie in the first quadrant. The parameter $\lambda = \frac{25}{26}$ and the curve is a hyperbola (see Figure 7.18(e)).

The following is a second method for solving Problem 2 using the model for this problem illustrated in Figure 7.19. The circle is tangent to $y = -1$ at A and to $x = -1$ at B and passes through C. Using the methods of Example 3.2.10, we find the matrix M of the projective transformation T sending P_1 to A,

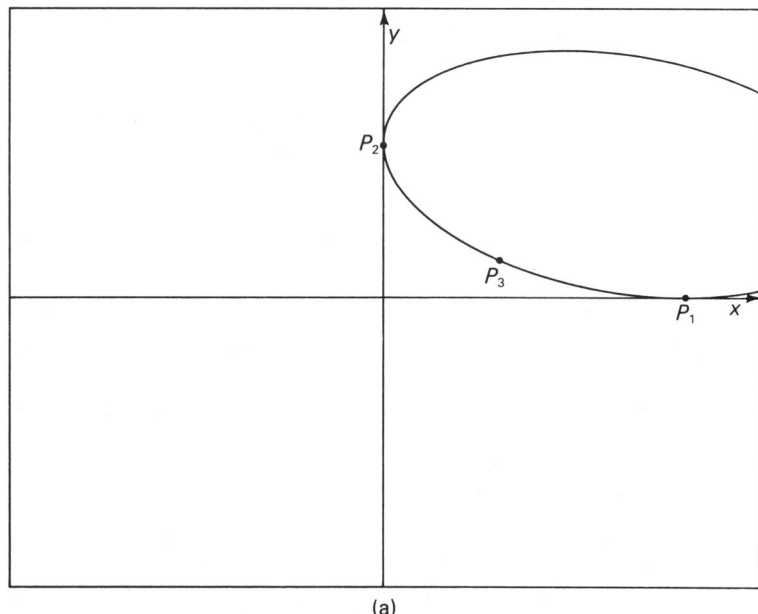

(a)

Figure 7.18 Solutions to the five problems in Example 3.3.

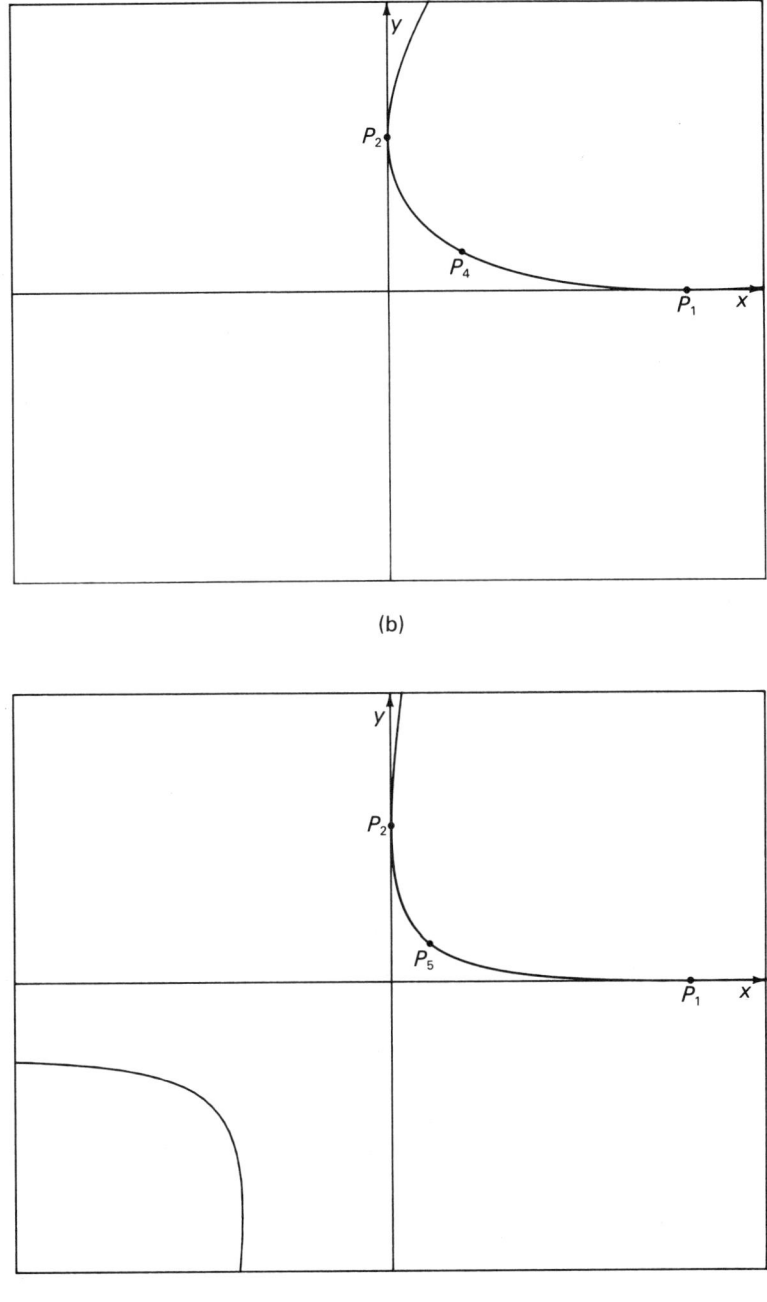

(b)

(c)

Figure 7.18 Continued.

Sec. 3 Additional Conic-Fitting Problems

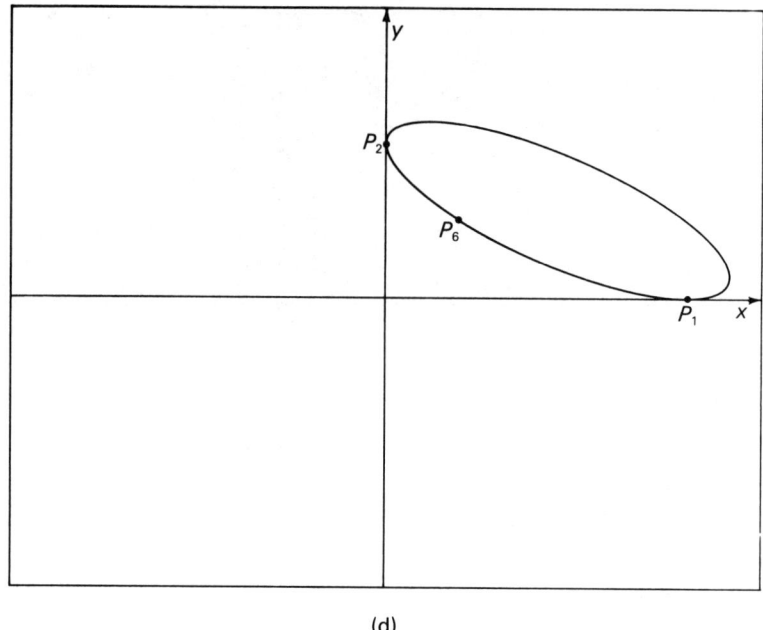

(d)

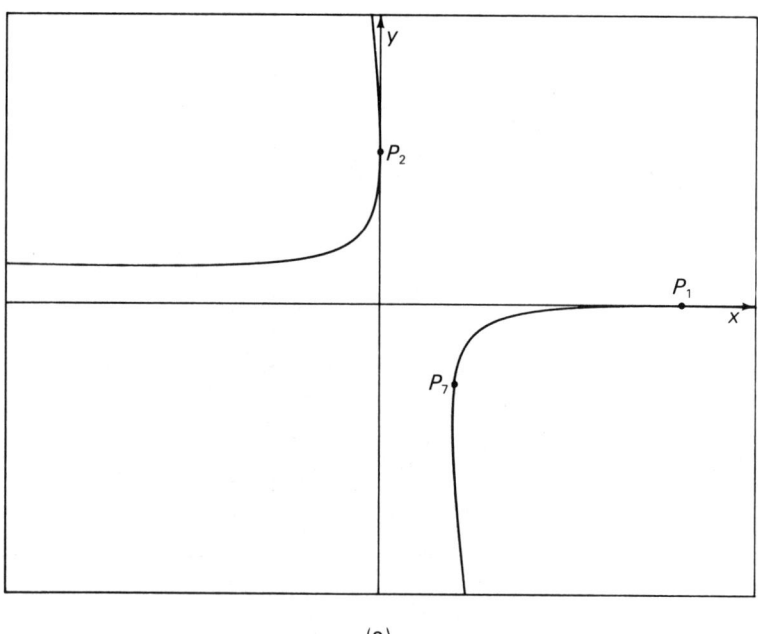

(e)

Figure 7.18 Continued.

384 Conics in Design Chap. 7

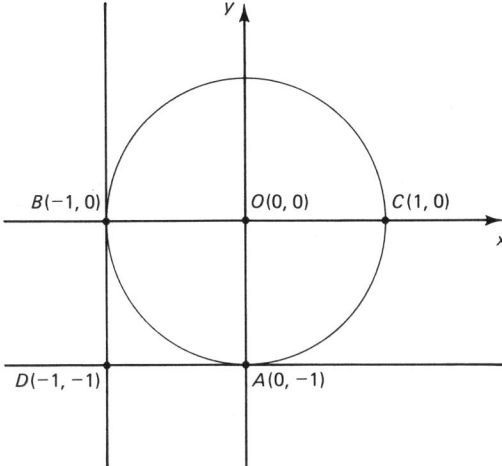

Figure 7.19 A model for a solution of problem 2.

P_2 to B, P_3 to C, and the intersection point P_4 of l_1 and l_2 to D. The transformation T^{-1}, represented by M^{-1}, maps A to P_1, B to P_2, C to P_3, D to P_4, the lines $y = -1$ and $x = -1$ to the lines l_1 and l_2, and the points of the circle to points of the conic that solves our design problem.

Let Q be the matrix of the quadratic form of the unit circle:

$$Q = \begin{pmatrix} 1 & 0 & 0 \\ 0 & 1 & 0 \\ 0 & 0 & -1 \end{pmatrix}$$

By Proposition 1.8, the equation of the conic that solves our problem is

$$\langle x, y, 1 \rangle M Q M^T \langle x, y, 1 \rangle^T = 0$$

The matrix M is, as usual, easier to find in two steps (see Figure 7.20). The matrix N_1 of the transformation T_1 that sends I_x to A, I_y to B, O to D, and U to C can be computed immediately: We find

$$N_1 = \begin{pmatrix} 0 & -2 & 2 \\ -1 & 0 & 1 \\ 2 & 2 & -2 \end{pmatrix}$$

The complete computation of M is illustrated by an example.

3.4 Example. We repeat (a) of Example 3.3. The matrix N_2 of the transformation T_2 that sends I_x to $P_1[8, 0, 1]$, I_y to $P_2[0, 4, 1]$, O to $P_4[0, 0, 1]$ and U to $P_3[3, 1, 1]$ is

$$N_2 = \begin{pmatrix} 24 & 0 & 3 \\ 0 & 8 & 2 \\ 0 & 0 & 3 \end{pmatrix}$$

The inverse of this matrix, up to a scalar multiple, is

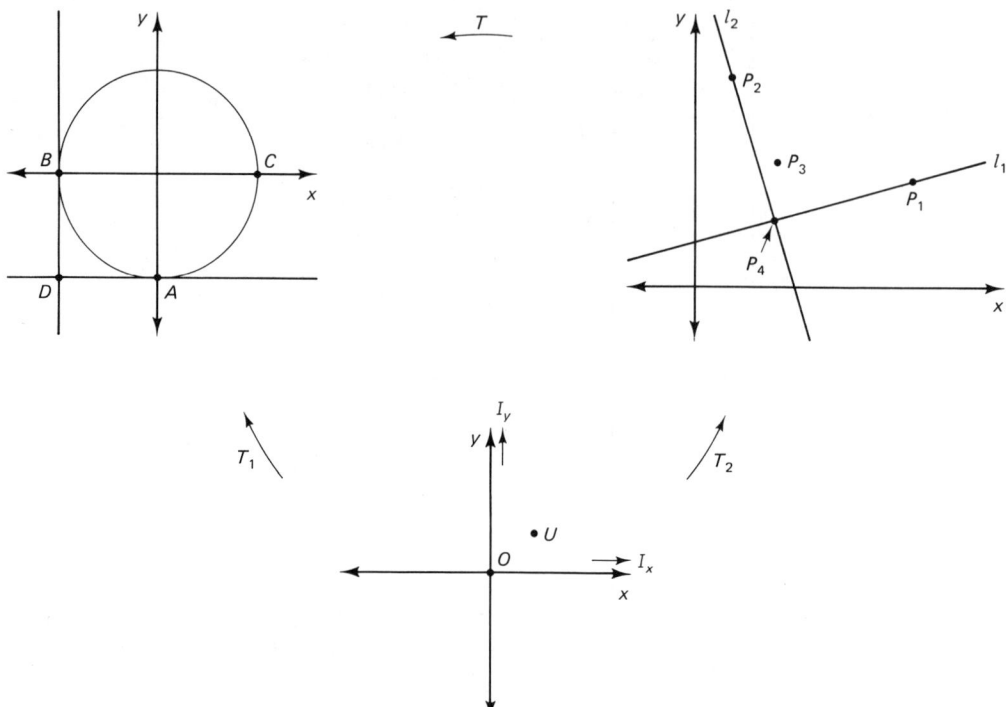

Figure 7.20 Steps in finding the matrix that transforms the circle to another conic.

$$N_2^{-1} = \begin{pmatrix} 1 & 0 & -1 \\ 0 & 3 & -2 \\ 0 & 0 & 8 \end{pmatrix}$$

Hence

$$M = N_2^{-1} N_1 = \begin{pmatrix} -2 & -4 & 4 \\ -7 & -4 & 7 \\ 16 & 16 & -16 \end{pmatrix}$$

Thus

$$MQM^T = \begin{pmatrix} 4 & 2 & -32 \\ 2 & 16 & -64 \\ -32 & -64 & 256 \end{pmatrix}$$

and the equation of the conic is

$$x^2 + xy + 4y^2 - 16x - 32y + 64 = 0$$

Problem 1: Find the equation of the conic that passes through four given points and is tangent to a given line at one of these points. (At least two of the points must not lie on the line, and if three of the points lie off the line, then no two of them may be collinear with the point of tangency.)

We use the method of Proposition 2.4 to find the equation of the conic tangent to l at P_1 and passing through P_2, P_3, and P_4 (see Figure 7.11). Assume P_2 and P_3 do not lie on l. Construct lines l_2 through P_1 and P_2, l_3 through P_1 and P_3, and l_4 through P_2 and P_3 (see Figure 7.21). Now proceed as before with degenerate conics C_1, composed of lines l and l_4, and C_2, composed of lines l_2 and l_3.

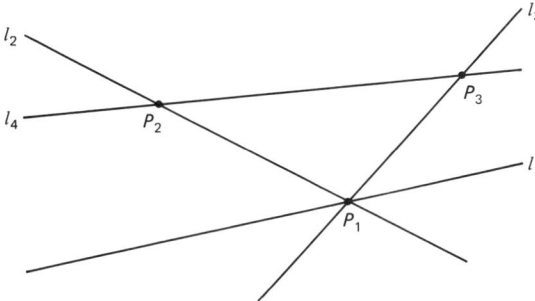

Figure 7.21 Construction lines for solving problem 1.

3.5 Example. We find the equation of the conic tangent to the x-axis at the origin P_1, which passes through $P_2(-1, 1)$, $P_3(1, 1)$ and $P_4(1, 2)$. We use the construction above and find homogeneous coordinates for the lines:

$$l_1 \begin{bmatrix} 0 \\ 1 \\ 0 \end{bmatrix}, \quad l_2 \begin{bmatrix} 1 \\ 1 \\ 0 \end{bmatrix}, \quad l_3 \begin{bmatrix} 1 \\ -1 \\ 0 \end{bmatrix}, \quad l_4 \begin{bmatrix} 0 \\ 1 \\ -1 \end{bmatrix}$$

Then we compute

$$Q_1 = \begin{pmatrix} 0 & 0 & 0 \\ 0 & 2 & -1 \\ 0 & -1 & 0 \end{pmatrix}, \quad Q_2 = \begin{pmatrix} 2 & 0 & 0 \\ 0 & -2 & 0 \\ 0 & 0 & 0 \end{pmatrix}$$

and

$$Q_\lambda = \begin{pmatrix} 2 - 2\lambda & 0 & 0 \\ 0 & -2 + 4\lambda & -\lambda \\ 0 & -\lambda & 0 \end{pmatrix}$$

The equation $\mathbf{p}_4 Q_\lambda \mathbf{p}_4^T = 0$ with $P_4[1, 2, 1]$ leads to $\lambda = \frac{3}{5}$ and the equation $2x^2 + y^2 - 3y = 0$. In standard form this equation is

$$\frac{x^2}{\frac{9}{8}} + \frac{(y - \frac{3}{2})^2}{\frac{9}{4}} = 1$$

(see Figure 7.22).

It can be verified (see Exercise 7) that if the points satisfy the technical conditions given in the problem statement, then the conic will satisfy the requirements. Certain initial configurations lead to a solution that is a degenerate conic (see Exercise 8).

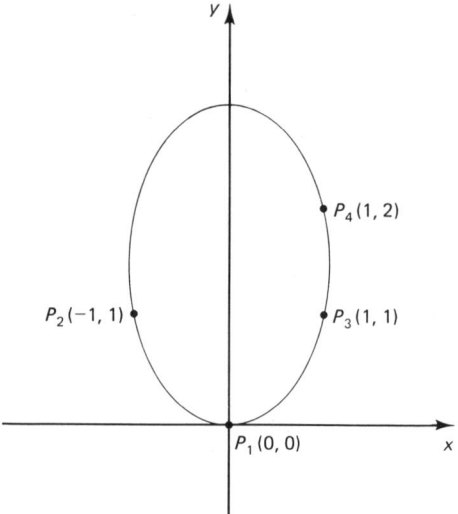

Figure 7.22 Solution to Example 3.5.

Problems 3, 4, and 5 can be solved using the method of Proposition 2.4 with line conics. The conics with which we have been working so far are *point conics*. For the remainder of this section we use the term *point conic* to avoid confusion with line conics.

3.6 Definition. A *line conic* is the set of all lines l in the projective plane whose homogeneous coordinates $\mathbf{l}$ (written as a column matrix) satisfy
$$\mathbf{l}^T Q \mathbf{l} = 0$$
for some symmetric 3×3 matrix Q.

The following proposition relates line conics and point conics in the nondegenerate case (see Figure 7.23).

3.7 Proposition. If Q is the matrix of a nondegenerate quadratic form, the set of all tangent lines to the point conic
$$\mathbf{p} Q \mathbf{p}^T = 0$$
is the line conic
$$\mathbf{l}^T Q^{-1} \mathbf{l} = 0$$
The points of the point conic are in one-to-one correspondence with the lines of the line conic. To the point P_0 of the point conic corresponds the tangent line $\mathbf{l}_0$ with homogeneous coordinates $Q\mathbf{p}_0^T$. To the line l_0 of the line conic corresponds the point of tangency P_0 with homogeneous coordinates $\mathbf{l}_0^T Q^{-1}$.

Proof. The tangent line $\mathbf{l}_0 = Q\mathbf{p}_0^T$ to the point conic at P_0 satisfies the equation of the line conic:
$$\mathbf{l}_0^T Q^{-1} \mathbf{l}_0 = (\mathbf{p}_0 Q) Q^{-1} (Q \mathbf{p}_0^T) = \mathbf{p}_0 Q \mathbf{p}_0^T = 0$$
Similarly, $\mathbf{p}_0 = \mathbf{l}_0^T Q^{-1}$ satisfies the equation of the point conic. ∎

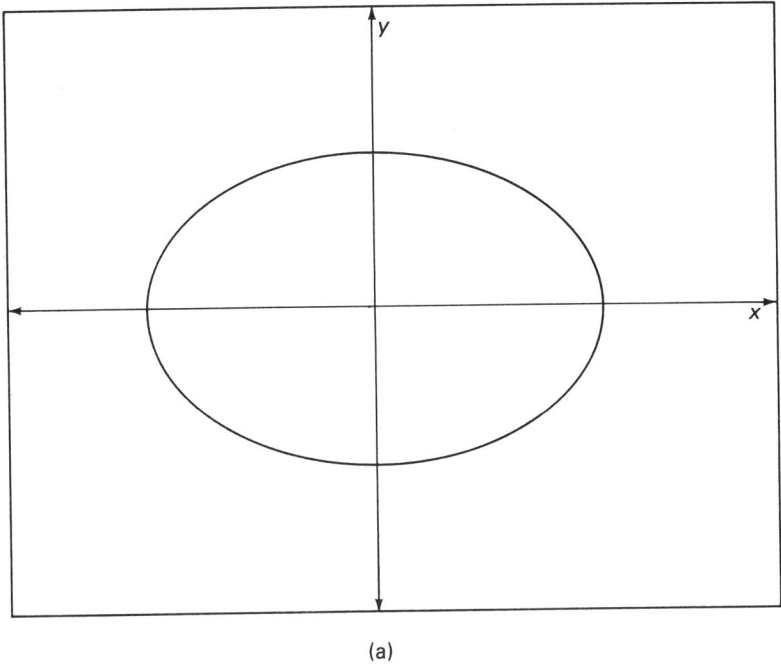

(a)

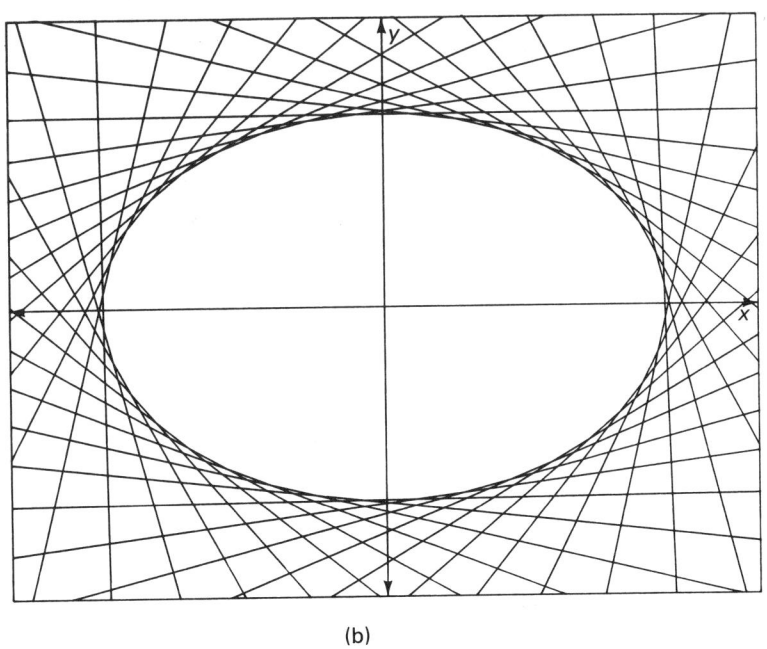

(b)

Figure 7.23 An ellipse as (a) a point conic and (b) a line conic.

Sec. 3 Additional Conic-Fitting Problems

Problem 3: Find the equation of the (point) conic that passes through two given points, is tangent to two given lines at these points and is tangent to one other given line. (The three lines may not be concurrent, and neither point may be the point of intersection of the first two lines.)

We find the equation of the (point) conic tangent to l_1 at P_1, tangent to l_2 at P_2, and tangent to l_3 (see Figure 7.13), using a construction dual to the first solution of Problem 2. Let P_3 be the point of intersection of l_1 and l_2. One line conic is determined by the points P_1 and P_2; it has matrix

$$Q_1 = \tfrac{1}{2}(\mathbf{p}_1^T \mathbf{p}_2 + \mathbf{p}_2^T \mathbf{p}_1)$$

The other line conic has matrix

$$Q_2 = \mathbf{p}_3^T \mathbf{p}_3$$

Now the coordinates of l_3 can be used to solve for λ in

$$Q_\lambda = \lambda Q_1 + (1 - \lambda) Q_2 = 0$$

The matrix of the point conic is Q_λ^{-1}.

3.8 Example. We find the equation of the conic tangent to the x-axis at $P_1(8, 0)$, tangent to the y-axis at $P_2(0, 4)$, and tangent to the line $x + y = 2$ (see Figure 7.24). The points P_1 and P_2 determine the matrix

$$Q_1 = \begin{pmatrix} 0 & 16 & 4 \\ 16 & 0 & 2 \\ 4 & 2 & 1 \end{pmatrix}$$

The point $P_3[0, 0, 1]$ determines the matrix

$$Q_2 = \begin{pmatrix} 0 & 0 & 0 \\ 0 & 0 & 0 \\ 0 & 0 & 1 \end{pmatrix}$$

From this we compute

$$Q_\lambda = \begin{pmatrix} 0 & 16\lambda & 4\lambda \\ 16\lambda & 0 & 2\lambda \\ 4\lambda & 2\lambda & 1 \end{pmatrix}$$

Since the equation $\mathbf{l}^T Q_\lambda \mathbf{l} = 0$ is satisfied by $l_3[1, 1, -2]$, $\lambda = -\tfrac{1}{2}$. Hence

$$Q_\lambda = \begin{pmatrix} 0 & -8 & -2 \\ -8 & 0 & -1 \\ -2 & -1 & 1 \end{pmatrix}$$

which is equivalent to

$$Q_\lambda = \begin{pmatrix} 0 & 8 & 2 \\ 8 & 0 & 1 \\ 2 & 1 & -1 \end{pmatrix}$$

The inverse of Q_λ is (up to a scalar multiple)

$$Q_\lambda^{-1} = \begin{pmatrix} -1 & 10 & 8 \\ 10 & -4 & 16 \\ 8 & 16 & -64 \end{pmatrix}$$

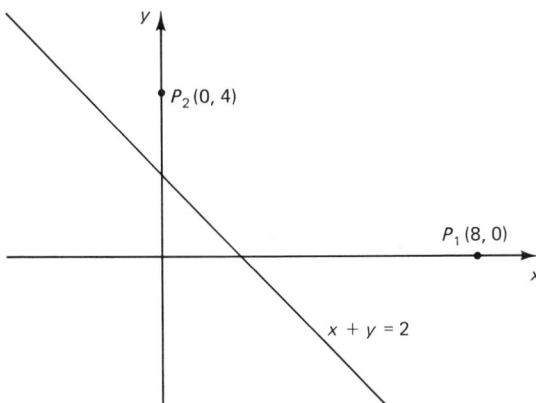

Figure 7.24 Data for Example 3.8.

The Cartesian equation of the point conic is, therefore,
$$-x^2 + 20xy - 4y^2 + 16x + 32y - 64 = 0$$
The discriminant is positive, so the curve is a hyperbola. The point of tangency of the line l_3 has homogeneous coordinates $l_3^T Q_\lambda = [4, 6, 5]$ and Cartesian coordinates $(\frac{4}{5}, \frac{6}{5})$. The conic is illustrated in Figure 7.25.

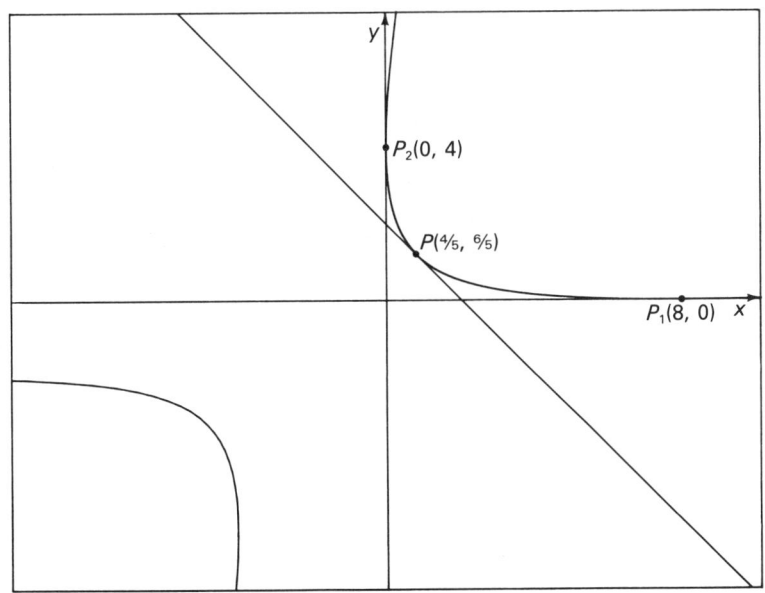

Figure 7.25 Solution to Example 3.8.

Problem 4 is solved by a line conic construction dual to the point conic construction used to solve Problem 1. Similarly, Problem 5 is solved by a line conic construction dual to the point conic construction used in Example 2.8. Note, however, that the conditions in Problems 4 and 5 are stronger than the corresponding conditions for point conics. For example, in Problem 5, requiring that no three lines may be concurrent is stronger than the condition used in Section 2 that no four points may be collinear. If three lines are concurrent, Problem 5 can be solved far enough to obtain the matrix Q_λ of a degenerate line conic, but since Q_λ is not invertible, we cannot find a corresponding point conic.

SECTION 3 EXERCISES

1. Find the equations of the tangent lines to the following conics at the indicated points.
   (a) $3x^2 + 5y^2 = 23$, $P(1, 2)$
   (b) $y = x^2$, $P(1, 1)$
   (c) $y = x^2$, $P[0, 1, 0]$
   (d) $9x^2 - 4y^2 = 32$, $P[2, 1, 1]$
   (e) $9x^2 - 4y^2 = 32$, $P[2, 3, 0]$

2. Find the equations of the conics that are tangent to the x-axis at $P_1(4, 0)$, tangent to the line $y = -4x$ at $P_2(-1, 4)$, and that pass through each point.
   (a) $P_3(1, 1)$
   (b) $P_3(3, 3)$
   (c) $P_3(2, -2)$
   (d) $P_3[1, 1, 0]$

3. Write a program that will take as input the data for Problem 2 and produce the desired equation, as well as graphing the conic. (*Hint:* Consider using the second solution of Problem 2. Find the matrices M and M^{-1}. Use M to produce the equation. Use M^{-1} in the Matrix command of Graph2D to draw the conic, using a circle as test pattern.)

4. In Problem 2, if the three points are collinear, what conic results from using the first solution? Does it solve the problem?

5. In Problem 2, what conic results from the first solution if one of the points of tangency is the point of intersection of the lines? Consider the example in which l_1 is the y-axis, l_2 is the x-axis, and the points are $P_1[0, 0, 1]$, $P_2[1, 0, 1]$, and $P_3[1, 1, 1]$. Does the conic solve the problem?

6. Find the equation $C_\lambda(x, y) = 0$ for the family of conics tangent to the x-axis at the origin P_1 and passing through $P_2[1, 1, 1]$ and $P_3[-1, 1, 1]$. Then find the equation of the conic satisfying the following conditions.
   (a) Passing through $P_4[0, 2, 4]$
   (b) Passing through $P_4[2, 0, 1]$
   (c) Passing through $P_4[1, 2, 1]$
   (d) Corresponding to $\lambda = \pm\infty$

   What value of λ determines the parabola $y = x^2$?

7. Verify analytically that the conic resulting from the solution to Problem 1 satisfies all the conditions in the statement of the problem.

8. In Problem 1, what conic results from the solution method in each of the following cases? Does the conic solve the problem? In each case the line l is the x-axis.
   (a) $P_1[0, 0, 1]$, $P_2[0, 1, 1]$, $P_3[1, 1, 1]$, $P_4[1, 0, 1]$

(b) $P_1[0, 0, 1]$, $P_2[0, 1, 1]$, $P_3[0, 2, 1]$, $P_4[1, 0, 1]$
(c) $P_1[0, 0, 1]$, $P_2[0, 1, 1]$, $P_3[0, 2, 1]$, $P_4[0, 3, 1]$
(d) $P_1[0, 0, 1]$, $P_2[0, 1, 1]$, $P_3[0, 4, 1]$, $P_4[1, 3, 1]$
(e) $P_1[0, 0, 1]$, $P_2[0, 1, 1]$, $P_3[1, 1, 1]$, $P_4[2, 1, 1]$

9. If $P[0, 0, 1]$ is a point and $Q = \mathbf{p}^T\mathbf{p}$ is the corresponding symmetric matrix, geometrically describe the degenerate line conic determined Q.

10. If $P_1[0, 0, 1]$ and $P_2[1, 0, 1]$ are two points and $Q = \frac{1}{2}(\mathbf{p}_1^T\mathbf{p}_2 + \mathbf{p}_2^T\mathbf{p}_1)$ is the corresponding symmetric matrix, geometrically describe the degenerate line conic determined by Q.

11. Find the equation and the graph of the point conic whose matrix is

$$Q = \begin{pmatrix} 1 & 0 & -1 \\ 0 & 1 & 1 \\ -1 & 1 & 1 \end{pmatrix}$$

Find the matrix of the corresponding line conic. Find the tangent lines at three points of the point conic and verify that these lines satisfy the equation of the line conic. Find the points of tangency of these lines and verify that they are the original points.

12. We saw in Proposition 3.7 that the point conic $\mathbf{p}Q\mathbf{p}^T = 0$ has, as tangent lines, the line conic $\mathbf{l}^T Q^{-1}\mathbf{l} = 0$. What is the relationship of the line conic $\mathbf{l}^T Q \mathbf{l} = 0$ to the point conic?

13. Find the equation of the conic that is tangent to $y = x$ at $P_1(1, 1)$, tangent to $y = -x$ at $P_2(-1, 1)$, and that is also tangent to the given line.
    (a) $y = 2$
    (b) $y = 0$
    (c) The ideal line.

14. Verify analytically that the conic resulting from the solution to Problem 3 satisfies all the conditions given in the statement of the problem.

15. Find a method for solving Problem 3 that uses a standard model, as in the second solution of Problem 2.

16. Find a solution method for Problem 4 that is dual to the solution of Problem 1.

17. Find the matrix Q_λ for the family of line conics that are tangent to the x-axis, l_1, at $P_1(1, 0)$ and are also tangent to the y-axis, l_2, and the line $y = 1$, l_3. Then find the equation of the line conic that is also tangent to the given line.
    (a) The line $x = 2$
    (b) The line $x = -1$
    (c) What line conic is determined if l_4 is the line $x + y = 1$?

18. Find a solution method for Problem 5 that is dual to that used in Example 2.8 for point conics.

19. Find a solution method for Problem 5 that uses determinants, as in Proposition 2.1.

20. Find the equation of the conic tangent to the five lines $x = -1$, $x = 1$, $x + y = -1$, $x - y = 1$, and $y = 2$. Also find the points of tangency.

21. Why is it important to understand the technical conditions that accompany the statements of all six conic-fitting problems (including the one discussed in Section 2)?

Bibliography

ADLER, C. F. *Modern Geometry, An Integrated First Course.* 2nd ed. New York: McGraw-Hill, 1967.

American Society of Photogrammetry. *Manual of Photogrammetry.* 4th ed. 1980.

ANDERSON, D. P. "Hidden Line Elimination in Projected Grid Surfaces." *ACM Transactions on Graphics* 1, no. 4 (October 1982): 274–288.

ANTON, H. *Elementary Linear Algebra,* 4th ed. New York: John Wiley, 1984.

BALLARD, D. H., and C. M. BROWN *Computer Vision.* Englewood Cliffs, N.J.: Prentice-Hall, 1982.

BARNARD, S. T., and M. A. FISCHLER. "Computational Stereo." *Computing Surveys* 14, no. 4 (December 1982): 553–571.

BALTRUŠAITIS, J. *Anamorphic Art.* New York: Harry N. Abrams, 1977.

BUTLAND, J. "Surface Drawing Made Simple." *Computer Aided Design* 11, no. 1 (January 1979): 19–22.

CARLBOM, I., and J. PACIOREK. "Geometric Projection and Viewing Transformations." *Computing Surveys,* 10, no. 4 (December 1978): 465–502.

ERNST, B. *The Magic World of M. C. Escher.* New York: Ballantine Books, 1976.

FAUX, I. D., and M. J. PRATT. *Computational Geometry for Design and Manufacture.* New York: Halsted Press, 1979.

FISHBACK, W. T. *Projective and Euclidean Geometry.* 2nd ed. New York: John Wiley, 1969.

FOLEY, J. D., and A. van DAM. *Fundamentals of Interactive Computer Graphics.* Reading, Mass.: Addison-Wesley, 1982.

GREENBERG, M. J. *Euclidean and Non-Euclidean Geometries, Development and History.* San Francisco: W. H. Freeman and Company Publishers, 1980.

Liming, R. A. *Mathematics for Computer Graphics,* Fallbrook, Calif.: Aero Publishers, 1979.

Newman, W. M., and R. F. Sproull. *Principles of Interactive Computer Graphics.* 2nd ed. New York: McGraw-Hill, 1978.

Strang, G. *Linear Algebra and its Applications.* 2nd ed. New York: Academic Press, 1980.

Sutherland, I. E., R. F. Sproull, and R. A. Shumacker. "A Characterization of Ten Hidden Surface Algorithms." *Computing Surveys* 6, no. 1 (March 1974): 1–55.

Wright, T. J. "A Two Space Solution to the Hidden Line Problem for Plotting Functions of Two Variables." *IEEE Transactions on Computers* TC22, no. 1 (January 1973): 28–33.

Wylie, C. R., Jr. *Introduction to Projective Geometry.* New York: McGraw-Hill, 1970.

Index

A

Absolute command, 192–93, 244
Abstract duality (*See* Duality, formal)
Adjustment:
   image, 97–100, 241–43, 246, 271
   object, 97–100, 191, 193–94, 201–2, 241–43, 246, 271
Affine coordinates, 126–27, 144–45
Aircraft frame, 306–7
Amount of perspective, 275–76, 295–96
Analytic duality (*See* Duality, analytic)
Analytic geometry, 5
Anamorphic art, 288, 290
Antipodal point, 16, 28, 34
Artificial horizon, 313–15
Artificial horizon indicator, 303
Aspect ratio, 185–89, 198, 200, 244
Asymptotes of a hyperbola, 62, 368
Attitude indicator, 303
Attitude of an aircraft, 307–8
Axiomatic method, 52
Axis of rotation, 140, 308
Axonometric projection, 289, 292
Azimuth angle, 353

B

Back clipping plane, 224–25, 296
Base line, 9, 12, 15–16
Best possible solution, 323–24

Billboarding, 184–85
Bump function, 269–70

C

Camera calibration, 322–23, 336–41
Ceiling emphasis, 275, 282
Center of perspectivity, 19, 41, 221–22, 241
Center of the object, 222, 241, 277, 282–85
Checkerboard, 3, 6, 14
Circle:
   intersection with a line, 366, 369
   as a model for a conic-fitting problem, 382–86
   as a nondegenerate conic, 365–66
   parametrization, 366
   tangent lines to, 366, 369
   as a topological model for the projective line, 29, 87, 150, 207
   twelve-point, 23
   unit, 359, 366
Clipping, 30–31, 177, 184, 200, 224–25, 296–302
Clipping plane, 224–25, 296
Collinearity:
   of points in projective space, 85, 90–92
   of points in the projective plane, 70–72
Column operation, 362
Command, 189
   absolute, 192–93, 244
   relative, 192–93, 244

397

Complete viewing operation, 177–84, 190–91, 194–96, 221, 226–38, 241–43
Concrete duality (*See* Duality, analytic)
Cone of vision, 224–25, 296
Conic, 358
    degenerate, 365, 368, 392–93
    through five points, 369–75
    through four points, tangent to one line, 377, 386–87
    intersection with a line, 366
    least squares fitting, 376
    line, 388
    nondegenerate, 365–66
    through one point, tangent to four lines, 379, 391–92
    parametrization, 366
    point, 388
    projective equivalence of, 359–62, 365–66
    tangent line to, 366–67, 379–80, 388–89
    tangent to five lines, 379, 391–92
    through three points, tangent to two lines, 377–78, 380–86
    through two points, tangent to three lines, 378, 390–91
Control points, 343–44
Coordinates:
    affine, 126–27, 144–45
    device, 177, 179–80, 191, 213–14, 221, 241
    general homogeneous:
        and the cross ratio, 154, 168–69
        in photogrammetric reconstruction, 346
        on the projective line, 149–52
        on the projective plane, 153–54
        on projective space, 154–55
        and projective transformations, 149–56, 159–61
    homogeneous, 55–61
        and data storage, 63
    normalized device, 184, 188–89, 199–200, 228
    spherical, 226, 237, 241, 258, 271
    standard homogeneous:
        of a line, 68–69
        of a plane, 82
        on the projective line, 60, 150, 160
        on the projective plane, 58–60
        on projective space, 60–61
    transformation of, 149–56, 159–61, 194–96
    world, 177, 180, 191, 213–14, 221, 241
Coplanarity:
    of lines in projective space, 94–95
    of points in projective space, 82–83
Counterclockwise rotation, 140–41
Cramer's Rule, 67–68
Critical value, 209, 248, 250, 252–53
Cross product of vectors, 70, 75, 229
Cross ratio, 43–48, 162–71
    of collinear points in the plane, 166–68
    of collinear points in three-space, 166–68
    and general homogeneous coordinates, 154, 168–69
    invariance under change of coordinates, 164–65
    invariance under projective transformation, 45, 164–66
    as a perspective scaling, 43, 154, 169
    in photogrammetric reconstruction, 344–47
    as a projective transformation, 45, 164
Cube:
    in one-point perspective, 33, 35–36, 249–51, 272
    in three-point perspective, 33, 37–38, 272, 276
    in two-point perspective, 33, 36–37, 272, 274
CurrentPosition, 192–93, 243–44

D

DefineCenterOfObject, 241
DefineCenterOfPerspectivity, 241
DefineCenterOfPerspectivitySpherical, 241
DefineDirectionOfProjection, 241
DefineViewingMechanism, 245, 254
DefineViewplaneDistance, 241
DefineViewplaneNormalVector, 241
DefineViewplaneReferencePoint, 241
DefineViewplaneUpDirection, 241
DefineViewport, 190, 241
DefineWindow, 190, 241
Degenerate conic, 365, 368, 392–93
Depth clipping, 224–25, 296–302
Desargues' Theorem, 73
Device coordinates, 177, 179–80, 213–14, 221, 241
Device independence, 184, 189, 199–200, 219
Dimetric projection, 289, 292
Direction:
    of a line, 35, 59
    of projection, 224, 241, 288, 291–93
    of a vector, 35, 59
Displacement:
    due to tilt and relief, 342–44
    radial, 282, 342–43
DisplayPicture, 202–3, 246
DisplayPictureByParametrizing, 209–10
DisplayPictureWithHiddenLines, 248–49
DisplayPictureWithTrap, 203
Display transformation, 180, 184, 195, 221, 223, 228, 241
DrawAbsolute, 192–93, 244
DrawFrom . . . To . . . , 191, 243
DrawInDCFrom . . . To . . . , 189
DrawRelative, 192–93, 244
Duality:
    abstract (*See* Duality, formal)
    analytic:
        planar, 69, 76–77
        spatial, 81–82, 84–85
    comparison of analytic and formal, 70, 82
    concrete (*See* Duality, analytic)

extrinsic and intrinsic, 76–77, 84–85
formal:
    planar, 40
    of propositions, 40, 49, 66, 73–74, 76, 79, 93
    spatial, 40–41
geometric (*See* Duality, analytic)
of line and point conics, 388–92

E

Elementary row, column operations, 362
Ellipse, 359, 362, 365, 389
Equation:
    of a conic, 358–59
    homogeneous, 76, 358
    in polar normal form, 318
    of a projective line, 63–65, 75–76
    of a projective plane, 78, 84
    system of (*See* System of equations)
Equivalence, projective, 360
Equivalence relation, 56
Equivalent systems of equations, 328
Error:
    in least squares, 324, 328–29
    in measurement, 334
    in reconstruction, 334, 339–40
Euclidean geometry, 1–2, 13–15, 19, 127
Euclidean perspectivity, 43, 62
Euclidean point, 55, 57–61
Euclidean properties, 19, 127
Euclidean transformation, 15, 19, 127, 140
Euler angles, 307–8, 316–17
Extended real number, 45, 63, 162
Extrinsic geometry, 13–14, 18, 35
Eye, 7–9, 19, 41–42 (*See also* Center of perspectivity)

F

Fiducial mark, 351
Fixed line, 147
Fixed point, 147
Floor emphasis, 275, 282
Foreshortening ratio, 288–89, 292–93
Formal duality (*See* Duality, formal)
Frame, aircraft and reference, 306–7
Front clipping plane, 224–25, 296
Function of two variables, 255–68
Fundamental Theorem of Projective Geometry:
    for lines, 43, 111–12, 150
    for planes, 48, 121–22
    for three-space, 126, 298

G

Gauss–Jordan elimination, 71–72
General homogeneous coordinates (*See* Coordinates, general homogenous)

GeneralImageTransformation, 243
GeneralObjectTransformation, 243
GeneralTransformation, 194
Geometric duality (*See* Duality, analytic)
Geometric properties:
    Euclidean, 19, 127
    projective, 19–20, 127
Geometry:
    analytic, 5
    Euclidean, 1–2, 13–15, 19, 127
    extrinsic, 13–14, 18, 35
    intrinsic, 13–14, 18, 35
    projective, 1–2, 18–20, 127
    synthetic, 5, 39–48, 52–53
GetPictureFile, 201–2, 246
Global scaling, 130, 144, 194, 242–43
GraphicsMode, 189
Graphing a function of two variables, 255–68
Graph2D, 202
Graph3D, 245
Grid plane, 306
Ground plane, 306

H

Hardware scaling, 214
Harmonic point, 24
Harmonic quadrad, 24–25, 170
Hidden line algorithm, 245, 255, 261
Hither clipping plane, 224–25, 296
Homogeneous coordinates (*See* Coordinates, homogeneous)
Homogeneous equation, 76, 358
Homogeneous system of equations, 326–27
Horizon, 11, 13, 22, 272, 275, 278–79, 282, 311, 313–14, 350–52 (*See also* Vanishing line)
    artificial, 313–15
Horizon plane, 11–12
Horizon point, 350
Hyperbola, 62, 365, 367–68

I

Ideal line, 18, 35, 96, 311, 367
Ideal plane, 34–35
Ideal point, 16–18, 33–35, 55, 57–61, 134, 153, 155, 209, 311
    of a conic, 62, 359, 367
Image adjustment, 97–100, 241–43, 246, 271
Image line, 41–42
Image plane, 7–9, 16, 41–42
Image reversal, 7–9, 14, 275
Incidence matrix, 128, 139–40
Inconsistent system of equations, 323, 326–27
Inhomogeneous system of equations, 326–27
Initialize2DimensionalGraphics, 190
Initialize3DimensionalGraphics, 240–41
Intersection:
    of a line and a conic, 366, 369

Intersection (*cont.*)
  of a line and a plane, 93
  of a line and the ideal line, 96, 209
  of three planes, 79–80, 125–26
  of two lines in projective space, 94–95
  of two lines in the projective plane, 66–67, 75, 125
Intrinsic geometry, 13–14, 18, 35
Inverse parametrization, 97–100, 108, 120, 221, 228–30, 241
Isocenter, 350
Isometric parallel, 350–51
Isometric projection, 289, 292

L

Lateral axis, 307
Lateral clipping, 225, 296
Least squares, 323–28
  fitting of conics and polynomials, 376–77
Left inverse of a matrix, 337
Library:
  three-dimensional, 240–44
  two-dimensional, 189–96
Liming's method, 372
Linear perspective, 8
Line conic, 388
Line of sight:
  principal, 222, 236–37, 282–83, 285, 287
  projector, 7, 288, 289, 291
Local scaling, 130, 144, 194, 242–43
Longitudinal axis, 307

M

Map plane, 343
Matrix, 202, 246
  elementary operations, 362
  incidence, 128, 139–40
  left inverse, 337
  point, 128, 138–40
  rank, 62
  right inverse, 71–72, 76, 83, 107–11, 119–21, 172–75, 229, 323–28, 332–34
  symmetric, 357–58, 363
  of a transformation (*See* Transformation)
Moebius band, 30–31, 211, 213
MoveAbsolute, 192–93, 244
MoveRelative, 192–93, 244
Multiview orthographic projection, 289, 291

N

Nadir point, 350
Nondegenerate conic, 365–66
Nonlinear perspective, 8, 15, 296
Nonorientable surface, 29–30, 77
Normalized device coordinates, 184, 188–89, 199–200, 228
NumberOfSteps, 209, 211

O

Object adjustment, 97–100, 191, 193–94, 201–2, 241–43, 246, 271
Object line, 41–42
Object plane, 9, 16, 41–42
Object reconstruction, 322–23, 329–34
Oblique projection, 288, 291, 293
One-point perspective, 22, 33, 35–36, 272–73, 282, 286
Orientation elements, 350–55
Orthographic projection, 288–89, 291–92
Overdetermined system of equations, 323, 326–27

P

Panning, 211
Paper strip technique, 355–56
Pappus' Theorem, 156–57
Parabola, 359, 365, 367
Parallel lines, 16–17, 33, 58, 145, 288
Parallel projection (*See* Projection, parallel)
Parametrization:
  of a circle, 366
  of a conic, 366
  of a family of conics, 372–74
  of a family of lines, 94
  of a family of planes, 95
  inverse, 97–100, 108, 120, 221, 228–30, 241
  of a line in projective space, 90–92, 115
  of a line in the projective plane, 86–90, 105–6
  of the line of intersection of two planes, 95–96
  nonlinear, 87
  nonuniqueness, 88–89, 106
  of a plane in projective space, 119
  of the projective line, 149–50
  of the projective plane, 153
  of projective space, 154–55
  of the viewplane, 97–100, 221, 223–24, 241
Parametrizing plane, 221, 241
Perpendicular lines, 145
Perspective, 42
  amount of, 275–76, 295–96
  linear, 8
  nonlinear, 8, 15, 296
PerspectivelyProject, 194
PerspectivelyProjectImage, 243
PerspectivelyProjectObject, 243
Perspective projection (*See* Projection, perspective)
Perspective transformation (*See* Transformation, perspective)
Perspectivity (*See* Transformation, perspective)
Photogrammetric reconstruction, 323, 341–55
Photogrammetry, 341–55
Photo plane, 341, 343
Pitch, 308

Pixel ruler, 333
Plot, 191, 242
PlotAbsolute, 192–93, 244
PlotInDC, 189
PlotRelative, 192–93, 244
Point conic, 388
Point matrix, 128, 138–40
Point of intersection (*See* Intersection)
Point of tangency, 388
Polar normal form, 318
Primitive command, 189
Principal line, 350–51
Principal line of sight, 222, 236–37, 282–83, 285, 287
Principal plane, 350–51
Principal point, 350
PrintPicture, 202, 246
Procedural command, 189
Projection:
   axonometric, 289, 292
   dimetric, 289, 292
   isometric, 289, 292
   multiview orthographic, 289, 291
   oblique, 288, 291, 293
   orthographic, 288–89, 291–92
   parallel, 42–43, 132, 224
     in a complete viewing operation, 235–36
     qualitative properties, 288–93
   perspective, 224
     in a complete viewing operation, 97–100, 221, 226–38, 241
     as a projective transformation, 131–37, 144, 146–47, 194, 202–10, 216–18, 243, 248
   trimetric, 289, 292
Projective equivalence, 360
   of conics, 359–62, 365–66
Projective geometry, 1–2, 18–20, 127
Projective line, 18, 35
   equation, 63–65, 75–76
   nonparametric description, 63–74
   parametric description, 86–96, 105–6, 115, 149–50
   topological model, 29, 87, 150, 207
   transformations, 41–43, 101–13, 124, 149–52
   through two points, 66, 75, 86–96, 125
Projective plane, 18, 26–31, 35
   decomposition, 59
   determined by a line of intersection and a point, 93
   equation, 78, 84
   nonparametric description, 77–85
   parametric description, 119, 153
   through three points, 79, 126
   topological model, 16–17, 26–31
   transformations, 41–43, 116–22, 125, 127–38, 153–54
Projective properties, 20, 127
Projective space, 33–35
   parametric description, 154–55
   transformations, 127, 138–44, 154–55

Projective three-space (*See* Projective space)
Projective transformation (*See* Transformation, projective)
Projectivity (*See* Transformation, projective)
Projector, 7, 288, 289, 291

Q

Quadratic form, 357–58
   simplified, 363

R

Radial displacement, 282, 342–43
Rank of a matrix, 62
Reconstruction, 322–23
   camera calibration, 322–23, 336–41
   error, 334, 339–40
   object, 322–23, 329–34
   photogrammetric, 323, 341–55
Rectification, 343–55
Reference frame, 306–7
Reflection, 130, 143, 145–46, 148
   in the ideal line, 138, 147
Reinitialize, 202, 246
Relative command, 192–93, 244
Representative, 56
Resolution, 186
Right-hand rule, 140
Right inverse of a matrix (*See* Matrix, right inverse)
Roll, 308
Rotate, 193–94
RotateImage, 243
RotateObject, 242
Rotation, 128–29, 140–42, 146–48, 193–94, 216–18, 242–43
   axis of, 140, 308
Row operation, 362
RP^1, 63
RP^2, 55–62
   decomposition, 56, 59, 127, 159
   not a vector space, 56

S

Scale, 194
ScaleImage, 243
ScaleObject, 242–43
Scaling, 130, 143–44, 146–47, 194, 242–43
   global, 130, 144, 194, 242–43
   hardware, 214
   local, 130, 144, 194, 242–43
   software, 213–14
Segment, 30–31, 206–9
Shear, 194
ShearImage, 243
Shearing, 130–31, 144, 146–47, 149, 194, 243
ShearObject, 243
Software scaling, 213–14
Solution, best possible, 323–24

Spherical coordinates, 226, 237, 241, 258, 271
Square, 20–24
Standard embedded projective line, 35
Standard embedded projective plane, 32
Standard homogeneous coordinates (*See* Coordinates, standard homogeneous)
Standard model for perspective, 32
Stereo pair, 247
Stereoplotter, 343
Swing angle, 353
Symmetric matrix, 357–58, 363
Synthetic geometry, 5, 39–48, 52–53
System of equations:
   equivalent, 328
   homogeneous, 326–27
   inconsistent, 323, 326–27
   inhomogeneous, 326–27
   overdetermined, 323, 326–27

T

Tangent line to a conic, 366–67, 379–80, 388–89
Test pattern, 127–28, 138–40
TextMode, 189
Three-point perspective, 33, 37–38, 272, 276–77
Tilt:
   in photogrammetry, 341–44, 354–55
   of the viewplane, 288–89
Topological model:
   Euclidean plane, 18–19, 26–27
   projective line, 29, 87, 150, 207
   projective plane, 16–17, 26–31
   projective space, 33–35
Topology, 26
Transform, 190, 206, 242
Transformation:
   of coordinates, 149–56, 159–61, 194–96
   display, 180, 184, 195, 221, 223, 228, 241
   Euclidean, 15, 19, 127, 140
   GeneralImageTransformation, 243
   GeneralObjectTransformation, 243
   GeneralTransformation, 194
   Matrix, 202, 246
   nonuniqueness of the matrix, 104–5, 113, 116, 118
   parallel projection (*See* Projection, parallel)
   perspective:
      analytic, 102, 113, 117, 123, 228
      synthetic, 18–19, 41–43
   PerspectivelyProject, 194
   PerspectivelyProjectImage, 243
   PerspectivelyProjectObject, 243
   perspective projection, 131–37, 144, 146–47, 194, 202–10, 212, 216–18, 241, 243, 248 (*See also* Projection, perspective)
   projective:
      analytic:
         by nonsingular matrix multiplication, 108–11, 119–21, 126–49, 153–55, 193–94, 242–43
         by singular matrix multiplication, 102–5, 117–18
      as a composite, 127, 138, 140
      and the cross ratio, 45, 164
      and generalized homogeneous coordinates, 149–56, 159–61
      synthetic, 19, 41–44
   of projective lines, 41–43, 101–13, 124, 149–52
   of projective planes, 41–43, 116–122, 125, 127–38, 153–54
   of projective space, 127, 138–44, 154–55
   rectification, 343–55
   reflection, 130, 143, 145–46, 148
      in the ideal line, 138, 147
   Rotate, 193–94
   RotateImage, 243
   RotateObject, 242
   rotation, 128–29, 140–42, 146–48, 193–94, 216–18, 242–43
   Scale, 194
   ScaleImage, 243
   ScaleObject, 242–43
   scaling, 130, 143–44, 146–47, 194, 242–43
   Shear, 194
   ShearImage, 243
   shearing, 130–31, 144, 146–47, 149, 194, 243
   ShearObject, 243
   Translate, 194
   TranslateImage, 243
   TranslateObject, 242–43
   translation, 129, 141, 143, 146–47, 194, 242–43
   viewing, 180, 184, 195, 221
Translate, 194
TranslateImage, 243
TranslateObject, 242–43
Translation, 129, 141, 143, 146–47, 194, 242–43
Trimetric projection, 289, 292
Twelve-point circle, 23
Two-point perspective, 22, 33, 36–37, 272, 274–75

U

Unit circle, 359, 366
Unit point, 150–55

V

Vanishing line, 11–12, 14, 16, 32–33, 272, 274, 276 (*See also* Horizon)
Vanishing point, 9–12, 14, 17, 20, 32–33, 38, 134, 137, 272, 274–76, 282, 311
Vertical axis, 307
Vertical photograph, 341

Viewing mechanism, 33, 220–22, 245, 254, 271, 277, 282
Viewing transformation, 180, 184, 195, 221
Viewplane, 221–24
    distance, 222, 241, 275–76, 279–80
    normal vector, 222, 241, 287–88
    parametrization, 97–100, 221, 223–24, 241
    reference point, 222–24, 241, 287–88
    up-direction, 223–24, 241, 282, 286, 288
Viewport, 177, 180–81, 184, 190–91, 201, 226, 241
View volume, 224–25, 296–97

W

Wall emphasis, 275, 278, 282, 284
Window, 177, 179, 181, 184, 190–91, 201, 223–24, 226, 241, 287–88

World coordinates, 177, 180, 213–14, 221, 241
WriteCharacter, 189
WriteString, 191

X

x-curve, 259

Y

y-curve, 259
Yaw, 308
Yon clipping plane, 224–25, 296

Z

Zooming, 211, 213